FLOW AND CONTAMINANT TRANSPORT IN FRACTURED ROCK

FLOW AND CONTAMINANT TRANSPORT IN FRACTURED ROCK

Edited by

Jacob Bear
Department of Civil Engineering
Technion — Israel Institute of Technology
Haifa, Israel

Chin-Fu Tsang
Earth Sciences Division
Lawrence Berkeley Laboratory
University of California
Berkeley, California

Ghislain de Marsily
Université Pierre et Marie Curie
Paris, France

Academic Press, Inc.
Harcourt Brace Jovanovich, Publishers
San Diego New York Berkeley Boston London Sydney Tokyo

ACADEMIC PRESS RAPID MANUSCRIPT REPRODUCTION

This book is printed on acid-free paper. ∞

Academic Press, Inc.
1250 Sixth Avenue, San Diego, California 92101-4311

United Kingdom Edition published by
Academic Press Limited
24–28 Oval Road, London NW1 7DX

Library of Congress Cataloging-in-Publication Data

Flow and contaminant transport in fractured rock / edited by Jacob Bear, Chin-Fu Tsang, Ghislain de Marsily.
p. cm.
Includes bibliographical references and index.
ISBN 0-12-083980-6
1. Groundwater flow. 2. Water, Underground–Contamination.
I. Bear, Jacob. II. Tsang, Chin-Fu. III. Marsily, Ghislain de.
GB1197.7.F548 1993
551.49–dc20 92-21209
CIP

PRINTED IN THE UNITED STATES OF AMERICA
93 94 95 96 97 98 MM 9 8 7 6 5 4 3 2 1

CONTENTS

3. Solute Transport through Fracture Networks

Leslie Smith and Franklin W. Schwartz

4. Stochastic Models of Fracture Systems and Their Use in Flow and Transport Modeling

Jean-Paul Chilès and Ghislain de Marsily

5. Tracer Transport in Fracture Systems

Chin-Fu Tsang

9. A Summary of Field Test Methods in Fractured Rocks

P. Jouanna

CONTRIBUTORS

Numbers in parentheses indicate the pages on which the authors' contributions begin.

J. Bear (1)
Department of Civil Engineering, Technion — Israel Institute of Technology, 32000 Haifa, Israel

Jean-Paul Chilès (169)
Bureau de Recherches Géologiques et Minières, 45060 Orléans Cedex 2, France

J. R. Gilman (267)
Marathon Oil Company, Littleton, Colorado 80160-0269

P. S. Huyakorn (396)
Hydrogeologic, Inc., Reston, Virginia 22090

P. Jouanna (437)
Université Montpellier II, Département Eau-Environment, Place Eugène Bataillon, 34095 Montpellier Cedex 5, France

H. Kazemi (267)
Marathon Oil Company, Littleton, Colorado 80160

Ghislain de Marsily (169)
Université Pierre et Marie Curie, 75272 Paris Cedex 5, France

T. N. Narasimhan (325)
Earth Sciences Division, Lawrence Berkeley Laboratory, University of California, Berkeley, California 94720

Ivars Neretnieks (39)
Department of Chemical Engineering, Royal Institute of Technology, S-100 44 Stockholm, Sweden

G. F. Pinder (396)
College of Engineering and Mathematics, University of Vermont, Burlington, Vermont 05401

Franklin W. Schwartz (129)
Department of Geology, Ohio State University, Columbus, Ohio 43210

Leslie Smith (129)
Department of Geological Sciences, University of British Columbia, Vancouver, British Columbia V6T 1W5, Canada

E. A. Sudicky (396)
University of Waterloo, Waterloo, Ontario N2L 3G1, Canada

Chin-Fu Tsang (237)
Earth Sciences Division, Lawrence Berkeley Laboratory, University of California, Berkeley, California 94720

J. S. Y. Wang (325)
Earth Sciences Division, Lawrence Berkeley Laboratory, University of California, Berkeley, California 94720

PREFACE

The question of how to deal with fractured rock domains has always been on the agenda of geohydrologists. Are Darcy's law and the theory of flow through porous media applicable to fractured rock aquifers, at least when the flow is assumed to be in the laminar flow range? And is the continuum approach applicable? The subject has also been investigated by reservoir engineers, this time in connection with multiphase flow, because many important petroleum reservoirs are in fractured rock formations. Of special interest are reservoirs composed of fractured porous rocks, in which the blocks surrounded by the network of fractures are porous. The permeability of such blocks is often rather low; but the porosity, and hence the storage capacity for fluids, is very high. This "double porosity" model for fractured porous rock domains was first introduced in the field of reservoir engineering.

In the past two or three decades, fractured rock domains have received increasing attention not only in reservoir engineering and hydrology but also in connection with geological isolation of radioactive waste. Locations in both the saturated and unsaturated zones have been under consideration. Such repositories are sources of heat and potential sources of ground water contamination. Thus, in addition to the transport of mass of fluid phases in single or multiphase flow, the issues of heat transport and mass transport of components have to be addressed.

A large number of articles on these subjects exist in the scientific and professional literature of a number of disciplines, including geology, hydrology, reservoir engineering, and environmental engineering. We feel that the time is ripe to put together the main ideas and methodologies found in the literature on flow and tracer transport in fractured rock domains, in the form of an edited book written by experts from various disciplines, for the benefit of researchers and practitioners. We have decided not to excessively edit or unify the chapters. This has the advantage that each chapter is relatively self-contained and can be studied independently of the others. Also, the reader may find it interesting to sample the varying styles of different authors in different disciplines.

The editors hope that the book will serve its purposes: to present the state of the art on flow and tracer transport in fractured rock domains as viewed by scientists working in different disciplines, to encourage practical field engineers and scientists to use the various methods suggested in this book, and to stimulate researchers to further advance the state of the art of this fruitful research area.

We would like to acknowledge the U.S. Department of Energy, Office of Environmental Restoration and Waste Management, for supporting the publication of this book. The assistance of Jim Miller, Loretta Lizama, Connie Silva, and Jean Wolslegel of the Technical Information Department, Lawrence Berkeley Laboratory, in word processing, English editing, and organizing the practical aspects of this book is gratefully acknowledged. Finally, the patience and understanding of Charles Arthur of the Academic Press through this whole process are most appreciated.

Jacob Bear
Ghislain de Marsily
Chin-Fu Tsang

December 1992

1

MODELING FLOW AND CONTAMINANT TRANSPORT IN FRACTURED ROCKS

J. Bear
Technion—Israel Institute of Technology
Haifa, Israel

The objective of this chapter is to present the basic models of transport in a fractured porous medium domain of two *extensive quantities*: mass of a fluid phase and mass of a component of a fluid phase. Actually, whenever the movement of a phase is involved, we consider also the transport of *momentum* of that phase, except that the *equilibrium equation* and Darcy's law are usually used to represent (a simplified version of) the averaged momentum balance equation of a solid phase and a fluid one, respectively.

1.1 INTRODUCTION

1.1.1 Some Basic Definitions

A *fracture* is part of the void space of a porous medium domain that has a special configuration: one of its dimensions—the *aperture*—is much smaller than the other two ones.

We shall use the term *fractured rock* (abbreviated: FR) when the blocks of rock surrounded by fractures contain no void space. We shall employ the term *fractured porous medium*, or *fractured porous rock* (abbreviated: FPR), when the blocks are porous. Thus, in a FPR, the void space if composed of two parts (Fig. 1.1.1): a network of fractures and blocks of porous medium. Since we are interested in the movement of a fluid phase in the fractures, we shall assume that the latter form an *interconnected* network.

For the purpose of this chapter, a *phase* is defined as a portion of space, whether connected or non-interconnected, that is separated from other such portions by a well defined surface, referred to as *interface* or *interphase boundary*. A phase is characterized by the fact that its behavior at all points can be described by *the same* set of state variables. There can be only a single gaseous phase in the void space, as different gases do not maintain a distinct interface between them. On the other hand, we may have a number of liquid phases, each occupying a well defined portion of the void

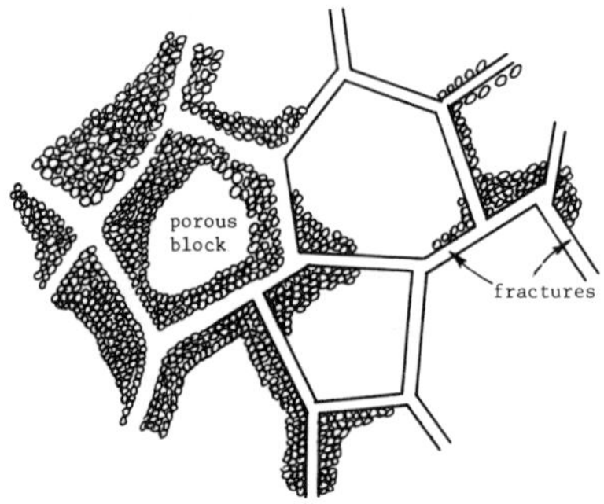

Figure 1.1.1. Schematic representation of a fractured porous medium.

space. We regard liquids as *immiscible*, if a distinct sharp interface is maintained between them, even when certain components contained in them do cross interphase boundaries in relatively small quantities, diffusing in both phases.

A phase may be composed of more than one component. A *component* is a part of a phase that is made up of an identifiable, chemical constituent, or an assembly of constituents, e.g., ions, or molecules.

The transport of a considered *extensive quantity* of a phase in a porous medium domain may take place through a single (fluid or solid) phase, through some of the phases present in the domain (possibly including the solid phase), or through all of them. In the first case, at least part of the (microscopic) domain occupied by that phase must be connected. In the last two cases, the transport of a considered extensive quantity may take place across the (microscopic) interphase boundaries separating the phases through which the transport occurs.

Since we consider fluid flow through the void space of a porous medium domain, we require that at least part of the void space be *interconnected.* Here we shall assume that the entire void space in a FR- (or a FPR)-domain is interconnected.

1.1.2 The Continuum Approach to Transport in Porous Media

In principle, the mathematical model that describes the transport of an extensive quantity of a—fluid, or solid—phase through a porous medium domain (= balance equations for the considered transported quantities, constitutive relations, initial conditions within the phase, and boundary conditions on the surface that bounds the phase), can be stated for every point within the considered phase. This description is said to be at the *microscopic level,* as we focus our attention on what happens at a (mathematical) point *within* a considered phase present in the domain. Although the transport problem can be stated as a *well posed model,* the latter cannot be solved at this level, since the detailed the geometry of the surface that bounds the phase is not known and/or is too complex to be described. Also, we cannot measure values of variables at points *within* a phase in order to validate a flow or transport model and to determine model parameters. As a consequence, the complete description and solution of a transport problem at the microscopic level is impossible.

To circumvent these difficulties, the transport problem is transformed from the microscopic level to a *macroscopic* one, at which the transport problem is reformulated in terms of averages of the microscopic values. The average values are measurable quantities. We refer to this approach, employed in many branches of science, as the *continuum approach.*

In the continuum approach, the real porous medium domain, consisting of two or more phases, is replaced by a *model* in which each phase is assumed to be present at *every point* within the entire domain. Each phase, thus, behaves as a continuum that fills up the *entire* domain. We speak of *overlapping continua*, each corresponding to one of the phases, or components. If the individual phases interact with each other, so do these continua. In fact, every extensive quantity of every phase, or of a component of a phase, is modelled as a continuum that occupies the entire domain.

For every point within each of these continua, average values of phase and component variables are taken over an *elementary volume* centered at the point, regardless of whether this point falls within the considered phase, or not. The averaged values are referred to as *macroscopic* values of the considered variables. By traversing the porous medium domain with a moving elementary volume, we obtain a *field* of the averaged values for every variable. These values are continuous and differentiable functions of the space coordinates and of time. This is the macroscopic level of description of transport phenomena.

By introducing the macroscopic (or continuum) level of description, we have circumvented the need to know the exact configuration of (microscopic) interphase boundaries, and obtained a description of processes in terms of measurable quantities (provided we make sure that our measuring device is designed to measure the averaged values). The price of this achievement is the introduction of various *coefficients* that reflect, at the macroscopic level, the effects of the microscopic configuration of interphase boundaries. The detailed structure of each coefficient depends on the model that is used to represent the *microscopic* reality. For a specific porous medium, *the numerical values* of these coefficients must be determined experimentally, in the laboratory, or in the field. Actually, these coefficients are *of the model* that we have selected for the problem on hand. However, following common practice, we refer to them as ones of the porous medium.

Accordingly, we shall start by introducing the *continuum approach* that leads to the macroscopic level of describing phenomena of transport in porous media.

A porous material domain is characterized by the fact that part of it is occupied by a *persistent solid phase*, called the *solid matrix*. The remaining part, called the *void space*, is occupied by one or more fluid phases. Another common characteristic of a porous medium domain is that the solid phase (and, hence, also the void space) *is distributed throughout it*. What we mean by this is that if we take sufficiently large samples of the porous material at different locations within the domain, we shall find in each of them both a solid phase and a void space. At the same time, if a sample centered at a point is to represent what happens at that point and at its close neighborhood, it is obvious that the size of the sample should not be too large. We shall refer to the volume of a sample that satisfies these conditions, as a *Representative Elementary Volume* (abbreviated REV) of the considered porous medium domain at the given point.

With these considerations in mind, we now define a *porous medium* as a *multiphase material body* characterized by the following features:

(a) A Representative Elementary Volume (REV) can be determined for it, such that no matter where we place it within a domain occupied by the porous medium, it will always contain both a persistent solid phase and a void space. *If such an REV cannot be found for a given domain, the latter cannot qualify as a porous medium domain.*

(b) The size of the REV is such that parameters that represent the distributions of the void space and of the solid matrix within it are statistically meaningful.

The quantification of the last requirement will not be considered here (see, among many others, Bear and Bachmat, 1990). We may summarize the guidelines for selecting the size of the REV for a given porous medium domain in the following way:

- The volume of the REV should be uniform in magnitude and shape throughout the considered porous medium domain.

- The resulting (volume) averaged values should be independent of the size, shape, and orientation of the REV.

- Using ℓ to denote the size of an REV (e.g., a diameter of a sphere), we require that

$$\ell \geq \ell_{\min} \gg d,$$

where d denotes a characteristic length of the microscopic geometry of the void space (e.g., the hydraulic radius). The length d represents the *scale of microscopic inhomogeneity of the void space.*

- Because the solid matrix is usually heterogeneous at the macroscopic level (e.g., porosity varies as a function of location), the size of the REV must also satisfy

$$\ell \ll \ell_{\max},$$

where $\ell_{\max}$ defines a domain around a point in a heterogeneous medium within which the (macroscopic) parameters that describe the void geometry may still be considered uniform, *within a prescribed error level.*

- Finally, we have to relate ℓ to the size of the considered domain. We require that

$$\ell \ll L,$$

where L denotes a characteristic length of the considered porous medium domain. This requirement ensures that the boundary region (of width ℓ) within which the continuum approach is not applicable will be small compared to the size of the domain itself.

In simpler language, the size of an REV must be much larger than the scale of microscopic heterogeneity, due to the presence of solid and void space, and much smaller than the *scale of the domain of interest.* Note that we have emphasized 'the domain of interest' (and not merely 'the domain'), as it is only the heterogeneity within the domain of interest (given a specific problem) that counts in determining the size of the REV (for the given specific problem); an aquifer may be large, but the domain of interest, say an advancing plume of a contaminant, may be a small fraction of it, and may vary with time.

A conceptual determination of the size of an REV for a porous medium domain is shown in Fig. 1.1.2. The determination is based on porosity as a geometrical property of the porous medium. We recall that the size of the REV is selected such that the macroscopic properties are independent of the domain of averaging. Using porosity as an example, this does not necessarily mean that the porosity must be uniform within the REV. To illustrate this point, consider the ratio $U_v(\mathbf{x})/U(\mathbf{x})$, where $U(\mathbf{x})$ is the volume of a sphere centered at an arbitrary fixed point, $\mathbf{x}$, within a considered domain, and $U_v(\mathbf{x})$ denotes the volume of the void space within $U(\mathbf{x})$.

Figure 1.1.2 shows the variations of the ratio U_v/U as U increases. For very small values of $U(\mathbf{x})$, the above ratio is one or zero, depending on whether $\mathbf{x}$ happens to fall in the void space, or in the solid matrix. As $U(\mathbf{x})$ increases, we note large fluctuations in the ratio U_v/U. However, as U continues to grow, these fluctuations are gradually attenuated, until, above some value $U = U_{\min}$, they decay, leaving only small fluctuations around some constant value. Further increase in U

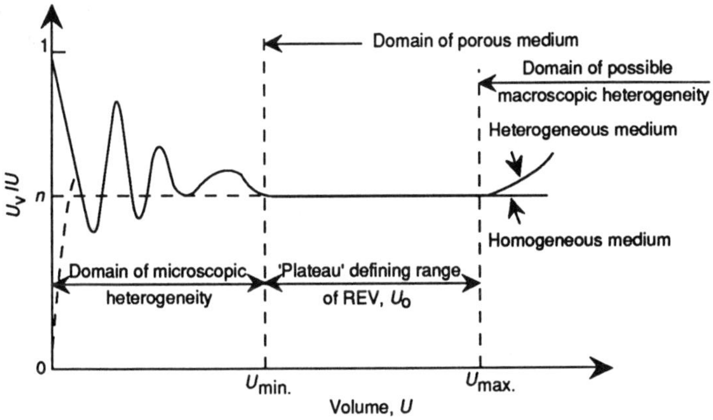

Figure 1.1.2. Variation of porosity, ϕ, in the neighborhood of a point as a function of the size of the averaging volume (after Bear and Bachmat, 1990).

may lead to additional deviations from the constant value. These result from inhomogeneity in the medium's porosity.

Bear and Bachmat (1990) examine the behavior of the $U_v/U \equiv \phi(U|_{\mathbf{x}})$ in the domain in which $\partial\phi/\partial U|_{\mathbf{x}} = 0$. They conclude that if the ratio $U_v/U \equiv \phi(U|_{\mathbf{x}})$ has a *plateau* within a given range of U, then $\phi(\mathbf{x}|U)$ is a linear function of $\mathbf{x}$, including the case ϕ = const. in that range. The opposite is also true, namely, that if the porosity, ϕ, varies linearly within U, or may be approximated as such, a plateau exists.

If $U(\mathbf{x})$ is further increased, say beyond some value $U = U_{max}$, we may observe a trend in the considered ratio, $U_v(\mathbf{x})/U(\mathbf{x})$, due to a systematic variation in the latter.

The Representative Elementary Volume is that volume, $U = U_o(\mathbf{x})$, within the range of $U_{min} < U < U_{max}$, that will make the ratio U_v/U independent of U, and hence a single valued function of $\mathbf{x}$ only. For any $U = U_o$ in that range, the ratio U_v/U represents the porous medium's *porosity*, ϕ, at $\mathbf{x}$.

In defining the REV, we required that macroscopic coefficients of the porous medium be independent of the volume of the REV. We considered porosity, ϕ, as a typical geometrical porous medium property. The above requirement should be extended to all relevant coefficients, requiring that a common REV be found for all of them.

Turning now to transport and other processes, where the state of each phase is specified by a set of relevant state variables, (e.g., densities), an analogous approach should be undertaken with respect to these variables. A range for REV (= domain of averaging) should be selected for each state variable, following considerations similar to those associated with the geometrical characteristics of the void space. In this way, the size, ℓ, of the REV corresponding to each state variable will be bounded by ℓ_{min} that depends on the spatial distribution of the microscopic values of that variable within the phase, and on ℓ_{max} that depends on the spatial variation of its macroscopic counterpart. The continuum description of the process involving them can be employed only if a common range of REV can be found for all of them. The same range should also be common with that associated with the configuration of the void space.

The requirement that across the REV, any *macroscopic property* (whether one of the void space, or of a state variable) *should vary linearly, or approximately so, justifies the assignment of the averaged values taken over the REV to the latter's centroid.*

An important consequence of the continuum approach and the introduction of the REV is related to field (and laboratory) measurements of state variables. In order to be comparable with calculated values, based on the continuum model, these measurements should also represent averages taken over the same REV that appears in the (mathematical) description of the considered transport phenomenon. The size of the 'window' of the measuring device should satisfy this requirement.

In the above discussion, each phase (gas, liquid or solid) is regarded as a continuum, overlooking its molecular structure. Actually, the averaging methodology employed for the passage from the microscopic level to the d macroscopic one, was originally introduced for the passage from the *molecular level* to the microscopic (phase continuum) one, with a Representative Elementary Volume used for averaging the behavior at the molecular level in order to obtain the behavior of the phase as a continuum.

Once we decide that the continuum approach is applicable to a given transport problem in a given domain (i.e., that an appropriate REV can be determined), we can construct the macroscopic transport model by averaging the microscopic one over the REV. As explained above, macroscopic coefficients, e.g., porosity and permeability, are created in the averaging process to reflect the effects of the microscopic configuration of interphase (including fluid-solid) boundaries.

1.1.3 Multicontinua Models

From the discussion above it follows that once an REV is determined for a porous medium domain, we may *model* the transport of every relevant extensive quantity of a phase, or of a component of one, as a continuum that occupies the entire domain. Sometimes, we shall find it *convenient* and/or *useful* to model a single fluid phase (as defined above) as two or even more *apparent phases*. Although apparent phases are not separated by a sharp interface, their different behavior justifies such a model. This follows from the definition of a model as a *selected, simplified version of a real system that approximately simulates the relevant excitation-response features of the latter.*

1.1.4 A Fractured Domain as a Continuum

We can now turn to the question of whether the continuum approach is applicable to fracture rock and fractured porous rock domains.

From the discussion presented thus far, it follows that the continuum approach is applicable to an FR-domain, and to an FPR- domain, as long as an REV can be determined for the specified problem and domain. For example, in an FR-domain, the scale of heterogeneity is the distance between fractures. If the domain is sufficiently large, an REV can be found. It size should be much larger than the spacing between fractures, and sufficiently smaller than the length characterizing the domain of interest. In many cases, this may mean quite a large REV. The question then arises as to how can state variables be measured in the field for such a large REV, and what is the error in measuring values by a device that averages over a volume that may be much smaller than the REV.

Figure 1.1.3 shows schematically, how the common REV can be determined for a fractured rock domain.

1.1.5 Scales of Problems

A *fractured porous medium* is defined as a portion of space in which the void space is composed of two parts (Fig. 1.1.1): an interconnected network of fractures, and blocks of porous medium. The entire void space is occupied by one fluid, or more. Such a domain can be treated as a *single continuum*, provided an appropriate REV can be found for it.

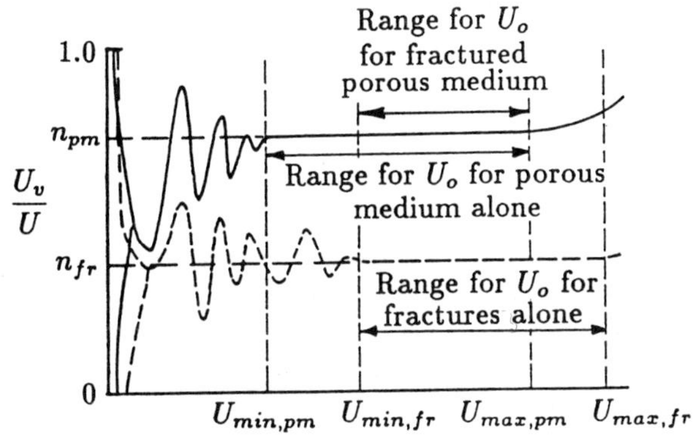

Figure 1.1.3. Definition of representative elementary volumes for a network of fractures and porous blocks (after Bear and Bachmat, 1990).

However, sometimes the fluid in the fractures behaves differently from that in the porous blocks. For example, when the fractures' widths (= apertures) are large, while the pores in the blocks are very small, practically all the flow takes place through the fractures, while most of the fluid storage is provided by the pore space of the blocks, due to the large porosity of the latter.

Under such conditions, it is convenient and useful to model the fluid in the void space as made up of two 'apparent fluid phases': one occupying the fractures and the other occupying the porous blocks, regarding each such phase as a continuum that occupies the *entire* domain. Exchange of transported extensive quantities between the two apparent phases (= continua) is assumed possible.

With the ideas expressed in the previous subsections, we may classify the various problems of flow and contaminant transport in fractured porous rock domains in the following way:

- **Zone 1: The very near field.** Interest is focused on fluid flow and contaminant transport within a single, well-defined fracture, possibly with transport into the porous blocks that bound it.

- **Zone 2: The near field.** Flow and contaminant transport is considered in a relatively small domain which contains a small number of well-defined fractures. The location and shape of the individual fractures must be known. However, the definition of the fractures may be statistical, and not deterministic. In the latter case, we envisage the use of the fractures' statistics in order to construct domains with fractures that are random realizations of the real system (which, as we recall, is unknown to us deterministically).

- **Zone 3: The far field.** Here transport may be visualized as taking place, simultaneously, in two *overlapping continua*: one composed of the fluid within the network of fractures, and the other involving the fluid within the porous blocks. Mass of the fluid phase and its components may be exchanged between the two continua.

- **Zone 4: The very far field.** The fluid in the entire fractured porous medium domain may be regarded as a single continuum. The geometrical properties of the void space reflect those of both the network of fractures and the porous medium blocks.

Figure 1.1.4 gives a schematic representation of the four zones.

The names: very near field, near field, etc., given to the various classes of problems are related to the *scale*, or size of the problem domains. Local problems, corresponding to flow and contaminant transport near sources and sinks, take place in the first and second zones. In the very close vicinity of a contaminant source, we represent the problem as a 'very near field' model. In fact, the very near field may consist of a single fracture intersecting a well. If the (flow and contaminated) domain of interest is larger, yet insufficient for the application of the continuum approach, we construct a 'near field' model, which requires information on *every* individual fracture. If the information about fractures is available only in the form of fracture statistics (say, of aperture, length, orientation, spacing, etc.), we may construct realizations (or conditional realizations) of the considered domain, model and study them, and derive a statistical descriptions of their behavior. However, these models are not practical when considering contaminant transport over large domains, as the inclusion of hundreds or thousands of fractures demands inordinately large amounts of detailed input data and computer storage. Such large zones are referred to as the 'far and very far zones'.

For larger-scale problems (in terms of spatial dimensions, e.g., involving a plume from a source of contamination), the double-continua approach of the 'far field' model may be used. The conceptualization of the two-continua approach, originally proposed by Barenblatt and Zheltov (1960) and Barenblatt *et al.* (1960), is also known as the *'overlapping continua'*, *'double continuum'*, or *'double-porosity'* model. In this conceptual model, the fractured porous medium domain is represented by two distinct, but interacting, subsystems: one consisting of the network of fractures and the other of the porous blocks. Each subsystem is visualized as a continuum occupying the entire investigated domain. Interaction phenomena between the two continua include the exchange of fluid and contaminants between fractures and porous blocks. Since the definition of two continua is required, one for the porous blocks and one for the fracture network, it follows that the application of this model assumes the existence of an REV which is common to both subsystems.

While Barenblatt and Zheltov (1960) and Barenblatt *et al.* (1960) suggested the use of two interacting continua, it is possible (and, under certain conditions, even preferable) to separate the network of fractures into two or more continua. Several different geological and environmental processes may contribute to the creation of fractures within a given domain. Depending on the nature of the formation, these mechanisms can produce a fracture network in which distinct subsystems of fractures can be identified (e.g., microfractures and larger ones) with regular, preferred, and/or random orientations, and with varying hydraulic conductivities and degrees of interconnectedness. Consequently, each subsystem possessing unique properties, may be considered as a distinct continuum; together they can be treated within the framework of a single model provided that there exists an REV common to all of them.

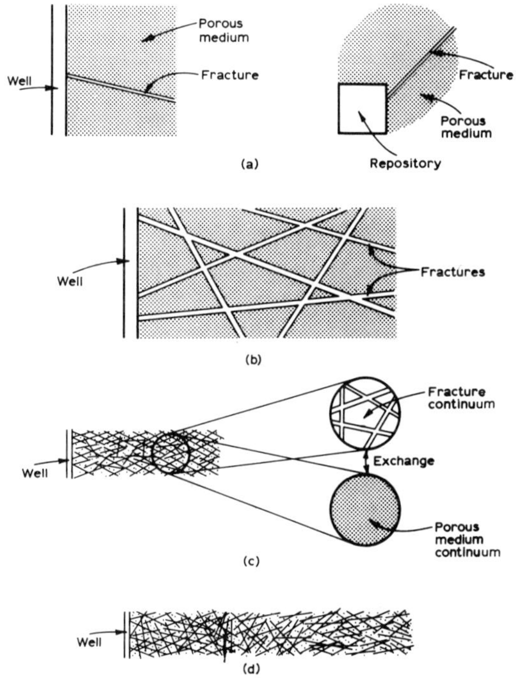

Figure 1.1.4. Schematic representation of the four zones of flow and contaminant transport in a fractured porous medium: (a) the very near field, (b) the near field, (c) the far field, and (d) the very far field (Bear and Berkowitz, 1987).

If very large distances and travel times are involved, we use the 'very far field' model, in which we regard the entire domain as a single continuum. This kind of model is applicable in cases where the system under consideration allows sufficient interaction between fluid and contaminants in the fractures and in the porous blocks, bringing the two systems to a local equilibrium at every (macroscopic) point. Such a system can be described by a model analogous to that of a regular porous medium. This approach has been shown by Mercer and Faust (1979) to be justified in some cases of flow and heat transport in fractured media.

Although at a number of places in this section we have made the *assumption* that the fractures are interconnected, this is not necessarily always the case in reality. The answer to the question of whether, in a given case, the fractures are interconnected or not depends to a large extent on the density of fractures and on the areal extent of the individual fractures (assuming that each fracture has a finite areal extent). It is obvious that as the density of fractures (i.e., the number of fractures per unit volume of fractured rock) become smaller, the chance of a fracture intersecting a neighboring one is reduced. The same is true as the fracture area becomes smaller. This subject has been investigated by using *percolation theory*. Researchers have attempted to determine the *percolation threshold*, viz., the density of fractures above which the connectivity of fractures is sufficient to enable flow through the network, at least through part of the fractures. Since connectivity of fractures *throughout* a domain (and not merely locally, within a subdomain) is also a function of the size of the domain, we may add one more requirement to the definition of an REV (if indeed we are interested in flow through the fractured rock domain), viz., that it will be sufficiently large so as to ensure connectivity of fractures.

Several authors (e.g., Robinson, 1984; Charlaix, 1985; Wilke *et al.*, 1985; Marsily, 1985) studied the issue of connectivity in fractured rock, making use of percolation theory. Charlaix (1984) suggested that at the threshold point, the density of fractures, n_{fr}, and the size of the fractures assumed to have the shape of discs of constant radius r_{fr}, are related by the expression

$$n_{fr}r_{fr}^2 = 0.15 \quad \text{to} \quad 0.30.$$

This expression is based on the assumption of random orientation of fractures in space.

With this in mind, we shall continue to assume that the network of fractures is interconnected.

1.2 FLOW MODELS

1.2.1 Flux in a Single Fracture

Any *motion equation*, i.e., an equation that relates the velocity of a fluid to driving forces, is actually a *momentum balance equation* for the fluid. The Navier-Stokes equation, often referred to as the *motion equation*, may serve as an example for a Newtonian fluid. This (momentum balance) equation describes the fluid's flow at the *microscopic level*, i.e., at points *within* the fluid continuum. When modeling transport phenomena at the *macroscopic level*, we need a macroscopic motion equation. This equation is obtained by averaging the corresponding microscopic one over the REV of the considered domain. The result is an averaged momentum balance equation that describes three- dimensional flow in a porous medium domain. Under certain simplifying assumptions, e.g., neglecting inertial effects and friction within the fluid (not at the fluid-solid interface!), the averaged momentum balance equation reduces to the well known *Darcy law*.

A fracture was defined above as a domain characterized by a width (= aperture) that is much smaller than any length of interest along the fracture's *axis surface* (see below). Hence, we *assume* that the flow in a fracture takes place essentially parallel to the fracture's axis, overlooking variations, say, of pressure, velocity, or solute concentration across the fracture's width. This implies that we are interested in the averaged behavior across the fracture. For example, we seek an expression for the averaged velocity at a point on the fracture's axis, rather than for the velocity variations over the aperture at that point, with the flow visualized as one in a three-dimensional domain. This goal is achieved, for a Newtonian fluid, by averaging the Navier–Stokes equation over the aperture of a fracture.

We shall start by introducing some definitions related to the geometry of a fracture. The discussion is confined to a nondeformable fracture, with a fracture aperture that varies in space in such a way that a *fracture axis surface* can be defined throughout the fracture. To obtain this surface, we consider a point on one of the walls that bound the fracture and determine the largest sphere that (1) can be placed inside the fracture, and (2) is tangent to the wall at that point. The centroid of this sphere is a point on the axis surface. We repeat this procedure for all points on the selected wall. The centroids of all spheres form the entire axis surface. Often, we approximate this surface as a plane (or as composed piecewise of planar segments). In what follows, we shall use the term fracture to part of a fracture that is characterized by such a planar segment (Fig. 1.2.1).

At every point on the fracture's axis plane, an orthogonal coordinate system, x', y', z', can be defined, with x', z' denoting a point in the axis plane. The stationary walls of the fracture can be described by the equations

$$F_1(x', y', z') = y_1' - f_1(x', z') = 0, \quad F_2(x', y', z') = y_2' - f_2(x', z') = 0, \tag{1.2.1}$$

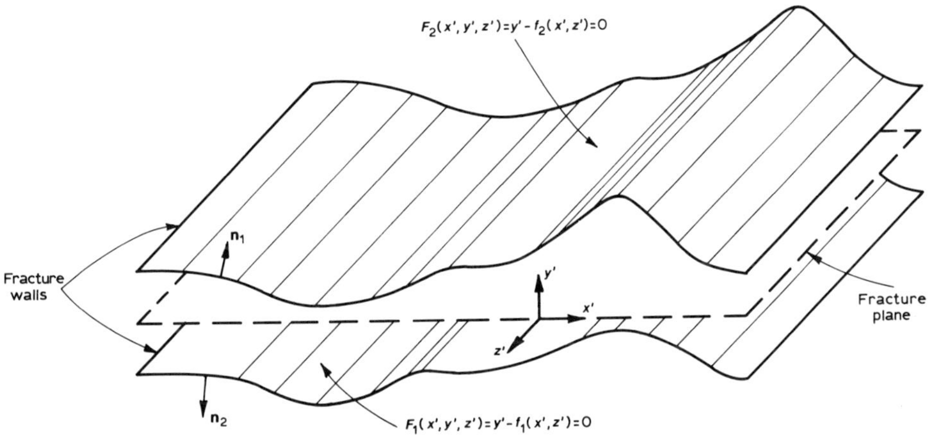

Figure 1.2.1. Nomenclature for a fracture geometry.

in which y' is the coordinate normal to the axis plane, and y_1' and y_2' are the values of y' on the walls of the fracture. From (1.2.1) it follows that the normal outwardly directed unit vector at a point on a fracture wall, is defined by

$$\mathbf{n}_m = \frac{\nabla F_m}{|\nabla F_m|}, \qquad m = 1, 2, \tag{1.2.2}$$

and the fracture aperture, $b = b(x', z')$, at any point on the axis surface is expressed by

$$b(x', z') = f_2(x', z') - f_1(x', z'). \tag{1.2.3}$$

We now turn to consider the flow within the fracture.

Assuming negligible pressure variations across the width of fracture (based on the assumption that the fracture's width is much smaller than its extent in the direction of the fracture surface), the point (i.e., microscopic) balance equations for the fluid's mass and linear momentum, can be averaged over the fracture width to produce averaged (or integrated) equations for two-dimensional flow in the fracture plane.

The three-dimensional balance equation for linear momentum of an incompressible fluid in a fracture, when combined with the mass balance one, takes the form

$$\rho \frac{\partial \mathbf{V}}{\partial t} + \rho \nabla \cdot (\mathbf{V}\mathbf{V}) + \nabla p - \rho \mathbf{g} - \mu \nabla^2 \mathbf{V} = 0, \tag{1.2.4}$$

where ρ and μ denote the fluid's density and dynamic viscosity, respectively, p is pressure, $\mathbf{V}$ is the fluid's mass weighted velocity, t is time, $\mathbf{g}$ $(= -g\nabla z)$ denotes the gravitational acceleration, and z is the vertical coordinate (positive upward).

Hubbert's potential, φ^*, for a compressible fluid, $\rho = \rho(p)$, is defined by

$$\varphi^* = z + \int_{p_o}^{p} \frac{dp'}{g\rho(p')}.$$

We shall approximate it by the piezometric head, φ, defined for a fluid of constant density, by

$$\varphi = z + \frac{p}{\rho g}. \tag{1.2.5}$$

Substituting (1.2.5) into (1.2.4), yields

$$\rho\frac{\partial \mathbf{V}}{\partial t} + \rho\nabla\cdot(\mathbf{V}\mathbf{V}) + \rho g\nabla\varphi - \mu\nabla^2\mathbf{V} = 0. \tag{1.2.6}$$

We now average (1.2.6) across the fracture width (normal to the fracture axis) to obtain a two-dimensional balance equation for linear momentum in the fracture plane

$$\int_{f_1}^{f_2}\left\{\rho\frac{\partial \mathbf{V}}{\partial t} + \rho\nabla\cdot(\mathbf{V}\mathbf{V}) + \rho g\nabla\varphi - \mu\nabla^2\mathbf{V}\right\}dy' = 0. \tag{1.2.7}$$

Since the limits of integration are independent of time, the first term in the integrand of (1.2.7) yields

$$\int_{f_1}^{f_2}\rho\frac{\partial \mathbf{V}}{\partial t}dy' = \rho\frac{\partial}{\partial t}\left(\int_{f_1}^{f_2}\mathbf{V}dy'\right) = \rho\frac{\partial(b\widetilde{\mathbf{V}})}{\partial t}, \tag{1.2.8}$$

where the average of a quantity A over a fracture width is defined

$$\widetilde{A}(x',z') \equiv \frac{1}{b}\int_{f_1}^{f_2}A(x',y',z')dy'. \tag{1.2.9}$$

By applying Leibnitz's rule (for taking a derivative of an integral) to the second term in the integrand of (1.2.7), we obtain

$$\begin{aligned}\int_{f_1}^{f_2}\rho\nabla\cdot(\mathbf{V}\mathbf{V})dy' &= \rho\nabla'\cdot\int_{f_1}^{f_2}\rho\mathbf{V}\mathbf{V}dy' - \rho\mathbf{V}\mathbf{V}|_{f_2}\cdot\nabla F_2 + \rho\mathbf{V}\mathbf{V}|_{f_1}\cdot\nabla F_1\\ &= \rho\nabla'\cdot(b\widetilde{\mathbf{V}\mathbf{V}}) - \rho\mathbf{V}\mathbf{V}|_{f_2}\cdot\nabla F_2 + \rho\mathbf{V}\mathbf{V}|_{f_1}\cdot\nabla F_1,\end{aligned} \tag{1.2.10}$$

where ∇' denoted differentiation only with respect to coordinates lying in the fracture $(x'z')$-plane. Let us define a velocity deviation, $\mathring{\mathbf{V}}$, such that

$$\mathbf{V}(x',y',z') \equiv \widetilde{\mathbf{V}}(x',z') + \mathring{\mathbf{V}}(x',y',z'), \quad \widetilde{\mathring{\mathbf{V}}} = 0. \tag{1.2.11}$$

We then have the relation

$$\widetilde{\mathbf{V}\mathbf{V}} = \widetilde{\mathbf{V}}\widetilde{\mathbf{V}} + \widetilde{\mathring{\mathbf{V}}\mathring{\mathbf{V}}}, \tag{1.2.12}$$

with $\rho\widetilde{\mathring{\mathbf{V}}\mathring{\mathbf{V}}}$ representing a *dispersive momentum* flux.

By introducing (1.2.12) into (1.2.10), we obtain

$$\begin{aligned}\int_{f_1}^{f_2}\rho\nabla\cdot(\mathbf{V}\mathbf{V})dy' &= \rho\nabla'\cdot(b\widetilde{\mathbf{V}}\widetilde{\mathbf{V}}) + \rho\nabla'\cdot(b\widetilde{\mathring{\mathbf{V}}\mathring{\mathbf{V}}})\\ &\quad -\rho\mathbf{V}\mathbf{V}|_{f_2}\cdot\nabla F_2 + \rho\mathbf{V}\mathbf{V}|_{f_1}\cdot\nabla F_1.\end{aligned} \tag{1.2.13}$$

The third term in the integrand of (1.2.7) is evaluated by

$$\int_{f_1}^{f_2}\rho g\nabla\varphi dy' = \rho g\nabla'(b\widetilde{\varphi}) - \rho g\varphi|_{f_2}\nabla F_2 + \rho g\varphi|_{f_1}\nabla F_1. \tag{1.2.14}$$

Applying Leibnitz's rule to the fourth term in the integrand of (1.2.7), gives

$$\int_{f_1}^{f_2} \mu\nabla^2\mathbf{V}\,dy' = \mu\nabla'^2(b\widetilde{\mathbf{V}}) - \mu\nabla'\cdot(\mathbf{V}|_{f_2}\nabla F_2 - \mathbf{V}|_{f_1}\nabla F_1) - \mu(\nabla\mathbf{V}|_{f_2}\cdot\nabla F_2 - \nabla\mathbf{V}|_{f_1}\cdot\nabla F_1). \tag{1.2.15}$$

Finally substituting (1.2.8), (1.2.13), (1.2.14) and (1.2.15) into (1.2.7), produces the averaged linear momentum balance equation in the fracture plane in the form

$$\begin{aligned}\rho\frac{\partial b\widetilde{\mathbf{V}}}{\partial t} &+ \rho\nabla'\cdot(b\widetilde{\mathbf{V}}\widetilde{\mathbf{V}}) + \rho\nabla'\cdot(b\widetilde{\mathring{\mathbf{V}}\mathring{\mathbf{V}}})\\ &- \rho\mathbf{V}\mathbf{V}|_{f_2}\cdot\nabla F_2 + \rho\mathbf{V}\mathbf{V}|_{f_1}\cdot\nabla F_1\\ &+ \rho g\nabla'(b\widetilde{\varphi}) - \rho g\varphi|_{f_2}\nabla F_2 + \rho g\varphi|_{f_1}T\nabla F_1.\\ &+ \mu\nabla'^2(b\widetilde{\mathbf{V}}) - \mu\nabla'\cdot(\mathbf{V}|_{f_2}\nabla F_2 - \mathbf{V}|_{f_1}\nabla F_1)\\ &- \mu(\nabla\mathbf{V}|_{f_2}\cdot\nabla F_2 - \nabla\mathbf{V}|_{f_1}\cdot\nabla F_1) = 0.\end{aligned} \tag{1.2.16}$$

The averaged mass balance equation takes the form

$$\frac{\partial(b\widetilde{\rho})}{\partial t} + \nabla'\cdot(b\widetilde{\rho}\widetilde{\mathbf{V}}) + (\rho\mathbf{V})|_{f_2}\cdot\nabla F_2 + (\rho\mathbf{V})|_{f_1}\cdot\nabla F_1 = 0. \tag{1.2.17}$$

With the assumptions of constant fluid density and stationary, nondeformable fracture walls employed above, (1.2.17) reduces to

$$\nabla'\cdot(b\widetilde{\rho}) - \mathbf{V}|_{f_2}\cdot\nabla F_2 + \mathbf{V}|_{f_1}\cdot\nabla F_1 = 0. \tag{1.2.18}$$

Substituting (1.2.18) into (1.2.16), yields

$$\begin{aligned}\rho\frac{\partial b\widetilde{\mathbf{V}}}{\partial t} &+ \rho b\widetilde{\mathbf{V}}(\nabla'\cdot\widetilde{\mathbf{V}}) + \rho\nabla'\cdot(b\widetilde{\mathring{\mathbf{V}}\mathring{\mathbf{V}}})\\ &- \rho\mathbf{V}\mathbf{V}|_{f_2}\cdot\nabla F_2 + \rho\mathbf{V}\mathbf{V}|_{f_1}\cdot\nabla F_1 + \rho g\nabla'(b\widetilde{\varphi})\\ &- \rho g\varphi|_{f_2}\nabla F_2 + \rho g\varphi|_{f_1}\nabla F_1\\ &+ \mu\nabla'^2(b\widetilde{\mathbf{V}}) - \mu\nabla'\cdot(\mathbf{V}|_{f_2}\nabla F_2 - \mathbf{V}|_{f_1}\nabla F_1)\\ &- \mu(\nabla\mathbf{V}|_{f_2}\cdot\nabla F_2 - \nabla\mathbf{V}|_{f_1}\cdot\nabla F_1)\\ &+ \rho\widetilde{\mathbf{V}}(\mathbf{V}|_{f_2}\nabla F_2 - \mathbf{V}|_{f_1}\nabla F_1) = 0.\end{aligned} \tag{1.2.19}$$

To further analyze (1.2.19), we shall consider the simple case of steady, unidirectional flow through a two-dimensional fracture bounded by the planar, parallel walls defined in Fig. 1.2.2. Furthermore, we assume that

- the dispersive momentum flux is much smaller than the advective one, i.e., $|\rho\widetilde{\mathring{\mathbf{V}}\mathring{\mathbf{V}}}| \ll |\rho\widetilde{\mathbf{V}}\widetilde{\mathbf{V}}|$, and
- across any aperture, the piezometric heads at the fracture walls satisfy $\varphi|_{f_2} \simeq \varphi|_{f_2}$ (the stronger condition, $\varphi|_{f_2} \simeq \varphi|_{f_2} \simeq \widetilde{\varphi}$, is required if the fracture walls are not assumed parallel).

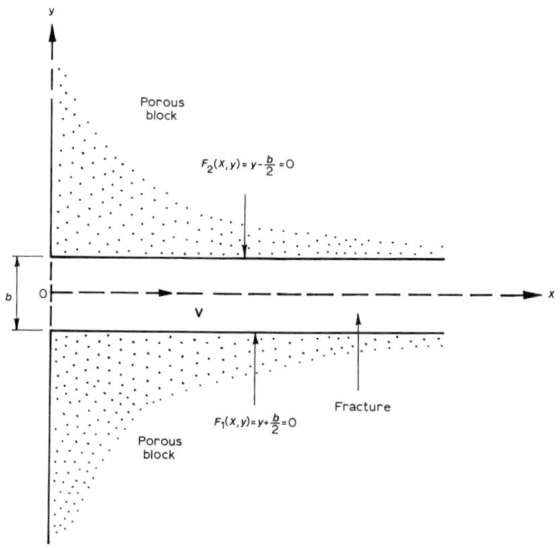

Figure 1.2.2. Fracture–porous block geometry in one-dimensional case.

Under these assumptions, and for steady flow, (1.2.19) reduces to

$$\rho g b \frac{d\widetilde{\varphi}}{dx} - \mu \left(\left. \frac{\partial V_x}{\partial y} \right|_{f_2} - \left. \frac{\partial V_x}{\partial y} \right|_{f_1} \right) = 0. \tag{1.2.20}$$

Although (1.2.20) was developed for steady flow, this restriction would have been unnecessary had we neglected inertial effects already at the microscopic level, i.e., for the constant density assumed here

$$\rho \left\{ \frac{\partial \mathbf{V}}{\partial t} + \mathbf{V}\nabla\mathbf{V} \right\} \simeq 0, \quad \text{and} \quad \nabla\cdot\mathbf{V} = 0.$$

If the fracture walls are stationary and impervious, and a no-slip condition (i.e., $V_x = 0$ at the walls) is imposed on them, then the velocity distribution across the fracture width will be parabolic, symmetric about the fracture axis (Lamb, 1945), with

$$V_x(y) = \frac{6\widetilde{V_x}}{b^2}\left(\tfrac{1}{2}b + y\right)\left(\tfrac{1}{2}b - y\right), \qquad \tfrac{1}{2}b \le y \le \tfrac{1}{2}b. \tag{1.2.21}$$

By differentiating (1.2.21), and substituting the result into (1.2.20), we obtain the average velocity in a fracture, in the form

$$\widetilde{V_x} = -\frac{\rho g}{\mu}\frac{b^2}{12}\frac{d\widetilde{\varphi}}{dx}. \tag{1.2.22}$$

Equation (1.2.22) can be rewritten in the form

$$\widetilde{V_x} = -K_{fr}\frac{d\widetilde{\varphi}}{dx}, \tag{1.2.23}$$

where K_{fr}, is the *hydraulic conductivity in the fracture*, defined by

$$K_{fr} = \frac{\rho g}{\mu}\frac{b^2}{12}. \tag{1.2.24}$$

In general, the hydraulic conductivity, K, and the permeability, k, are related to each other by the expression

$$K = -\frac{\rho g}{\mu} k. \tag{1.2.25}$$

Hence, the permeability *in* the fracture, k_{fr}, is defined as

$$k_{fr} = \frac{b^2}{12}. \tag{1.2.26}$$

The total discharge through a fracture, $\mathbf{Q}_{fr}$, is expressed by

$$\mathbf{Q}'_{fr} = b\widetilde{\mathbf{V}} = -\frac{\rho g}{\mu}\frac{b^3}{12}\nabla\widetilde{\varphi} = -T_{fr}\nabla\widetilde{\varphi}, \tag{1.2.27}$$

where the prime indicates a vector in the fracture plane, and

$$T_{fr} = -\frac{\rho g}{\mu}\frac{b^3}{12} = K_{fr}b \tag{1.2.28}$$

denotes the *transmissivity of a fracture.* Most researchers refer to T_{fr}, defined by (1.2.28), as the *hydraulic conductivity of the fracture.* The above expressions are valid for any parallel wall fracture, independent of orientation.

Equation (1.2.28), often referred to as the *cubic law,* has been derived by many authors for the flow in fractures with parallel walls. Witherspoon *et al.*, (1980) have demonstrated the validity of the cubic law in a fracture where the walls are not parallel (in fact the walls have contact points) and the aperture is being reduced by the application of stress.

In the development presented above, it was implicitly assumed that the fracture wall is *smooth,* and that the flow is in the laminar range. This is, obviously, not the situation in reality. Lomize (1951) investigated the influence of fracture roughness on the (laminar) flow rate. He suggested the relationship

$$\Psi = \frac{96}{\text{Re}},$$

where Ψ, expressed by

$$\Psi = \frac{96}{\text{Re}}\left[1 + 6.0\left(\frac{\epsilon}{2b}\right)^{3/2}\right], \quad \text{valid for} \quad \frac{\epsilon}{b} > 0.0065,$$

is a friction factor, Re is the Reynolds number, ϵ is the fracture roughness, and b is the aperture. In general, laminar flow between parallel plates is expressed as

$$\Psi = \frac{96}{\text{Re}}f, \quad \text{or} \quad Q = \frac{1}{f}\frac{b^3}{12\mu}\mathcal{J},$$

with $f = 1$ for smooth walls, and $f > 1$ for rough ones.

Louis (1974) suggested formulæfor walls with various degrees of roughness in both laminar and turbulent flow regimes. Sayers (1990) studied the case of partially closed fractures. In his model, a fracture is envisioned as a planar region in which separation and contact points both exist.

Let us now return to (1.2.19), and consider the case of fracture walls that are assumed permeable. Then, with the other assumptions used to derive (1.2.20), we obtain

$$\begin{aligned} \rho g b\frac{d\widetilde{\varphi}}{dx} &- \mu\left(\left.\frac{\partial V_x}{\partial y}\right|_{f_2} - \left.\frac{\partial V_x}{\partial y}\right|_{f_1}\right) + \rho\widetilde{V_x}\left(V_y|_{f_2} - V_y|_{f_1}\right) \\ &- \mu\frac{\partial}{\partial x}\left(V_y|_{f_2} - V_y|_{f_1}\right) = 0. \end{aligned} \tag{1.2.29}$$

Assuming that the leakage into or out of the fracture is equal on both fracture walls, and uniform over the fracture length (i.e., $V_y|_{f_2} = V_y|_{f_1} = q_\ell$, equation (1.2.29) reduces to

$$\rho g b \frac{d\widetilde{\varphi}}{dx} - \mu \left(\left. \frac{\partial V_x}{\partial y} \right|_{f_2} - \left. \frac{\partial V_x}{\partial y} \right|_{f_1} \right) + 2\rho \widetilde{V_x} q_\ell = 0, \tag{1.2.30}$$

where q_ℓ denotes the rate of leakage into the wall.

We now assume that the velocity distribution across a fracture is described by (1.2.21), and that the no-slip condition exists also in the case of flow in a fracture with leakage through the walls. This assumption is, in fact, valid only when the flux through the fracture is much larger than the leakage into the fracture walls. Differentiating (1.2.21), and substituting the result into (1.2.30), yields a 'modified' average velocity in the fracture

$$\widetilde{V_x} = -\frac{\rho g b^2}{\mu + 2b\rho q_\ell} \frac{d\widetilde{\varphi}}{dx}, \tag{1.2.31}$$

to be compared with (1.2.22). Thus, for a fracture imbedded in a permeable host rock, the hydraulic conductivity in a fracture is also a function of the magnitude of the leakage through the fracture walls. If q_ℓ is negligible, or $\mu \gg b\rho q_\ell/6$, the conductivity in the fracture can be approximated by (1.2.24).

In addition to the observation made above that fracture walls are not smooth, the aperture of a fracture varies. The ideal model of 'parallel plates' does not exist in reality. Moreover, at points and areal segments within a fracture, the aperture may disappear altogether, as adjacent blocks come into direct contact.

Several authors have studied the effects of a variable aperture on the flow in a fracture. For a fracture approximated as a series of m discrete segments with different apertures (Fig. 1.2.3a), Wilson and Witherspoon (1974) expressed the 'effective aperture', b_{eff}, as

$$b_{\text{eff}}^3 = \frac{\sum_{i=1}^m \ell_i}{\sum_{i=1}^m (\ell_i / b_i^3)}, \tag{1.2.32}$$

where ℓ_i denotes the length of a fracture segment of aperture b_i.

This resultant harmonic-type mean is obtained by solving a problem analogous to that of determining the equivalent hydraulic conductivity of a layered soil, with flow normal to the layers (see, for example, Bear, 1979). For an aperture that varies continuously along the fracture (Fig. 1.2.3b), i.e., $b = b(s)$, equation (1.2.32) is replaced by

$$b_{\text{eff}}^3 = \int_0^L \left(\int_0^L \frac{ds}{[b(s)]^3} \right)^{-1} ds, \tag{1.2.33}$$

where L is the total fracture length. In both cases, we replace b in (1.2.24) and (1.2.28) by b_{eff}.

It is also possible to introduce a statistical distribution of apertures. Then

$$T_{fr} = \frac{1}{12} \frac{\rho g}{\mu} \ell \sum_{k=1}^{\infty} f_k b_k^3, \tag{1.2.34}$$

or

$$T_{fr} = \frac{1}{12} \frac{\rho g}{\mu} L \int_0^\infty b^3 f(b) db, \tag{1.2.35}$$

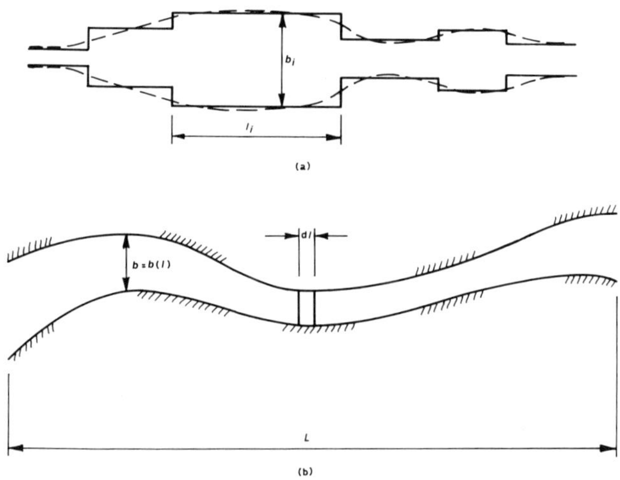

Figure 1.2.3. Fracture of variable aperture: (a) discrete aperture variation, (b) continuous aperture variation.

where b_k is the central value of aperture in the kth interval, f_k is the number of segments of width b_k in the kth interval, ℓ denotes the length of the segments, and $f(b)$ is the frequency distribution. For flow parallel (Q_1') and normal (Q_2') to these changes, we have (Fig. 1.2.4)

$$Q_1' = \frac{b_{\text{eff}}^3}{12}\frac{\rho g}{\mu}\mathcal{J}_1 = \frac{\rho g m}{12\mu \sum_{i=1}^{m}\left(\frac{1}{b_i^3}\right)}\mathcal{J}_1, \tag{1.2.36}$$

where $\mathcal{J} \equiv \nabla'\tilde{\varphi}$ denotes the hydraulic gradient, and

$$Q_2 = \frac{\rho g \ell \sum_{i=1}^{m} b_i^3}{12\mu}\mathcal{J}_2. \tag{1.2.37}$$

1.2.2 Flux in an Ordered Fracture System

The *parallel plate model* for a single fracture presented above, can be extended to various types of multiple fracture systems, by considering regular families of parallel fractures. Consider the simplest case of parallelepiped fractured rock domain intersected by a single family of m fractures of equal aperture, b (Fig. 1.2.5a), oriented parallel to the flow direction. The total discharge, Q, and specific discharge, q, through a cross-section this domain having a width L, and a height of unit length normal to the flow direction, are given by

$$Q = m\frac{b^3}{12}\frac{\rho g}{\mu}\mathcal{J}, \quad q \equiv \frac{\mathbf{Q}}{L} = \frac{m}{L}\frac{b^3}{12}\frac{\rho g}{\mu}\mathcal{J} \equiv K_f\mathcal{J}, \tag{1.2.38}$$

with

$$K_f = \frac{mb^3}{12L}\frac{\rho g}{\mu} = \phi_f\frac{b^2}{12}\frac{\rho g}{\mu}, \qquad \phi_f = \frac{mb}{L}, \tag{1.2.39}$$

denoting the hydraulic conductivity and the porosity of the fractured rock, respectively.

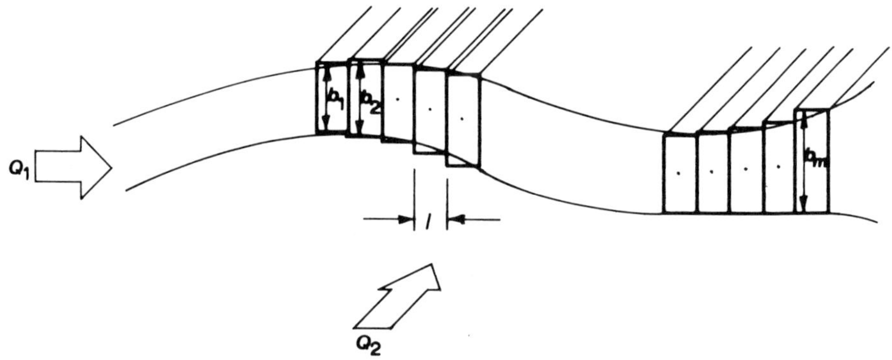

Figure 1.2.4. Flow in a fracture of variable aperture width.

If the fractures are of varying apertures, b_i, $i = 1, 2, ..., m$, then

$$Q = qL = \frac{1}{12}\frac{\rho g}{\mu}\left(\sum_{i=1}^{m} b_i^3\right)\mathcal{J} \equiv K_f L \mathcal{J}. \tag{1.2.40}$$

Then

$$K_f = \frac{1}{12L}\frac{\rho g}{\mu}\sum_{i=1}^{m} b_i^3 = \frac{\phi_f}{\sum_{i=1}^{m} b_i}\frac{1}{12}\frac{\rho g}{\mu}\sum_{i=1}^{m} b_i^3. \tag{1.2.41}$$

Consider a network composed of two orthogonal families of parallel fractures in the cross-sectional area of $L \times L$ shown in Fig. 1.2.5, normal to the flow direction. The variable apertures of the two families of fractures are denoted by b_i, $i = 1, 2, ...m_1$, and b_j, $j = 1, 2, ..., m_2$. The total discharge and hydraulic conductivity are determined by superposition

$$Q = Q_{m_1} + Q_{m_2} = qL^2 = \frac{1}{12}\frac{\rho g}{\mu}\left(\sum_{i=1}^{m_1} b_i^3 + \sum_{i=1}^{m_2} b_j^3\right)\mathcal{J}. \tag{1.2.42}$$

For such a system,

$$K_f = \frac{1}{12L^2}\frac{\rho g}{\mu}\left(\sum_{i=1}^{m_1} b_i^3 + \sum_{j=1}^{m_2} b_j^3\right), \qquad \phi_f \approx \frac{\sum_{i=1}^{m_1} b_i + \sum_{j=1}^{m_2} b_j}{L^2}. \tag{1.2.43}$$

We note that in (1.2.42) and (1.2.43), the flow through the fracture junctions is counted twice. However, it seems reasonable to assume that this will have only a very small effect on the calculated discharge and porosity.

Let us now consider the case in which the blocks surrounded by the fractures are pervious, with a permeability k_{pb}. For a fractured porous rock composed of parallel fractures of a constant aperture, the total flux, q_{FPR}, is expressed by

$$q_{\text{FPR}} = \frac{\rho g}{\mu L}\left(\frac{mb^3}{12} + k_{pb}(L - mb)\right)\mathcal{J}. \tag{1.2.44}$$

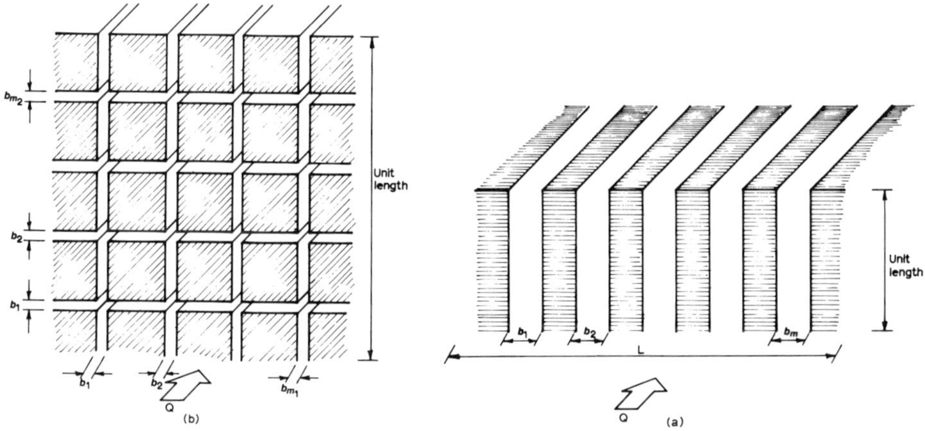

Figure 1.2.5. Multiple fracture system: (a) one family of parallel fractures, (b) two orthogonal families of parallel fractures.

Since $mb \ll L$, we have

$$q_{\text{FPR}} \approx \frac{\rho g}{\mu} \left(\frac{mb^3}{12L} + k_{pb} \right) \mathcal{J} = \frac{\rho g}{\mu} \left(\frac{\phi_f b^2}{12} + k_{pb} \right) \mathcal{J}. \tag{1.2.45}$$

In the entire development presented above, it has been implicitly assumed that the fracture walls are impervious, or that the leakage across them between the fractures and the porous blocks, is negligible. Otherwise, the development should have been based on (1.2.31).

1.2.3 Flux in an Assembly of Randomly Oriented Fractures

Consider a network of randomly oriented fractures imbedded in an impervious rock. The motion equation of a Newtonian fluid in such a porous domain can be obtained by averaging the Navier-Stokes equation over the void space contained in an REV, with a no-slip condition on the fluid-solid surface. In the laminar range of flow, the inertial effects are neglected in the Navier-Stokes equations.

The averaging over the void space can also be performed in two steps: first over the aperture width, b, in an segment of an individual fracture that has a planar axis surface (or can be approximated as such), and then over all the elementary areas of all the fractures within an REV.

Let us consider a simple example of this two-step averaging procedure. With the nomenclature of Fig. 1.2.6, let $\mathcal{J}$ denote the average hydraulic gradient at a point within the fractured rock domain. The average driving force *in the fracture plane*, $\mathcal{J}^*$, is

$$\mathcal{J}^* = \mathcal{J} - (\mathcal{J} \cdot \boldsymbol{\nu})\boldsymbol{\nu}.$$

Alternatively, we can express $\mathcal{J}^*$ by its components $\mathcal{J}_1^*$, $\mathcal{J}_2^*$ and $\mathcal{J}_3^*$, in the x_1, x_2 and x_3 directions,

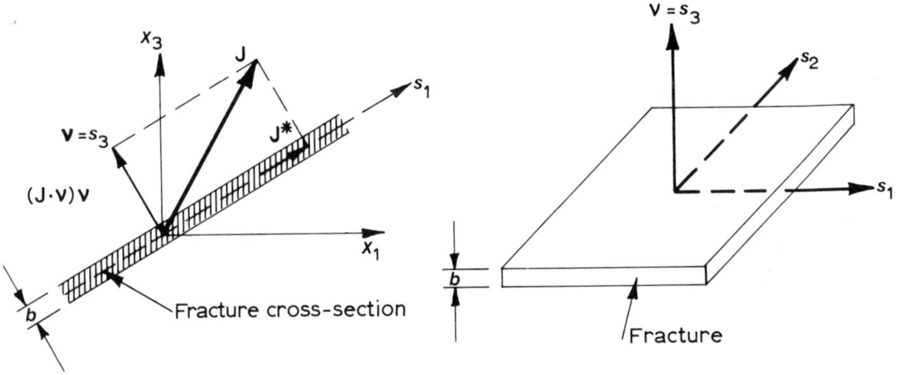

Figure 1.2.6. Nomenclature for flow in a fracture element.

respectively, so that

$$\mathcal{J}_j^* = \sum_{i=1}^{3} \mathcal{J}_i(\delta_{ij} - \nu_i\nu_j), \qquad i,j = 1,2,3, \tag{1.2.46}$$

where the $\mathcal{J}_i's$, $(i = 1,2,3)$, are components of the averaged hydraulic gradient, $\mathcal{J}$, in the domain, and δ_{ij} denotes the *Kronecker delta.* Hence, the average velocity at a point *in a fracture*, $\widetilde{\text{V}}$, is given by

$$\widetilde{V_j} = K_{fr}\mathcal{J}_j^* = K_{fr}(\delta_{ij} - \nu_i\nu_j)\mathcal{J}_i, \qquad i,j = 1,2,3, \tag{1.2.47}$$

where K_{fr} is the hydraulic conductivity *in an individual fracture* defined by (1.2.24), and we have invoked the *summation convention.* The expression $K_{fr}(\delta_{ij} - \nu_i\nu_j)$ is a second rank symmetric tensor. Equation (1.2.47) was also developed by Snow (1969).

To obtain the hydraulic conductivity of the FR-domain, we have now to average (1.2.47) over the entire void space within an REV. Since the velocity has already been averaged across a fracture, the second averaging, mentioned above, has to be performed over the total area of fractures within the REV. Integrating (1.2.47) in this manner, and dividing by the volume of the REV, gives

$$\begin{aligned} \overline{V_j} &\equiv \frac{1}{U_o}\int_{(U_{ov})} V_j\,dU = \frac{1}{\Sigma A}\int_{(\Sigma A)} \widetilde{V_j}\,dA \\ &= \frac{1}{\Sigma A}\int_{(\Sigma A)} K_{fr}(\delta_{ij} - \nu_i\nu_j)\mathcal{J}_i\,dA, \end{aligned} \tag{1.2.48}$$

where U_o denotes the volume of the REV, U_{ov} denotes the volume of the void space within U_o, ΣA $(= U_{ov}/ < b >)$ represents the total area of the fracture surface elements within U_o, and we have used the symbol $< .. >$ to denote the average over the total surface area of the fractures.

Recalling that $\mathcal{J}$ already denotes an average value, we can rewrite (1.2.48) in the form

$$\overline{V_j} =< K_{fr}(\delta_{ij} - \nu_i\nu_j) > \mathcal{J}_i. \tag{1.2.49}$$

The overall specific discharge, q_j, in the fracture rock domain, is given by

$$q_j \equiv \phi\overline{V_j} = \phi_{fr}\overline{K_{fr}(\delta_{ij} - \nu_i\nu_j)}\mathcal{J}_i \equiv \left(K_{fr}^*\right)_{ij}\mathcal{J}_i, \tag{1.2.50}$$

where

$$\left(K^*_{fr}\right)_{ij} \equiv \phi_{fr} < K_{fr}(\delta_{ij} - \nu_i\nu_j) >,$$

with $\phi_{fr} =< b > \Sigma A/U_o$, is now the hydraulic conductivity of the fractured rock with randomly oriented fractures. We note that, depending on the distribution of orientations of the unit vector $\boldsymbol{\nu}$, the permeability may exhibit the behavior of an anisotropic porous medium. In the latter case. One can determine the *principal directions* of the anisotropic permeability.

Often, the network of fractures is composed of a number of sets of fractures, each composed of parallel (say, equally spaced) fractures having a specified orientation in space. The various sets are not necessarily orthogonal to each other. In performing the integration over the total surface area of all fractures within an REV in (1.2.48), we may replace the integral by a sum of integrals, one for each set of fractures (for which the direction of the unit vector is a constant). We may also combine the discussion in Subs. 2.2.2 above, with the discussion here. We can then relate the anisotropic nature of a fractured rock domain to the orthogonal or non-orthogonal sets of parallel fractures.

Maini and Hocking (1977) in two dimensions, and Feuga (1981) in three dimensions, presented expressions for the anisotropic hydraulic conductivity of a fractured rock domain with a number of sets of parallel fractures.

It may be of interest to note that by making use of the notion employed above in the averaging procedure, that a segment of a fracture is a fracture that has its own aperture and permeability, segments with zero (or almost zero) aperture can be included. However, when dealing with multiphase (e.g., air-water) flow, we have to take care of the fact that due to capillary phenomena, fracture segments with smaller apertures will be occupied by the wetting fluid (water in the case of an air-water system), while the non-wetting fluid will occupy (and flow through) the segments with larger apertures (noting that smaller apertures are, in general, located around contact points and segments). Since we deal here only with single phase flow, we shall not pursue this issue any further (see Chapter 5 on unsaturated flow of this volume).

When the network of randomly oriented fractured is imbedded in a porous rock, we may employ the conclusions of Subs. 2.2.2 in order to derive the permeability of an FPR-domain.

1.2.4 Mass Balance Equations in a Single Fracture

We consider flow in a single fracture. The microscopic differential balance equation for any transported extensive quantity can be averaged over the aperture to produce an averaged equation for two-dimensional transport of the considered extensive quantity in the fracture plane. In doing so, we usually assume negligible variations in the relevant intensive quantities *across* the fracture's aperture.

We shall focus our attention first on the motion of mass of a fluid phase. As above, we shall assume that an 'axis surface', or an 'axis plane' can be defined for the considered fracture.

We start from the microscopic differential mass balance equation for the fluid, in the form of

$$\frac{\partial \rho}{\partial t} = -\nabla \cdot (\rho \mathbf{V}). \qquad (1.2.51)$$

Note that since $\mathbf{V}$ denotes the mass weighted velocity of the fluid, there is no need to introduce the diffusive mass flux of the fluid. We have assumed that no distributed sources of fluid mass are present in the fracture. The mass balance equation can always be modified to accommodate such sources.

By averaging this equation across the aperture, b, (normal to the latter's axis), and with the nomenclature of Fig. 1.1.1, we obtain

$$\int_{f_1(x',y',z')}^{f_2(x',y',z')} \left[\frac{\partial \rho}{\partial t} + \nabla\cdot(\rho \mathbf{V})\right] dy' = 0. \tag{1.2.52}$$

Noting that the limits of integration are assumed to be independent of time, and employing Leibnitz' rule, we rewrite (1.2.52) in the form

$$\frac{\partial}{\partial t} b\tilde{\rho} + \nabla'\cdot(b\widetilde{\rho \mathbf{V}}) - (\rho \mathbf{V})|_{f_2}\cdot\nabla F_2 + (\rho \mathbf{V})|_{f_1}\cdot\nabla F_1 = 0. \tag{1.2.53}$$

We now introduce the relation

$$\widetilde{\rho \mathbf{V}} = \tilde{\rho}\tilde{\mathbf{V}} + \widetilde{\mathring{\rho}\mathring{\mathbf{V}}}, \tag{1.2.54}$$

in which $\mathring{\rho}$ and $\mathring{\mathbf{V}}$ are deviations from the average, defined by

$$\mathring{\rho}(x', y', z') = \rho(x', y', z') - \tilde{\rho}(x', z'), \qquad \tilde{\mathring{\rho}} = 0,$$

and

$$\mathring{\mathbf{V}}(x', y', z') = \mathbf{V}(x', y', z') - \tilde{\mathbf{V}}(x', z'), \qquad \tilde{\mathring{\mathbf{V}}} = 0.$$

Assuming that the dispersive flux of the total mass in (1.2.54) is much smaller than the advective one, i.e., $|\tilde{\rho}\tilde{\mathbf{V}}| \gg |\widetilde{\mathring{\rho}\mathring{\mathbf{V}}}|$, equation (1.2.53) reduces to

$$b\frac{\partial}{\partial t}\tilde{\rho} + \nabla'\cdot(b\tilde{\rho}\tilde{\mathbf{V}}) - (\rho \mathbf{V})|_{f_2}\cdot\nabla F_2 + (\rho \mathbf{V})|_{f_1}\cdot\nabla F_1 = 0. \tag{1.2.55}$$

If we now assume steady averaged flow through a fracture of constant aperture (Fig. 1.2.2), equation (1.2.55) reduces to

$$b\left(\frac{\partial \tilde{\rho}}{\partial t} + \frac{\partial \tilde{\rho}\widetilde{V_x}}{\partial x} + \frac{\partial \tilde{\rho}\widetilde{V_z}}{\partial z}\right) - (\rho V_y)|_{f_2} + (\rho V_y)|_{f_1} = 0. \tag{1.2.56}$$

For a fluid of constant density, (1.2.56) is further reduced to

$$b\left(\frac{\partial \widetilde{V_x}}{\partial x} + \frac{\partial \widetilde{V_z}}{\partial z}\right) - (V_y)|_{f_2} + (V_y)|_{f_1} = 0. \tag{1.2.57}$$

The last two terms on the l.h.s. of (1.2.57) represent influx and efflux of fluid through the two fracture walls. For a fracture imbedded in an impervious rock, these two terms vanish.

1.2.5 Mass Balance Equations in a Double Porosity Model

In Subs. 2.1.3 we have introduced the *double porosity model* for a fractured porous rock domain. In such a model, we visualize the void space of the fractures as a continuum (occupied by one or more fluids), while the void space within the blocks is regarded as another contimuum which is *occupied by the same fluid, or fluids*. The two void-space continua may exchange fluid (or fluids) mass between them at every (macroscopic) point within the considered domain. The transport of other extensive quantities, e.g., mass of a component, may also take place within each of the two continua, with a possible exchange between them. Needless to add that a double porosity model requires that a common REV be found for the network of fractures and for the porous medium in the blocks.

Many authors developed models of fluid flow based on the two- continua (or double-porosity) approach introduced by Barenblatt and Zheltov (1960). Among many others, we may mention Warren and Root (1963), Odeh (1965), Kazemi *et al.* (1969), Streltsova (1976), Da Prat *et al.* (1981).

In the present subsection, we shall consider the mass balance of a single fluid phase that occupies the void space of both the fractures and the porous blocks. We shall employ subscripts s, fr and pb to denote the solid, the fluid in the fractures and that within the porous blocks, respectively.

The microscopic mass balance equation is (1.2.51). Let us rewrite this equation, once for the fluid in the fractures and once for the fluid in the void space of the porous blocks, in the form

$$\frac{\partial \rho_{fr}}{\partial t} = -\nabla\cdot\rho_{fr}\mathbf{V}_{fr}, \tag{1.2.58}$$

$$\frac{\partial \rho_{pb}}{\partial t} = -\nabla\cdot\rho_{pb}\mathbf{V}_{pb}. \tag{1.2.59}$$

The fluid in the fractures within an REV, is surrounded by the fracture-block interface, $\mathcal{S}_{fr,pb}$. The latter consists of a fracture-solid interface, $\mathcal{S}_{fr,s}$-surface and a fracture-fluid surface, $\mathcal{S}_{fr,f}$-surface. Hence, the averaged mass balance for the fluid within the fractures is given by

$$\begin{aligned}\frac{\partial \theta_{fr}\overline{\rho_{fr}}^{fr}}{\partial t} &= -\nabla\cdot\theta_{fr}\left(\overline{\rho_{fr}}^{fr}\overline{\mathbf{V}_{fr}}^{fr} + \overline{\mathring{\rho}_{fr}\mathring{\mathbf{V}}_{fr}}^{fr}\right)\\ &\quad - \frac{1}{U_o}\int_{S_{fr,f}} \rho_{fr}(\mathbf{V}_{fr} - \mathbf{u})\cdot\boldsymbol{\nu}_{fr}\, dS\\ &\quad - \frac{1}{U_o}\int_{S_{fr,s}} \rho_{fr}(\mathbf{V}_{fr} - \mathbf{u})\cdot\boldsymbol{\nu}_{fr}\, dS,\end{aligned} \tag{1.2.60}$$

where $\theta_{fr}(= U_{ofr}/U_o)$ denotes the volumetric fraction of volume of fractures within the REV, $\mathbf{u}$ denotes the speed of displacement of an $\mathcal{S}_{fr,s}$- or $\mathcal{S}_{fr,f}$-surface, and $\mathring{(..)}_{fr}$ denotes here the deviation of $(..)_{fr}$ from its intrinsic phase average over the REV. The last term in (1.2.60) vanishes because the fracture-solid interface, $\mathcal{S}_{fr,s}$ is a material surface with respect to the fluid's mass. The remaining surface integral, over the fracture-fluid surface in the void space, $\mathcal{S}_{fr,f}$ expresses the rate at which fluid mass is transported, by advection from the fractures to the porous blocks, at a point within the domain (per unit volume of FPR-domain). Within the fracture network, we have mass transport by advection and dispersion. However, in most cases of practical interest, the fluid's density is almost constant, so that the dispersive flux of the fluid's mass may be neglected.

The averaged mass balance for the fluid within the porous blocks is given by

$$\begin{aligned}\frac{\partial \theta_{pb}\overline{\rho_{pb}}^{pb}}{\partial t} &= -\nabla\cdot\theta_{pb}\left(\overline{\rho_{pb}}^{pb}\overline{\mathbf{V}_{pb}}^{pb} + \overline{\mathring{\rho}_{pb}\mathring{\mathbf{V}}_{pb}}^{pb}\right)\\ &\quad - \frac{1}{U_o}\int_{S_{pb,f}} \rho_{pb}(\mathbf{V}_{pb} - \mathbf{u})\cdot\boldsymbol{\nu}_{pb}\, dS\\ &\quad - \frac{1}{U_o}\int_{S_{pb,s}} \rho_{pb}(\mathbf{V}_{pb} - \mathbf{u})\cdot\boldsymbol{\nu}_{pb}\, dS,\end{aligned} \tag{1.2.61}$$

$\theta_{pb}(= U_{opb}/U_o)$ is the volumetric fraction of the void space within the blocks $(= \phi - \theta_{fr})$, and denotes deviation of $(..)_{pb}$ from its intrinsic phase average. The $\mathcal{S}_{pb,s}$-surface is a material one with respect to the fluid's mass. Neglecting the dispersive fluxes in both the blocks and the fractures, and denoting the rate at which fluid mass moves from the void space in the fractures to the blocks, per unit volume of porous medium, by $f^{m}_{fr\to pb}$, we may rewrite the two balance equations in the form

$$\frac{\partial \theta_{fr}\overline{\rho_{fr}}^{fr}}{\partial t} = -\nabla\cdot\theta_{fr}\overline{\rho_{fr}}^{fr}\overline{\mathbf{V}_{fr}}^{fr} - f^{m}_{fr\to pb}, \tag{1.2.62}$$

$$\frac{\partial \theta_{pb}\overline{\rho_{pb}}^{pb}}{\partial t} = -\nabla \cdot \theta_{pb}\overline{\rho_{pb}}^{pb}\overline{\mathbf{V}_{pb}}^{pb} + f^m_{fr\to pb}, \tag{1.2.63}$$

$$f^m_{pb\to fr} = \frac{1}{U_o}\int_{S_{pb,f}} \rho_{pb}(\mathbf{V}_{pb} - \mathbf{u})\cdot\boldsymbol{\nu}_{pb}\, dS = -f^m_{fr\to pb}. \tag{1.2.64}$$

Barenblatt *et al.* (1960) constructed (1.2.62) and (1.2.63) from macroscopic mass balance considerations. However, they developed the mass exchange term from heuristic arguments only.

Let us assume that

$$\overline{\rho_{fr}}^{fr} \equiv \overline{\rho_{pb}}^{pb} \equiv \rho.$$

Then, with $\overline{p_{fr}}^{fr} \equiv p_{fr}$ and $\overline{p_{pb}}^{pb} \equiv p_{pb}$, we can write the flux equations for the fluid in the fractures and in the porous blocks continua in the forms

$$\theta_{pb}\overline{\mathbf{V}_{pb}}^{pb} = -\frac{k_{pb}}{\mu}\left(\nabla p_{pb} + \rho g \nabla z\right), \tag{1.2.65}$$

$$\theta_{fr}\overline{\mathbf{V}_{fr}}^{fr} = -\frac{k_{fr}}{\mu}\left(\nabla p_{fr} + \rho g \nabla z\right). \tag{1.2.66}$$

For the exchange term, Barenblatt *et al.* (1960) suggested

$$f^m_{fr\to pb} = C_{\text{FPR}}\frac{K_{pb}\rho A^2}{\mu}(p_{fr} - p_{pb}), \tag{1.2.67}$$

where A is the area of fracture–porous block interface, per unit volume of fractured porous medium ($\approx 2\theta_{fr}/ < b >$), with $< b >$ representing the average fracture aperture, and C_{FPR} denoting a dimensionless shape factor of the fractured porous medium.

For a compressible fluid, the density variations may be approximated by

$$\rho = \rho_o[1 + \beta_p(p - p_o)], \tag{1.2.68}$$

where β_p denotes the fluid's compressibility at constant temperature, and p_o denotes a reference pressure.

For a deformable porous rock, in which

$$\theta_{fr} = \theta_{fr}(p_{fr}, p_{pb}), \qquad \theta_{pb} = \theta_{pb}(p_{fr}, p_{pb}),$$

we may write

$$\begin{aligned} d\theta_{fr} &= \alpha_{11} dp_{fr} - \alpha_{12} dp_{pb} \\ d\theta_{pb} &= \alpha_{21} dp_{fr} - \alpha_{22} dp_{pb} \end{aligned} \tag{1.2.69}$$

where

$$\begin{aligned} \alpha_{11} &= \frac{\partial \theta_{fr}}{\partial p_{fr}} \qquad \alpha_{12} = \frac{\partial \theta_{fr}}{\partial p_{pb}}, \\ \alpha_{21} &= \frac{\partial \theta_{pb}}{\partial p_{fr}} \qquad \alpha_{22} = \frac{\partial \theta_{pb}}{\partial p_{pb}} \end{aligned} \tag{1.2.70}$$

are coefficients of porous rock compressibility.

In spite of the deformation of the fracture porous rock as a whole, we assume that the volume of solids, per unit volume of fracture porous rock remains unchanged, viz.

$$1 - \theta_{fr} - \theta_{pb} = \theta_s = \text{const.}$$

Hence,

$$\alpha_{11} = -\alpha_{21}, \qquad \alpha_{12} = -\alpha_{22}.$$

Under these conditions, and for an isotropic fracture porous rock, we insert the flux equations into the mass balance equations, and assume that the effect of gravity is much smaller than that of the pressure gradient.

For flow in the network of fractures, we obtain

$$\begin{aligned} \frac{k_{fr}}{\mu}\nabla^2 p_{fr} &= (\alpha_{11} + \beta_p\theta_{fr})\frac{\partial p_{fr}}{\partial t} + \alpha_{12}\frac{\partial p_{pb}}{\partial t} \\ & C_{fpr}\frac{k_{pb}A^2}{\mu}(p_{fr} - p_{pb}) - \frac{\beta_p k_{fr}}{\mu}(\nabla p_{fr})^2. \end{aligned} \tag{1.2.71}$$

For flow in the porous blocks, we obtain

$$\begin{aligned} \frac{k_{pb}}{\mu}\nabla^2 p_{pb} &= (\alpha_{22} + \beta_p\theta_{pb})\frac{\partial p_{pb}}{\partial t} + \alpha_{21}\frac{\partial p_{fr}}{\partial t} \\ & C_{fpr}\frac{k_{pb}A^2}{\mu}(p_{fr} - p_{pb}) - \frac{\beta_p k_{pb}}{\mu}(\nabla p_{pb})^2. \end{aligned} \tag{1.2.72}$$

The last term on the r.h.s. of the last two equations is very small and may be neglected. Assuming that

$$|\alpha_{12}dp_{pb}| \ll |\alpha_{11}dp_{fr}|, \qquad |\alpha_{21}dp_{fr}| \ll |\alpha_{22}dp_{pb}|, \tag{1.2.73}$$

$$|\nabla\cdot\rho\theta_{pb}\overline{\mathbf{V}_{pb}}^{pb}| \ll \left|\frac{\partial(\theta_{pb}\rho)}{\partial t}\right|, \qquad |\nabla\cdot\rho\theta_{fr}\overline{\mathbf{V}_{fr}}^{fr}| \gg \left|\frac{\partial(\theta_{fr}\rho)}{\partial t}\right|, \tag{1.2.74}$$

we obtain the approximate mass balance equation for flow in the network of fractures in the form

$$k_{fr}\nabla^2 p_{fr} + C_{\text{FPR}}k_{pb}A^2(p_{fr} - p_{pb}) = 0. \tag{1.2.75}$$

We note that in this equation we have only flow in the fractures and exchange of fluid between fractures and porous blocks.

For the porous blocks, the approximate mass balance equation takes the form

$$(\alpha_{22} + \beta_p\theta_{pb})\frac{\partial p_{pb}}{\partial t} - C_{\text{FPR}}\frac{k_{pb}A^2}{\mu}(p_{fr} - p_{pb}) = 0. \tag{1.2.76}$$

Here we note only fluid exchange between fractures and porous blocks, and elastic storage in the porous blocks.

By eliminating p_{pb} from (1.2.75) and (1.2.76), we obtain another approximate mass balance equation for the fluid in the fractures, in the form

$$\frac{\partial p_{fr}}{\partial t} - \frac{k_{fr}}{k_{pb}}\frac{1}{C_{\text{FPR}}A^2}\frac{\partial}{\partial t}(\nabla^2 p_{fr}) = \chi\nabla^2 p_{fr}, \tag{1.2.77}$$

where

$$\chi = \frac{k_{fr}}{\mu(\alpha_{22} + \beta_p\theta_{pb})}. \tag{1.2.78}$$

We note that the expression $\alpha_{22} + \beta_p\theta_{pb}$ plays here the role of *specific storativity* of the porous block, where most of the water is stored (compare with the same definition as used in flow through porous media).

As the size of the blocks decreases, the area, A, increases, and the second term on the l.h.s. of (1.2.34) vanishes. If also $\beta_p\theta_{pb} \gg \alpha_{22}$, and $k_{fr} \ll k_{pb}C_{\text{FPR}}A^2$ (corresponding to a reduction in the dimensions of the porous blocks), we obtain the mass balance equation for flow in the fractures in the simplified form

$$\beta_p\theta_{pb}\frac{\partial p_{fr}}{\partial t} = \frac{k_{fr}}{\mu}\nabla^2 p_{fr}. \tag{1.2.79}$$

This equation states that the flow takes place only in the fractures, while the change in fluid storage is due to the compressibility of the fluid in the porous blocks.

We note here that (1.2.67) represents just one possible way of expressing the exchange of fluid between the fracture and the porous block continua. Three alternative flow models have been developed and presented in the literature. Equation (1.2.67), first proposed by Barenblatt and Zheltov (1960) and Barenblatt *et al.* (1960) assumes quasi-steady leakage, and an exchange that is proportional to the difference between the average pressures in the fractures and in the porous blocks. The coefficient of proportionality, assumed a constant, is approximately correlated to the porous block geometry.

Duguid and Lee (1977) presented a slightly different double porosity model. Their two balance equations are:

$$\begin{aligned}
(1-\phi_{pb})\phi_{fr}\beta_p\frac{\partial p_{fr}}{\partial t} + (1-\phi_{pb})\phi_{pb}\beta_p\frac{\partial p_{pb}}{\partial t} &= -\nabla\cdot\mathbf{q}_{fr} + f_{pb\to fr}/\rho,\\
(1-\phi_{fr})\phi_{pb}\beta_p\frac{\partial p_{pb}}{\partial t} + (1-\phi_{fr})\phi_{fr}\beta_p\frac{\partial p_{fr}}{\partial t} &= -\nabla\cdot\mathbf{q}_{pb} - f_{pb\to fr}/\rho,
\end{aligned} \tag{1.2.80}$$

where the fluxes are expressed by

$$\begin{aligned}
\mathbf{q}_{fr} &= -\frac{\mathbf{k}_{fr}}{\mu}\cdot\left(\rho\frac{\partial \mathbf{q}_{fr}}{\partial t} + \nabla p_{fr}\right),\\
\mathbf{q}_{pb} &= -\frac{\mathbf{k}_{pb}}{\mu}\cdot\nabla p_{pb}),
\end{aligned} \tag{1.2.81}$$

the pressures are relative to the initial one, and it is assumed that $\phi_{fr} \ll \phi_{pb}$, while $k_{pb} \ll k_{fr}$. In their model, the rate of exchange of fluid between the porous blocks and the fractures is expressed as

$$f_{pb\to fr} = \frac{\alpha^{p\star}}{\mu}\left[(p_{pb}-p_{fr}) + 2\sum_{n=2}^{\infty}\left[(-1)^n p_{pb} - p_{fr}\right]\exp\left(\frac{-k^m n^2\pi^2 t}{2L^2\mu\phi_{pb}\beta_p}\right)\right],$$

in which

$$\alpha^{p\star} = \frac{4k_{pb}\phi_{fr}\rho}{\pi(b/2)L}$$

is the transfer coefficient for fluid mass from porous blocks to fractures, and L is a characteristic half dimension of a block. The term expressing the rate of exchange used here is similar to that used by Bredehoeft and Pinder (1970) for a leaky aquifer.

In unsteady leakage, the exchange is determined by solving the flow equation in the individual porous blocks, in an attempt to improve the correlation of the coefficient to porous block geometry and to the pressure distribution within the blocks. To date, two idealizations of the porous block geometry have been considered: prismatic blocks, and spherical blocks.

Overviews of these various formulations are provided by Huyakorn *et al.* (1983) and Sauveplane (1984). Obviously, all models are limited by the selected simplified porous block geometry, and by the assumption that all blocks are of uniform size.

Warren and Root (1963) employed a model which is essentially the same as (1.2.75) and (1.2.76), except that they did not neglect the effect of fluid storage in the fractures, related to $\partial p_{fr}/\partial t$ in (1.2.71). They presented an analytical solution for flow to a well producing at a constant rate from an infinite or finite horizontal reservoir. Kazemi *et al.* (1969) solve a similar problem. They solved for the spatial as well as the temporal variation in pressure. Da Prat *et al.* (1981) solved the problem of flow to a well producing at a constant pressure.

Without presenting their actual solution, it may be interesting to summarize the main features of the Warren and Root (1963) solution:

- Initially, fluid is removed primarily from the fractures. This is a consequence of the fact that the permeability of the fractures is much higher than that of the porous blocks. The case of impervious blocks reduces to that of a homogeneous single continuum with fracture permeability. During this phase, a straight line is obtained when the pressure decline in a well producing at a constant rate is plotted against the logarithm of time.

- Gradually, flow at an increasing rate takes place from the porous blocks to the fractures. This appears as a curve of variable slope on the above plot.

- Then equilibrium is reached between the two continua. This appears on the above plot as a straight line having the same slope as that describing the pressure decline during the initial period, but now the line is displaced parallel to itself. This indicates that the entire fractured porous rock behaves as an equivalent, single, homogenized continuum. The equality of the slopes during the first and third stages indicates that the permeability in the latter is primarily that of the fractures.

The solution derived by Warren and Root (1963), requires the knowledge of the coefficient C_{FPR} appearing in the transfer function. This coefficient depends on the configuration of the blocks. Warren and Root assumed that the coefficient is a constant that can be determined from tests. Crawford *et al.* al. (1976) describe such tests. In a nonhomogeneous material, magnitude of this coefficient will vary in space. However, Braester (1984) showed that even when this coefficient is taken as a constant (as in the case of uniform regular porous blocks), the pressure decline is not sufficiently sensitive to the block size. Braester, therefore, questions the usefulness of the solution.

1.2.6 Mass Balance in a Single Continuum Model

An important conclusion from the features discussed above is that the flow in a fractured porous rock differs from that in an ordinary porous medium only during the initial stages of transient flow, and only in the vicinity of a producing well. Beyond a relatively short time period, the flow regime is identical to that of a single continuum (Zone 4). In that single continuum, the flow can be described by a single mass balance equation for the fluid within the void space. We can obtain such equation by regarding the FPR as a continuum that has a porosity and a permeability like an 'ordinary' porous medium. We may obtain the equation also by adding (1.2.62) and (1.2.63). The terms expressing the exchange of mass between fractures and blocks will be eliminated.

Usually, it is assumed that the two subsystems–the fractures and the porous blocks–are in equilibrium from the point of view of pressure, so that they are everywhere under the same pressure. There is no net exchange of fluid between the two subsystems. The permeability of the FPR is then the sum of the two permeabilities.

1.3 CONTAMINANT TRANSPORT MODELS

1.3.1 General Considerations

Like flow, the transport of contaminants (= components of a fluid phase) in a fractured porous rock may take place in both the network of fractures and in the porous blocks.

The discussion on contaminant transport (= transport of the mass of a component of a fluid phase) in a fractured rock (FR), or in a fractures porous rock (FPR) domain, should be based on considerations which are similar to those underlying the discussion on flow (= the transport of mass of a fluid phase). The different scales of problems are essentially described by the four zones presented in Sec. 2.1. In fact, as explained there, if an appropriate REV can be found, a fractured rock domain will behave as an 'ordinary' porous medium also with respect to contaminant transport. However, depending on the spacing between fractures, the REV may be rather large. This means that the scale of heterogeneity resulting from the presence of fractures may be large. This, in turn, will affect continuum coefficients that depend on this scale, such as dispersivity.

We may, thus, have:

- Problems involving the transport of contaminants in a single fracture, say from an injection well. In a fractured porous rock, the contaminant, as it moves in the fracture, may spread also into the porous material bounding it (Zone 1).
- Problems involving a small number of fractures imbedded in an impervious or in a pervious rock. In the latter case, the contaminants will spread from the fractures to the adjacent porous material (Zone 2).

In both cases, the continuum approach to the description of the transport problem is not applicable.

- Problems involving a large number of fractures imbedded in a pervious or impervious rock, such that the continuum approach is applicable to the entire domain (as a single continuum model (Zone 4), or as a double porosity one (Zone 3)).

In some cases, the permeability of the porous blocks is very low relative to that of the fracture network, so that transport of contaminants in such block (as well as the exchange of contaminants between the blocks and the fractures) is by the mechanism of molecular diffusion only. This is a relatively slow mechanism. Contaminants transported in the fractures in an FPR domain will diffuse into the porous blocks. Then, while the contaminants may be flushed out of the fractures relatively quickly, say as part of clean-up operations, the porous blocks will only slowly release the contaminants into the fractures.

In a fractured rock domain, regarded as a continuum, or in the continuum of fractures in a double porosity model, the total flux of a component is made up of advection, dispersion and diffusion (similar to the fluxes in an 'ordinary' porous medium. However, the dispersivity may be rather large, while molecular diffusion in the network of fractures may be neglected (relative to the advective flux).

Similar to contaminant transport in an 'ordinary' porous medium, sources and sinks of a considered contaminant within a fluid phase may be of two kinds:

- Sources that are due to production of the contaminant by processes that take place within the fluid (per unit volume of the domain). Decay is considered a negative source (= a sink). Chemical reactions and radioactive (and other types of) decay may serve as examples. We often consider point injection and withdrawal of fluids and contaminants as belonging to this group, although they are actually conditions on domain boundaries.

- Sources that are due to exchange between the considered phase and other phases across their (microscopic) interphase boundaries. Adsorption is an example. In the averaged balance equation of the contaminant, this exchange appears as a source or a sink (per unit volume of domain).

In a fractured rock, adsorption can take place only on the (relatively small) surface area of the fractures. However, in a fractured porous rock, adsorption may take place also (actually primarily) on the very large solid surface area within the porous blocks.

1.3.2 Component Transport in a Single Fracture

We recall that at the microscopic level, transport takes place by advection and diffusion only. In principle, like flow, the transport within a fracture is three-dimensional. Because of the special geometry of the flow domain within a fracture, we may simplify the model describing flow in a fracture by transforming it (as we did for the flow in Subs. 2.2.1) into one that describes flow in a two-dimensional domain–that of the fracture axis.

Assuming no sources or sinks to be present, the microscopic differential mass balance equation for a contaminant at any point within a fracture is given by

$$\frac{\partial c}{\partial t} + \nabla\cdot(c\mathbf{V} - \mathcal{D}\nabla c) = 0, \tag{1.3.1}$$

where $c(x', y', z', t)$ denotes the contaminant's concentration (= mass of contaminant per unit volume of fluid), and $\mathcal{D}$ denotes the coefficient of molecular diffusion of the contaminant in the considered fluid. We have assumed that the concentration of the contaminant is rather low, so that the fluid's density remains unaffected by changes in c.

Referring to the geometry of Fig. 1.2.1, we average (1.3.1) across the fracture aperture, b, to obtain

$$\int_{f_1}^{f_2} \left[\frac{\partial c}{\partial t} + \nabla\cdot(c\mathbf{V} - \mathcal{D}\nabla c)\right] dy' = 0, \tag{1.3.2}$$

Following the methodology employed in Subs. 2.2.1, and assuming that the aperture, b, is independent of time, the resulting integrated equation takes the form

$$\begin{aligned} b\frac{\partial \tilde{c}}{\partial t} &+ \nabla'\cdot\lfloor b(\tilde{c}\tilde{\mathbf{V}} - \mathbf{D}\cdot\nabla'\tilde{c} - \mathcal{D}\nabla'(\tilde{c})\rfloor \\ &- (c\mathbf{V}) - \mathcal{D}\nabla c)|_{f_2}\cdot\nabla F_2 + (c\mathbf{V}) - \mathcal{D}\nabla c)|_{f_1}\cdot\nabla F_1 = 0 \\ &+ \nabla'\cdot[\mathcal{D}(c|_{f_2}\nabla F_2 - c|_{f_1}\nabla F_1 - \tilde{c}\nabla' b) \end{aligned} \tag{1.3.3}$$

where, $\tilde{c}(x', z', t)$ is the average concentration across a fracture, and $\widetilde{\mathring{c}\mathring{\mathbf{V}}}$, defined by

$$\widetilde{c\mathbf{V}} \equiv \tilde{c}\tilde{\mathbf{V}} + \widetilde{\mathring{c}\mathring{\mathbf{V}}}, \tag{1.3.4}$$

is the *dispersive flux*, introduced in the process of averaging. It is expressed as

$$\widetilde{\mathring{c}\mathring{\mathbf{V}}} = -\mathbf{D}\cdot\nabla'\tilde{c}, \tag{1.3.5}$$

where $\mathbf{D}$, a second rank symmetric tensor, is the coefficient of dispersion in a single fracture.

We note that (1.3.3) requires information on both c and ∇c on the fracture boundaries, F_1 and F_2. We also recall that f_2 and f_1 are related to the aperture by (1.2.3).

If we assume that the concentrations across a fracture satisfy

$$c|_{f_1} = c|_{f_2} = \widetilde{c},$$

the last term on the l.h.s. of (1.3.3) vanishes. If the fracture walls are parallel, it is sufficient to assume that $c|_{f_1} = c|_{f_2}$.

Thus, for the case of parallel walls (Fig. 1.2.2) for which $\nabla F_1 = \nabla F_2$, and with the assumptions listed in the development of the one- dimensional balance equation for mass transport in a fracture, (1.3.3) reduces to

$$\begin{aligned} b\left[\frac{\partial \widetilde{c}}{\partial t} + \frac{\partial}{\partial x_i}(\widetilde{c}\widetilde{V}_i) - \frac{\partial}{\partial x_i}\left(D_{hij}\frac{\partial \widetilde{c}}{\partial x_j}\right)\right] \\ - \left(cV_y - \mathcal{D}\frac{\partial c}{\partial y}\right)\Big|_{f_2} + \left(cV_y - \mathcal{D}\frac{\partial c}{\partial y}\right)\Big|_{f_1} = 0, \quad x_i, x_j = x, z, \end{aligned} \tag{1.3.6}$$

where $\mathbf{D}_h$ ($\equiv \mathbf{D} + \mathcal{D}\boldsymbol{\delta}$) denotes the *coefficient of hydrodynamic dispersion*, with $\boldsymbol{\delta}$ denoting the unit tensor.

In the one-dimensional case (in the x-direction only), the coefficient of hydrodynamic dispersion, in analogy to dispersion in an ordinary porous medium, can be expressed by

$$D_{hxx} \equiv D_{xx} + \mathcal{D} \equiv a_L \widetilde{V_x} + \mathcal{D}, \tag{1.3.7}$$

where a_L may be referred to as the *longitudinal dispersivity of the fracture*. In the more general two- dimensional flow in the fracture, if we assume that the fracture is 'isotropic' (say, it has the same behavior in all directions within a fracture), both longitudinal and transversal dispersivities in a fracture can be defined, in analogy to those in an ordinary isotropic porous medium. One can also assume an analogous expression for the relationship between the coefficient of dispersion and these dispersivities.

For a constant ρ and nondeformable fracture, the fluid mass balance equation, (1.2.57), reduces to $\nabla' \cdot \widetilde{\mathrm{V}} = 0$. Incorporating this statement in (1.3.6), reduces the latter to

$$\frac{\partial \widetilde{c}}{\partial t} + \widetilde{V}_i \frac{\partial \widetilde{c}}{\partial x_i} - \frac{\partial}{\partial x_i}\left(D_{hij}\frac{\partial \widetilde{c}}{\partial x_j}\right) - q_c|_{f_2} + q_c|_{f_1} = 0, \tag{1.3.8}$$

where

$$q_c \equiv \frac{1}{b}\left[(c - \widetilde{c})V_y - \mathcal{D}\frac{\partial c}{\partial y}\right] \tag{1.3.9}$$

denotes the contaminant flux through the wall (as viewed from the fracture side). If the matrix permeability is assumed small so that no advection occurs, the flux is only by diffusion.

Equation (1.3.8) describes two-dimensional transient contaminant transport in a fracture. In this equation, the unknown variable is $\widetilde{c}$.

The last two terms of (1.3.8) represent contaminant mass flux (by advection with the fluid and by diffusion) exchanged between the fracture and the porous block through the fracture walls. When the permeability of the block is sufficiently low, no exchange by advection takes place, and we set $V_y = 0$. There may still be exchange by diffusion. In a fracture rock domain, the source and sink terms vanish.

At the fracture-porous block interface, continuity of the flux normal to the boundary requires that there be no jump in the normal component of the total flux of the contaminant, i.e.,

$$(cV_y + \mathcal{D}\frac{\partial c}{\partial y})|_{fr} = \phi_{pb}(cV_y + \mathcal{D}^*\frac{\partial c}{\partial y})|_{pb}, \quad (V_y)|_{fr} = \phi_{pb}(V_y)|_{pb}, \tag{1.3.10}$$

where $\mathcal{D}^*$ takes into account the tortuosity of the void space in the porous block. We may use this condition to rewrite (1.3.10) in terms of porous block variables and parameters. We often assume that the boundary between the porous block and the fracture behaves like a 'well mixed zone' boundary (Bear and Bachmat, 1990), so that the flux normal to the boundary on the fracture's side is given by $\tilde{c}\mathbf{V}_y$.

Thus, (1.3.8) involves conditions on the boundary between the fracture and the porous block. These conditions must be specified, or calculated by other means.

The effects of contaminant adsorption and decay have not been included in this development.

The boundary conditions (1.3.9) couple the flow and contaminant transport in the fracture with those in the adjacent porous rock. The complete problem involves two variables: the concentration $\tilde{c}(x', z', t)$ in the fracture, and that in the porous block, $c_{pb}(x', y', z', t)$. the model involves two mass balance equations for the contaminant: one for the fracture and the other for the porous block. Appropriate boundary conditions have to be specified on the common boundary, and the two equations, because they are coupled, have to be solved simultaneously.

Several papers dealing with the transport of contaminants in a single fracture have been published in recent years. Most of them assume that the flow in the fracture is laminar, that the aperture is a constant with a planar axis surface, and that diffusion is the sole transport mechanism within the porous blocks bounding the fracture.

As a simple example, let us consider a vertical fracture with a constant aperture, b. The fracture axis is the xz-plane. We shall assume that the fracture's axis is also a plane of symmetry, so that it will be sufficient to model the half space $y \geq 0$. We assume that the permeability of the porous block is very low such that no advection takes place within it. The porous block is homogeneous and nondeformable. The balance equation for the mass of a contaminant in the fracture, obtained from (1.3.8), is

$$\frac{b}{2}\left(\frac{\partial \tilde{c}}{\partial t} + \tilde{V}_i \frac{\partial \tilde{c}}{\partial x_i} - \frac{\partial}{\partial x_i} D_{hij} \frac{\partial \tilde{c}}{\partial x_j}\right) + q_c = 0, \quad i, j = x, z, \tag{1.3.11}$$

where, expressed now as a flux on the porous medium side, we have

$$q_c = -\phi_{pb} \mathcal{D}^* \frac{\partial c}{\partial y}. \tag{1.3.12}$$

For the diffusion in the porous block, we write

$$\frac{\partial c}{\partial t} - \mathcal{D}^* \nabla^2 c = 0, \qquad y \geq \frac{b}{2}, \tag{1.3.13}$$

where the Laplacian operator is now in three dimensions. In this set of equations, the variables are

$$\begin{aligned} c &= c(x, y, z, t), \quad y \geq \frac{b}{2}, \\ \tilde{c} &= \tilde{c}(x, z, t), \quad y \leq \frac{b}{2}. \end{aligned}$$

Initial and boundary conditions have to be specified on the the external boundary of the fracture domain. Similarly, initial conditions have to be specified on the external boundaries of the porous rock domain. For the porous block, we often assume a semi-infinite domain, with zero concentration at infinity. Sometime we consider a case of two parallel fractures with a porous block between them. then, because of symmetry, it is sufficient to investigate half the distance between the fractures. The condition there is one of no diffusive flux normal to the boundary (a Neumann condition).

In addition, we need conditions at the common boundary–the fracture wall at $y = b/2$. We often assume there that $\widetilde{c} = c$.

Analytical solutions for contaminant transport in a single fracture, with advection in the fracture and diffusion in the porous block, were published, among others, by Neretnieks (1980), and Grisak and Pickens (1981). Since the studied situation is analogous to that of heat transfer by convection over a flat plate, solutions followed those of Carslaw and Jaeger (1959).

Tang *et al* (1981) solved a model with advection, dispersion and diffusion in the fracture, and diffusion in the porous block. They took into account adsorption and radioactive decay. For a fracture in the z-direction, and a porous medium in the yz- plane, their model consists of the following equations:

Mass balance equation for the transport of contaminant in the fracture

$$\frac{\partial \widetilde{c}}{\partial t} + \widetilde{V_z}\frac{\partial \widetilde{c}}{\partial z} - D_h\frac{\partial^2 \widetilde{c}}{\partial z^2} + \frac{2}{2}q_c + \frac{2}{b}f_{fr\to s} + \lambda\widetilde{c} = 0, \tag{1.3.14}$$

where it was assumed that $\partial V_z/\partial z = 0$, λ is a decay coefficient, q_c is the leakage into the porous block expressed by (1.3.12), and $f_{fr\to s}$ expresses the rate at which the considered contaminant leaves the fluid in the fracture to be adsorbed on the solid wall of the latter. It can be expressed by considering the mass balance for the contaminant on the solid wall.

Mass balance equation for the contaminant on the solid fracture wall

$$\frac{\partial c_{sfr}}{\partial t} = \frac{2}{b}f_{fr\to s} - \lambda c_{sfr}, \tag{1.3.15}$$

where c_{sfr} denotes the mass of the contaminant adsorbed on the solid wall per unit area of the latter. Assuming that adsorption can be described by the *linear equilibrium isotherm*

$$c_{sfr} = K_{dfr}\widetilde{c}, \tag{1.3.16}$$

where K_{dfr} is a *partitioning coefficient.* The next step is, usually, to insert (1.3.16) into (1.3.15), use the latter as a definition of $f_{fr\to s}$, and use the latter in (1.3.14).

Mass balance for the contaminant in the porous block. With c denoting the contaminant's concentration, this equation takes the form

$$\frac{\partial c}{\partial t} - \mathcal{D}^*\left(\frac{\partial^2 c}{\partial y^2} + \frac{\partial^2 c}{\partial z^2}\right) + \lambda c + \frac{f_{f\to s}}{\phi_{pb}}\widetilde{c} = 0, \tag{1.3.17}$$

where $f_{f\to s}$ denotes the rate at which the contaminant leaves the fluid in the void space to be adsorbed on the solid matrix. Again, assuming adsorption to be described by a linear equilibrium isotherm, we have:

$$\phi\frac{\partial c_{spb}}{\partial t} = f_{f\to s} - \phi_{pb}\lambda c_{spb}, \qquad c_{spb} = K_{dpb}c, \tag{1.3.18}$$

where c_{spb} denotes the mass of the contaminant adsorbed on the solid wall per unit mass of the solid matrix, and where K_{dpb} is a *partitioning coefficient* in the porous block. Again, the next step is to insert (1.3.18) into (1.3.14).

Tang *et al.* (1981) obtained an analytical solution for a case in which the diffusion in the z-direction in the porous block was neglected.

Sudicky and Frind (1982) solved the same model for a family of parallel fractures imbedded in a porous rock. Because of symmetry, midway between the fractures, the condition $\partial c/\partial y = 0$ is imposed. A similar problem was solved by Hodgkinson *et al.* (1983). They (1984) also presented an analytical solution for the problem of a radionuclide chain which decays during transport in a single fracture imbedded in a porous rock. The model takes into account adsorption in the fracture and in the block, but assumes no advection (and no dispersion) in the latter.

Rasmuson and Neretnieks (1981) added dispersion in the porous block to the model of Grisak and Pickens (1981) and considered a block of finite size. They considered the case of three orthogonal families of fractures. However, to simplify the solution, they replaced each cubic porous block by a sphere which has a finite capacity to absorb the contaminant.

We consider two continuæ: one composed of the porous blocks which form a continuous porous medium domain, and the other composed of the network of interconnected fractures imbedded in the first one. The respective porosities are θ_{pb} and θ_{fr}.

We may apply the same procedure as that applied to the mass of a phase also to the mass of a component of the fluid phase. Denoting the average concentration of the component in the network of fractures and in the porous blocks by c_{fr} and c_{pb}, respectively, the two (macroscopic) balance equations for the mass of the component take the form

$$\frac{\partial \theta_{fr} c_{fr}}{\partial t} = -\nabla \cdot \theta_{fr} \left(c_{fr} \mathbf{V}_{fr} + \mathbf{J}^{*\gamma}_{fr} + \mathbf{J}^{\gamma}_{fr} \right) - f^{\gamma}_{fr \to pb}, \tag{1.3.19}$$

$$\frac{\partial \theta_{pb} c_{pb}}{\partial t} = -\nabla \cdot \theta_{pb} \left(c_{pb} \mathbf{V}_{pb} + \mathbf{J}^{*\gamma}_{pb} + \mathbf{J}^{\gamma}_{pb} \right) - f^{\gamma}_{pb \to fr}, \tag{1.3.20}$$

where $\mathbf{J}^{*\gamma}_{\beta}$ and $\mathbf{J}^{\gamma}_{beta}$ denote the dispersive and diffusive mass fluxes of the component, respectively, relative to the mass weighted velocity of the fluid in the fractures ($\beta = fr$) and in the porous blocks ($\beta = pb$), and $f^{\gamma}_{fr \to pb}$ denotes the rate of transfer of the mass of the γ-component from the fractures to the blocks per unit volume of porous medium.

Often, the rate of component transfer, $f^{\gamma}_{fr \to pb}$, is assumed proportional to the difference between the average concentrations in the two continua, viz.,

$$f^{\gamma}_{fr \to pb} = \alpha^{c*}_{fr} (c_{fr} - c_{pb}), \tag{1.3.21}$$

where α^{c*}_{fr} is a *transfer coefficient*, often regarded as a constant proportional to the coefficient of molecular diffusivity in the porous blocks, to the area of contact between the two continua, and inversely proportional to some characteristic distance between the fractures and the interior of the blocks. However, the assumption that this coefficient is a constant is rather questionable, especially in the case of large blocks. This observation stems from the fact that the characteristic distance to the point within a block to which the average value may be attributed, varies continuously as the component diffuses into the block, or out of it, especially in large blocks.

An analogous model can be written for the transport of heat in a fractured porous medium.

List of Main Symbols

a_L	Longitudinal dispersivity of isotropic porous medium.
a_T	Transversal dispersivity of isotropic porous medium.
b	Aperture of fracture.
c	Concentration of a component (= mass of component per unit volume of phase).
D	Coefficient of dispersion of mass of a component.
$\mathcal{D}$	Coefficient of molecular diffusion of a component in a fluid phase.
$\mathcal{D}^*$	Coefficient of molecular diffusion of a component in a porous medium.
$\mathbf{D}_h$	Coefficient of hydrodynamic dispersion of a component.

f As subscript, symbol denoting fluid.
$f^E_{\alpha\to\beta}$ Rate of transfer of E from the α-phase to the β one, across their common microscopic interface per unit volume of porous medium.
fr As subscript, a symbol denoting a fracture.
g Gravity acceleration.
h As subscript, symbol denoting hydrodynamic dispersion.
$\mathcal{J}$ Hydraulic gradient.
$\mathbf{k}$ Permeability.
$\mathbf{K}$ Hydraulic conductivity.
$\mathbf{K}_{fr}$ Hydraulic conductivity in a fracture.
K_d Partitioning coefficient.
ℓ Characteristic size of REV.
L Characteristic size of domain.
p Pressure.
pb As subscript, symbol denoting porous block.
pm As subscript, symbol denoting porous medium.
$\mathbf{Q}$ Discharge.
Re Reynolds number.
s As subscript, symbol denoting solid.
$\mathcal{S}_{\alpha\beta}$ Surface of contact of α-phase with all other phases
t Time.
$\mathbf{T}_{fr}$ Transmissivity of a fracture.
$\mathbf{u}_\alpha$ Velocity of surface (e.g., of $\mathcal{S}_{\alpha\beta}$).
U Volume.
U_o Volume of domain of REV.
v As subscript, symbol indicating void space.
$\mathbf{V}$ Velocity.
x Horizontal coordinate.
$\mathbf{x}$ Position vector.
y Horizontal coordinate.
z Vertical coordinate (positive upward).

Greek letters

α As subscript, symbol for an α-phase.
$\alpha^{E\star}$ Transfer coefficient of E.
β As subscript, symbol for a β-phase. A symbol for all other phases, except α).
β_p Coefficient of fluid compressibility at constant pressure.
γ As superscript, symbol denoting a γ-component.
θ_α Volumetric fraction of an α-phase.
λ Coefficient of radioactive decay.
μ Dynamic viscosity of a fluid phase.
$\boldsymbol{\nu}_\alpha$ Outward unit vector to a surface.
ρ Mass density.
φ Piezometric head.
ϕ Porosity.

Special symbols

$\overline{(..)}$ Average, volume average, or phase average of (..) $(=\frac{1}{U_o}\int_{\mathcal{U}_o}(..)\,dU)$.
$\overline{(..)}^{\alpha}$ Intrinsic phase average of (..) $(=\frac{1}{U_{o\alpha}}\int_{\mathcal{U}_o}(..)\,dU)$.
$\mathring{G}$ Deviation of G from its intrinsic phase average, $\overline{G}^{\alpha}$, over an REV.
$\widetilde{(..)}$ Average of (..) over the thickness of a fracture.
$\overline{\overline{(..)}}$ Average of (..) over thickness of thin domain.
$\hat{(..)}$ Deviation of (..) from $\overline{\overline{(..)}}$.
$\nabla\cdot\mathbf{A}$ Divergence of a vector **A** (≡ div **A**).
∇A Gradient of a scalar A (≡ grad A).

References

Barenblatt, G. I., and Zheltov, I. P. Fundamental equations of filtration of homogeneous liquids in fissured rocks. *Soviet Dokl. Akad. Nauk.*, **13**(2):545–548, 1960.

Barenblatt, G. I., Zheltov, I. P., and Kochina, I. N. Basic concepts in the theory of seepage of homogeneous liquids in fissured rocks. *Soviet Appl. Math. Mech. (P.M.M.)*, **24**(5):852–864, 1960.

Bear, J. *Hydraulics of Groundwater.* McGraw-Hill, New York, 569pp., 1979.

Bear, J., and Bachmat, Y. *Introduction to Modeling of Transport Phenomena in Porous Media.* Kluwer Academic Publishers, Dordrecht, 553pp., 1990.

Bear, J., and Berkowitz, B. Groundwater Flow and Pollution in Fractured Rock Aquifers. In *Developments in Hydraulic Engineering* (ed. P. Novak), Elsevier Appl. Science Publ., **4**:175-238, 1987.

Braester, C. Influence of block size on the transition curve for a drawdown test in a naturally fractured reservoir, *J. Soc. Pet. Eng.*, October, pp. 494-504, 1984.

Bredehoeft, J. D., and Pinder, G. F. Digital analysis of areal flow in multiaquifer groundwater systems–a quasi-three- dimensional model, *Wat. Resour. Res.*, **6**(3):883-888, 1970.

Carslaw, I. S., and Jaeger, J. C. *Conduction of Heat in Solids*, 2nd ed., Oxford Univ. Press, London, 1959.

Charlaix, E. A criterion for percolation threshold in a random array of plates, *Solid State Communications*, **50**(11):999-1002, 1985.

Crawford, C. E., Hagedorn, N. R., and Pierce, N. E. Analysis of pressure build-up test in a naturally fractured reservoir, *J. Pet. Tech.*, December, pp. 1295-1300, 1976.

Da Prat, G., Cinco-Ley, H., and Ramey, H. J. Jr. Decline curve analysis using type curve for two-porosity system, *J. Soc. Pet. Eng.*, June, pp. 354-362, 1981.

Duguid, J. O., and Lee, P. C. Y. Flow in fractured porous media, *Wat. Resour. Res.*, **13**(3):558-566, 1977.

Feuga, B. Détermination des directions principales et de l'anisotropie de permeabilité d'un milieu rocheux fracture à l'aide de levés de fracturation. Approche théorique et premiéres application. *Rep. Rech. Géol. Min.*, 81 SGN 497 GEG, Orléans, 1981.

Grisak, G. E., and Pickens, J. F. An analytical solution for solute transport through fractured media with matrix diffusion, *J. Hydrol.*, **52**:47-57, 1981.

Hodgkinson, D. P., Lever, D. A., and England, T. H. *Mathematical Modeling of Radionuclide Transport Through Fractured Rock Using Numerical Inversion of Laplace Transforms: Application to INTRACOIN Level 3*, Theoretical Physics Div., AERE Harwell, Oxfordshire, 1983.

Huyakorn, P. S., Lester, B. H., and Faust, C. R. Finite element techniques for modeling groundwater flow in fractured aquifers, *Wat. Resour. Res.*, **19**(4):1019-1035, 1983.

Kazemi, H., Seth, M. S., and Thomas, G. W. The interpretation of interference tests in naturally fractured reservoirs with uniform fracture distribution, *J. Soc. Pet. Eng.*, **9**:463-472, 1969.

Lamb, H. *Hydrodynamics*. Dover, New York, 563pp., 1945.

Louis, , C. Introduction à l'hydraulic des roches. *Bull. Bur. Rech. Géol. Min.*, Ser. 2, Sect. III, No. 4, 1974.

Maini, T., and Hocking, G. Examination of the feasibility of hydrologic isolation of a high level waste repository in crystalline rocks. Session on Geologic Disposal of High Level Radioactive Waste. *Ann. Meet. Geol. Soc. Am.*, Seatle, 1977.

Marsily, G. de. Flow and transport in fractured rocks: connectivity and scale effects. *Proc. Int. Assoc. of Hydrogeologists: Hydrogeology of rocks of low permeability*, Tucson, AZ., **17**:267-277, 1985

Mercer, J. W., and Faust, C. R. Geothermal reservoir simulation, 3: Application of liquid- and vapor-dominated hydrothermal modeling techniques to Wairakei, New-Zealand, *Wat. Resour. Res.*, **15**(3):653-671, 1979.

Neretnieks, I., Diffusion in the rock matrix: an important factor in radionuclide retardation?, *J. Geophys. Res.* **85**(B8):4379-4397, 1980.

Odeh, A. S. Unsteady- state behavior of naturally fractures reservoirs, *J. Soc. Pet. Eng.*, **3**:60-64, 1965.

Rasmuson, A., and Neretnieks, I. Migration of radionuclides in fissured rock: The influence of micropore diffusion and longitudinal dispersion, *J. Geophys. Res.* **86**(B5):3749- 3758, 1981.

Robinson, P. C. *Connectivity, flow and transport in network models of fractured media.* Ph. D. thesis, Oxford. Rep. TP 1072, Atomic Energy Research Authority, Harwell, UK, 1984.

Sauveplane, C. Pumping test analysis in fractured aquifer formations: state of the art and some perspectives. In *Groundwater Hydraulics* (Rosenshein, J. S., and Bennet, G. D., eds.), Water Resources Monograph Series, Amer. Geophys. Union, Washington D. C., **9**, 1984.

Sayers, C. M. Fluid flow in a porous medium containing partially closed fractures, *Transport in Porous Media*, 1990. ???

Snow, D. T. Anisotropic permeability of fractured media, *Wat. Resour. Res.*, **5**:1273-1289, 1969.

Sudicky, E. A., and Frind, E. O. Contaminant transport in fractured porous media: analytical solution for a system of parallel fractures, *Wat. Resour. Res.*, **18**(6):1634-1642, 1982.

Streltsova, T. D. Hydrodynamics of groundwater flow in fractured formation, *Wat. Resour. Res.*, **12**:405- 414, 1976.

Tang, D. H., Frind, E. O., and Sudicky, E. A. Contaminant transport in fractured porous media: analytical solution for a single fracture, *Wat. Resour. Res.*, **17**(3):555-564, 1981.

Warren, J. E., and Root, P. J. The behavior of naturally fractured reservoirs, *J. Soc. Pet. Eng.*, September, 245-255, 1963.

Wilke, S., Guyon, E., and Marsily, M. de. Water penetration through rocks: test of a tridimensional percolation description. *J. Int. Assoc. Math. Geol.*, **17**(1):17-24, 1985.

Wilson, C. R., and Witherspoon. P. A. Steady state flow in rigid networks of fractures. *Wat. Res. Res.*, **10**(2):328- 335, 1985.

Witherspoon, P. A., Wang, J. S. Y., Iwai, K., and Gale, J. E. Validity of cubic law for fluid flow in a deformable rock fracture, *Wat. Resour. Res.*, **16**(6):1016-1024, 1980.

2

SOLUTE TRANSPORT IN FRACTURED ROCK — APPLICATIONS TO RADIONUCLIDE WASTE REPOSITORIES

IVARS NERETNIEKS
Department of Chemical Engineering
Royal Institute of Technology
S-100 44 Stockholm, Sweden

2.1 BACKGROUND AND INTRODUCTION

In recent years interest has increased considerably in the area of flow and transport in low permeability fractured rock. One important reason for this is that many countries are seriously considering siting final repositories for nuclear waste in such environments, at depths ranging from a few tens of meters, for low and intermediate level waste, to 500 m or even a kilometer or more for high level waste.

There is considerably less information and experience on depths below a few hundred meters than at shallower depths. To assess whether a repository is sufficiently isolated, information in several fields is needed. The flow rate and flow distribution at repository depth will strongly influence the rate of dissolution of many radionuclides. The flow paths and velocities will influence their travel time. A longer residence time will allow the radionuclides more time to decay. Over time, axial dispersion will dilute the species, but will also allow a fraction of the nuclides to travel faster. Channeling has the same effect. Transverse dispersion will cause dilution but also exchange species between fast and slow flow paths. For those nuclides that sorb on fissure surfaces and/or diffuse into the rock matrix, the frequency of water-conducting channels and the extent of their exposed surface directly influences the contact area between flowing water and rock.

In Canada, Switzerland, and Sweden large national research programs are underway to study radionuclide transport in crystalline rocks. Finland also intends to site their repository in similar rocks and has a research program directed towards this end. Other countries, including France, Japan, Spain, the United Kingdom, and the USA, have or have had research programs oriented toward crystalline rocks. In addition, an international field research program organized by the OECD/NEA in the Stripa mine in Sweden is in its third phase.

FLOW AND CONTAMINANT TRANSPORT IN FRACTURED ROCK

The investigations are aimed at assessing the long-term safety of nuclear waste repositories, where the time scales involved range from hundreds of years to many millions of years because several of the radionuclides have very long half-lives. The research is aimed at understanding the processes that determine the flowpaths and the flow rates of water in fractured rock masses. At least as important is the need to assess the release rates, the transport capacity, and rate of transport of the radionuclides. Solute transport is very much affected by the chemistry of the waters and by reaction processes involving exchange of dissolved and suspended species with the rock that is in contact with the mobile water.

The water moving in the fractures in the rock may transport dissolved species such as radionuclides from a repository for radioactive waste. The assessment of how much and at what rate the nuclides are carried by the moving water is usually based on calculations of the flow rate of the water in the rock, its velocity and pathways, and on the retardation of the radionuclides by physical and chemical interactions with the rock.

In three recent safety analyses for nuclear waste repositories KBS–3 (1983) in Sweden, NAGRA (1985) in Switzerland, and SFR (Neretnieks, 1988) in Sweden, the water flow rate calculations were based on conventional hydrological models that apply Darcy's law to a medium that was assumed to be well-described by a porous medium equivalent. The results from such calculations were flow rates and pathways in all parts of rock that were included in the model.

These flow rates and flow paths were used in the transport models. These models describe the movement of the radionuclides with the water. However, the radionuclides do not move with the velocity of the water in general. Their velocity will deviate from the average velocity because of variations in the water velocity and the mixing of different portions of the water. Because of these processes, which are summarized in the term "dispersion," some portion of the nuclides may move faster and some slower than the average velocity. In extreme cases, when the pathways are few and far between, some fast pathways may dominate the transport process. This is sometimes called "channeling."

Most of the nuclides are expected to be cationic or neutral in the waters present in the crystalline rocks investigated. They will adsorb and/or ion-exchange on the negatively charged surfaces of the rock minerals. These processes may considerably retard the radionuclides which, in some instances, can be expected to move many orders of magnitude slower than the water. For a given flow rate the retardation will be greater for a nuclide if there are more exposed surfaces with which the nuclides interact. Therefore, one of the crucial questions is how much fracture surface the flowing water encounters. In addition to interaction with the fracture surface, the radionuclides may diffuse into the rock matrix and sorb onto the inner surfaces of the rock matrix. The inner surfaces are larger by many orders of magnitude than the fracture surfaces in contact with the flowing water. If the inner surfaces are accessed, the retardation may increase considerably.

It is usually assumed that the fractured rock, even at depth, is sufficiently fractured to allow an averaging of the properties in a volume of rock such that it is meaningful to assign an average value of the property to a "point" in the rock. The properties of interest for flow are the hydraulic conductivity and the porosity. The size of the volume over which the averaging is performed is called a "representative elementary volume" (REV). This is the basis for the continuum models. It is assumed that the REV is considerably smaller than the rock volume studied. A mass balance equation into which Darcy's law is incorporated is used, together with the appropriate boundary conditions, to calculate the flow rate and flow directions at "all" points in the rock mass of interest. Known variations of properties in space, such as the presence of fracture zones and other large-scale features, are readily included. In principle only the amount of available information and the size of the computer limits the amount of detail that can be included in the model calculations. A detailed map of the flow rates, flow paths, and, if needed, the actual flow velocities, once obtained, can then be used directly in the transport models. This was done in KBS-3 (1983) and NAGRA (1985).

When there is channeling, the rates of flow in the different pathways are assessed and the transport calculations are done for the individual pathways. This approach was taken in the SFR (Neretnieks, 1988) repository for low and intermediate level reactor waste, where the site was shallower and the flow paths shorter.

This chapter starts with an overview of some concepts of flow and transport in fractured rock that have not yet become textbook material. It is based on general observations and large scale field investigations. Some general features of fractured rocks are discussed in relation to flow and transport of solutes. These concepts and ideas form the background for the design of many of the experiments that will be discussed later. Finally, some recent laboratory and field experiments are described, and their interpretation is discussed in relation to the concepts discussed earlier. The experiments described have been selected because they have given new insights or data. Several other experiments are mentioned only briefly.

2.2 SOME CONCEPTS OF FLOW OF WATER IN FRACTURED ROCK

2.2.1 Large Scale Flow

The flow in large volumes of fractured rock is usually assumed to take place in a pseudo homogeneous porous medium. Also, Darcy's law is often assumed to be valid in sparsely fractured rock. The scale at which a fractured rock has enough fractures to become representative of the rock mass depends on the frequency, orientation, length, other geometrical properties and permeability of the individual fractures and fracture clusters.

Recently five potential sites for a spent fuel repository were investigated in Sweden in great detail (KBS-3, 1983). They cover gneissic and granitic rocks. Aerial photographs and geophysical measurements were used to locate fracture zones over areas on the scale of tens of kilometers. On the more local scale tens of bore holes intersecting the fracture zones were used for mapping them in depth and for obtaining fracture and hydrological data. The widths of the fracture zones range from a few meters to hundreds of meters.

It was found that rock blocks between fracture zones on the order of several cubic kilometers could be found in which a repository could be sited. Rock blocks with a horizontal projected area smaller than a square km were not uncommon, but considerably larger blocks seem to be scarce. Figure 2.2.1 shows the interpreted fracture zones around the Gideå site and Fig. 2.2.2 shows a more localized view of the same site. The other sites are similar as regards fracture zones. Up to 15 deep (approximately 700 m), diamond drilled holes were investigated in each site in addition to many shallower holes (approximately 100 m). Fracture mapping of the cores show fracture frequencies varying from less than 1 to 4 fractures per meter at depths below a few hundred meters. This includes closed fractures. Double packer tests at distances of 5 and 10 m, and in some holes at 2-3 m, show that only a fraction of the visible fractures are conducting water. At

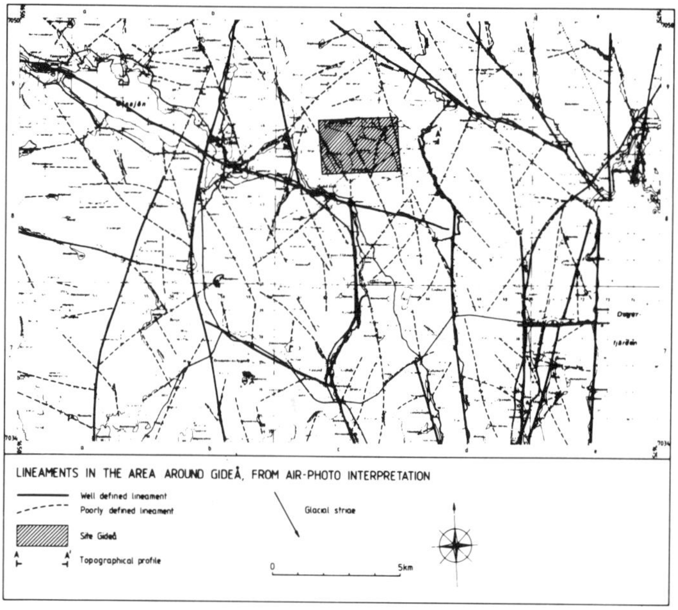

Figure 2.2.1. Interpreted lineaments in the area around Gideå.

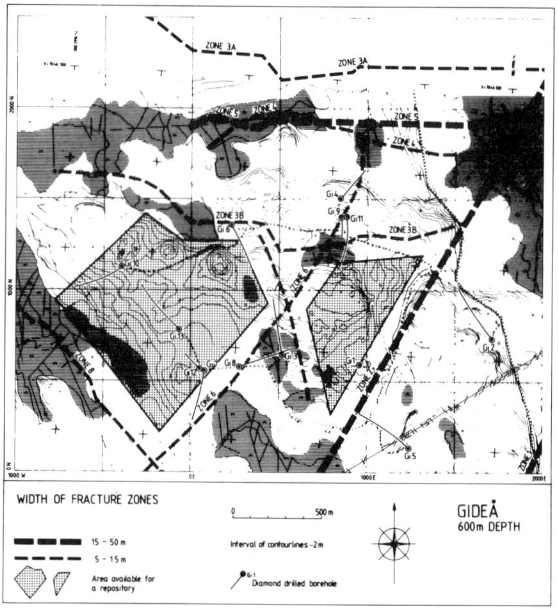

Figure 2.2.2. Location of fracture zones at 600 m depth in Gideå. Available area for possible repository at 500 and 600 m depth is indicated.

the Finnsjön and Sternö sites the distance between water-conducting fractures below 200 m was between 5 and 10 m. This includes even those fractures where the lowest conductivity (above the measurement limit) was obtained. Fractures and fracture zones with high conductivities have been found even at the largest depths. Similar observations have been made in deep boreholes in Switzerland that reach down to more than 1500 m (NAGRA 1985). Because of thick sedimentary overburdens in the Swiss investigated sites, surface mapping has not been possible.

Figure 2.2.3 shows a plot of hydraulic conductivity data from Kamlunge (KBS-3, IV). The hydraulic conductivities of the fracture zones were about an order of magnitude higher than that of the rock mass.

Within the Stripa project (Abelin et al., 1987), the 3D migration study has recently been finished. In this study tracer movements in a larger rock mass are observed. For this purpose a drift, 75 m long, with two side arms 12.5 m each, has been excavated at the 360-m level in the granitic rock of the rock laboratory. The rock is saturated with water and there is a natural inflow of water to the drift. This experiment will be described in more detail later. At present, only some observations that support the general concepts are outlined.

The natural flow to the upper half of the drift has been monitored in considerable detail by gluing on approximately 375 plastic collector sheets about 2 m × 1 m in size. The water flow seeping into every sheet is monitored. The water flow was found to be very unevenly distributed over the full 100 m of drift sections, covering a monitored area of more than 800 m^2. The total flow rate is less than 1 l/h. A third of the water flow takes place in about 2% of the area covered. In about 70% of the area no water flow has been found to take place.

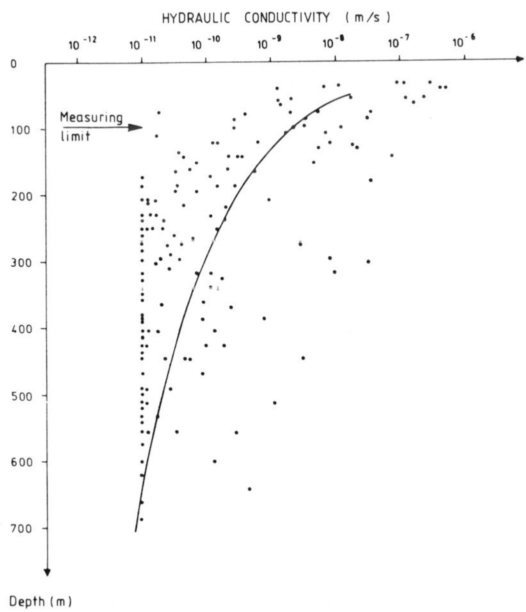

Figure 2.2.3. Correlation between hydraulic conductivity and depth for the rock mass.

A similar large scale flow and tracer experiment in central France at Fannay Augere was performed in a much more permeable rock than in Stripa (Cacas et al., 1990a,b). The length of the tunnel section used for the experiments was 100 m. The tracers arrived in three distinct locations in that section.

2.2.2 Flow in Individual Fractures

The flow in individual fractures is often assumed to be like flow between two parallel and smooth plates. In this case, the relation between flow and pressure gradient is governed by the "cubic law." This has been shown experimentally for fresh fractures in granite (Witherspoon et al., 1980). However, many natural fractures are old and weathered, and filled with precipitated minerals or clays, and cannot be expected to obey the cubic law. The pressure drop may be determined more by the physical size of the infilling material.

Analysis of tracer experiments between two boreholes in crystalline rock in three different experiments in Sweden indicate that the equivalent smooth plate fracture opening is different from that obtained from the residence time data of the conservative tracers. (Moreno et al., 1983, 1989). Similar results were obtained by analyzing flow and solute movement in three different natural fractures intersecting tunnels in the granite in the Stripa mine (Abelin et al., 1985). In this investigation, it was also found that the same fracture had very different hydraulic properties at points intersected by boreholes as little as 0.5 m apart. These observations indicated that the fractures were closed or very tight (mostly) with less frequent open parts between the areas where the two fracture surfaces are in contact. Observations in drifts and tunnels strengthen this impression. (Neretnieks, 1987a,b).

The observed preferential pathways in a fracture plane may have any of several causes. In a fracture that has uneven surfaces, and where the aperture varies, the water will seek out the easiest paths for the prevailing gradient. The paths will change as the direction of the gradient changes (Tsang et al., 1988; Moreno et al., 1988). Such paths may be clogged or opened by chemical and erosion processes so that they become permanent in the fracture plane. Such chemical changes are frequently observed in old crystalline rocks, where several generations of fracture filling materials are often found. Prominent dissolution channels have been found in fracture zones in Switzerland (Nagra, 1985). In two investigations (Neretnieks, 1987a; Abelin et al., 1987 SFR) it was found that fracture intersections are often high-flowrate conduits.

So far, fracture zones have not been investigated in the same detail as the regular rock mass. There are indications that some fracture zones may have properties that are quite variable (KBS-3, 1983; Black et al., 1987). On the other hand, in an experiment in a very prominent and very highly conductive fracture zone at Finnsjön, in Sweden, the hydraulic properties were within an order of magnitude of each other in several bore holes (Ahlbom and Smellie, 1989).

An ongoing investigation in Stripa aims to specifically measure the hydraulic conductivity variation in fracture planes on the scale of 2 m. First results (Abelin et al., 1988) indicate that there are very strong variations in conductivity in the plane of the fractures.

The form of the channels and specifically the contact surface between the mobile water and the rock is of prime importance for the interaction of dissolved species with the rock. The interaction may be purely physical, for instance, physical adsorption and diffusion into the rock matrix of solute molecules, or it may be chemical, for instance, ion exchange or precipitation.

2.3 SOME CONCEPTS OF TRANSPORT AND INTERACTION OF DISSOLVED SPECIES IN FRACTURED ROCK

Solutes that are dissolved in water will be carried by the moving water but will also move independently by various other mechanisms. Small molecules or ions diffuse in a concentration gradient and can move from one "stream tube" to another. Different water volumes move with different velocities and may mix at more or less regular intervals. A regular type of mixing may be described as a random process of the same type as molecular diffusion and is often described as such by the so called Fickian dispersion.

With only advection and dispersion active, the classical advection-dispersion equation results. It has been used extensively to describe tracer movement in porous media. It is easily modified to account for instantaneous chemical reaction with linear or nonlinear equilibria and can also easily accommodate reaction rates. The same equation can be used also to describe transport in parallel slits and in channels.

Ideally, hydrologic tracers are not supposed to react chemically with the solid material, but some naturally occurring tracers do, such as C-14, and the effect cannot be neglected. Also, in cases where it is of interest to describe the movement (and retardation) of reactive species, such as chemical waste and many radionuclides, the chemical reactions are an integral part of the problem. In field experiments kinetic effects often influence the results and must be accounted for in the interpretations.

The dissolved species may also experience kinetic effects caused by physical processes. One such process that may have a very large impact for solute transport in fractured rock is the diffusion in and out of zones in which the water is moving so slowly that it can for practical purposes be assumed to be stagnant. Such stagnant zones can be expected in fractures with uneven surfaces and with fracture-filling materials. In rocks with a connected matrix porosity, the accessible pore volume of the matrix can be very much larger than the mobile water volume in the fracture. Species that have access to the pore water will then have a residence volume and residence time determined by the sum of the water volume in fractures with flow and the water volume accessed in stagnant areas. Because the stagnant zones are reached by diffusion, the volume of stagnant water accessed is dependent on the residence time of the flowing water. Rock fractures often seem to have preferential channels where the water flows. Channels in the same fracture may never meet or mix their water for considerable distances. The mixing required for the process to become one of hydrodynamic dispersion in the sense necessary for it to be modeled as a Fickian process, may not occur to a sufficient extent over the distances of interest. The situation may then be better described as stratified flow or channeling.

The properties of the medium can vary considerably in crystalline fractured rock. Fracture apertures and thus the transport capacity of the individual fractures are known to span several orders of magnitude. The porosity of and diffusivity in the rock matrix has similar variability. Fracture coating and filling materials also may vary considerably in both type and amount. In fracture zones the block sizes vary from very small particles up to blocks of considerable size. The size of the blocks will strongly influence the amount of stagnant volume accessible in a given contact time.

The details of the mathematical formulations and their solutions will not be discussed here. These have been discussed in some detail in the literature (Neretnieks, 1985). Instead, some potential effects of channeling, matrix diffusion and sorption will be discussed in more detail.

2.3.1 Dispersion and Channeling

The advection-dispersion formulation implies that the dispersivity D_L is constant, independent of the distance. This is expected to be true for great distances when the fluid in the different fractures has been repeatedly mixed. In fractured rock at depths greater than a few hundred meters the fissures are far apart and long distances are needed for mixing. As long as the channels do not intersect or only a few of them intersect, and incomplete mixing has taken place, the transport may not be described well by an advection-dispersion model with constant dispersivity D_L. Most tracer experiments in fissured rock have been performed over distances of less than 50 m, which may not be sufficient to allow for frequent mixing.

The impact of the matrix diffusion for experiments involving short residence time should be small if laboratory diffusion data on crystalline rock are used for the estimation. In the Finnsjön field test (Moreno et al., 1983) the condition of the fracture surface is not known. The regression analysis of the breakthrough curve indicated that the matrix diffusivity was much higher than that of intact granite. The matrix diffusion would account for most of the pulse spreading if the fissure surfaces are more porous than the granite further away from the surface. Experiments in granite at Studsvik in Sweden (Landström et al., 1978) over 22 and 51 m distance were reported to give dispersion lengths of 6.1 and 7.7 m, respectively, when matrix diffusion is not accounted for. Webster et al. (1970) reported a dispersion length of 134 m for a migration distance of 538 m in fractured crystalline rock. The analysis does not account for matrix diffusion effects. Lallemand-Barres and Peaudecerf (1978) report on a confidential study in granite, where a dispersion length of 0.8 m was obtained over a 11.8 m migration distance. Gelhar (1987) compiled a number of investigations which give dispersion data of varying quality. Some of these investigations were performed in fractured rocks. These data also indicate that the dispersion coefficients increase with travel distance.

Figure 2.3.1 shows some data on dispersion (Neretnieks, 1985). The dispersion length increases with migration distance. This compilation of data is approximate because different methods have been used to determine the dispersivity.

An apparent increase in dispersion length can be caused by channeling and by various interaction mechanisms with the rock, e.g., matrix diffusion. Neretnieks (1983) has investigated a pure channeling model and shown that when pure channeling occurs, the apparent dispersion length is proportional to distance and never becomes constant. In the same work it is also shown that matrix diffusion may contribute considerably or even totally dominate the spreading of a tracer pulse when contact times are long. A large dispersion may dilute a contaminant plume and it has been argued that low dispersion cases therefore are of more interest, because the contaminant arrives in higher concentration. Because radionuclides decay, this is not necessarily the case for radionuclide transport. An early arrival of some portion of the plume may not have had time to decay and, although diluted, may have retained a considerably higher activity than that portion of the plume that travels with longer residence time.

In view of the low frequency of fissures in the deeper regions of the rock, it is not clear that the mixing between channels is frequent enough to let the dispersivity become constant and independent of distance before the flow intersects fractures or fracture zones with other properties.

The non-connected channel models will not account for mixing between channels. In a recent model (Cacas et al., 1990a,b) the fractures are assumed to have channels connecting the centers of

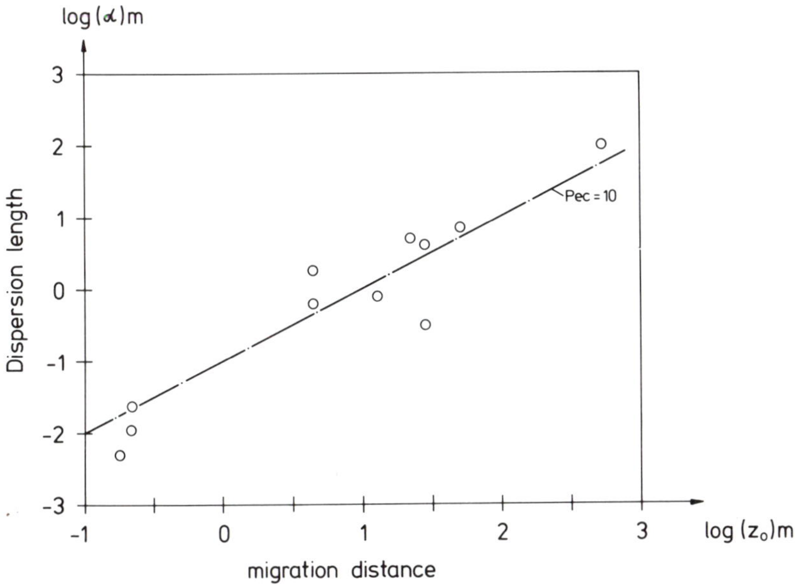

Figure 2.3.1. Dispersion in fractured rock as function of observation distance.

two intersecting fractures. The fracture network will, from the flow and transport point of view, form a network of connected tubes. This approach permits the channels to mix their water at intersections and can thus bridge the difference between an advection-dispersion concept and that of non-interacting channels. To calculate tracer transport, the volume of the channels is needed in addition to the hydraulic properties. For interacting species, the contact surface must also be known.

In other fracture network models the whole fracture plane is assumed to be available for the mobile water (Long et al., 1985). Other approaches have also been used where the individual fracture is made up of regions with different apertures (Tsang et al., 1988). Such fractures can then be combined in networks (Robinson, 1984, 1986; Derschowitz et al., 1985).

There has been a lack of data to test the various model concepts.

2.3.2 Matrix Diffusion Effects

Most rocks, even dense crystalline rocks such as granites and gneisses, have microscopically small fissures between the crystals. These fissures constitute a pore system, containing water, that is at least partly interconnected. Small molecules are much smaller than the microfissures and can diffuse in and out of this pore system. The inner surfaces in the rock matrix are many times (thousands and more) larger than the surfaces in the fractures in which the water flows, and the water volume in the microfissures is often much larger than the water volume in fractures. Over a long time scale, molecular diffusion may have an important effect in exchanging solutes between the mobile water in the fractures and the stagnant or near-stagnant water in the interior of the rock matrix. Cathles (1974) found in a field tracer experiment that the small molecule chloride did not arrive in the monitoring hole when colloidal silica particles did. This was interpreted as an effect of matrix diffusion, where the small molecule was taken up by the porous matrix while the colloids, with a much smaller diffusivity, did not.

Gneisses and granites in Swedish precambrian rock have been found to have a continuous pore system consisting of the microfissures between the crystals in the rock matrix. The porosity in this pore system varies between 0.06% and 1% (Skagius and Neretnieks, 1986a) for the rock matrix. Similar results have been obtained by other investigators (Brace, 1965; Bradbury, 1982; and NAGRA, 1985), Fracture minerals and rock in crushed zones have higher porosities, with values between 1 and 9% having been measured. Substances dissolved in the water can diffuse into this pore system and sorb on the inner surfaces. Figure 2.3.2 shows the concept of penetration and sorption in the porous rock matrix.

The penetration depth increases with time. Non-sorbing species penetrate far into the matrix, while sorbing nuclides are retarded, because they also have to fill up the sorption sites before they migrate further.

The penetration depth can be calculated with the aid of Fick's laws for diffusion, if the diffusivity is known, and by taking retardation due to sorption into account.

The size and form of the contact area will have a strong influence on the rate of interaction between the solutes and the rock matrix. Studies have been performed in which rock has been modeled as blocks or spheres of equal size, as blocks of unequal and irregular sizes, and as circular or banded channels (Neretnieks and Rasmuson, 1984; Rasmuson and Neretnieks,

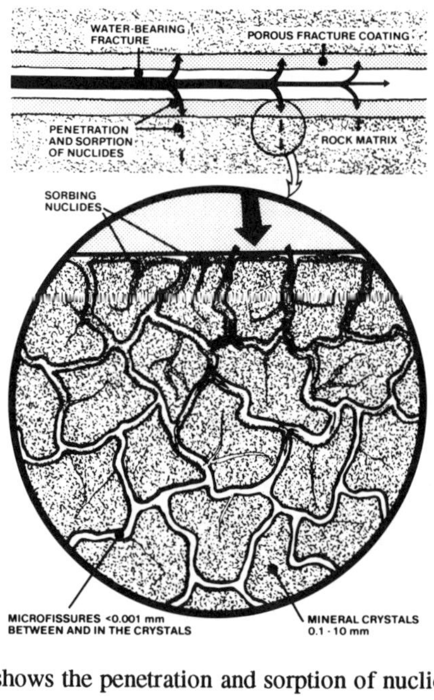

Figure 2.3.2. The figure shows the penetration and sorption of nuclides in the microfissures in the rock matrix.

1986a,b). The rate of interaction was found to depend primarily on the size of the wetted surface, but also on the form of the blocks and channels in cases where the range of diffusion is larger than the channel dimensions.

2.3.3 Sorption and Other Interactions

The minerals of crystalline rocks have a considerable cation exchange capacity. They also exhibit a capacity to form surface complexes with anions as well as cations. The capacity to bind different ions to the surface depends on the pH, the concentration of many of the other ions, and on the form the species have in the solution. Many cations form complexes with the anions and other complexing agents in the water. The complexed metal ions, which would have a positive charge if not complexed, may be positively charged, neutral species, or even negatively charged species, depending on the composition of the water.

The component may be bound (adsorbed, ion exchanged) to the surface of the minerals of the rock, or it may form its own minerals by precipitation. It may also react irreversibly by mineralization into some very stable solids.

The different complexes will of course have different affinity to the mineral surfaces. Because of the complexity of the situation, it is often convenient to summarize all the effects into a term "sorption coefficient" or K_d. This expresses how much of a component is bound to the surfaces of the rock and how much is in the water at equilibrium. One can neglect which form (complexed or non-complexed) the various species containing the component have in the solution as well as in the bound state. K_d is the ratio of concentrations of the component in the two phases.

The component may simultaneously be a part of several species (complexes) which are sorbed in different ratios. K_d often varies a great deal with changes in the composition of the groundwater (pH, Eh, concentration of complexing agents etc.). It is therefore a very imprecise entity, but because of its simplicity it has been found to be very useful in mathematical modeling and in transferring information from chemists to modelers.

Adsorption and ion exchange are often summarily called sorption, to denote fast reversible reactions of the dissolved components with the surface of the rocks. These processes are treated by the K_d concept. Processes such as precipitation and irreversible mineralization are usually treated as such.

2.3.4 A Quantitative Illustration of Matrix Diffusion Effects

The well known advection-diffusion equation is shown in (2.3.1). To it has been added a fourth term that accounts for the exchange of the solute with the matrix of rock by diffusion over the fracture surface. The fourth term consists of three parts: the concentration gradient in the matrix, the effective diffusivity in the matrix D_e, and the surface-to-volume ratio, a, of the fracture surface in contact with the mobile water to the volume of the mobile water in the rock. The equation can be applied to a variety of cases with different geometries, including flow in a porous medium with porous particles, flow in a fracture in a porous rock where the water in the rock matrix is stagnant, flow in channels, etc.

$$\frac{\partial C}{\partial t} = -u_f \frac{\partial C}{\partial z} = D_L \frac{\partial^2 C}{\partial z^2} + a\, D_e \left.\frac{\partial C_p}{\partial x}\right|_{x=0} \tag{2.3.1}$$

The movement by diffusion in the rock matrix is shown in (2.3.2). The right-hand term contains the apparent diffusivity D_a, which for diffusion in the pore water can be expressed as the ratio of the pore diffusivity D_p to the retardation coefficient R_d, which accounts for sorption effects.

$$\frac{\partial C_p}{\partial t} = \frac{D_p}{R_d} \frac{\partial^2 C_p}{\partial x^2} = D_a \frac{\partial^2 C}{\partial x^2} \tag{2.3.2}$$

R_d depends on the sorption coefficient K but also to a very large extent on the matrix porosity, which for small values may make the retardation coefficient (equation 2.3.3) very large.

$$R_d = 1 + K\left(1 - \varepsilon_p\right) / \varepsilon_p \tag{2.3.3}$$

It has been shown (Neretnieks, 1980) that for long contact times the matrix diffusion will have an important impact on the mobility of sorbing solutes and also that nonsorbing solutes such as tritiated water would potentially move considerably slower than the water velocity, making it difficult to utilize natural radionuclides to assess the water velocity (Neretnieks, 1981). It was shown by Neretnieks (1980), and Rasmuson and Neretnieks (1980), that for long contact times the velocity of the water may have a negligible impact on the velocity of a solute. The velocity of the latter may be influenced only by the ratio of the flow rate of the water to the accessed sorption capacity of the rock. These effects must be accounted for when designing and interpreting tracer tests.

To illustrate the effects quantitatively a simple example is given below, based on a case where there is no hydrodynamic dispersion but only advection, matrix diffusion and sorption in the matrix. For a step injection of a solute in a stream of water that is flowing in a medium consisting of rock blocks, where penetration depth is small compared to size of the blocks so that the geometry of the blocks can be approximated by flat surfaces, the solution to (2.3.1)-(2.3.3) is:

$$C / C_o = \mathrm{erfc}\left(\frac{B a t_w}{2\left(t \quad t_w\right)^{1/2}}\right) \tag{2.3.4}$$

The value a is the wetted surface per volume of mobile water, B is defined in (2.3.6) and accounts for the uptake in the rock matrix by diffusion and sorption. Even if there is no sorption, the uptake into the stagnant water in the matrix influences the movement of the solute, as can be seen from (2.3.6), which gives B for no sorption.

$$B = \left[D_p \varepsilon_p \left(\varepsilon_p + \left(1 - \varepsilon_p\right) K\right)\right]^{1/2} \tag{2.3.5}$$

For $K = 0$, B becomes

$$B = \left(D_p\right)^{1/2} \varepsilon \tag{2.3.6}$$

For small porosities and high sorption, B is given by equation (2.3.7).

$$B = \left(D_p \varepsilon_p K\right)^{1/2} \tag{2.3.7}$$

A simple criterion may be constructed from (2.3.4) to show when matrix diffusion becomes important as a retardation mechanism. It is formulated in the following way: "Matrix diffusion is important when, for a step input, the resulting breakthrough curve has a C/C_o equal to 0.5 at double the water residence time t_w." The argument in (2.3.4) must equal 0.477 for the left-hand side to become 0.5, and so it is found that for $t_w = 0.91/(B \cdot a)^2$, matrix diffusion will become important.

Figure 2.3.3 shows the responses to a finite pulse, calculated by the advection-dispersion equation and by the advection-dispersion-matrix diffusion equations for a case with a water residence time of 1 year. Rock property data typical for Swedish granites have been used. The responses differ considerably. If, in a field experiment, the matrix diffusion curve was evaluated using the advection-dispersion model, erroneous water residence times and dispersivities would be obtained. The use of the obtained data and the advection-dispersion model for extrapolation to other cases might give very misleading results. Maloszewski and Zuber (1985) have arrived at the same conclusion.

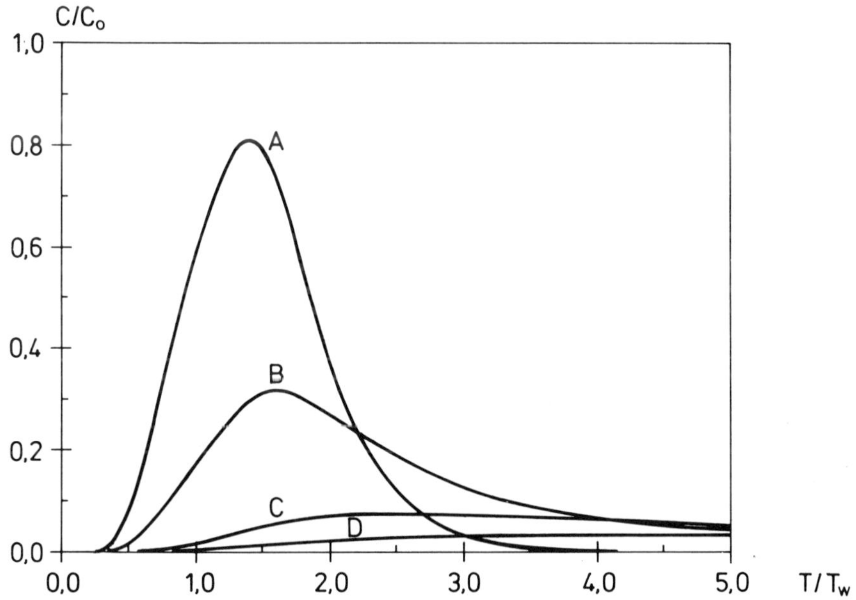

Figure 2.3.3. Response to a pulse in a 0.1 mm fracture. Influence of porosity and diffusivity. A: No matrix diffusion, B: D_p=2.5·10^{-11} m^2/s, ε_p= 0.002. C: D_p=4·10^{-11} m^2/s, ε_p= 0.005. D: D_p=10·10^{-11} m^2/s, ε_p= 0.002. Water residence time and injection pulse duration are 1 year.

2.3.5 Impact of Channeling on Transport

For media with largely unconnected channels, the advection-dispersion equation may not be a good descriptor in the sense that the "dispersivity" of the overall system is governed by velocity differences between channels and not by the attenuating influence of mixing. However, the transport in the individual channels may still be described well by the advection-dispersion model. Radionuclides transported in fast channels that mix their water only a little or not at all with other channels will have a shorter time in which to decay. In fast channels there will also be less time for the species to diffuse into the matrix. This may have very strong effects, as was found in a recent analysis of potential radionuclide migration from a repository for low and intermediate waste (Neretnieks, 1988).

2.4 EMERGING CONCEPTS OF CHANNELING

The notion of channeling seems to be very old. It has, however, seldom been described in the hydraulic literature. To water well drillers, miners and other underground workers it is well known that the water flow is unevenly distributed in the rock and that, often, unexpected "channels" are encountered when mining a new drift.

The following discussion describes some emerging concepts of channeling in an individual fracture.

A fracture is made up of two surfaces that are wavy and rough. The two surfaces are in contact with each other in some points but are at a distance from each other at other points. The openings in the fracture are potential channels. These are only potential conduits because they have to be connected to other open sections in order to form a continuous network.

When a hydraulic gradient is imposed over a fracture with a variable aperture, the water will seek out the easiest pathways. If there is a random but large variation in the transmissivities in different places, the water will seek out a tortuous path, always avoiding those sections where the transmissivity is low. In such a case most of the water may choose to flow along one path because it is the path with least resistance for that specific gradient.

Recently several model studies have been made on flow and transport in fractures that have variable apertures (Smith et al., 1987; Moreno et al., 1988; Tsang et al., 1988). Although based on very limited field data, the results indicate that considerable channeling is to be expected in such fractures.

It is possible that the flowing water will influence the transmissivity of the pathway in several ways. The pathways may be clogged due to filtering of particles or due to precipitation of minerals, or they may become larger due to erosion and dissolution. New pathways may open up due to small shearing movements in the fracture. All of these processes are known to occur. There have also been several observations that indicate that fracture intersections form easy pathways (Abelin et al., 1987; Neretnieks, 1987b; Moreno and Neretnieks, 1988).

The models indicate that there is a tendency for some pathways to carry much more water than others. In the limit, one or a few pathways will dominate. Mixing between the pathways would be small and a tracer introduced in one channel may not reach the other channel even after considerable travel distances. There will, then, be distinct channeling from the transport point of view.

There have been very few observations of channeling in a fracture plane (Bourke, 1987; Abelin et al., 1987,1988), and even less is known of what happens at fracture intersections. When fracture planes intersect, a channel may either connect to a closed or practically closed portion in the other fracture, or to some open portion. In the latter case the channel will continue over more than one fracture. At fracture intersections several distinct possibilities exist. The fracture intersection may act as a conduit, effectively connecting all channels from one fracture to the other. Another possibility is that the intersection has no flow or transport properties of its own, so that if a channel from one fracture does not meet a channel in the other fracture, there is no connection. A third possibility also exists: that the fracture intersection has a low transmissivity, effectively controlling (or hindering) the flow from one fracture plane to another.

In the first case there will be mixing "pools" at every fracture intersection. This will tend to average out concentration differences between channels. In the second case there will be considerably fewer channels which are connected from fracture to fracture. As more fractures are involved the tendency will be for fewer channels to traverse them all. In the latter case there will, in the limit, be no water flow.

No information is available today that would make it possible to distinguish between the cases. Probably, all three cases exist in real rocks. It is more a question of how common the cases are.

When there is no mixing between channels they can be treated as a bundle of individual conduits. It would then be sufficient to characterize the properties and the variation of the properties of the channels. With such information it would be possible to statistically generate a bundle of channels (or many bundles) and investigate their flow and transport properties. This is a very attractive notion because channels are essentially one-dimensional features from the flow and transport point of view. It would in this case not be necessary to make 2-dimensional, much less 3-dimensional calculations of water flow and transport. A series of simple 1-dimensional calculations would suffice. This has been explored by theoretical calculations (Neretnieks, 1983; Tsang et al., 1987).

Most likely, it is not possible to make such a radical simplification, but it is nonetheless a very interesting approach because it is so simple and has such attractive properties. It is also probable that the fewer and the more widely spaced the channels are, the better such an approach is. If there are only few mixing points along the flowpath, it would also be possible to use the recently developed mixing rule methods (Rasmuson, 1985) to determine the impact of mixing on the transport in a bundle of channels.

2.5 DATA NEEDS AND APPROACHES TO OBTAIN DATA

In Table 2.5.1 the most important data needed for assessing radionuclide escape rates to the biosphere are indicated and commented. The importance of the mechanism and data is based on experiences from the safety analyses: KBS-3 (1983), NAGRA (1985), SFR (1987) and WP-Cave (1988). The indicated importance may not be generally applicable for repositories of vastly different design.

Table 2.5.1. Data needed for radionuclide transport modeling.

Variable		Importance for near field	Importance for far field Nonsorbing	Sorbing
Q	Flow rate of water	**VL**	**VL**	**VL**
u	Water velocity	VL	VL	Negl
z	Travel path length	-	**VL**	**VL**
D_L	Hydrodynamic dispersion*	-	L	S
D_e	Diffusion in rock matrix	-	S	VL
K_d	Sorption in rock matrix	**VL**	-	**VL**
a	wetted surface	S	S	**VL**
Redox conditions, pH, complexing agents		**VL**	-	**VL**
Channel aperture, porosity		L	L	Negl
Channel frequency		L	S	**VL**
Channel flow rate distribution		L	L	**VL**

* in individual pathways. For overall dispersion see "Channel flow rate distribution."

VL very large importance in the context
L large importance in the context
S small importance in the context
Negl negligible in the context
\- not applicable in the context
VL Very large importance for the overall safety assessment and final results.

The flow rate Q and its variability are assessed by hydraulic modeling that uses data on hydraulic conductivity K_p and its variations in the rock, topography, and knowledge of the location and properties of fracture zones. These data are obtained by geologic and hydraulic investigations such as down-hole conductivity measurements and surface mapping, and by geophysical methods. Most of the investigation techniques give no detailed information on channel frequency, wetted surface, dispersivity, or exchange with the matrix.

The water velocity u is obtained from Q and information on flow porosity. The latter entity is not known well for fractured crystalline rock. Good investigative techniques are lacking to determine flow porosity of large rock bodies. For sorbing nuclides in the far field this is uninteresting because it does not appreciably influence their velocity. For the nonsorbing nuclides it does not seem to be very critical in the safety assessments, because the travel times are so short anyway that no appreciable nuclide decay (I-129, C-14) will take place.

The travel path length z is obtained from the hydraulic modeling and is a very important entity. It has the same impact on the radionuclide transport as the wetted surface a and flow rate Q.

The hydraulic dispersion D_L is not well-understood for fractured rock. The dispersivities evaluated from available field tests seem to increase with distance, which should rule out the use of the advection-dispersion model. The model is often used anyway to mimic the dispersion behavior. The dispersion in individual channels seems not to have a large impact. The velocity differences between channels dominate the overall "dispersion."

The flow rate, velocity and wetted surface may vary considerably between channels. The rather new concept of channeling in the context of radionuclide transport in fractured rock is far from fully understood, but there are strong indications both experimentally in the field and theoretically that it may have a most profound influence on the RTD (residence time distribution) of the nuclides, nonsorbing as well as sorbing. For the sorbing nuclides, channeling is very important because channels with high flow rates may have a very small wetted surface relative to the flow rate in the channel.

Well-tested and accepted methods do not exist at present to investigate channeling.

Sorption on the minerals of crystalline rock has been studied extensively. There is a fair understanding of the processes involved and confidence in the data is reasonably high. The uptake in the rock matrix by diffusion in the micropore waters is fairly well understood, except for the surface diffusion of some nuclides. Practically all investigations have been performed in the laboratory and there is a need to confirm the findings on the field scale and under natural stress and chemical conditions.

The matrix diffusion and sorption on the inner surfaces of the rock is by far the most important retardation mechanism for the sorbing nuclides. The sorption of the nuclides on fracture filling and coating materials is practically negligible compared to the sorption in the matrix. This is because there is so much more material to sorb on in depth in the matrix as compared to the thin layer on the surface.

Redox conditions, pH, and the presence of complexing agents very strongly influence the solubility and sorption properties of the nuclides and thus directly influence the nuclide mobility. Minerals in the rock such as ferrous iron minerals, silica and carbonates play a dominating role in determining the water chemistry. Iron minerals, for example, which are abundant in the matrix of the rock, may neutralize even quite large disturbances caused by the intrusion of surface waters or by radiolysis.

Most of the data on these phenomena have been obtained in the laboratory and there is a clear need to make field experiments and observations. Transport modeling using coupled geochemical and transport models will play an important role in interpreting the experiments.

Channel aperture and porosity influence the migration of the nonsorbing nuclides, but this seldom influences the maximum rate of escaping nonsorbing nuclides, because the important ones will not have time to decay anyway.

The channel geometry and the wetted surface have a very strong influence on the RTDs of the sorbing nuclides. The wetted surface has the same influence as the flow rate in a channel and as the travel distance.

Very little is known at present on channel geometry and wetted surface. Some first investigations have been made but there is a great need for more field observations, investigations, and theoretical deliberations.

The channel flow rate distribution has a very strong influence on the RTD of the sorbing nuclides. A channel with a high flow rate not only carries a nuclide in large amounts; it may also have a small surface to sorb on and for the nuclide to diffuse into the matrix. A "fast" channel, then, may allow a certain portion of the sorbing nuclide to be minimally retarded by the sorption because the flow rate of water is large compared to the wetted surface in this channel.

Very little is known at present on channel flow rate distribution. Some first investigations have been made, but there is a great need for both more field observations and investigations as well as theoretical deliberations.

Available information suggests that the variability of flow rates and wetted areas may sometimes be more important than the average flow rate along the flow path.

Information on flow rate distribution is not normally obtained in conventional hydraulic investigations. Special investigation methods must be developed and tested, and present channeling models must be further developed.

The channel frequency has a large impact on the near field release. If there are only a few channels there may be many radioactive waste storage canisters that in practice will not have access to mobile water. On the other hand some canisters may be subject to large flow rates if they are intersected by a channel with a high flow rate.

The views expressed above are based on the experiences of more than 10 years working with safety analyses, and devising and applying models for calculating radionuclide transport from repositories for radioactive waste. In the course of this work many concepts have changed as new observations and data have become available. It has been noted that a lot of effort has gone into hydraulic and geologic investigations in the field. It has been possible to use only a very little of that information to assess the migration of radionuclides. On the other hand, there has been a very noticeable lack of information in some areas. The wetted surface is still poorly known. The understanding of channeling is far from complete. The surface diffusion phenomena are not so well understood that it is possible to confidently predict how a solute will diffuse in the rock matrix.

To assess the transport of radionuclides, those processes that are of importance for the migration of the nuclides should be put in the foreground. The hydraulic parameters make up only one part of the important parameters.

In the following discussion, some recent experiments are described that have set out to investigate various aspects of the flow and transport in fractured crystalline rock.

2.6 DIFFUSION IN THE ROCK MATRIX

2.6.1 Overview

The rock matrix of even dense crystalline rocks has been observed to be porous. The porosity, if connected in depth, could potentially act as a sink for radionuclides and other dissolved substances in the mobile water in the fractures (Neretnieks, 1980). Positively charged or neutral species will sorb on the mineral surfaces of the pores and microfractures of the matrix.

Matrix porosity, diffusivity in the pore waters, and sorption properties of some sorbing species that do not sorb very strongly can be and are measured in the laboratory by simple experiments. If the data are to be used for predictions of transport over much larger time and space scales, however, the laboratory experiments must be supplemented by other experiments and observations. Some key issues which have been addressed to varying extents in various investigations are

i) Connectivity of the pore system over long distances.

ii) Questions of constrictivity: Will larger molecules be less mobile than small, even accounting for the smaller diffusivity in water?

iii) Effects of surface charge: Many mineral surfaces are negatively charged and may, by electric repulsion, "constrict" the pores for anions.

iv) Sorption mechanisms and mobility: Are the sorbed cations immobilized or are they mobile even in the "sorbed" state?

v) What influence does rock stress have on the porosity of the rock and diffusivity of the solutes?

vi) Have the de-stressed samples taken to the laboratory been irreversibly changed?

vii) Will fracture coating and filling material allow diffusing species to access the underlying matrix?

viii) Do the transport and sorption properties change over long periods of time?

This section will describe some of the theoretical background, techniques used, and experiments performed.

Concept of Transport in a Porous Rock Matrix

Crystalline rocks are made up of small crystals ranging in size typically from 0.1 mm to many millimeters. There are microfractures in and between the crystals which make up a connected pore system. Some pores are dead-ended. Granites and gneisses have been found to have low hydraulic conductivity and transport by flow can usually be neglected compared to that by diffusion. Transport by flow in the rock matrix will not be discussed in this section.

When the pores of the matrix are filled with water, the solutes will migrate by molecular diffusion. The movement will be restricted at narrow passages and the path length will be longer than the physical distance in the rock because the pores form tortuous pathways.

In steady state, transport will only take place in the connected pore system. During the unsteady state phase the dead-end pores will also contribute by acting as sinks and/or sources. The dissolved solutes move by diffusion in the pore water. Sorbed species may be bound by mechanisms that make them immobile in the sorbed state, in which case they can only move if they are desorbed first, or they may be mobile, as when they are attracted by diffuse electric forces which make the water near the surface more concentrated in the "sorbed" species without actually binding them.

In steady state, the transport can be described by Fick's first law

$$N = -D_p \varepsilon_t \frac{\partial C}{\partial x} \tag{2.6.1}$$

ε_t is the transport porosity, D_p is the diffusivity in the pore water and can be related to the diffusivity in unconfined water, D_w, by the relation

$$D_p = D_w \frac{\delta_D}{\tau^2} \tag{2.6.2}$$

δ_D is the constrictivity and τ is the tortuosity.

The product $\varepsilon_t \cdot \delta_D/\tau^2$ is called the formation factor F_f.

If all solutes were influenced in the same way by the pore system, then it would suffice to measure the formation factor for one solute. This is not always the case, but F_f can still be a useful entity for comparison purposes.

When the transport is instationary and the solute accumulates or is depleted from the pore system, Fick's second law and the law of conservation of mass is used to describe the transport and accumulation of a solute in the pore system. Assuming that transport takes place only in the water in the pores, that the diffusion coefficient is constant, and that the sorption is reversible, the following equation results:

$$(\varepsilon + K(C))\frac{\partial C}{\partial t} = D_p \varepsilon_t \frac{\partial^2 C}{\partial x^2} \tag{2.6.3}$$

ε is the total porosity, including dead-end pores, and $K(C)$ is the volumetric sorption coefficient, which may be concentration-dependent.

These equations, together with the initial and boundary conditions, often suffice to describe the transport and accumulation of solutes in a porous medium. For many species the sorption coefficient can be assumed to be constant.

Ion mobility in an electrical field can be used to determine diffusivity of charged solutes and is used to determine the formation factor, which is directly obtained from the ratio of electrical conductivity of the solution-filled sample to the conductivity of the same solution.

There are indications that some sorbed species may be mobile in the "sorbed" state (Skagius and Neretnieks, 1988). This is sometimes called surface diffusion. The process is not well-understood (Rasmuson and Neretnieks, 1983). Then, parallel to the transport in the pore water, there will be transport of the sorbed species. The rate of surface transport can be written

$$N = -D_s \frac{\partial C_s}{\partial x} \tag{2.6.4}$$

C_s is the concentration of the sorbed substance in the solid and is, in general, concentration-dependent. $D_s = D_s(C_s)$. As sorption is fast, equilibrium is assumed to exist between C_s and C.

$$C_s = K(C) \tag{2.6.5}$$

The total transport rate under steady state conditions, then, is the sum of the transport by pore diffusion and by surface diffusion.

$$N = N_{pore} + N_{surf} = -D_s A \frac{\partial C_s}{\partial x} - D_p \varepsilon_t A \frac{\partial C}{\partial x} \tag{2.6.6}$$

For instationary transport the equivalent expression to Fick's second law is

$$\varepsilon\frac{\partial C}{\partial t}+\frac{\partial C_s}{\partial t}=\frac{\partial}{\partial t}\left(D_s(C_s)\frac{\partial C_s}{\partial x}+D_p\varepsilon_t\frac{\partial C}{\partial x}\right) \tag{2.6.7}$$

Experimental Approaches

Different experimental techniques are used to determine the transport properties. In the "through diffusion" experiment the rate of transport through a section of rock is monitored (Garrels et al., 1974; Skagius and Neretnieks, 1986a; Bradbury and Green, 1985) as the solute moves from a vessel with high concentration to a receiving vessel with low concentration. The product of the pore diffusivity and the transport porosity $D_p\varepsilon_t$ is obtained. It is called the effective diffusivity D_e. The early instationary part of the experiment can be used to determine the capacity of the rock for holding the solute, $e+K$. For nonsorbing species this is equal to the total porosity. In systems with long dead end pores, not all these are equilibrated during the early time period (Bradbury and Green, 1986). This problem will decrease in thicker samples. For the time scales of interest for assessing radionuclide transport in the field, the water in the dead end pores will be in equilibrium with the transport pores.

Different solutes, both charged and uncharged and of different molecular weight, have been used in the diffusion experiments. Colloids have also been used to determine their possible blocking of the pore structure (Bradbury and Green, 1986). Diffusion in fracture coating materials has also been measured (Skagius and Neretnieks, 1984, 1986a; Bradbury and Green, 1986).

The porosity can be determined by several other methods. Weighing the sample wet and dry gives the total porosity, as does leaching a solute out of a previously saturated sample. Pore sizes and pore size distributions have been measured by mercury porosimetry, BET sorption and microscoping. For large samples and for sorbing solutes the through diffusion method is very time-consuming and strictly instationary methods have been applied, to reduce the time. Immersion of tablets and larger pieces of rock samples in traced solutions has been used (Skagius, 1986). The decrease of the bath composition is then utilized to determine the transport properties and the final steady state to determine equilibrium sorption. Profiling of the samples was used by Skagius and Neretnieks (1988) to get profiles of Cesium and Strontium in granites and gneisses.

Electrical conductivity measurements of rock samples saturated with highly saline solutions have been used successfully to determine the formation factor (Skagius and Neretnieks, 1986b). This technique has been used on meter-length samples (Atkinson and Titchell, 1985).

Through-diffusion and electrical conductivity measurements have been performed in pressure cells in the laboratory to determine the influence of stress on porosity and transport properties (Brace et al., 1968; Skagius and Neretnieks, 1986b; Bradbury and Green, 1986; Gilling et al., 1987).

Diffusion experiments in undisturbed rock at depth under natural stress conditions have been performed in granite (Birgersson and Neretnieks, 1987; Birgersson, 1988) over time periods of more than 3 years. Attempts have also been made to measure concentration profiles of salt water intrusion in granite over time scales of over 30 years in a large block immersed in salt water (Jefferies, 1988).

2.6.2 Results from Diffusion Experiments with Nonsorbing Substances in Crystalline Rocks

The following section gives a more detailed description of some of the experiments with nonsorbing tracers and the results obtained.

Skagius and Neretnieks (1986a) made measurements in granites and gneisses from 8 different locations in Sweden. The samples were taken from depths down to more than 500 m. The diffusing substances used were iodide, Cr-EDTA and uranin. The sample thickness was 10 mm. Several of the samples were taken adjacent to fractures and had fracture-coating materials.

In Fig. 2.6.1 the effective diffusivities of iodide in the granites, gneisses and fissure coating materials are plotted versus the porosities determined by the leaching method. Results obtained by Bradbury et al. (1982) for iodide diffusion in different granites from the United Kingdom are also presented in the figures for comparison. The logarithmic mean values are marked in the figures. The effective diffusivity in the granites (logarithmic mean value = $22.0 \cdot 10^{-14}$ m^2/s and arithmetic mean value = $25.2 \cdot 10^{-14}$ m^2/s) is higher than in the gneisses (logarithmic mean value = $5.1 \cdot 10^{-14}$ m^2/s and arithmetic mean value = $9.2 \cdot 10^{-14}$ m^2/s). The mean values of the porosity determined by the leaching method are also higher for the granites (logarithmic = 0.24% and arithmetic = 0.26%) than for the gneisses (logarithmic = 0.13% and arithmetic = 0.15%).

Figure 2.6.1 also includes values for fissure-coating materials. The effective diffusivity of iodide in the fissure-coating materials is of the same order of magnitude or higher than in the granites and the gneisses. Those samples that have an effective diffusivity that is of the same order of magnitude as the granites and the gneisses have, however, much higher porosity values.

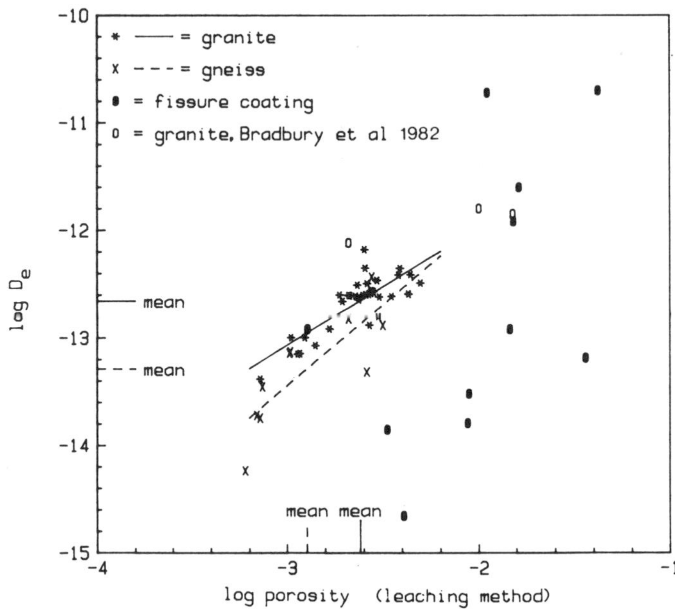

Figure 2.6.1. Effective diffusivity of iodide in the rock materials versus the porosity.

This could be due to a higher "storage" porosity in the fissure-coating material, or to a lower pore diffusivity in the fissure-coating material compared with the granites and the gneisses.

The effective diffusivities for Cr-EDTA and uranin typically are an order of magnitude lower than those for iodide. For iodide, values of the formation factors were mostly in the range $1\text{-}20 \cdot 10^{-5}$, with only a few outliers, and about a factor of 5 lower for the other two solutes. Skagius and Neretnieks (1986b) also measured formation factors, using the electrical conductivity method. Potassium iodide was used as the ionic conductor. Formation factors in the range $1\text{-}30 \cdot 10^{-5}$ were found, with only a few outliers. Variations in porosity and diffusivity between samples taken at neighboring positions from the same core typically give values that differ by 2-3 times.

These results indicate that the electric measurement technique gives comparable results to the direct diffusivity measurements. The large difference in formation factors for iodide on one hand and Cr-EDTA and uranin on the other hand indicate that the latter two bulkier molecules are considerably more influenced by the pore structure.

Bradbury and Green (1985, 1986) measured diffusivities in 4 British and 1 Canadian granites, using iodide as the diffusing substance. They found effective, or in their terminology intrinsic, diffusivity values ranging from $2 \cdot 10^{-14}$ to $130 \cdot 10^{-14}$ m^2/s. Formation factors were found to lie between $1.2 \cdot 10^{-5}$ and $80 \cdot 10^{-5}$. Porosities determined from the instationary part of the diffusion experiment were between 0.06% and 1.2%. Bradbury and Green (1985) also used technetium in the form of pertechnetate as the diffusing substance and found diffusivities comparable to those for iodide. The rock capacity factor $\varepsilon + K$, however, was found to be considerably larger, indicating that the pertechnetate sorbs or reacts somewhat in the solid.

Fracture coating materials and weathered surfaces were found to have higher porosities and diffusivities than "good rock." The increases in diffusivity ranged from 20-200 times and the porosity from 10-100 times.

The change in transport properties with distance from the fracture surface has been measured by Skagius and Neretnieks (1986a) and Bradbury and Green (1986). In both investigations it was found that both porosity and diffusivity dropped off with increasing distance from the surface. The decrease with distance strongly depends on the degree of weathering.

Skagius (1986) measured the electrical conductivity of samples of different lengths up to 30 cm long. She found that the specific conductivity decreased by a factor of about two when the sample thickness increased from 10 mm to 50 mm and that a further increase in length did not change the properties further. Atkinson and Titchell (1985) measured electrical conductivity of granite cores up to 1.25 m in length. They found very similar results.

At expected radioactive waste repository depths, the rock is exposed to high stresses caused by the large overburden of rock. When drill cores are taken up from the ground this overburden no longer exists. As a result there might be an increase in the porosity of the rock samples. The effective diffusivity measured in rock samples under atmospheric pressure in the laboratory would then be higher than the effective diffusivity in the rock in situ.

To simulate the stress that may exist in the bedrock at great depths, diffusion experiments with iodide and electrical resistivity measurements in rock materials under mechanical stress have been performed by Skagius and Neretnieks (1986b) with stresses up to 35 MPa, and by Bradbury and Green (1986) up to 16 MPa. In the first-mentioned experiment, which covered 6 different granites, the formation factor decreased 1.2 to 2 times for an increase in stress of 5 MPa and then decreased more gradually. At 30-35 MPa the formation factor has decreased by 1.6 to 4 times.

Bradbury and Green (1986) found a faster initial decrease and a leveling off at 5 MPa for one granite.

Figure 2.6.2 shows a plot of the concentration at the low concentration side versus time for iodide diffusion through a piece from Svartboberget. The experiment was started with the piece under atmospheric pressure. After 35 days the pressure was increased to 33 MPa.

It is seen from the slope that there is a marked decrease in diffusivity when the pressure is increased.

In some of the electrical conductivity measurements the loading stress was cycled several times. Hysteresis effects are clearly seen, and both somewhat higher and somewhat lower values were found at the end of a cycle. A second cycle closely followed the first in the experiment in which the cycling was repeated (Fig. 2.6.3).

The formation factors calculated from the diffusion experiments under stress are in fair agreement with the formation factors from the resistivity measurements at higher pressures.

The results from Skagius and Neretnieks measurements are summarized in Figs. 2.6.4 and 2.6.5. The results from the various methods are compared. Figure 2.6.4 shows the diffusivities of iodide determined in granites and Fig. 2.6.5 shows the diffusivities of iodide determined in gneisses. The open bars represent the diffusivities in samples under stressed conditions and the filled bars represent the diffusivity in samples under atmospheric pressure.

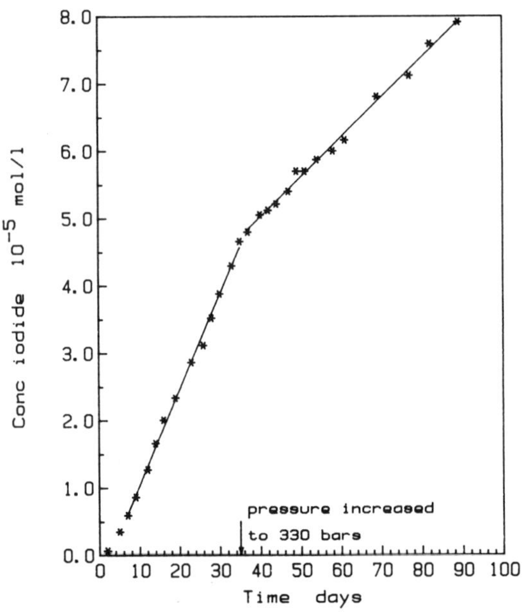

Figure 2.6.2. Concentration of iodide versus time, diffusion through biotite gneiss from Svartboberget.

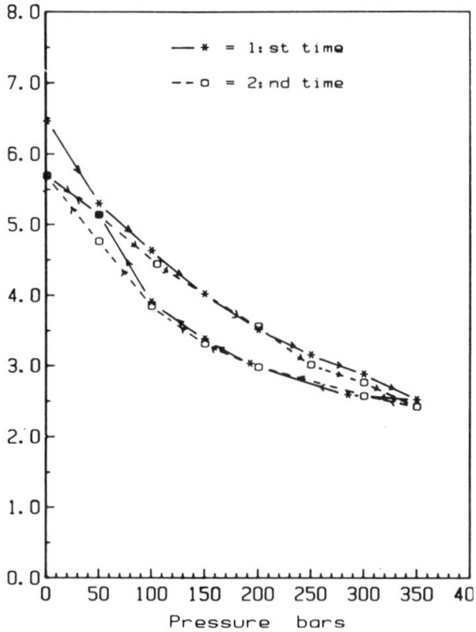

Figure 2.6.3. Formation factor × 10^5 versus pressure for a granite sample from Finnsjön.

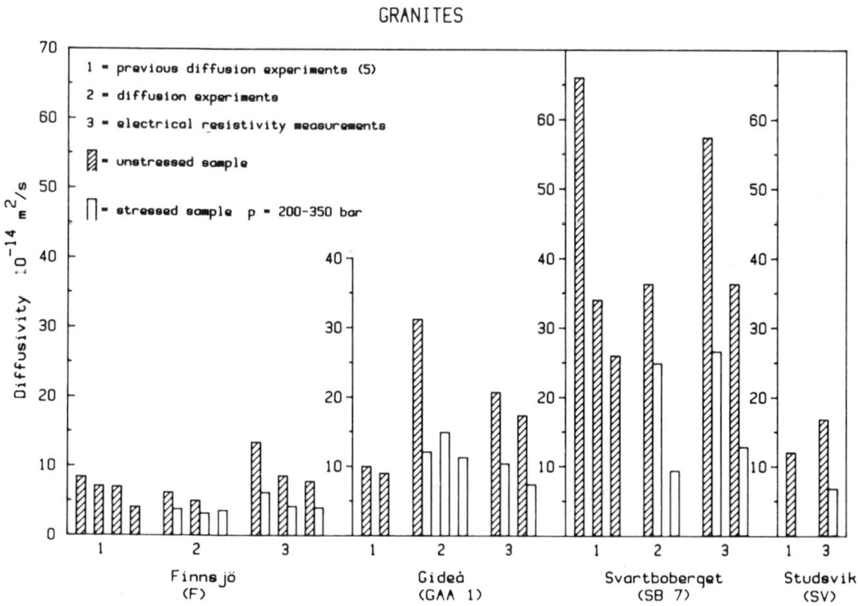

Figure 2.6.4. Effective diffusivities in granites.

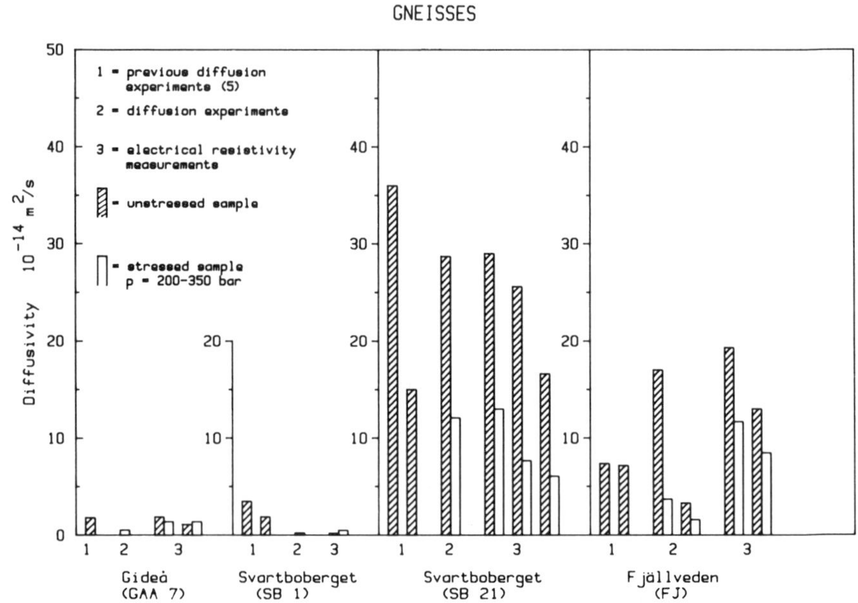

Figure 2.6.5. Effective diffusivities in gneisses.

The relative values of formation factors in samples at 30-35 MPa stress and samples at atmospheric pressure obtained in the diffusion experiments are in fair agreement with those determined from the electrical resistivity measurements for the same rock materials. Electrical resistivity measurements can thus be used to give approximate values of the diffusivity. The advantage of electrical resistivity measurements is that the experimental time is much shorter than in the diffusion experiments.

The diffusivity and the formation factor in samples at 30-35 MPa was typically a factor of 2 lower and in no case lower than about 1/5 of the value in unstressed samples.

2.6.3 Diffusion Experiments with Sorbing Substances in Crystalline Rocks

In the simplest case it may be assumed that transport takes place only in the pore water and that the sorption equilibrium is reversible and linear. Equation (2.6.3) then simplifies to

$$\alpha \frac{\partial C}{\partial t} = D_p \varepsilon_t \frac{\partial^2 C}{\partial x^2} \tag{2.6.8}$$

where α is the rock capacity factor $\varepsilon + K$ and $K = C_s/C$ is the volumetric sorption coefficient for linear sorption.

For systems in which the above assumptions are valid, it would be sufficient to determine the formation factor F_f, the diffusivity in water of the solute, and the sorption coefficient to fully define the system.

Several experiments have set out to test the validity of these assumptions experimentally.

Bradbury and Stephen (1985) measured the effective diffusivity and sorption properties of technetium as pertechnetate, cesium and strontium in a sandstone using a through diffusion technique. The sandstone had about 10% porosity. They found good agreement between the diffusivities for the three solutes. Sorption equilibrium measurements using a batch technique showed that the isotherms were weakly nonlinear for technetium and cesium, giving a Freundlich exponent between 0.71 and 0.94. The isotherm was linear for strontium. However, it was found that the sorption capacity could be up to one to two orders of magnitude higher in batch experiments with crushed rock and coupons than in the through-diffusion experiments on machined coupons. The reason for the differences between the results for coupons in the two different experiments is not known. The methods of evaluating the results are different, however. Batch experiments are evaluated by making a mass balance on the amount of tracer lost from the bath. The through-diffusion experiments were evaluated from the time lag of the breakthrough curve. This method gives the capacity of the transport pores and does not account for the possible capacity of the "storage pores." Bradbury et al. (1982) found that for nonsorbing species storage pores had a noticeable influence on the total capacity of the rock. No investigations so far seem to have addressed the influence on storage capacity in this type of instationary experiment.

The higher sorption values for the crushed rock, with sizes <0.1 mm, can be understood because the crushing will induce new microfractures and expose fresh surfaces to weathering. The sorption capacity for particles of 1-2 mm was comparable to that of the coupons, which were 5 mm thick.

Skagius et al. (1982) made sorption and diffusion measurements on two different granites. The technique used was in-diffusion into crushed particles with 6 differently sized fractions ranging from 0.1 to 5 mm. The sorption isotherm for strontium was found to be linear; that for cesium was slightly nonlinear with a Freundlich exponent of about 0.6 for both granites.

The sorption capacity was found to be practically independent on the particle size if account was taken of the larger outer surface for the smaller particles. For the smallest particle size (0.1-0.12 mm) the sorption on the outer surface accounted for 11-15% for cesium and strontium, respectively, on the Finnsjön granite and 40-46% on the Stripa granite. For particle sizes larger than 1 mm the contribution from the outer surface was negligible.

The in-diffusion experiments were performed with the batch technique, where the bath concentration was monitored for more than 10,000 h. Figure 2.6.6 shows the concentration time curve for two experiments together with the fitted curve. The fitted curve is the solution to the instationary diffusion equation (2.6.3) with the appropriate boundary conditions. The sorption coefficient and the effective diffusivity are obtained simultaneously from the fit.

Skagius (1986) made another set of batch in-diffusion experiments into 5 mm thick tablets on Finnsjön and Stripa granites. Results for Finnsjön granite are illustrated in Fig. 2.6.7. The sorption coefficients were up to three times lower in the machined tablets. The effective diffusivity was very strongly dependent on the particle size. Figure 2.6.8 shows the diffusivities obtained for particles.

Skagius and Neretnieks (1988) made sorption and diffusion measurements, using cesium and strontium, by the in-diffusion and the through-diffusion methods. Biotite gneiss from Fjällveden was used. After contact times of up to 1.5 years the tablets were profiled by grinding off thin sections. Sorption and diffusion values were determined both from the concentration profiles in the samples and from the time lag of the breakthrough curves. Sorption coefficients determined from measured uptake in the rock were about half those determined from the time lag of the

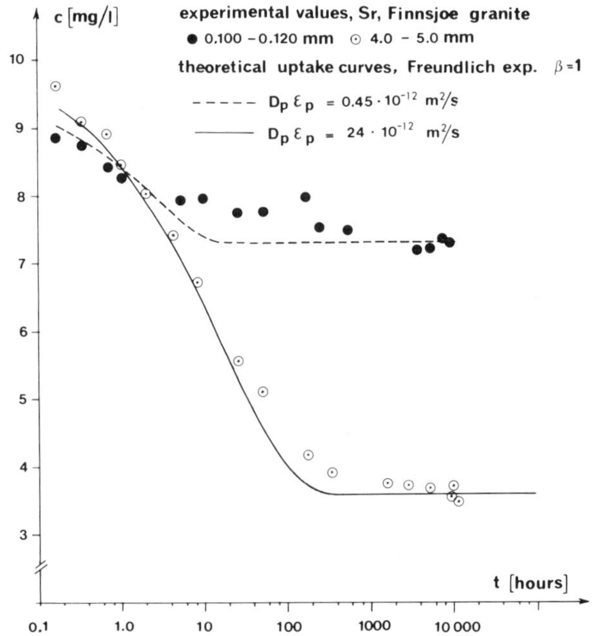

Figure 2.6.6. The fit of the diffusion model to the experimental uptake curve for strontium on Finnsjön granite.

breakthrough curve. This contrasts strongly with the large differences obtained by Bradbury and Stephen (1985), who found up to two orders of magnitude differences.

Attempts by Skagius and Neretnieks (1988) to evaluate the effective diffusivity, assuming only pore diffusion, failed because the "pore diffusivity" from the slope of the breakthrough curve was much higher than can be explained by pore diffusion only. The pore diffusivity of cesium would have to be larger by at least a factor of two over the diffusivity in pure water. This is nearly a hundred times more than expected from the formation factor measured with iodide. Furthermore, the experimental and theoretical breakthrough curves had distinctly different shapes for early points in time. A good fit could be obtained by assuming that at least part of the transport is by surface diffusion. Indeed, most of the transport must then be by surface diffusion, and the transport by pore diffusion is practically negligible.

Figure 2.6.9 summarizes the Skagius (1986) results as formation factors obtained for different solutes. It is seen that the formation factor is very much higher for cesium and strontium than values obtained from electrical conductivity measurements and from iodide measurements. This has been interpreted as indicating the presence of a transport mechanism in addition to transport in pore water only.

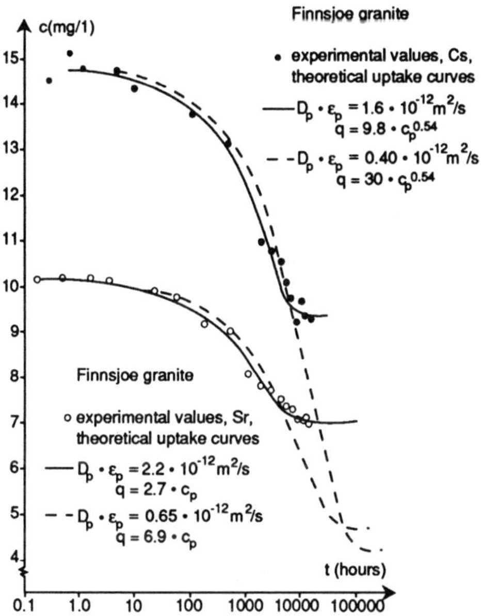

Figure 2.6.7. The fit of the diffusion model to the experimental uptake curve for cesium and strontium on Finnsjön granite.

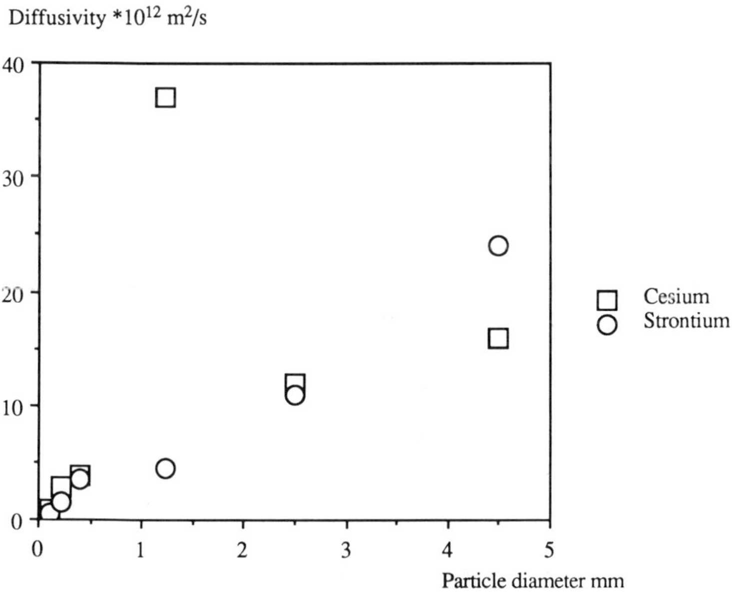

Figure 2.6.8. Effective diffusivity of cesium and strontium for different particle sizes.

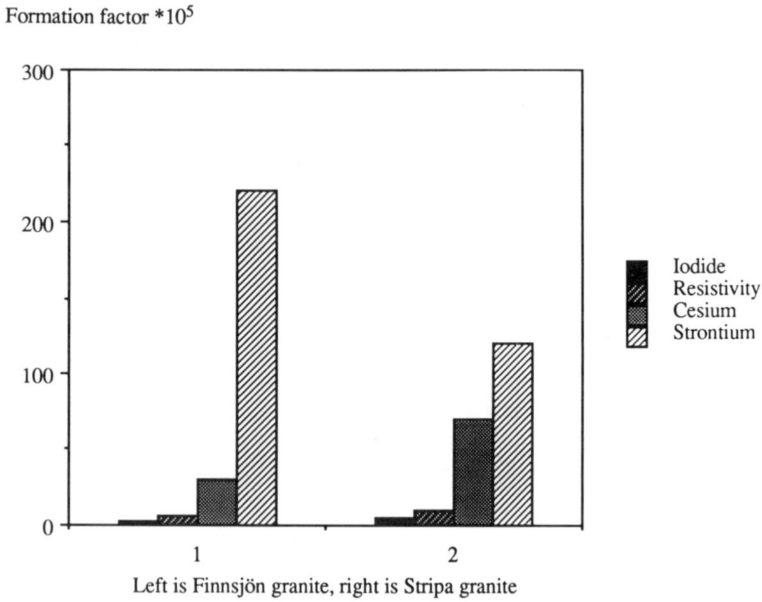

Figure 2.6.9. Formation factor for Finnsjön and Stripa granite.

2.6.4 Field Investigations

A Field Diffusion Experiment in Stripa

To test if the matrix of granites at great depths has a connected porosity, a set of three experiments has been conducted. The experiments have been carried out in "undisturbed" rock under its natural stress environment, at the 360 m level in the Stripa Mine. The rock is subject to nearly the same conditions as the rock surrounding a nuclear waste repository as proposed in the Swedish concept.

A mixture of three nonsorbing (conservative) tracers, uranin, Cr-EDTA, and I^-, were injected into the granitic rock matrix for three different time periods: about 3 months, about 6 months, and about 3.5 years. The subsequent overcorings of the injection holes showed that the tracers had in some cases migrated at least about 400 mm (measuring limit) into the rock matrix for the experiment with the longest injection time. It could also be seen that there were substantial differences in migration distance into the rock matrix for samples taken fairly close to each other.

The results from all three experiments showed that the three tracers had migrated through the disturbed zone close to the injection hole, through the fissure-coating material, and a distance into the "undisturbed" rock matrix (Birgersson, 1988).

The experiment was designed in the following way. Close to a drift or a borehole the rock stresses will be considerably different from those far away from the disturbance. In order to make the experiment in undisturbed rock, 3 large holes were drilled with lengths of more than two tunnel diameters (> 10 m). In the bottom of the large holes, thin holes (20 mm) extending 2-3 m were drilled (see Fig. 2.6.10). At distances larger than about 2 tunnel or hole diameters the rock is essentially undisturbed. The small holes were packed off and filled with tracer solutions, which were kept under somewhat higher than ambient hydraulic pressure. The tracers flowed and diffused out into the rock surrounding the hole.

The first hole was overcored after 3 months, the second after 6 months and the third after 3.5 years. In the two longer experiments, called II and III, additional holes were drilled parallel to the first after the injection period to reach rock at distances up to 40 cm from the injection hole.

Figure 2.6.11 shows the overcoring arrangement of experiment II. Figure 2.6.12 shows the coring-out arrangement of experiment III. The three large holes (146 mm) made it possible to study the concentration profile as far as about 400 mm outward from the injection hole. The three additional small holes (76 mm) made it possible to study the concentrations in other directions.

The cores were sectioned and sample cores were drilled from the sections. The sample cores were leached and the tracer concentrations of the leachates were determined. The number of sampling cores obtained was about 200 in experiment I, about 650 in experiment II, and about 1800 in experiment III.

Experiment I was a pilot experiment and gave only limited data on the migration distances. However, it showed that the tracers penetrate through quite thick fracture coating materials under the natural stress conditions (Birgersson and Neretnieks, 1982a,b). Experiment II has also been reported earlier (Birgersson and Neretnieks, 1983, 1984).

A rock stress measurement was performed in the hole of experiment I at 15.5, 16.0, and 17.5 m depth from the floor of the drift. The stress distribution agreed well with what is regarded as natural stress in the Stripa Mine.

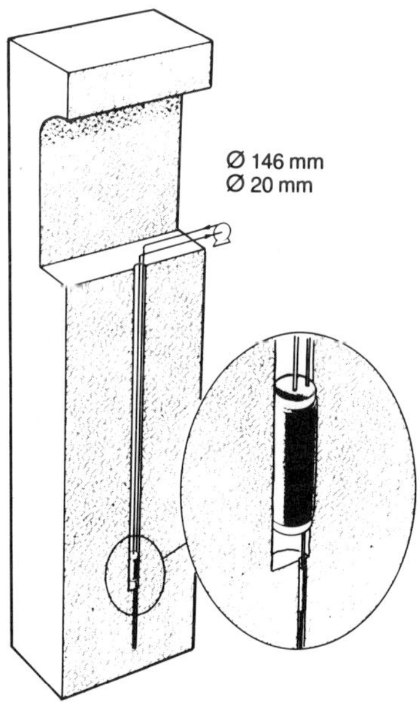

Figure 2.6.10. Drilling dimensions and packer positions.

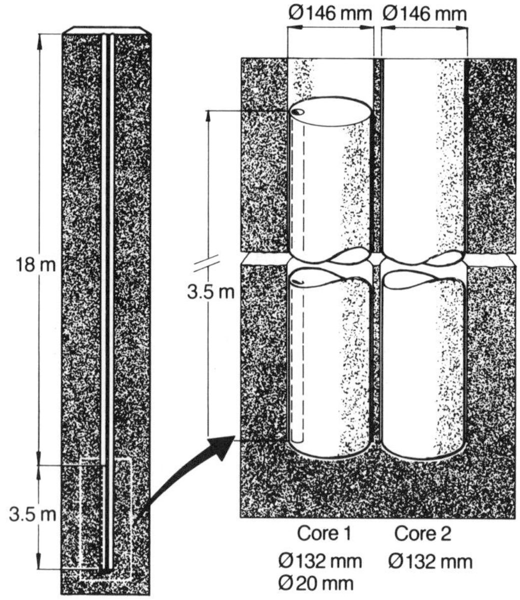

Figure 2.6.11. Drilling arrangement, Part II.

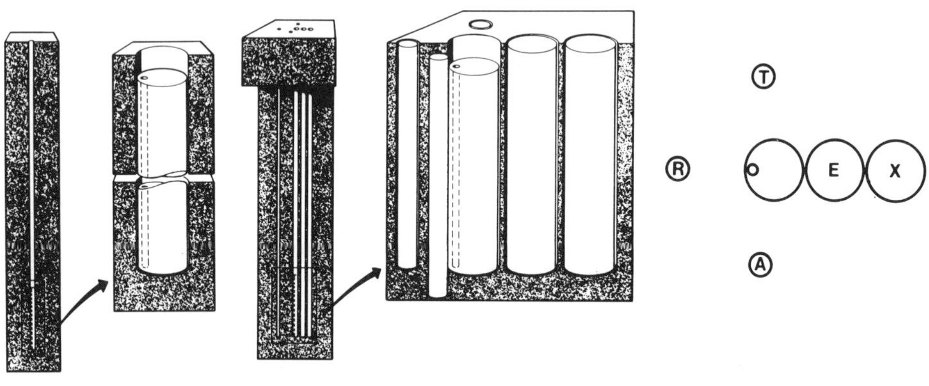

Figure 2.6.12. Part III. Drilling arrangement and notation of the cores.

The porosity was measured for every individual sampling core by comparing the weight difference between the wet and dry core. The porosity was found to be between 0.12-0.51%, with an average close to 0.3%.

The pore size distribution was measured using three different methods: mercury penetrometry, the BET-method (using nitrogen gas) and by electron microscope. The pore size distribution for 5 samples were determined by this method. Figure 2.6.13 shows an example of the results from the electron microscope.

The direction of the microfissures were determined in three samples. The method used was the DSA-method (differential strength analysis). No specific microfissure direction could be found in any of the three samples.

The diffusivity was also measured on 4 samples by the method used by Skagius and Neretnieks (1986a). Iodide pore diffusivities varied between 90 and $200 \cdot 10^{-12}$ m^2/s and uranin gave values of $3\text{-}10 \cdot 10^{-12}$ m^2/s.

The hydraulic conductivity was determined for 5 samples. It was found to be $0.2\text{-}2 \cdot 10^{-13}$ m/s at confining stresses of 15-30 MPa.

Some examples of concentration profiles are shown in Figs. 2.6.14, 2.6.15, 2.6.16, 2.6.17, and 2.6.18. Together they illustrate how, for experiment III, the penetration distances varied with tracer, with direction and with depth. Figure 2.6.19 illustrates the concentration of iodide at a distance of 200-250 mm from the injection hole. This plot also shows strong variations in the migration with location.

The observed differences in migration distance at different depths could be caused by differences in porosity (ε_p) and differences in the migration parameters (K_p and D_p).

The porosity was measured for the individual sampling cores. Figure 2.6.20 illustrates the mean value of the porosity (the standard deviation) for each sampling piece from core 1 and core 2 in experiment II. The variations are small and cannot alone explain the differences in migration distances.

The values of the migration parameters have been evaluated by comparing the experimentally obtained concentration profiles with theoretical concentration profiles obtained by solving the advection-diffusion equation for the appropriate geometry (Birgersson and Neretnieks, 1983). These approximate values of D_p and K_p are summarized in Table 2.6.1 for experiment II.

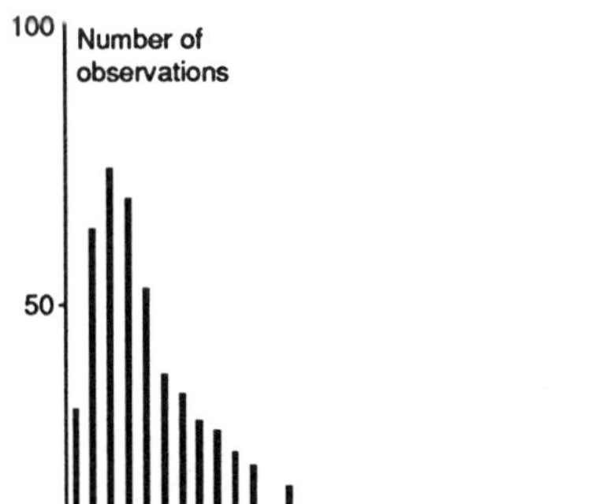

Figure 2.6.13. Pore size distribution for sample 25:9. Results obtained by electron microscope.

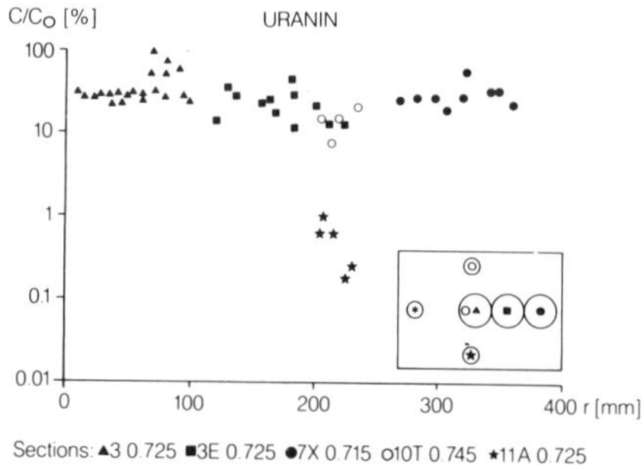

Figure 2.6.14. Part III. Uranin concentrations vs distance at 0.725 m depth.

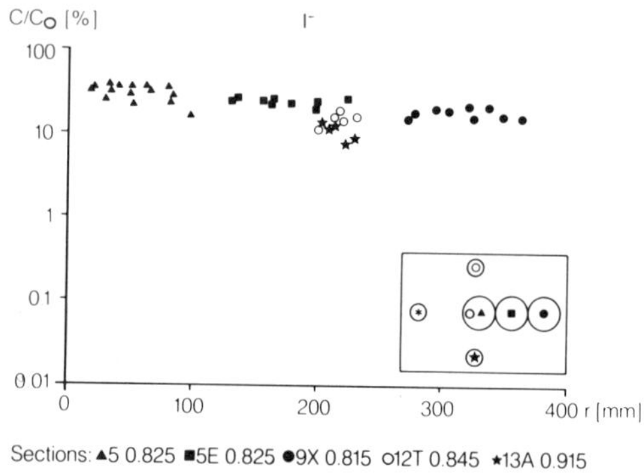

Figure 2.6.15. Part III. I^- concentration vs distance at 0.825 m depth.

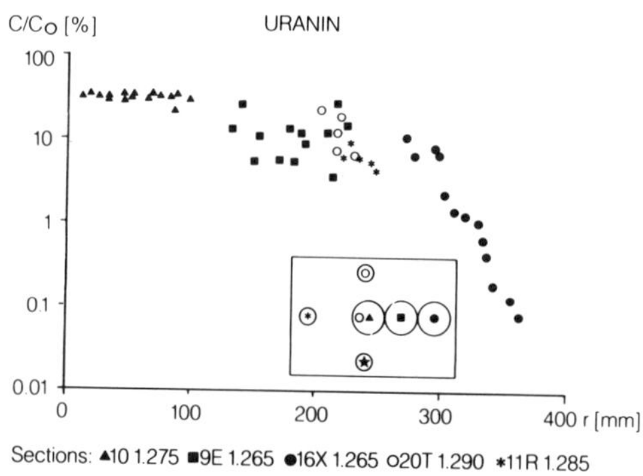

Figure 2.6.16. Part III. Uranin concentration vs distance at 1.275 m depth.

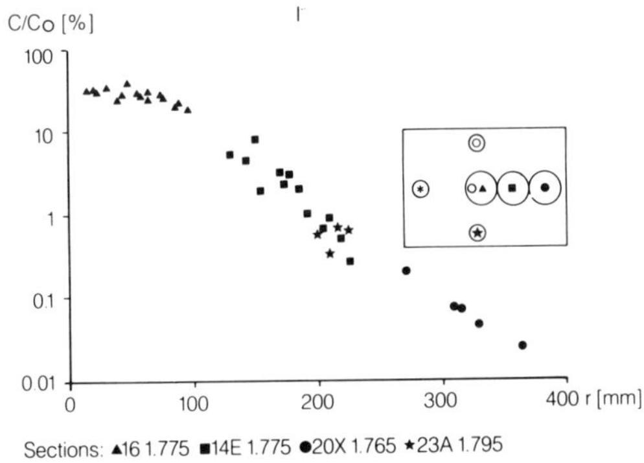

Figure 2.6.17. Part III. I^- concentration vs distance at 1.775 m depth.

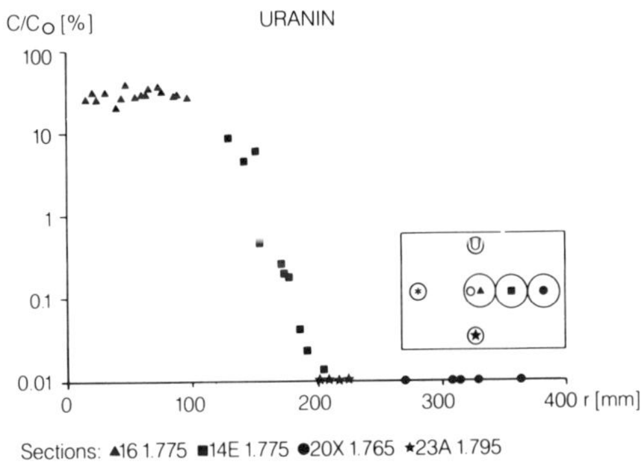

Figure 2.6.18. Part III. Uranin concentration vs distance at 1.775 m depth.

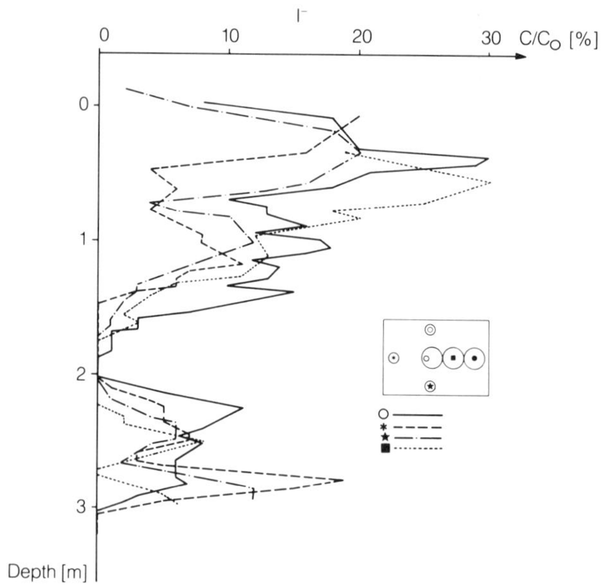

Figure 2.6.19. Average iodide concentration 200-250 mm from the injection hole.

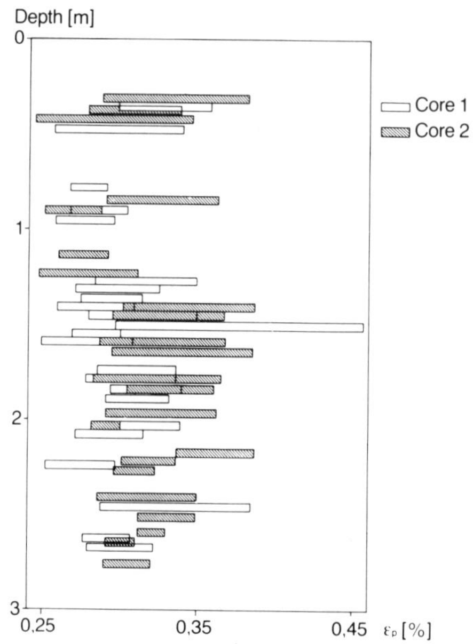

Figure 2.6.20. Experiment II. Core 1 and core 2. The porosity (± the standard deviation) vs depth.

Table 2.6.1. Approximate values on D_p and K_p for different depths.

Sampling piece	Depth (m) along little hole	D_p $(m^2/s) \cdot 10^{10}$	K_p $(m/s) \cdot 10^{13}$
1-2	0.36-0.48	>1	>2-5
3-9	0.78-1.41	0.5	0.1
10-13	1.46-1.59	0.05	<0.1
14-20	1.74-2.24	1	1
21-22	2.62-2.67	>1	>2-5

Table 2.6.1 shows that D_p and K_p can vary by an order of magnitude for sampling pieces separated by just a few tens of centimeters.

No obvious correlation between the penetration depths and porosity, pore size distribution or microfissure direction was found, although there was a tendency toward greater penetration distances in regions where the pores were larger.

The reason for the considerable differences in migration distance for sampling pieces separated by just a few tens of centimeters in depth or at the same depth but in different directions seems to be variations in the migration parameters, K_p and D_p, due to inhomogeneities in the rock matrix. It was not possible to measure these properties on the individual small samples as was done for the porosity. The few measurements of diffusivity and hydraulic conductivity made from the same cores show a considerable variation.

The diffusivity and hydraulic conductivity were evaluated at the locations illustrated in Figs. 2.6.21 and 2.6.22. The evaluation was made by fitting the obtained concentration profiles to the advection-diffusion equation for radially symmetric migration. It was not possible to follow the concentration profile as far out into the rock matrix as needed for the evaluation of D_p and K_p at the other locations. The difference in behavior between the tracers is also clearly illustrated.

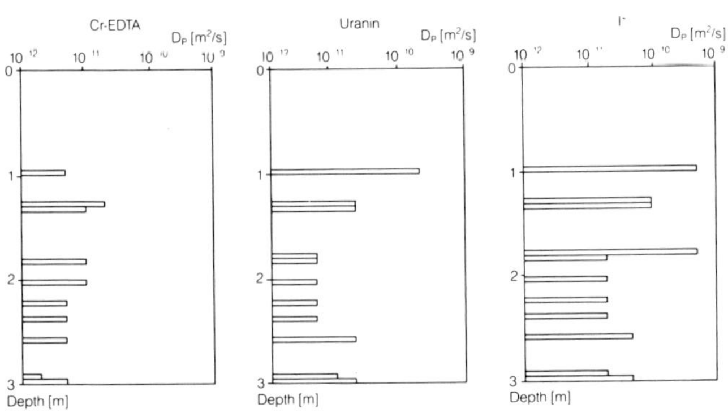

Figure 2.6.21. Experiment III. D_p vs depth in the cores for the three tracers.

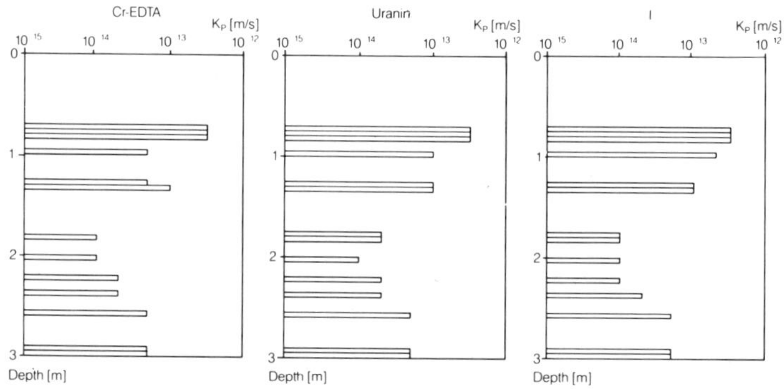

Figure 2.6.22. Experiment III. K_p vs depths in the cores for the three tracers.

Iodide has diffusivity values that are one to two orders of magnitude larger than Cr-EDTA, while the diffusivity for uranin is somewhere between these two tracers. The pore diffusivities vary between $2 \cdot 10^{-12}$ to $5 \cdot 10^{-10}$ m^2/s.

This difference between the tracers was also seen in the laboratory experiment. These values are of the same order of magnitude as those obtained in situ and are illustrated in Fig. 2.6.23. The difference in stress levels between the laboratory and the in situ measurements are expected to make the laboratory data about a factor of 2 higher (Skagius and Neretnieks, 1986b).

The hydraulic conductivities obtained in situ and those obtained in the laboratory experiment are of the same order of magnitude (see Fig. 2.6.24).

It was concluded from the investigations that the pore system in virgin undisturbed rock is open and connected. This was found to be the case for at least 40 cm. The diffusive and hydraulic properties vary by from one to several orders of magnitude even at close locations.

The changes in migration properties at different locations and in different directions could not be explained by the laboratory experiments performed concerning porosity, pore size distribution or direction of microfissures.

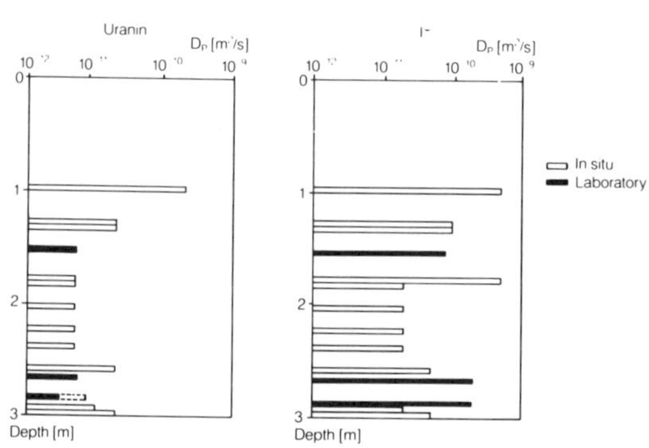

Figure 2.6.23. Comparison between diffusivities obtained in the laboratory and in situ.

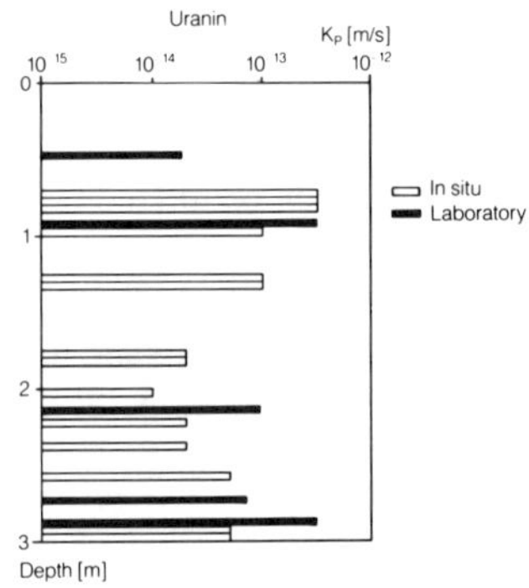

Figure 2.6.24. Hydraulic conductivity vs depth in core. From in situ conditions and laboratory experiments.

The diffusivities and hydraulic conductivities obtained in situ compared well with those obtained in the laboratory experiments.

Salt Water Diffusion Profiles in a Granite Block

Jeffries (1988) has reported on the investigation of a salt water profile in a block of granite rock which was immersed in salt water 30 years ago. The block of granite was 1.5·1·10.6 m and was taken from the Carnmenellis pluton in Cornwall. It was used to construct a water break in Falmouth in 1956. After 30 years of immersion in salt water it was retrieved and samples were taken by coring through the block in the shortest direction.

The core was sectioned and leached of its dissolved anions, Cl^-, Br^- SO_4^{2-}, and F^-. The concentration of the ions in the leachates was measured and concentration profiles were obtained. Cl^- and Br^- concentrations were 15 to 40 times higher than those in fresh granite samples obtained from the same quarry. The profiles were flat and only lower bounds of the diffusivities could be determined. The pore diffusivity was estimated to be larger than $9 \cdot 10^{-11}$ m^2/s. Measured values on samples of fresh unweathered rock were found to be at least 3 times lower.

The SO_4^{2-} and F^- profiles had sharp concentration peaks about 10 cm from the surface that were interpreted to be the result of weathering of the rock, which contains minerals with these species. It was suggested that these species are slightly sorbed on the mineral surfaces and that the movement of these species is slightly retarded in relation to non-interacting species. Diffusivities and sorption coefficients could not be determined unequivocally from the data but the assessed apparent diffusivities were found not to be unreasonable.

2.7 FLOW AND TRANSPORT IN INDIVIDUAL FRACTURES

2.7.1 Laboratory Experiments

A set of laboratory tracer experiments have been performed using natural fractures. The aim of the experiments was to study sorbing tracer retardation by sorption and matrix diffusion in natural fractures.

Nonsorbing tracers—tritiated water and a large organic molecule, lignosulphonate—were used to evaluate the fracture aperture and dispersive properties. The sorbing tracers cesium and strontium were used to study the retardation effects.

In the first set of experiments (Neretnieks et al., 1982) the rock used was a 30-cm-long granitic drill core, 20 cm in diameter, taken from the Stripa mine at a depth of 360 m below ground level. The core had a natural fracture which ran parallel to the axis. The cylindrical surface of the drill core was sealed to prevent any water leaving the rock except through the outlet end of the fissure. The granite cylinder was mounted between two endplates containing inlet and outlet channels (Fig. 2.7.1).

Artificial groundwater with a tracer was fed to the upper channel. The effluent was analyzed for tracer concentrations. The tracers were introduced, either as a step up or as a step up followed by a step down after a suitable amount of tracer had been introduced.

Figure 2.7.2 shows the breakthrough curves for tritiated water for different flow rates.

The figure shows plateaus at C/C_o = 0.7-0.8, indicating that there are at least two channels. Similar curves were obtained with the other nonsorbing tracer – the negatively charged lignosulphonate ion. Neglecting the presence of two channels, the best fit to the advection-dispersion equation gives Peclet numbers ranging between 8 and 27, with an average of 14.2. The dispersion length was 25 mm. The average fracture aperture calculated from the residence

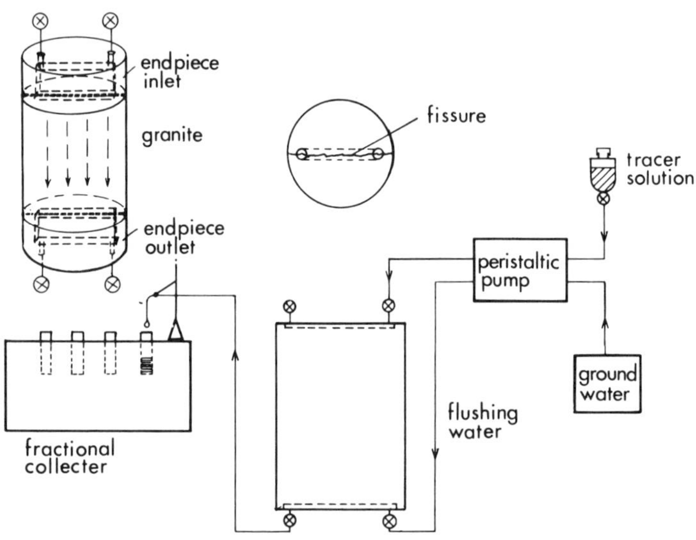

Figure 2.7.1. Experimental setup.

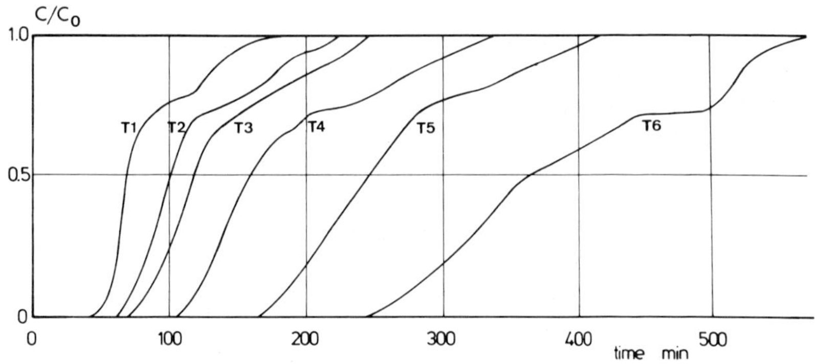

Figure 2.7.2. Experimental breakthrough curves for THO.

time was 0.18 mm. The same data were also analyzed with the advection-channeling model, assuming that the breakthrough curve is caused by different velocities in a multitude of independent channels. With a log normal (base 10) distribution, the logarithmic standard deviation was found to be 0.094 on the average.

Strontium and cesium were run in the same core. A typical breakthrough curve for strontium is shown in Fig. 2.7.3.

Whereas the tritiated water and the lignosulphonate ion were predicted and found to show negligible effects of penetration of the matrix, strontium and cesium were predicted to be strongly influenced by sorption within the rock matrix and by sorption on the surface of the fracture. The predicted and the experimental results compare reasonably well in all 13 runs with sorbing tracers. Figure 2.7.3 gives one example. Attempts to fit the experimental results with only surface sorption or with only matrix diffusion and inner sorption in the rock matrix were not successful. Simultaneous surface sorption and matrix diffusion with sorption in the rock matrix had to be assumed to account for the experimental results. The two curves with $K_d\rho_p$=20 and 40 in Fig. 2.7.3 show attempts to fit only with matrix diffusion.

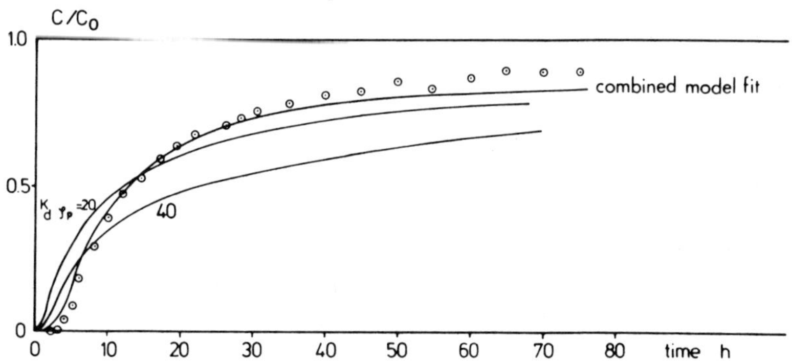

Figure 2.7.3. Experimental strontium breakthrough curve and curve predicted by advection-channeling-matrix diffusion model.

In a later set of experiments (Moreno et al., 1985) two more cores were used, 18.5 and 27 cm long. More refined models were used that also accounted for the dispersion effects of the inlet and outlet end pieces. Peclet numbers of 20 and 15 and fracture apertures of 0.14 and 0.13 mm were found. Predicted and experimental results on diffusivities and sorption equilibria for strontium differed much more for these cores than for the first core. The obtained results are, however, within the wide range of diffusivities and sorption data found for crystalline rocks and coating and alteration materials (Skagius and Neretnieks, 1985a).

The diffusion and sorption data obtained, D_e=10^{-12} m^2/s and $K_d\rho_p$=135 and 10 for strontium and cesium, respectively, agree with independent measurements (Skagius et al., 1982).

Table 2.7.1 summarizes the fracture properties and the data obtained for the sorbing species. The cores were obtained from the Stripa mine.

These investigations indicate that the transport of sorbing species in natural fractures is influenced by both surface sorption and by diffusion into the rock matrix. The fractures used were natural fractures with a thin coating, and this may influence the transport. The values obtained for the sorption coefficients and diffusivities are within the range of values obtained by other measurements.

Table 2.7.1. Fracture properties and the data obtained for the sorbing species on Stripa granite.

		Core 1	Core 2	Core 3
Reference		(Neretnieks et al., 1982)	(Moreno et al., 1985)	
l	length (cm)	30	18.5	27
W	width (cm)	20.0	10.0	10.0
δ	Aperture (mm)	0.18	0.14	0.13
Pe	Peclet number	12	20	15
Strontium:				
K_a	surface sorption coeff (mm)	1	0.12–0.31	0.14
$D_eK_d\rho_p$	(m^2/s)	$135\cdot10^{-12}$	$10\text{-}101\cdot10^{-12}$	$16\cdot10^{-12}$
Cesium:				
K_a	surface sorption coeff (mm)	1		
$D_eK_d\rho_p$	(m^2/s)	$10\cdot10^{-12}$		

Skagius et al. (1982) obtained values of K_a for both strontium and cesium on crushed rock pieces of 1 mm. For strontium Skagius (1986) obtained values of $D_eK_d\rho_p$ = $1\text{-}2\cdot10^{-12}$ m^2/s for tablets.

2.7.2 Field Experiments

Stripa 2D Experiment

This experiment was aimed at investigating, in situ, the flow and transport in a natural fracture with migration distances of 5 and 10 m. This is more than one order of magnitude larger than the distances used in laboratory experiments. The experiment was done by injecting tracers in the naturally flowing water in a fracture and by collecting the tracers as they emerged in the drift. In addition, the fracture was excavated after the experiment and its surfaces monitored for the sorbing tracers. The experiment was performed in the Stripa mine at a depth of 360 m.

Figure 2.7.4 shows a schematic illustration of the fracture with one of the injection holes and several sampling holes. Figure 2.7.5 shows the two main fractures in the actual site. Fracture 2 was used for the test.

The fracture was characterized hydraulically by pressure pulse tests in 5 holes intersecting the fracture and by measurements of natural flow rates and hydraulic heads. Figure 2.7.6 shows the location of the injection holes.

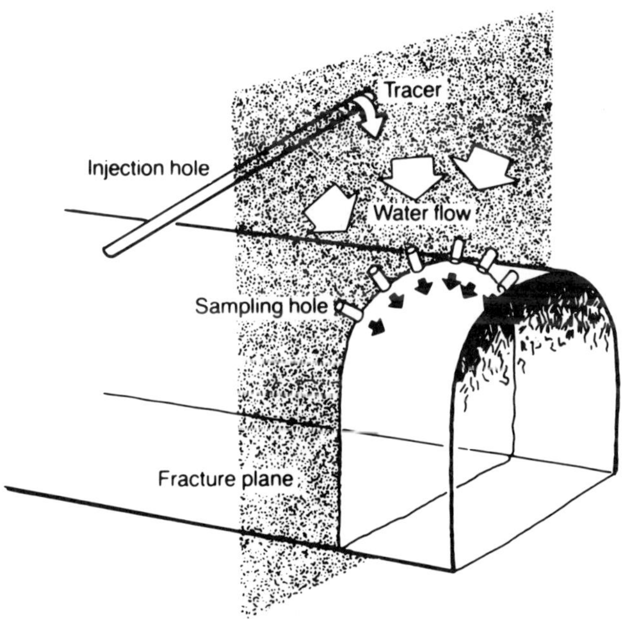

Figure 2.7.4. Schematic illustration of tracer injection.

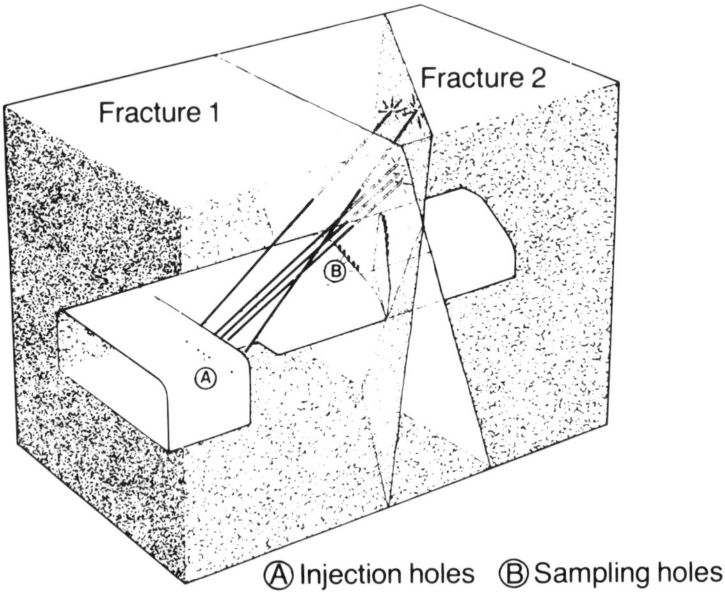

Figure 2.7.5. Schematic illustration of the two fractures intersecting the drift.

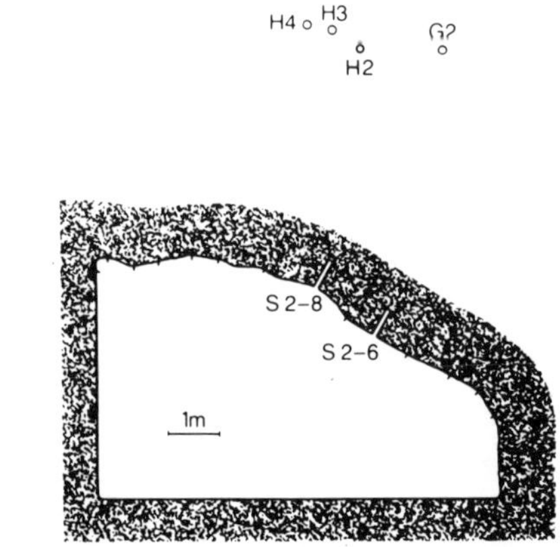

Figure 2.7.6. Location of injection and some collecting holes in fracture No. 2.

Figure 2.7.7 shows the natural flow rates into the collection holes of fractures No. 1 and No. 2. Water flow rates and hydraulic heads were also measured in the fracture of the pilot investigation.

Tracers were injected over a 6 month period. Four different injection points were used. In one of the injection points, sorbing tracers, Cs, Sr, Eu, Nd, Th, U, as well as nonsorbing tracers, iodide and uranin, were injected simultaneously.

All water emerging from the fracture was collected under anoxic conditions to reduce the risk of precipitation of ferric iron hydroxides and other redox sensitive elements.

After the injection stopped the fracture was excavated. Small sample cores were drilled perpendicular to and into the fracture surface. These small cores were ground off in sections and each section analyzed for sorbed tracers.

The seven nonsorbing tracers injected at the four different points H1, H2, H3 and H5 were monitored in the waters collected from the different collection points (see Figs. 2.7.6 and 2.7.7).

In hole H2 the tracers iodide and uranin were injected, the latter on two different occasions. Both uranin and iodide were found in collection holes S2-6 and S2-8. Uranin was also found in very small amounts in S2-9 and S2-10, two sampling points to the right of S2-8 at 0.7 and 1.4 m distance (see Fig. 2.7.7). Eosin from H3 was found in S2-6 and S2-8 whereas albumin, also from H3, was not found at all.

The sorbing tracers were injected in hole H2 continuously for 6 months. None of the sorbing tracers were found in the collected waters. This was expected, except that Sr had been predicted to have a fair chance of arriving within the year-long monitoring period.

A total of 210 sample cores were drilled into the excavated fracture surface. The small sample cores were ground down in thin sections of approximately 0.2 mm, starting from the surface of the fracture. The ground-off samples were analyzed for the sorbing tracers.

Figure 2.7.8 gives Cs concentrations on the fracture near the injection hole. The increased concentration can be traced for a distance of about 0.5 m, which is the end of the cored-out fracture in this direction.

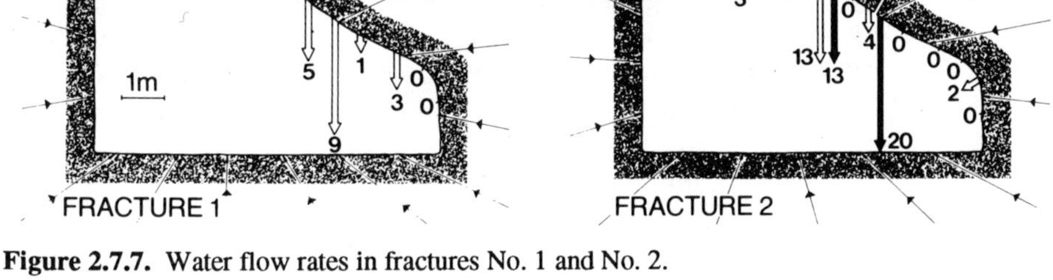

Figure 2.7.7. Water flow rates in fractures No. 1 and No. 2.

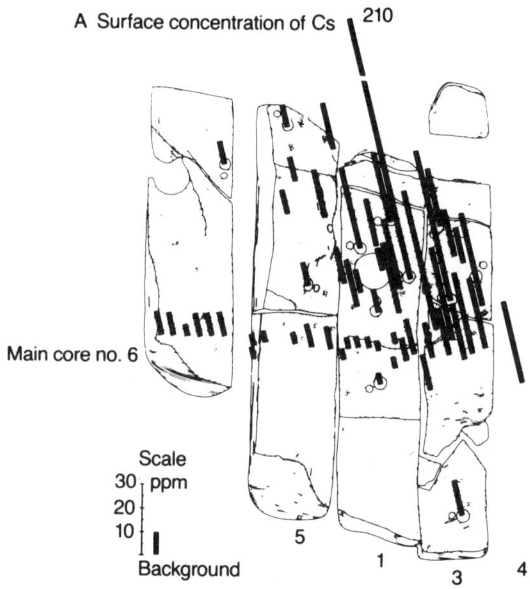

Figure 2.7.8. Examples of surface concentrations of cesium on fracture A. The injection hole is seen in core 1, a third of the way from the top.

Figure 2.7.9 shows the Cs concentration varying with depth at the different locations.

Figure 2.7.10 shows an example of the variation with depth of Cs, Eu, Nd, Th and U at a point about 50 cm "below" the injection point.

The data available made it possible to determine the fracture aperture in three different ways: by mass balance, by the laminar flow assumptions in a slit using pressure drop and flow rate data, and from laminar flow and residence time data.

For the ideal parallel, smooth fracture with laminar flow the three apertures should be the same.

The results depend on whether the tracer solution migrates without mixing or is mixed with the naturally flowing water at the injection point. If the injected tracer stream is carried by and flows with the velocity of the water in the fracture, the tracer residence time will be the same as that of the water. This case is called the mixing case. If on the other hand the tracer solution is injected in a flow path where it moves more or less independently of the water in the rest of the fracture the tracer residence time may not reflect the residence time of the water. This is called the non-mixing case. If the widths of the pathways in both cases are the same the fracture aperture for the much smaller tracer stream will be much smaller than that needed for all the water. Both radial and linear flow assumptions were tested in the analysis of the experiments.

The results are given in Tables 2.7.2 and 2.7.3.

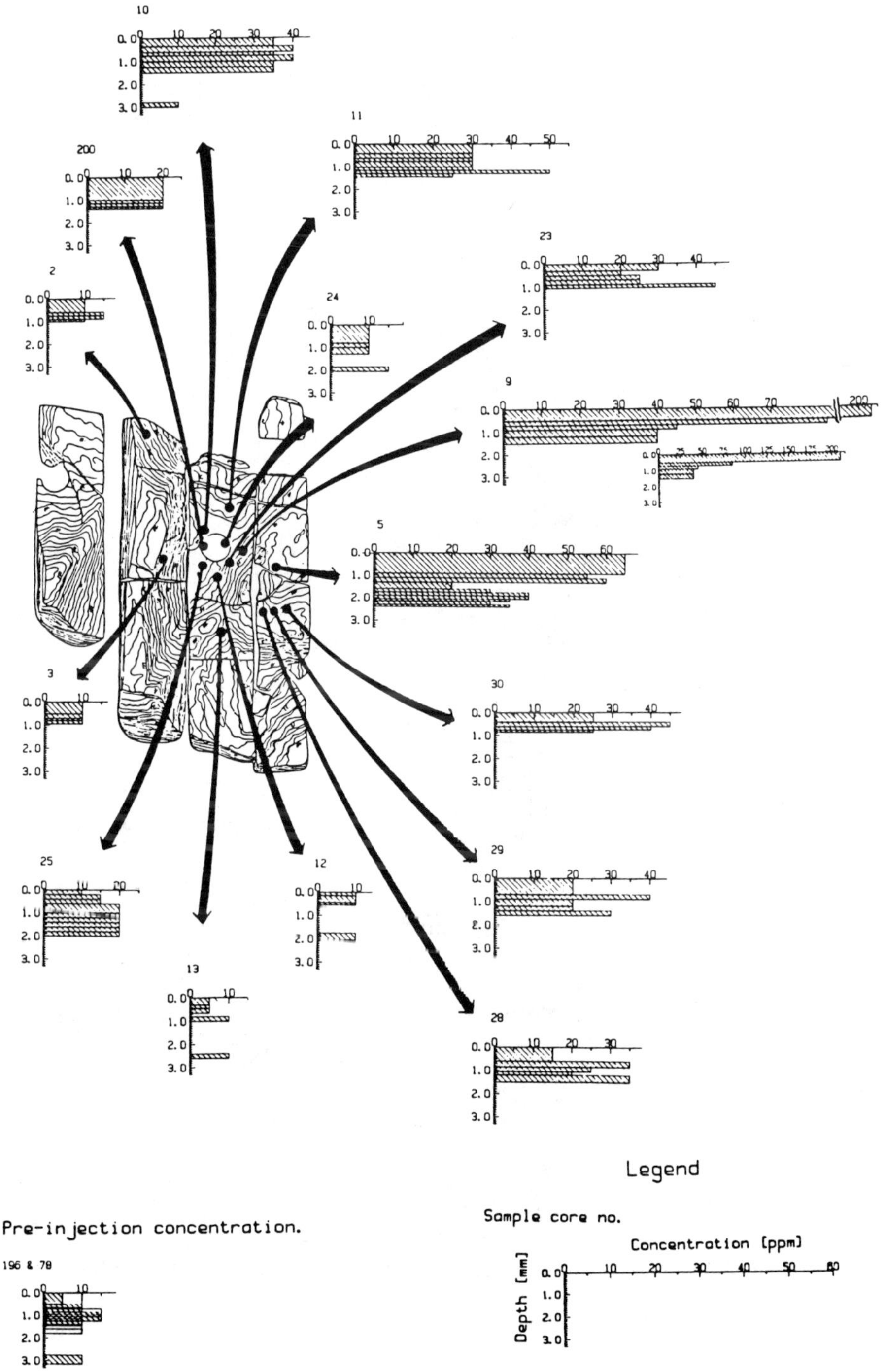

Figure 2.7.9. Variation of cesium with depth at different locations in fracture A.

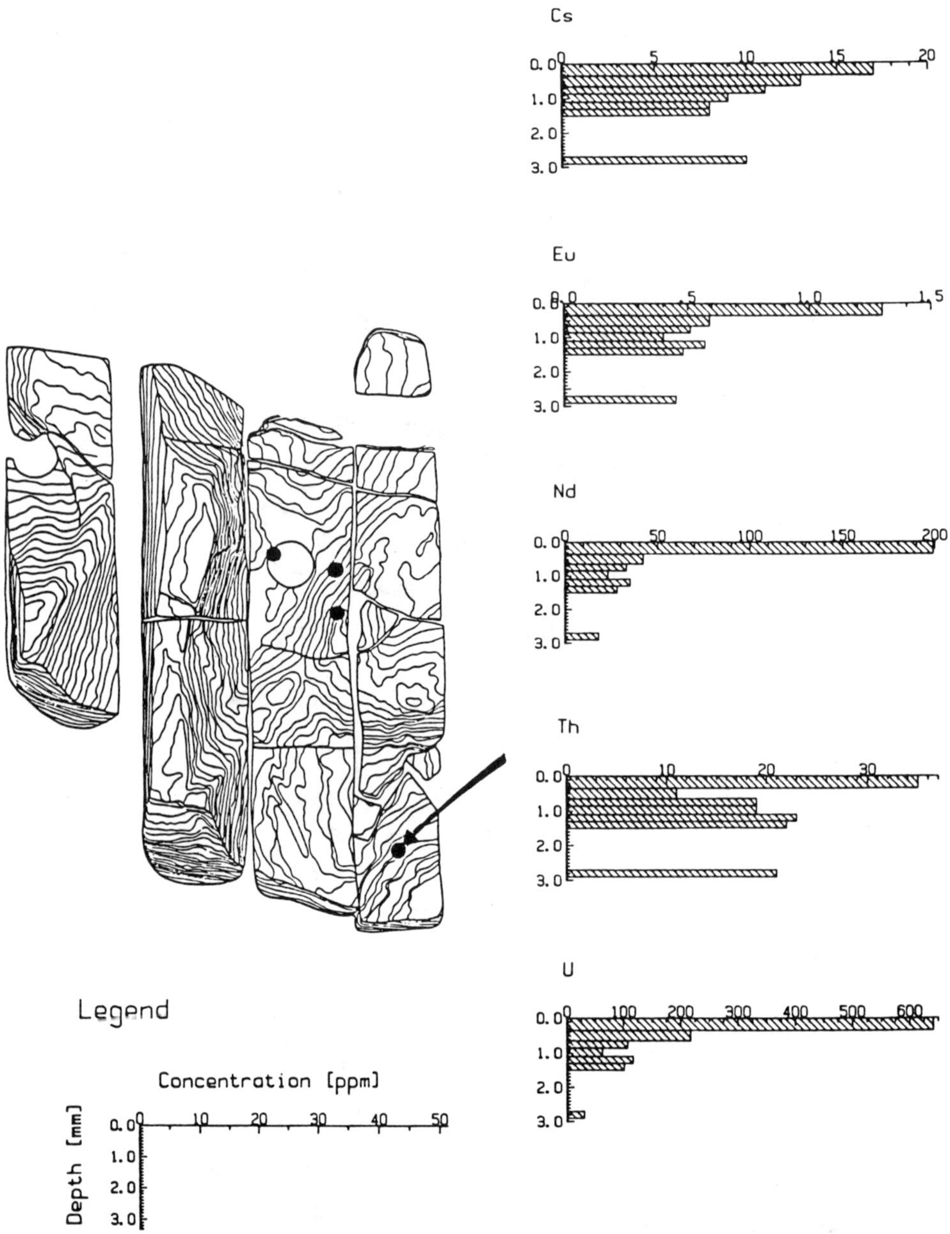

Figure 2.7.10. Variation of tracer concentration with depth in sample core No. 142 in fracture A.

Table 2.7.2. Fracture apertures (μm) assuming mixing at injection point.

Sampling hole	linear			radial		
	δ_f mass bal	δ_l friction	δ_c cubic law	δ_f mass bal	δ_l friction	δ_c cubic law
S2-6	240	1.1	6.7	130	1.2	5.5
S2-8 (100)	280	2.0	10	150	2.1	8.5
S2-8 (500)	1400	0.9	10	750	0.9	8.5
4[a]	2200	1.3	16	1200	1.4	13
5[a]	1500	0.9	11	820	0.96	9.1

[a]Two sampling holes used in a pilot investigation in another fracture.

Table 2.7.3. Fracture apertures (μm) assuming mixing at collection point.

Sampling hole	linear			radial		
	δ_f mass bal	δ_l friction	δ_c cubic law	δ_f mass bal	δ_l friction	δ_c cubic law
S2-6	60	1.1	4.7	43	1.2	3.8
S2-8 (100)	28	2.0	4.6	15	2.1	4.0
S2-8 (500)	140	0.9	4.6	75	0.9	4.0
4[a]	20	1.3	3.2	10	1.4	2.6
5[a]	90	0.9	4.3	50	0.96	3.6

[a]Two sampling holes used in the pilot investigation

A single fracture with an aperture corresponding to the one calculated from the frictional losses (δ_l) could not carry all the water entering the sampling holes. There have to be several of these fractures within the fracture plane to obtain water flow rates equal to those measured at sampling holes S2-6 and S2-8. The actual number is obtained by dividing the mass balance aperture by the frictional loss aperture (δ_f/δ_l).

Three different model concepts were tested against experimental data: the advection-dispersion surface sorption model, the advection-dispersion matrix diffusion model, and the channeling model with no intermixing between channels.

Fitting of the actual models to the experimental breakthrough curves for the nonsorbing tracers was done to obtain residence times and dispersion information. The residence times for the pathway emerging at collection hole S2-6 was found to be 280–350 h for all three models fitted. For the pathway to S2-8, the residence times were between 530 and 710 h. An extremely low value of 45 h was found in one fitting attempt where the effect of matrix diffusion was unrealistically large. Dispersion values corresponding to Peclet numbers of 1–3 were typical, resulting from the very long tails.

Figures 2.7.11 and 2.7.12 show the fit between the three different models and the experimental data. As can be seen in the figures it is possible to get a good fit with all the models. It is not possible to select between the models based only on the results of this experiment. Some of the mechanisms and their parameters have to be determined independently.

Uranin injected in hole H2 gave the highest concentrations and the shortest residence times: on the order of 300–700 h, depending on model. The other tracers arrived in very low concentration and after longer times. Elbenyl injected in hole H1 at about 9.5 m distance just reached the detection limit after 1500 h, when the experiment was closing down.

Iodide injected in hole H2 together with uranin behaved very similarly to uranin.

Eosin injected in the hole H3 was found in both S2-6 and S2-8. The breakthrough curves showed two peaks indicating at least two different pathways. The residence times were longer from H3 than from H2. Figure 2.7.13 shows the breakthrough curves of uranin and eosin plotted in the same diagram. It should be noted that the two injection holes are at practically the same distance from the collection point S2-8. This observation, together with the fact that hardly any tracer could be injected in hole H4, strongly suggests that the hydraulic and transport properties are quite variable within a fracture.

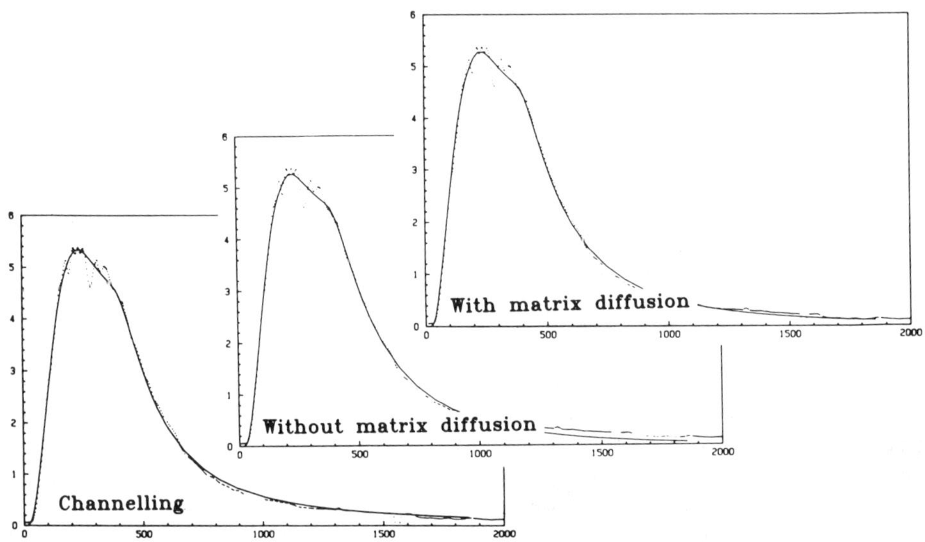

Figure 2.7.11. Best fit between models and experimental data at sampling hole S2-6. Left figure: channeling without matrix diffusion. Middle figure: advection-dispersion. Right figure: advection-dispersion with matrix diffusion.

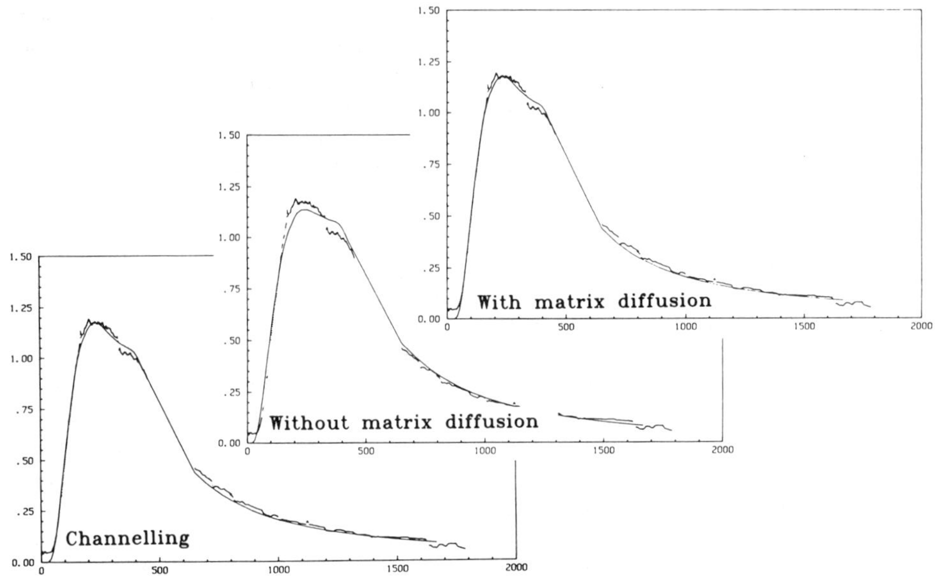

Figure 2.7.12. Fit between models and experimental data at sampling hole 2-8.

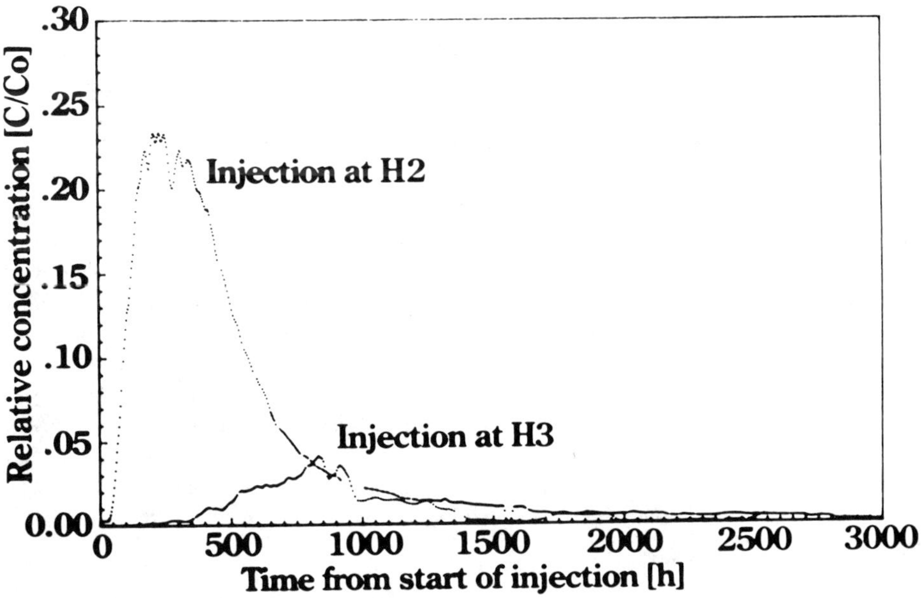

Figure 2.7.13. Breakthrough curves at sampling hole S2-8 for injection at point H2 and point H3

The results of the fitting show that the spreading of the tracer pulses is very great. If it is interpreted as hydrodynamic dispersion as in the advection-dispersion model, Peclet numbers on the order of 1–2 are obtained. This means that the dispersion length equals between half and the full travel distance.

The attempt to fit the advection-dispersion-matrix diffusion model gave similar results when laboratory data on the diffusivity were used. The matrix diffusion may have been noticeable in the pathway H2 to S2-6 but not in the pathway H2 to S2-8. The fitting of the channeling model gave similar results.

It is clear from the results that none of the models gives a unique interpretation of the experiments. There are indications, as for the eosin runs from H3, that there are at least a few different major flowpaths. In such circumstances attempts to fit models which basically assume one flowpath will lead to bad fits and seemingly high dispersion numbers.

The results show that the assumption of where the tracer enters the main stream of the water is crucial to the interpretation of the fracture aperture (mass balance). This assumption cannot be checked with the information available.

Whichever mixing case is the more correct, it is evident that the cubic law cannot give a good estimate of the aperture that determines the residence time (the mass balance aperture).

The sorbing tracers Cs, Sr, Eu, Nd, Th, U were injected in hole H2 together with the uranin. These stable nuclides already exist in the rock and groundwater, and the added concentrations will be superimposed on those in the system. Both water samples and rock samples were therefore taken before the injection started.

The sorbing tracer concentrations on the fracture surface show a very scattered picture (Figs. 2.7.9 and 2.7.10).

It was not possible to reconcile this information with the assumption of smooth, even pathways.

In this field experiment an attempt was made to investigate flow and tracer behavior in a natural fracture in granitic rock. The experiment was designed on the basis that the fracture would have a variable water flow in different parts, known as channeling effects. A considerable number of injection and collection points were therefore used. It was expected that the sorbing tracers would not arrive at the collection points and that the fracture would have to be excavated after the experiment. The variability of flow rate in the fracture was evident and proved to be more important than originally anticipated. It was found that the volumetric fracture aperture was very different from that calculated from the transmissivity. This may be caused by fracture-filling materials in a natural fracture. It was concluded that the fractures investigated did not nearly behave as parallel-walled fractures, either from the point of view of pressure drop or with reference to the uneven flow. Channeling was very pronounced.

The sorbing tracers were found to be strongly retarded and to have penetrated the matrix of the rock as anticipated from laboratory experiments. Because channeling effects made it impossible to monitor all possible surfaces for the sorbing tracers, no mass balance calculation for these was possible. The concentration profiles extending into the rock matrix for the sorbing tracers near the injection point had concentrations and penetration depths within the expected ranges.

Chalk River Experiments

A set of tracer experiments have been performed in a single fracture in a monzonitic gneiss rock body at Chalk River in Canada. In one radially diverging test the distance was 50 m and covered several pathways (Novakowski et al., 1985a). Hydraulic tests were made and resulted in cubic law apertures of 16-61 μm. The estimates from the tracer tests gave similar apertures.

Another experiment in the same location, but involving only one fracture, was run as injection withdrawal tests over a distance of 10.6 m. Novakowski et al. (1985b) analyzed these tests using an advection diffusion model and the assumption that the flow field was a dipole. The hydraulic formation gave cubic law apertures of 60 μm and the tracer tests gave values of 510 μm. Dispersion lengths obtained by curve fitting were found to be 1.4 m.

In a third experiment in the same rock body, five tests were made, four injection withdrawal tests and one radially converging test. The distances were 12.7 and 29.8 m. Both injection-withdrawal tests and radially converging tests were made with pulse injection of tracer. In the latter test the nonsorbing tracers $^{82}Br^{-}$ and a dye, sodium fluorescein, were used. The fracture was located at a depth of about 35 m (Raven et al., 1988).

The breakthrough curves had peak arrival times of between 2 and 6 h for the shorter distance and about 60 h for the longer distance. Hydraulic head differences for the injection withdrawal tests were 2.35–4.90 m and 5.95 m for the short and long paths, respectively.

These tests were analyzed using both the conventional advection-dispersion model and dipole flow field, when appropriate, but also with a model which accounts for exchange of solute with stagnant water volumes adjacent to the mobile water. The transfer rate between the mobile and the stagnant water was assumed to be proportional to the concentration difference and the volume of stagnant water was assumed to be finite. The model was called "the transient solute storage model." Two additional fitting parameters were needed, the transfer coefficient and the volume of stagnant water. Because of the short residence times the effects of matrix diffusion were ruled out.

The advection-dispersion model gave very bad fits to the experimental data, especially in the tails. The transient solute storage model gave excellent fits, with the fraction of mobile water ranging between 0.35 and 0.65 for the shorter flow path. For the longer flow path, where the residence times was more than one order of magnitude larger, all the water could be modeled as mobile. This was interpreted as the result of all stagnant water having been equilibrated with the mobile water.

Cubic law apertures ranging between 110 and 190 μm were obtained from hydraulic tests and the tracer tests gave values of 95-115 μm. This is somewhat surprising because hydraulic (cubic law) apertures tend to be smaller than residence time apertures. The reported apertures do not seem to include the presence of the stagnant water volume and as this was found to be about as large as the mobile water volume, the total volume available for a solute will be about twice as large once the solute has accessed it. Accounting for this effect the residence time aperture will be larger than the hydraulic aperture.

2.8 LARGE SCALE FLOW AND TRACER EXPERIMENTS

2.8.1 Stripa 3D Investigation

In the Stripa experimental mine a drift was excavated at the 360 m level below ground in order to study water flow and tracer migration in low permeability fractured granite. The broader aim was to understand and quantify transport processes relevant to the safety of a final repository for high level radioactive waste (Abelin et al., 1987).

The drift, which is located in water saturated rock, has a natural inflow of water. The water flow rate to the drift, whose total length was 100 m, was monitored by collecting the water in about 375 plastic sheets, which were glued onto the ceiling and upper part of the walls. More than 700 m^2 of the drift were covered. This is half the total surface of the drift. In addition, the water inflow to the lower part of the drift was measured by a ventilation test. The water inflow to the 140 m long access tunnel was also measured by a ventilation experiment. Hydraulic heads were measured in three vertical holes, each 70 m long, extending upwards from the drift. Water flows were monitored for more than 2 years.

The three vertical holes were completely sealed off except for nine 2.5 m long sections at distances varying from 10 to 55 m from the drift. Nine different tracers were injected for nearly two years into these sections. Later, two more tracers were added to the injected water in one of the previous locations.

The tracers were monitored in all sheets carrying water and in some locations in the floor of the drift where water was seeping out. Many of the tracers were strongly colored dyes and were looked for in other nearby drifts and galleries as well as in waters sampled in other places in the mine. Six of the tracers arrived in measurable concentrations in many of the water bearing sheets. In all, 167 tracer breakthrough curves suitable for interpretation were obtained. Average travel times were between 2000 and 7000 hours. Tritium was also looked for in some sheets and was found in two of them.

The recovery of the tracers varied between 2.8% and 65% for 5 of the tracers. One of the tracers had a recovery of 0.002%, four were not found at all, and one was recovered in a considerable but unknown quantity. The tracers that were not found were, as a rule, injected at the farthest distances.

Very few investigations have been made regarding flow and transport properties in deep-lying crystalline rocks. There are no well-developed techniques available to investigate the properties of interest. The flow patterns and mechanisms are not well known and the investigative techniques must be developed so that many conceivable results can be accommodated with the experiments.

Before the experiment, the flow rates, velocities, channeling frequencies and geometry of pathways were known only to within several orders of magnitude. Nor could the possible variations in the properties be known or reasonably assessed. To accommodate possible large variations of these properties some special techniques were developed. The basis for the whole experiment was the development of the large scale plastic sheeting technique, which permitted a very detailed monitoring of the water flow rate and tracer path distribution. Another special aspect of the experiment was the use of as many different tracers as possible, 9+2 tracers, in order to investigate variations in the transport properties of the rock. The choice of different injection distances was made with dual purposes. First, because of the a priori unknown water velocity, a large difference in the migration distance would ensure that at least some of the tracers would arrive during the course of the experiment. Second, different migration distances along essentially the same flow path may give information on how the dispersivity is influenced by the migration distance. This is an open question of some importance.

The size of the wetted surfaces is unknown. If it is large, small molecular weight tracers might diffuse into the micropores of the matrix and to a large extent be withdrawn from the mobile water (Neretnieks, 1980). This might cause the tracers to be retarded and diluted to such an extent that detection in the collected water in the drift would be difficult. Tracers of different molecular weight were used and a high-molecular-weight tracer was synthesized especially for this purpose.

Experimental

The experimental drift was excavated starting about 100 m from the nearest drifts and extending 75 m due north, further into the undisturbed granite. Figure 2.8.1 shows an overall view of the mine adjacent to the experimental drift. Figure 2.8.2 shows the plastic sheeting arrangement.

Water flow rates were monitored at various locations and during various time periods. A pilot hole was drilled into the region where the test was to be located, and water inflow rates were measured at 7.5 m intervals. As soon as the site had been fracture-mapped and photographed, the upper part was covered with plastic sheets, which allowed a high resolution in water inflow and tracer monitoring. The water inflow monitoring continued for 26 months. During this time period the injection holes were drilled, measured, and sealed, and the injection of tracers was completed. The water inflow rates to the three vertical injection holes were measured in 2-m sections. In addition to the detailed inflow rate in each sheet, the total water inflow rate was monitored by a ventilation experiment. In this experiment the total inflow rate to the access drift was also monitored.

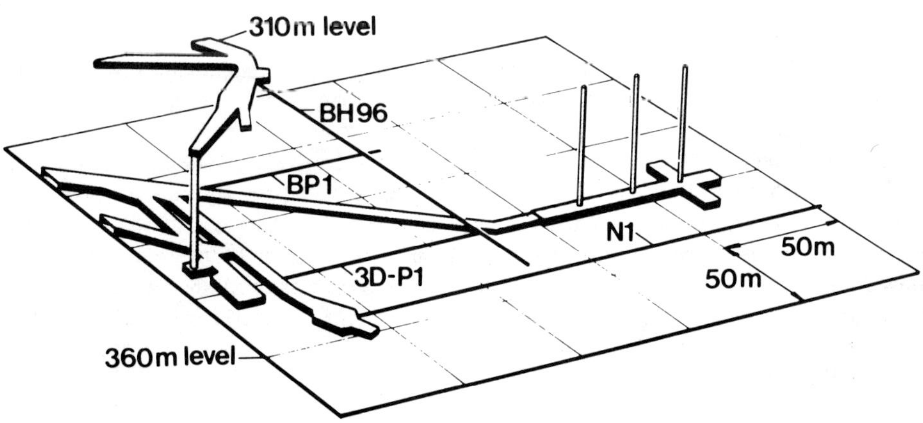

Figure 2.8.1. Overall view of the mine adjacent to the 3D test site.

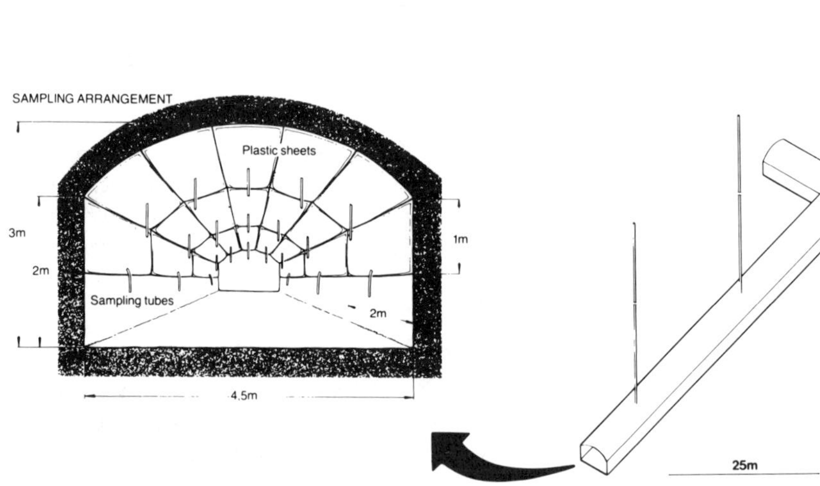

Figure 2.8.2. Dimensions of the 3D test site.

Water samples were collected in test tubes every 16 hour from all sampling sheets with a high enough water flow rate. Data exist on the breakthrough for 7 of the 11 injected tracers, which were found in about 65 different sampling sheets. Tritium analysis has been performed on water samples from both the site and from the injection holes.

Results

The upper part of the test site, approximately 700 m^2 (which is half the surface area of the test site) was divided into about 350 sampling areas, each with an area of 2 m^2. Each sampling area was covered with a plastic sheet. More sheets were added during the experiment at potential points of interest, so that there were approximately 375 sheets at the end of the experiment. The total water inflow to the covered area was 0.7 l/h. Of the 375 sampling sheets 145 gave measurable amounts of water, and 50% of the total inflow came from approximately 3% of the covered area. The wettest sheet carried about 10% of the total inflow. The water inflow rate distribution from the "undisturbed" rock (prior to the drilling of the three injection holes) was monitored for 6 weeks. The results from these measurements are presented in Fig. 2.8.3, where it can be seen that the water inflow is concentrated in a few areas with large dry areas in between.

The duration of the ventilation experiment was half a year. The total water inflow rate into the uncovered part of the test site and access drift was found to be 4 l/h. Of this measured total water inflow 2 l/h came from the uncovered part of the test site. Summing up all the measured inflow into the test site gives a total measured inflow of 3.1 l/h. This compares well with the water inflow rate of 3 l/h into the corresponding part of the pilot hole that was drilled prior to excavating the drift.

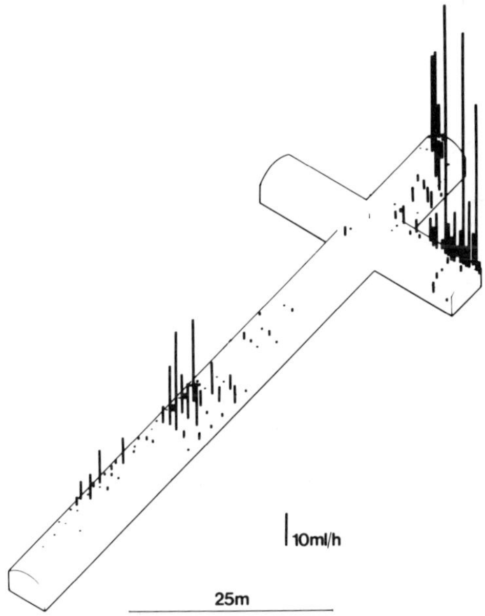

Figure 2.8.3. Water inflow rates into the test site before drilling of the injection holes.

The water flow rates and tracer occurrences along the two drifts were compared with the following fracture characteristics per surface area within 2 m sections, along the two drifts:

- total fracture lengths
- total zone lengths
- number of fracture intersections
- presence of different fracture filling materials
- fracture length within the 4 fracture sets.

The first three fracture characteristics for both the individual fractures and fracture zones are given in Fig. 2.8.4 and can be compared with the water flow rates for the main drift and the arm given in the same figure. In the histograms for the number of intersections, only the intersections concerning individual fractures are included, and not those concerning zones. Comparisons between the flow rates and histograms for individual fractures and fracture zones for the four fracture sets and fracture fillings were also attempted. No obvious correlation between these fracture characteristics and the measured water flow rates is evident. On the other hand, the areas that have a higher number of fracture intersections have higher water flow rates. This was thought to be a significant finding, and it is in accordance with observations in another investigated site in Sweden.

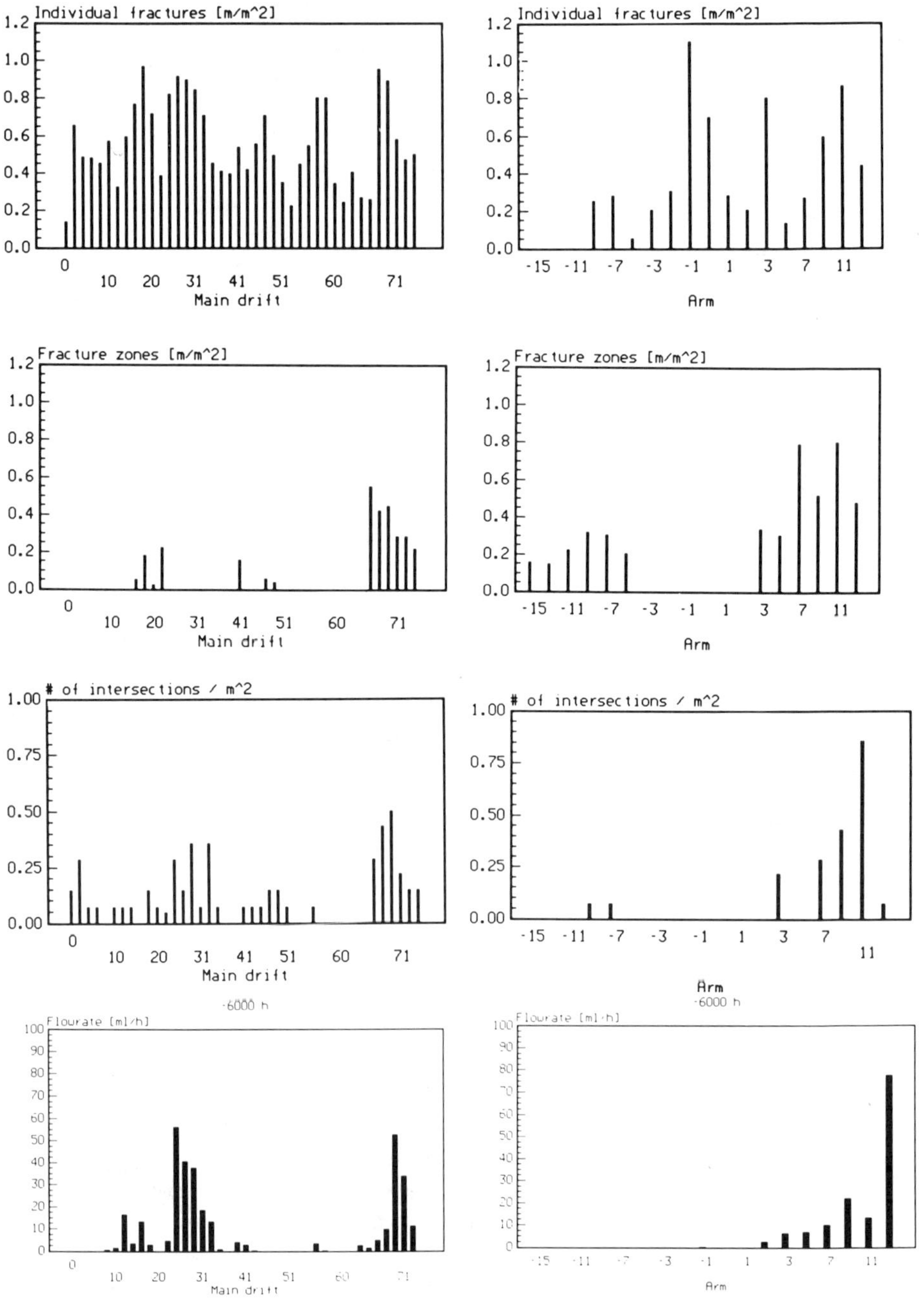

Figure 2.8.4. Fracture length, fracture zone length, and number of single fracture intersections per m for main drift and arm. Water flow rates for main drift and arm before drilling of the injection holes.

The tracers were injected continuously at distances ranging from 12 to more than 55 m for more than 20 months. Water entering the test site was sampled for tracers during the time of injection plus an extra 6 months after the end of injection. Injection flow rates varied between 1 and 20 ml/h.

An overview of tracer breakthrough is given in the following figures. Figure 2.8.5 shows the test site from the side and to which parts of the test site the different tracers have migrated.

It can be seen in Fig. 2.8.5 that tracers from 6 of the 9 zones have emerged into the test site. During the last period of reduced sampling a seventh tracer was found, which is illustrated with dashed lines in Fig. 2.8.5. This occurrence might be due to the preparations for Stripa Project Phase III.

No tracers were found in the right part of the arm, which is the area with the highest water inflow rates. The right arm is also closest to the injection zones in hole III, but the tracers injected at these zones emerged at the central part of the main drift.

Figure 2.8.6 shows a view from above of where tracers were found. Only those sampling areas with "high" water flow rates i.e. connected to fractional collectors, are included in the figure. Figure 2.8.7 shows the locations where the individual tracers emerged. As can be seen in Fig. 2.8.7 all tracers emerge in the central part of the main drift over a length of 35 meters.

Figure 2.8.8 gives an example of breakthrough curves for the different tracers. More than 160 such curves were obtained with up to a thousand individual measurements each.

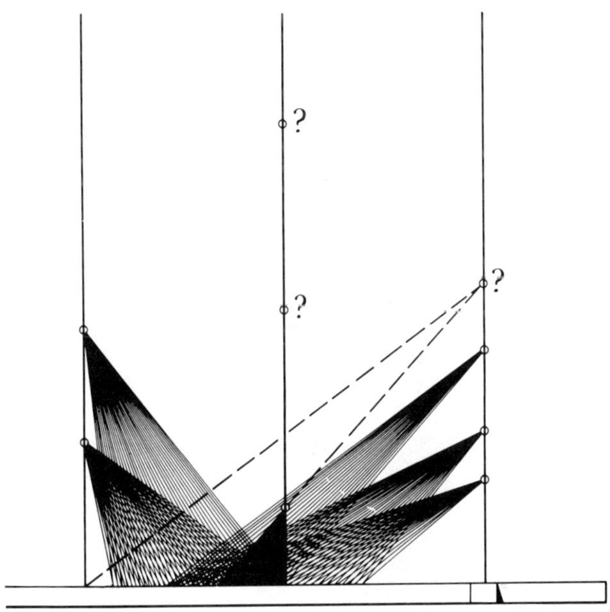

Figure 2.8.5. Areas in the test site where the different tracers have emerged.

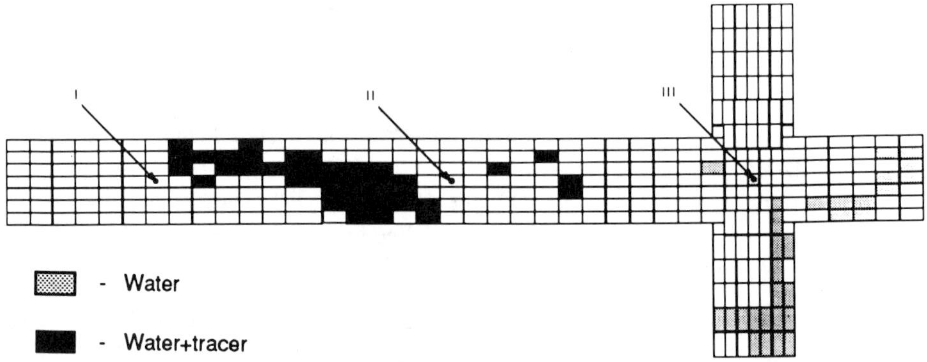

Figure 2.8.6. Areas where tracers have been found in the test site.

Tritium is often used as an environmental tracer. Infiltrating rain water contains about 10-20 TU (tritium units), at present. During the nuclear bomb tests in the 1950's and 1960's considerable amounts of tritium were released to the atmosphere. The tritium content of the rain was increased considerably and peak values of more than 500 TU were measured in the period 1963-1967. The tritium content of water before the bomb tests was on the order of 5 TU. Tritium has a half-life of 12.3 years. Therefore, "pre-bomb" tritium should have decayed to about a tenth of the initial concentration of those times i.e. to values below 1 TU. If values higher than this are found, then it is an indication that the surface water has reached the sampling points in less than about 30 years.

Tritium samples were taken in May 1985 and in June 1986 at some locations in the drifts and in two of the injection zones in hole III.

The results indicate that in at least two sheets where the tritium concentration was 5.1 and 5.7-7.3 TU, the surface water has had a residence time of less than 30 years.

Evaluation and Interpretation of Experimental Results

The tracer tests were analyzed using different models (see Table 2.8.1). Listed are some items that are of special interest in evaluating the different models:

- water travel time (flow porosity)
- dispersion length
- matrix diffusion effects
- wetted area
- channeling characteristics.

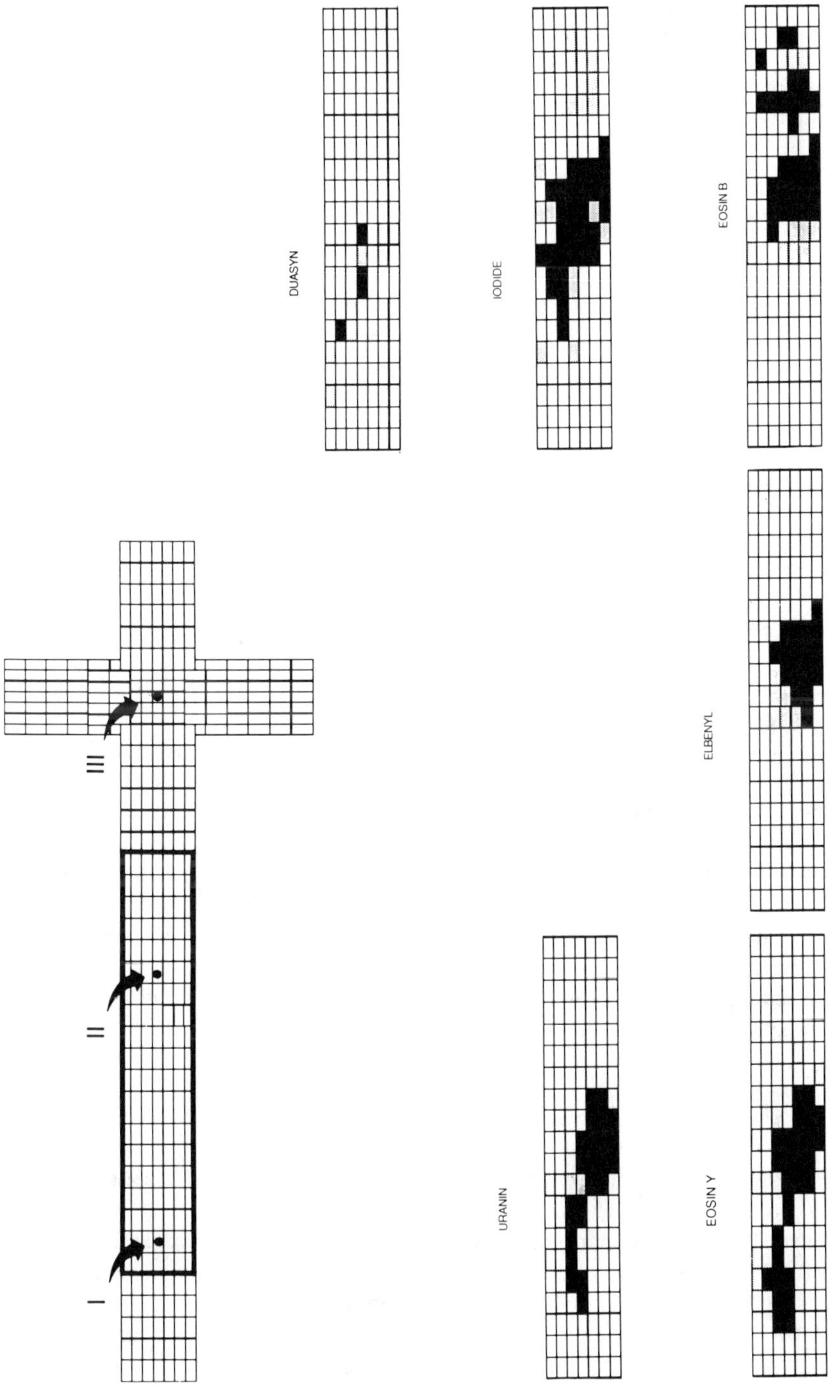

Figure 2.8.7. Tracer occurrence in the main drift for the individual tracers.

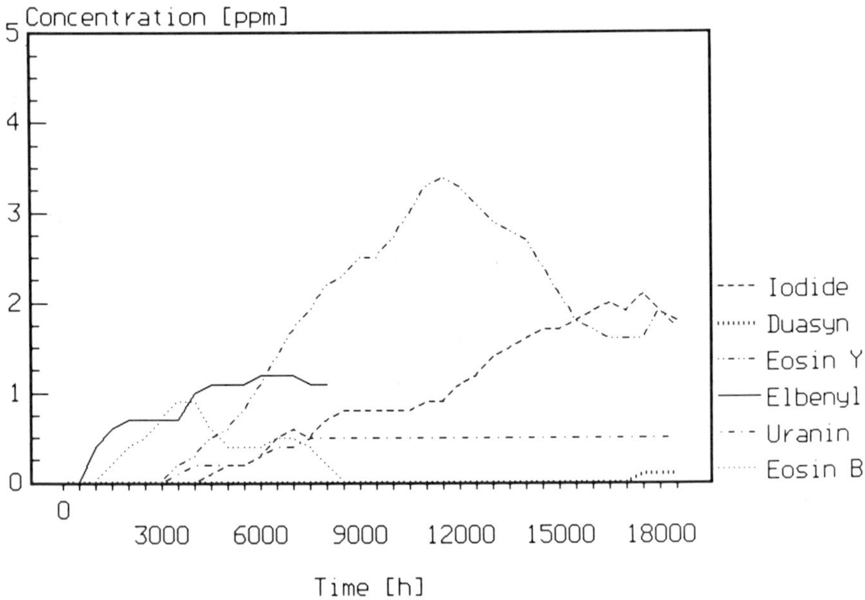

Figure 2.8.8. Example of different breakthrough curves.

Models that have more parameters may describe more mechanisms and thus may give better agreement with the experimental results. Of the models used, the simplest models contain only three parameters, e.g., travel time, dispersivity and dilution factor, and the more complex models include four independent parameters. By including more parameters it is easier to obtain better fits without actually increasing the physical meaning of the obtained parameter values. However, a good fit does not imply that the mechanisms which the model is based on actually are active at all. Therefore several models have been analyzed with different mechanisms which may give similar results, e.g. the spreading of a tracer pulse may be caused by hydrodynamic dispersion, channeling, matrix diffusion, or some other causes which are not included in the model. The three mentioned mechanisms will in many circumstances lead to a similar spreading of a pulse and cannot be distinguished from one another by just fitting one or a few experimental curves in a model. In order to separate out the various mechanisms, independent information is needed.

Models Used for Fitting Tracer Concentration Curves. Table 2.8.1 below summarizes the models used and the parameters which are obtained with each model.

The theoretical models were fitted to the experimental results by a least squares nonlinear fitting process.

First, all the selected individual curves were fitted with the AD model (advection-dispersion model). Of the nearly 170 different curves fitted many had high dilution factors and contained very small amounts of the tracer in comparison to the other curves.

After fitting "all" the tracer breakthrough curves, a set of 5 selected curves was chosen for each tracer. The selected curves accounted for a significant amount of the tracer. These curves were studied in more detail with the ADD model and AC model.

Table 2.8.1. Models used and parameter values determined.

Model	Parameters
Advection-dispersion (AD)	t_w, Pe, DF*
Advection-channeling (AC)	t_w, σ_l, DF
Advection-dispersion-matrix diff. (ADD)	t_w, Pe, A,** DF
Advection-channeling-matrix diff. (ACD)	t_w, σ_l, D_D, DF

*Dilution factor **Includes parameters to describe matrix diffusion effects.

The results from the model fits indicate that the dispersion for many of the breakthrough curves was very high. There are many conceivable reasons for high dispersion values. One is the spreading of the tracer pulse by diffusion of the tracer into and out of stagnant or nearly stagnant volumes of water. Such volumes of water were known to exist in the porous matrix of the rock as well as in the fracture itself, and can be accessed by molecular diffusion from the flowing water in the channels. The ADD model was used to investigate this possible cause of dispersion and also to investigate the matrix diffusion. The matrix diffusion mechanism may be an important cause of withdrawing the tracer from the flowing water and into the rock, resulting in less than full recovery even after very long collecting time periods.

The injection flow varied with time and for this reason a mean value for the injection flow (or mass of tracer injected in a given time) could not be used. The breakthrough curve of variable injection can be calculated using the convolution theorem, which requires an integration over time.

The breakthrough curves were first fitted using the advection-dispersion model. The curves that were near the detection limit were often erratic and were not used. Curves in which a plateau was not reached were not used either. In total, 167 individual breakthrough curves were fitted. In these nonlinear least-square fits three parameters were fitted: the Peclet number, the residence time, and a dilution factor. Preliminary results showed that most of the curves have a very low Peclet number (high dispersion). The model used was not valid for Peclet numbers lower than three (Sauty, 1980). For this reason the Peclet number was limited to values greater than 4.0. It was observed that the dispersion was low for the runs with the tracer iodide. The runs with the tracers eosin B and uranin showed high dispersion. The agreement between the fitted curves and the experimental data was good for the runs with the tracers eosin Y and iodide. The agreement was very bad for the runs with the tracer uranin. Figures 2.8.9-2.8.13 show some typical results of the fitting.

In Fig. 2.8.9 it is shown that the theoretical curve does not reproduce the form of the experimental points very well. There are two distinct dips in the experimental points at 5000 h and at 15000 h which are not seen in the fitted curve. The fit gives a residence time of 7045 h and a Peclet number of 5.1. A visual observation of the experimental points shows that the fitted curve cuts over the two early peaks. Inspection of the other curves for eosin B shows that the form of the response curves are quite different in some cases. This indicates that there is more than just advection and hydrodynamic dispersion occurring which influences the responses along these pathways.

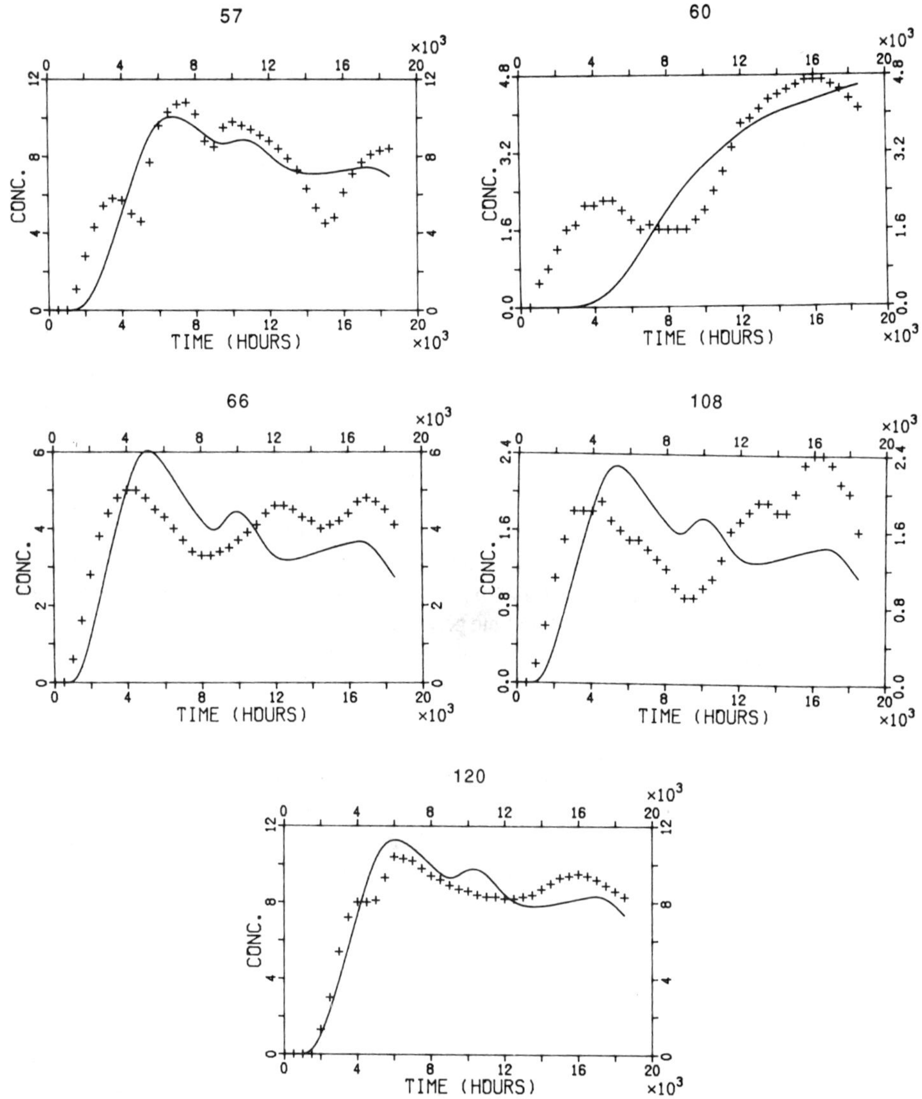

Figure 2.8.9-13. Eosin B breakthrough curves from selected sheets. The full line gives a fitted curve by the AD model and the crosses show selected experimental points. The figures are numbered from left to right, starting in the upper left corner.

Because a large number of the curves available contributed to a very small amount of the total recovered tracer, five curves for each tracer were chosen for continued analysis. These curves were chosen because they represented an important part of the tracer recovered during the observation time or because they represented the different zones where the tracer was found.

From these results representative values for the transport parameters are determined and are shown in Table 2.8.2. The tracers eosin B and uranin show a very high dispersion. The Peclet numbers for the tracer iodide were in the neighborhood of 30.

The ADD model requires the determination of four parameters: the Peclet number, the water residence time, a parameter which takes into account the diffusion into the rock matrix, and the dilution factor. It was found that the parameter that accounts for the matrix diffusion effects could not be determined by this method because it influences the curves in a way similar to dispersion, and the two processes could not be distinguished sufficiently with the data available. Another approach to determining matrix diffusion effects is discussed later.

When the flow rate, the residence time, and the travel distance are known, the flow porosity may be evaluated. This was done by assuming that the flow is radially converging.

Other assumptions also had to be made. Thus, only the flow rate of the water with the tracer was used to calculate the residence time of that tracer, and the flow area was taken to be within that section of the covered drift where tracer was found. The path length was taken to be the shortest distance between drift and injection hole. The uncertainties involved in these assumption can easily lead to uncertainties of a factor of two in the determined "average" porosity. It may even be questioned whether it is meaningful to try to assign an average porosity to so limited a portion of a heterogeneous rock.

Figure 2.8.14 shows a plot of the porosity as a function of the distance to the injection point. There is a tendency for the porosity to decrease with increasing distance.

In the determination of the porosity there is an underlying assumption that the residence time measured with the tracers represents that of the main flow in which the tracers were subsequently found as they emerged in the drift. The average individual injection flow rates for the six tracers are in the range of 0.45 to 8.94 ml/h. The sum of the flow rates for these six recovered tracers is 24.4 ml/h. Compare to this the total flow rate to the covered part of the drift, which ranged from

Table 2.8.2. Representative values for the parameters obtained from the advection-dispersion model.

Tracer	Peclet number	Residence time (hours)	Distance from injection (m)
Eosin B	4.0	6 000	31
Uranin	4.0	5 000	35
Elbenyl	5.0	2 000	11
Eosin Y	5.0	7 000	24
Iodide	30	8 000	41

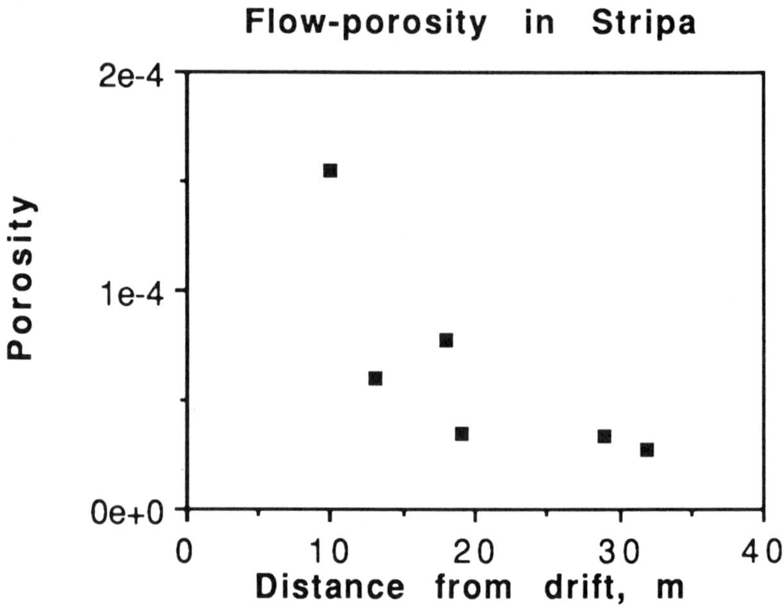

Figure 2.8.14. Porosity versus distance from drift.

600-700 ml/h. The injection flow rates of these tracers made up 3-4% of the flow rate to the drift. The tracers were recovered in the flow to the central part of the drift, where the flow was only about 150-200 ml/h. The dilution factor of the individual tracers is, then, between 22 and 396.

If the mixing of the tracer solution with the main water flow has occurred near the drift, then the paths taken by the tracers may not be representative of the paths taken by the water in which the tracers were found. In extreme cases, the porosity of the rock through which the tracers have migrated might be smaller by a factor equal to the dilution factor.

The recovery for the different tracers was calculated by integrating the breakthrough curves multiplied by the respective flow rates. The cumulative amount of tracer that reached the drift as a function of time was also calculated (see Fig. 2.8.15). The recoveries of the tracer varied from 2.8% for uranin to 65.8% for elbenyl. The tracers rose bengal and phloxine were not found at all. Only 0.002% of duasyn was recovered. The low value for duasyn was because its first arrival time was about 15,000 hours. Bromide showed a recovery of 153%. This impossible value for bromide was caused by errors in the analysis due to the presence of iodide. Therefore, the bromide data are not used in recovery calculation.

The results do not indicate that the distance from the injection point to the sampling area was the sole factor influencing the recovery. For example, iodide had a recovery higher than eosin B even though eosin B was injected at a shorter distance.

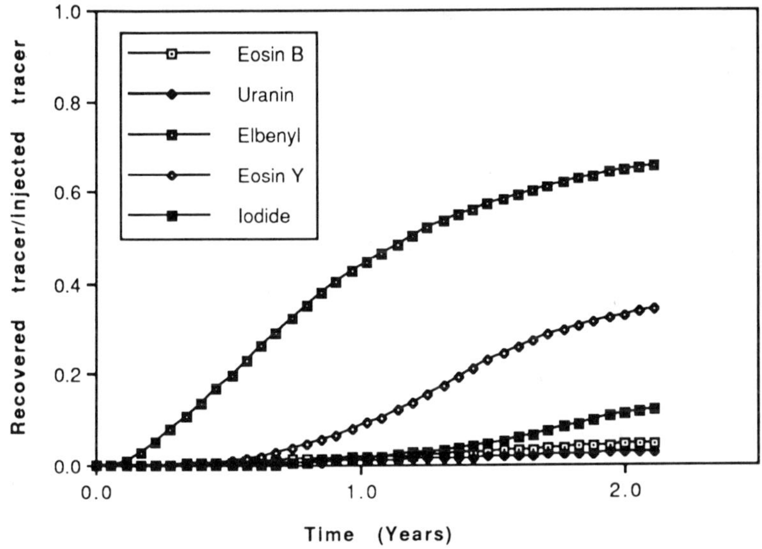

Figure 2.8.15. Recovered tracer/total injected tracer versus time.

Table 2.8.3 indicates that there is a trend in which the recovery of the tracers decreases with increasing injection distance. There are several possible causes of less-than-full recovery of the tracers, as follows:

1. the tracer has partly moved into pathways that do not lead to the experimental drift;

2. the tracer is still traveling in the mobile water;

3. the tracer has reacted "irreversibly" or been degraded;

4. the tracer has moved into stagnant volumes of water by molecular diffusion.

Cause 1 is known to have occurred for one tracer, eosin Y, which has been found in a drift 150 m distant. All tracers have been looked for in nearby drifts and holes but only the one mentioned has been found. The probability and extent of cause 1 can therefore not be assessed at present. Cause 2 is certainly true to some degree but can be accounted for. Cause 3 is not probable because extensive sorption tests in the laboratory over long periods of time have not revealed that the tracers react or degrade. Cause 4 is known to occur as diffusion into the stagnant water in the matrix in the rock and is also suspected to occur as diffusion into the stagnant water in the fractures.

The following discussion explores the possibility of matrix diffusion as a cause for the low recovery.

It was projected (Neretnieks, 1980) that small, readily mobile molecules, when dissolved in water, would migrate into the water in the micropores of the matrix and would thereby be withdrawn from the moving water. Subsequently a series of laboratory measurements have been

Table 2.8.3. Some data on tracer residence times and recoveries.

Tracer	Injection distance (m)	Residence time[a] (h)	Porosity[b] × 10^5	Section length (m)	Recovery (%)
Elbenyl(II)	9-11	2000	15.5	10	65.8
Bromide(III)	12-14	2500	4.5	31	-
Eosin Y(I)	17-19	6000	6.9	27	34.2
Eosin B(III)	18-20	3000	3.4	22	4.8
Iodide(III)	28-30	7000	3.0	31	12.4
Rose bengal(II)	33-35	-	-	-	-
Uranin(I)	31-33	5500	2.0	23	2.8
Duasyn(III)	36-38	>15000	-	-	0.002
Phloxine(II)	55-57	-	-	-	0
STR7(I)	31-33	-	-	-	0
Fluoride(I)	31-33	-	-	-	0

[a]Approximate average values estimated from the fits with AD, AC and ADD models.
[b]Determined from flow rate and residence time.

performed on the porosity and diffusivity of small molecules in granites and gneisses. Typical porosity values are 0.05-0.5% for intact rock. Higher values are often found near fracture surfaces. Effective diffusivities, D_e, are typically in the range 10^{-14} to 10^{-12} m^2/s. Pore diffusivities are higher by two to three orders of magnitude in comparison to the effective diffusivities ($D_e = D_p\ \varepsilon_p$, where ε_p is the matrix porosity). As an illustration of the potential capacity and accessibility of the matrix water to the molecules, consider that the water volume contained in a fracture of thickness 0.05 mm is as large as the water volume in a 100-mm-thick slab of rock adjacent to the fracture if the rock has a porosity of 0.05%. Consider the possible dilution effect: if the fracture were filled with a tracer solution with concentration C and the solution were given time to equilibrate with the non-contaminated water in the pores of a 50-mm-thick rock slab on either side of the fracture (total rock slab on both sides of the fracture is 100 mm), then the concentration in the fracture and in the pores would be 0.5 C. If the fracture were emptied of the water, the recovery would be 50% and the rest of the tracers would reside in the porous rock matrix. The time for this process can also be estimated. From the diffusion equation (Bird et al., 1960) it is found that the penetration thickness η for this case can be determined by

$$\overline{\eta} = \sqrt{D_p} \tag{2.8.1}$$

For a pore diffusivity D_p of 10^{-10} m^2/s the time to completely penetrate 50 mm is 0.8 years. The duration of the experiment was more than 2 years and the water residence times for the tracers were in the time interval 0.2 to 0.8 years or more. It thus seems reasonable that this effect may be of importance if the fracture apertures are small enough.

The fracture apertures can be estimated if the flow porosity and the fracture frequency is known. The flow porosity was estimated from data on residence times and total flow rates. The porosities ranged between $2 \cdot 10^{-5}$ and $15 \cdot 10^{-5}$. For a given fracture spacing (parallel, fully open fractures), the flow porosity ε_f is δ/S (the aperture divided by the fracture spacing). For a flow porosity of $5 \cdot 10^{-5}$ and a fracture aperture of 0.05 mm, the spacing is 1 m.

The above data are in the range of possible values. It thus seems worthwhile to explore this possibility further.

The specific surface a' was determined by comparing the experimental recovery and the recovery obtained from model calculations. The model calculations were made using the advection-dispersion-matrix diffusion model to fit the recovery. The obtained range of values are given in Table 2.8.4, column 3, using the matrix porosity and diffusivity data obtained in the laboratory diffusion measurements (Abelin et al., 1987). The porosity obtained from the diffusivity measurements is about 10 times lower than that obtained by weighing a wet and dry sample. The higher values may be more appropriate for the long contact times in the tracer experiment. If the higher values are used, the surface a' obtained is smaller by about a factor of three. It should be noted that the matrix porosity and diffusivity measurements were made on samples of rock not directly adjacent to a fracture with flowing water. Skagius and Neretnieks (1986a) found that the rock directly adjacent to fractures almost always has considerably higher porosity and diffusivity than rock further from the fracture. Porosities of 1 to more than 7% were found in the rock adjacent to the fracture. Diffusivities were mostly larger and sometimes several orders of magnitude larger in the rock adjacent to the fracture than in rock at some distance from the fracture surface. If the fractures in contact with the tracers in the field experiment have 2% porosity and a diffusivity of $5 \cdot 10^{-14}$ m^2/s, the specific surface needed to account for the nonrecovery of tracers is much smaller. Column 4 gives these v alues of $\acute{a}$.

Table 2.8.4. Compilation of obtained specific surfaces.

Tracer	Injection distance (m)	Specific surface (m^2/m^3)	Specific surface (m^2/m^3)
Column 1	2	3	4
Eosin B	19	14-27	1.4-2.7
Uranin	32	5.1-12	0.5-1.2
Elbenyl	10	15-20	1.5-2.0
Eosin Y	18	4.7-7.7	0.5-0.8
Iodide	29	0.49-1.1	0.22-0.5

In column 3, the large specific surfaces for diffusion into the rock matrix, a', obtained for the dyes are unreasonable because they imply that spacing of conducting fractures should be $2/a'$ m. For iodide this would mean spacing of 1–2 meters and for the dyes 0.07–0.4 meters. This seems to contradict the observations in the drift, where considerably fewer than 100 large water conducting fractures were observed over a length of 100 m. Because only a fraction of the fractures conduct water, the fracture spacing must be larger than 1 m in the rock near the drift. Unless the diffusivity and porosity of the rock adjacent to the water conducting channels is considerably greater (so as to give the values in column 4), matrix diffusion effects alone cannot explain the low recoveries if the large fractures alone are assumed to carry most of the flow.

In a recent mapping of fractures along 5 boreholes, near the experimental drift, whose total hole length was about 900 m, Gale (1988) found typical values of 2-4 fractures per m on the average. Locally, 10 or more fractures per m were found. These observations include all visible fractures in the bore cores also, which are very short fractures. Trace length measurements showed that most of the fractures have trace lengths less than 1.5 m. This would give specific surfaces of 4-8 m^2/m^2 and more than 20 m^2/m^2 locally if a parallel pattern is assumed. For a cubic or random pattern 3 times higher values are obtained.

Matrix diffusion can possibly account for most or even all of the nonrecovery of tracers if the rock adjacent to the fracture surfaces where water is flowing is more porous than rock further in or if a high percentage of the fractures are conducting.

It cannot be resolved at present which are the more correct values for diffusivity and porosity. A much more extensive sampling of rock near fracture surfaces is needed to do this.

Another possible cause for the low recovery of tracers is based on the concept that the water flows in channels that are in contact with practically stagnant pools of water. The tracer in the flowing water in the channel may diffuse into the nearby stagnant water and thus be withdrawn from the mobile water. Figure 2.8.16 illustrates this process.

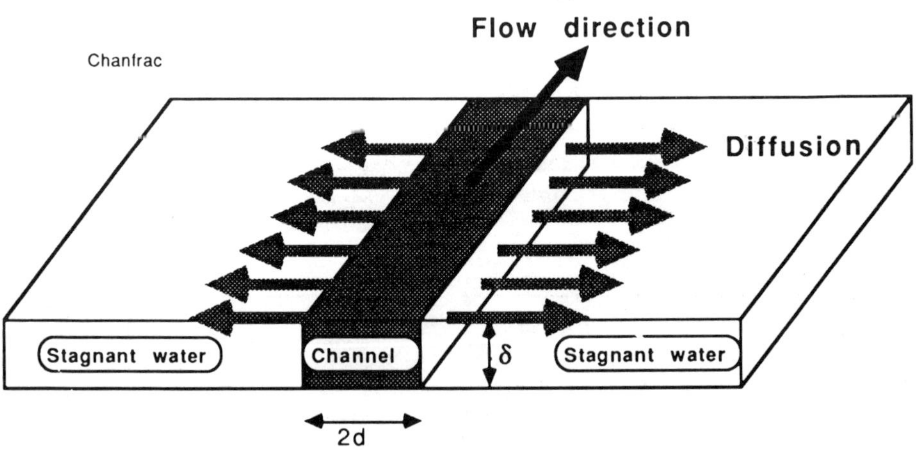

Figure 2.8.16. A fracture cross section with a channel containing flowing water is shown. In contact with the flowing water are stagnant volumes of water into which the tracer is carried by diffusion.

The aperture of the channel is δ and the width of the channel is $2d$. The species diffuse into the stagnant water. For a channel with a small width, in comparison to the diffusion distance, it may be assumed that the concentration of the species across the width of the channel is constant. Therefore, the previously used equation for diffusion into stagnant water in the rock matrix can also be used to describe this case. The effective diffusivity will then be equal to the diffusivity in unconfined water ($D_e=D_w$) and the equivalent of the matrix porosity is equal to unity.

The diffusivity D_w of the dyes in water is on the order of 10^{-9} m^2/s and for iodide it is $4 \cdot 10^{-9}$ m^2/s (Skagius, 1986). The obtained channel widths, $2d$, are summarized in Table 2.8.5. They are based on an average value of a' from the previous calculations.

The channel widths obtained are smaller or, at most, of the same magnitude as the penetration depths into the stagnant water for a contact time of one year. Therefore, the assumption that the concentration over the channel width is constant is not grossly violated.

The above analysis gives no indications of how large the apertures of the channels are. An estimate of the sum of the apertures of the channels, assuming that they all have the same widths, can be made in the following way. The sum of all the cross section areas of the channels is denoted as $2d\ (\Sigma\delta)$. The total cross section area of rock, A_o, that these channels intersect at the face of the drift was utilized to evaluate the flow porosity, ε_f. The porosity is the ratio of the cross section of the channels to the cross section of the area containing the channels.

$$\varepsilon_f = (\Sigma\delta)\cdot 2d / A_o \qquad (2.8.2)$$

From the above expression the sum of the channel apertures ($\Sigma\delta$) can be obtained. The values range from 19 mm to 567 mm and are given in Table 2.8.5.

The tracers arrived in many of the sheets. If every sheet on the average had a few channels, several tens of channels would determine the sum of apertures ($\Sigma\delta$). So for 50 channels this would mean a 3-mm individual channel aperture for eosin B, slightly more than 1 mm for uranin and elbenyl, more than 10 mm for eosin Y, and a 0.4-mm aperture for iodide. Most of these values are considerably larger than what the visual observations indicated.

These results are based on the assumption that flowpaths are straight and that tracers mix into the main flow of water. Flowpaths may easily be up to 2 times larger and, as discussed in relation

Table 2.8.5. Estimated equivalent channel widths 2d.

Tracer	ε_f flow porosity $\times 10^5$	a' (central value) (m^{-1})	$D_e\ \varepsilon_p$ $\times 10^{16}$ (m/s)	Channel width ($2d$) (m)	Sum of apertures ($\Sigma\delta$) (m)
Eosin B	3.4	20	0.1	0.034	0.154
Uranin	2.0	8	0.1	0.050	0.064
Elbenyl	15.5	17	0.1	0.182	0.060
Eosin Y	6.9	6	0.1	0.023	0.567
Iodide	3.0	0.8	2	0.335	0.019

to dilution factors, the uncertainties as to where mixing takes place can cause more than one order of magnitude overestimation of the porosity and thus of the sum of channel apertures.

The diffusion into stagnant pools of water cannot alone explain the loss of tracer. The diffusion into the rock matrix may also not be the sole cause for the loss of tracer. The tracers may have taken paths which are directed to some other drifts in the mine even though they are much further away. This has been observed in one instance.

Eosin Y was found to have migrated in substantial quantities to a gallery about 150 m away from the injection point even though the sheeted drift was located only 18 m away. This indicates that the assumption that all the injected tracer mass moves towards the experimental drift is not valid for one tracer and may not be for the other tracers. Therefore, the analysis based on this assumption may determine specific surfaces and channel widths that are too large.

If most of the tracer moved in the direction of the drift, the results of the calculations indicate that both matrix diffusion and diffusion into stagnant pools of water may play a role. There are probably other mechanisms which could also cause the recovery to be small. One possibility is that the injection, although small, may be large compared to the total flow rate when viewed on a local scale. A dipole type of flow may then develop, where some of the flow moves away from the drift for some distance before returning to the drift. This would produce a lengthening of the concentration curve tails and a lowering of the tracer recovery during a certain observation time.

Determination of Hydraulic Conductivity. The hydraulic conductivity of the rock adjacent to the drift was evaluated based on the assumption that the rock mass is a homogeneous porous medium and that Darcy's law is valid. Some simplifying assumptions needed to be introduced because the head boundary conditions were not known exactly, because of the presence of drifts and tunnels in the old mine. For a homogeneous porous medium the simplifying assumptions would introduce only minor errors and are generally accepted in most other cases involving study of the properties of fractured rocks.

The hydraulic head was measured in the three vertical injection holes. Flow rates were measured in about 375 sheets in the upper half of the drift. In addition, ventilation experiments were conducted to measure the inflow to the lower half of the drift and to the access drift.

Table 2.8.6 gives the flow rates in the different parts of the drifts.

Using the flow rate data from Table 2.8.6 and the head data, the hydraulic conductivities of the different parts of drifts were obtained. These values are summarized in Table 2.8.7.

Table 2.8.6. Flow rates in the different parts of the drifts.

Part of drift	Flow rate (l/h)	Length of drift (m)
Plastic sheets	0.7	100
Bottom part	2.4[a]	100
Access drift	2	115

[a]Covered parts of floor + ventilation

Table 2.8.7. Hydraulic conductivities in the Stripa 3D drifts.

Location		Conductivity × 10^{11} m/s	
		Steep gradient	Low gradient
Experimental drift			
	Ceiling + upper sides	0.56	1.07
	Floor + lower sides	1.9	3.6
	Average	1.3	2.3
Access drift		0.7	1.3

The flow rates varied considerably over the length of the drift and so did the hydraulic conductivities. The data in Table 2.8.7 were averaged over more than 100 m of drift. The variations within the experimental drift are shown in Fig. 2.8.17 where 10 m averages are shown. It is observed that on a 10 m scale the hydraulic properties of the rock are far from constant. The hydraulic conductivity in the access drift and the experimental drift differ by a factor of 2. This may indicate that the 100 m scale is not large enough to obtain reasonably constant average values.

Flow Rate and Tracer Distribution. The tracers arrived at quite varying rates, concentrations and flows of water to the different sheets. The mass flow rate to a sheet is determined by the

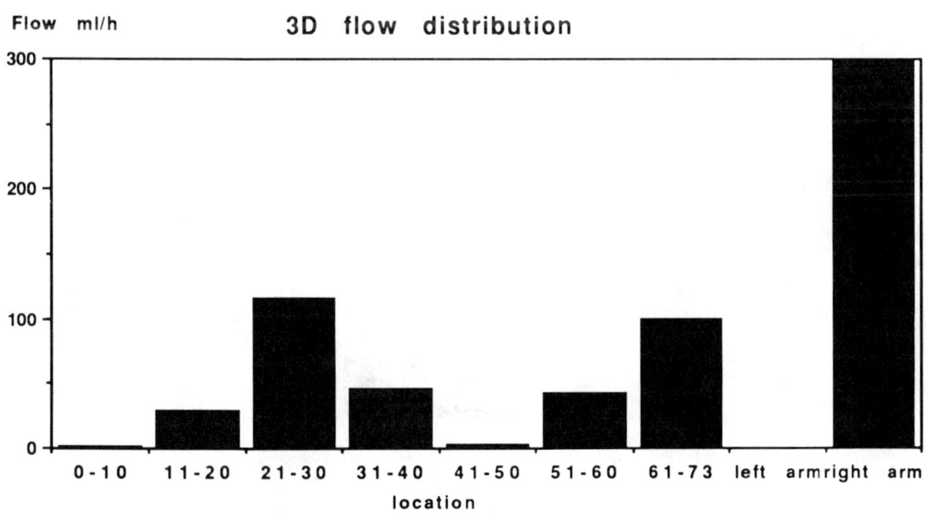

Figure 2.8.17. Flow rate in the experimental drift averaged over 10 m.

product of the flow rate and concentration. The differences in mass flow rates to the different sheets reveal the existence of any preferential paths. Figures 2.8.18 through 2.8.22 show how the tracer recovery is divided among different sheets for the five tracers eosin B, uranin, elbenyl, eosin Y, and iodide. The bars in the histograms show how much of a tracer was recovered in a given range of mass recovery. The right-hand bar in Fig. 2.8.18 shows that there were four sheets (the figure above the bar) that had a tracer recovery in the range 400-800 mg. The sum of the tracer that arrived at these four sheets is given by the height of the bar.

For all the tracers, 3 to 5 sheets out of 27 to 52 sheets carried more than half the tracer.

The distribution of flow rates in the different sheets show that a few spots carried most of the water. This is an indication of channeling. The dimensions of the channels are not known from these observations, but Fig. 2.8.4 indicates that the larger flow rates are observed in those parts of the drift where many fracture intersections exist. This may be an indication that the water flows along fracture intersections, where the channels are very narrow.

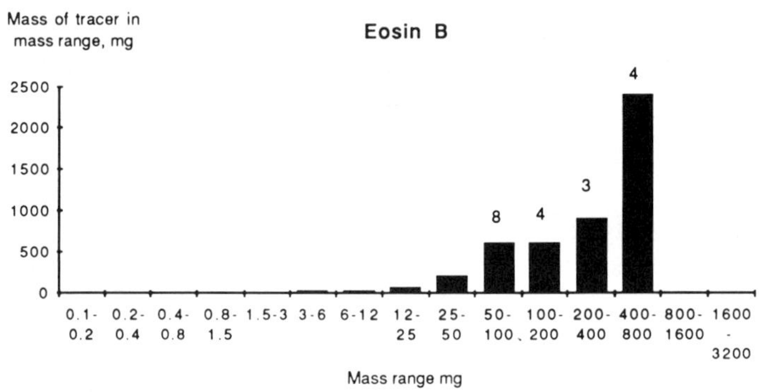

Figure 2.8.18. Histogram over the recovery of eosin B in different mass ranges. 43 sheets carried this tracer in measurable quantities.

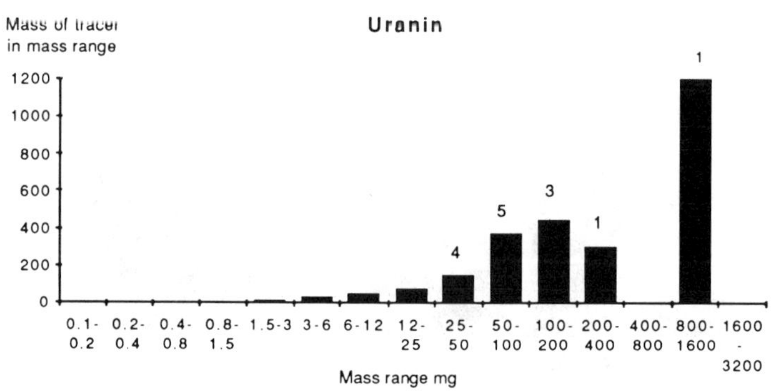

Figure 2.8.19. Histogram over the recovery of uranin in different mass ranges. 45 sheets carried this tracer in measurable quantities.

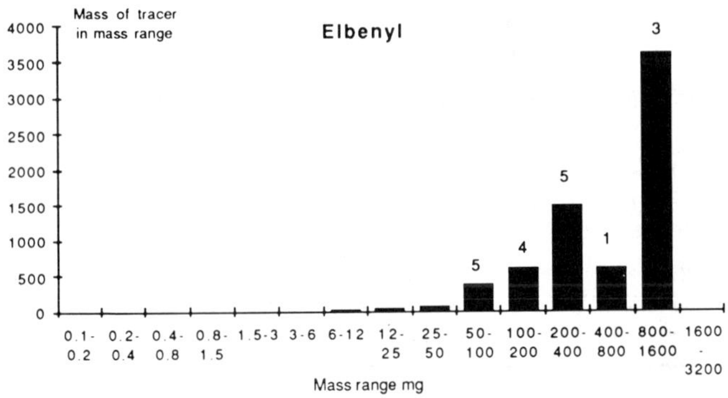

Figure 2.8.20. Histogram over the recovery of elbenyl in different mass ranges. 27 sheets carried this tracer in measurable quantities.

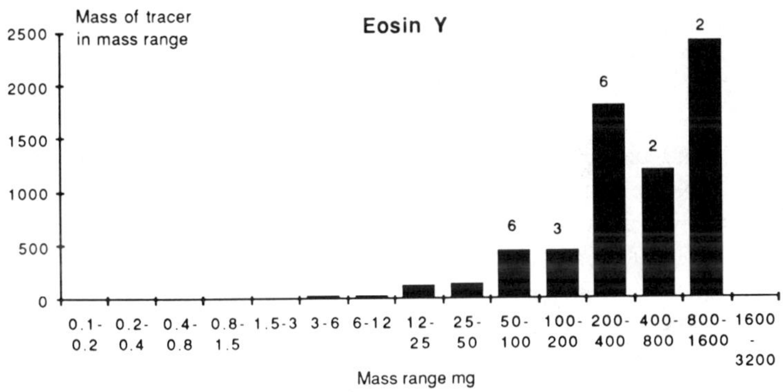

Figure 2.8.21. Histogram over the recovery of eosin Y in different mass ranges. 46 sheets carried this tracer in measurable quantities.

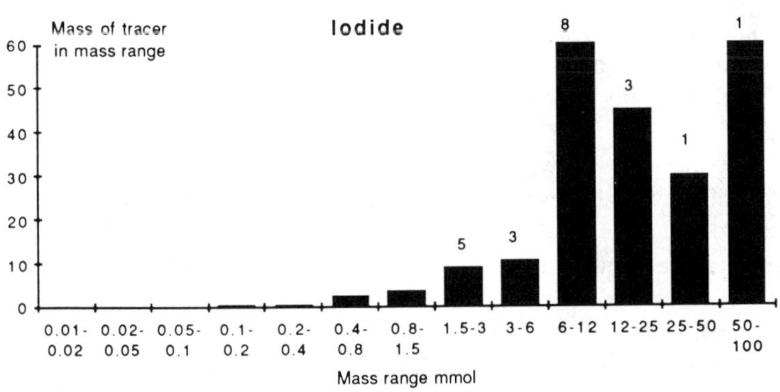

Figure 2.8.22. Histogram over the recovery of iodide in different mass ranges. 52 sheets carried this tracer in measurable quantities.

The tracers were spread out among many sheets. It is clear that the tracers were not confined entirely to one or a few channels but spread considerably. However, a small fraction of the sheets carried a large fraction of the tracer.

The flow rate distribution over the experimental drift shows clearly that adjacent sheets may carry quite varying flow rates. One of the 375 sheets carried 10% of the water. Often the sheets adjacent to a sheet that carried a considerable flow rate was dry or nearly dry. The size of the channels in the fractures were not measured in the experimental drift but observations in other tunnels and drifts in Sweden indicate that channel widths vary from a few centimeters to many tens of centimeters or even a meter or more (Neretnieks, 1987).

The distribution of tracers over the different sheets showed behavior similar to the flow rate distribution. One tracer, duasyn, which was injected in hole III at 37 m distance from the drift, was first detected in a sheet at the opposite end of the drift. The water where it was found contained very little or none of the other tracers. One conclusion is that duasyn must have passed over the main area of sheets, where the other tracers were found, without being drawn down into those sheets. The sheets where this tracer was subsequently found had another pattern of travel than that of the other tracers. Duasyn must have traveled in a pathway or pathways that bypassed the system of pathways that the other tracers traveled. Eosin Y, which was injected in hole I at 18 m from the drift, had a recovery of 34%. Neither this tracer nor any other tracers were found in any other waters in the mine area that were among those sampled at intervals during the experimental time. After the sampling was finished in December 1986 and the mining operations were starting for the next phase of the Stripa program, eosin Y was found in a newly excavated gallery about 150 m away from the injection point. The amounts found were estimated to be on the order of several tens of percent of the total injected mass of this tracer. There had been no indications either in the hydraulic head measurements or any other observations that the tracer would travel in this direction.

Summary of Main Results, Discussion and Conclusions

The single most prominent feature of this investigation was the use of a multitude of plastic sheets in a long underground drift. The use of a drift located well below the water table ensured that the water flow would mainly be directed towards the drift from practically all the tracer injection points. The use of a 100 m long drift with about 375 plastic sheets made it possible to monitor the flow rate and tracer transport in detail, which cannot be done by other known techniques such as "between hole tracer tests." Most of the information obtained with the present technique, such as variability of conductivity and channeling, is difficult or impossible to obtain with other methods.

Samples were taken for tritium analysis in several sheets and at other locations on two occasions. Tritium was found in two of the six sampled sheets. Had the samples been taken from the mixed water of the sheets, the tritium concentration would have been below the detection limit. The presence of tritium in the waters indicates the presence of fast, nearly isolated channels of short extent. Such channels probably would have gone unnoticed had not the plastic sheeting technique been used.

The porosity was found to decrease with distance from the drift. The values for the rock far from the drift were found to be on the order of $2 \cdot 10^{-5}$ to $7 \cdot 10^{-5}$. The porosity in the 10 m nearest the drift was found to be $15.5 \cdot 10^{-5}$. This indicates that the presence of the drift has increased the porosity.

The water residence times that were used to estimate the porosity varied considerably between sheets. Also the residence times for major flow paths and averaged flow paths were quite variable. These and other observations indicate the presence of channeling. This result underscores the need to use the average properties with considerable caution. The presence of tritium in some sheets, for example, indicates that there is some part of the porosity that is not well-connected to other parts of the rock porosity.

The models used do not describe some of the important causes of dispersion and attempt to force the parameter values to account for processes which they cannot explain. This is apparent in the fitted concentration curves. The model results cannot describe the several ups and downs in the experimental curves. Furthermore, the experimental curves in adjacent sheets for the same tracer are in many cases quite dissimilar. This indicates either that there must be very large differences in the values of the dispersivities and velocities in the different pathways to the different sheets or that there are mechanisms that cause these variations that have not been modeled. If the former is true then there are strong differences between adjacent pathways, a condition which would be termed channeling. If there are unmodeled processes, the variations imply that one or more of these processes is the transport mechanism in several but not all of the pathways. This possibility is supported by other information, such as the presence of tritium in some sheets and the very large variations in the ratios between tracer concentrations in adjacent sheets. If regular mixing were constantly occurring then the hydrodynamic dispersion model would apply, and the curves and the concentration ratios in adjacent sheets would behave in a much more regular way.

The results show that for the tracers to have been taken up by the rock matrix, fracture spacings (equivalent water-carrying fractures) for the dyes of between 7 and 40 cm and for iodide of between 50 and 100 cm are needed. These values are probably much too small if only the 100 prominent fractures visible in the drift of 100 m are assumed to be active conductors. The number of fractures with lengths of a few meters or less is orders of magnitude larger, and if these fractures actively participate in the conduction of the flow, the loss of tracer is easily accounted for by matrix diffusion effects. The diffusivity data used were for intact rock. It is known from other measurements that the rock and fracture coating or alteration materials adjacent to water conducting fractures may be much more porous and have much higher diffusivities. If this is the case then the fracture spacings needed are between 1 and 10 m, and matrix diffusion effects may still have played an important role in these experiments even if only the larger fractures conduct most of the flow.

The diffusion into stagnant water in the fracture was also explored. For uranin, elbenyl, and iodide this effect may contribute noticeably, but it probably has a small effect for eosin B and eosin Y. This latter effect is more speculative than the matrix diffusion because many assumptions must be made regarding the geometry and quantity of the stagnant water zones.

Similar effects may be caused by slow flow into semi-stagnant pools of water. There is even less information available for speculations on this process.

The results of the tracer experiments and the tritium measurements give strong support to the notion that a non-negligible portion of the flow takes place in channels which have little contact with other main channels. This cannot be treated by the models that have been applied in the analysis of the experiment. Probably, a fruitful avenue would be to try to incorporate the variability in the models. Stochastic models are available for porous media and some attempts to apply them to this type of rock have been made recently (Gelhar 1987b; Moreno et al., 1987a,b;

Tsang et al., 1987). These models need considerable amounts of data to obtain the desired statistical properties. They also must be devised so as to incorporate the correct processes.

The mechanisms that govern diffusion into stagnant pools of water, matrix diffusion, the frequency of mixing, and especially the non-mixing of channels need further study.

2.8.2 Fannay-Augeres Investigation

Overview

A large scale flow and tracer experiment has been performed in fractured granite in central France at Fannay-Augeres (Cacas et al., 1990a, b). The fractured structure of the rock and observations of strongly channeled flow and solute transport, both in the fractured planes and globally, have led to the development and application of a model whose structure should be able to account for these observations. The model is a stochastic network model where the flow in every fracture is assumed to follow one-dimensional channels from fracture to fracture. The geometric properties of the fractures, such as sizes, orientations and frequencies, are determined from local observations on walls, on cores, and in drillholes. Hydraulic properties are determined from short-interval injection tests by calibrating the injections to a sufficiently large number of simulated realizations. A set of assumptions must be made as to the form of the conductivity distribution, placement of channels in the fracture plane, volume of the channels, dispersion in the channels, mixing at channel intersections, and more. Some of the assumptions can be and have been tested.

Experimental

The flow and tracer experiments were performed in a section of granite tunnel 100 m in length. The tunnel is located at a depth of 150 m. The rock is highly fractured and there is a steady flow of water into the drift. Water flow rate into the drift was measured continuously both before and after the ten 50-m long measurement holes were drilled. The flow rate varied from 5.1 to $11 \cdot 10^{-5}$ m^3/s before the drilling of the holes and dropped to $3.7 \cdot 10^{-5}$ m^3/s after the holes were drilled. The decrease in flow rate was attributed to partial desaturation. The holes were located in three planes as shown in Fig. 2.8.23. Fracture orientation and fracture density data were obtained from the holes and the cores. Fracture trace lengths were obtained from a survey of the eastern wall of the tunnel section. Hydraulic injection tests were made in all holes using packed off lengths of 50 m, 10 m (50 sections) and 2.5 m (185). The flow rates in the 2.5 m sections were found to be well described by a log normal distribution. Piezometric heads were monitored in sections 685 m in length in the 10 boreholes at distances from 5.5 to 47.5 m from the drift. Of all the injection tests in holes, one third were below the detection limit $3 \cdot 10^{-7}$ m^3/s.

Tracer tests were performed by injecting 6 different tracers in 10 different locations divided between two of the boreholes. The tracers were collected in three different locations. One location was at the ceiling below the two injection holes. The other two locations were in the drainage ditch, 25 and 50 m downstream from the injection holes. Tracers were collected for up to 150 days. Of the 10 injected tracers 2 did not arrive in the drift in measurable concentrations, one had a multi-modal breakthrough curve, and two injections failed. Five tracers were thus successfully recovered and could be used for model comparisons.

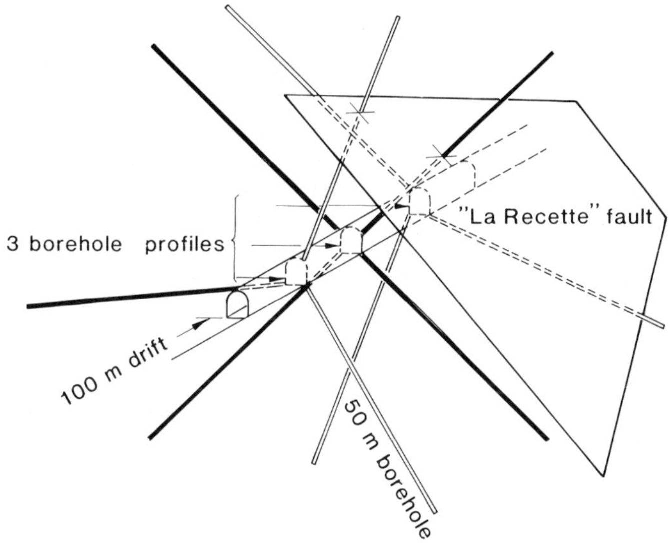

Figure 2.8.23. Position of drift and boreholes.

Some samples which were taken in two other boreholes at distances of 5, 12 and 18 m from the drift showed that the tracers had a component of movement that was parallel to the drift. Recovery of the 5 successful tracers varied between 5 and 45%. First arrival times were 2.5 and 3.5 h to the nearest collection point in the ceiling and up to 570 h to the furthest collection point. One tracer moved a longer distance in a 3 to 6 times shorter time than another tracer injected in the same hole but at less than half the distance. It was found that the tracers emerged into the drift at three main locations.

At this scale the medium and pathways are very heterogeneous and cannot be described by a homogeneous continuum.

Conceptual Model

The basis of the model is that the flow and transport take place in a network of fractures. The fracture locations, size and orientations are assumed to be random, although the fractures are allowed to be grouped in different sets. In this specific model it is assumed that the fractures are circular disks. In most fracture network models (Rouleau, 1984; Robinson, 1984; Andersson, 1987; Long and Billaux, 1987) it is assumed that the fractures can be modeled as parallel plates and thus have a uniform conductivity in the fracture plane. There are several observations that go contrary to this assumption (Gentier, 1986; Neretnieks, 1987; Tsang, 1987; Bourke 1987), indicating channeling.

In this model it is assumed that the flow takes place in a channel from the center of the fracture to the line of intersection with the next fracture. The channel in a fracture plane is connected with the channel in the intersecting fracture. A fracture will then have a quantity of channels connecting its center to the other fractures equal to the number of intersecting fractures.

The hydraulic resistance of a path connecting the center of two fractures will be the sum of its two channels. For simulations of steady state flow rates this information is sufficient.

For the simulation of the transport of nonreactive tracers, additional information on the volume and dispersive properties of channels is needed. The volume will be the sum the volumes of the channels. Further, it is necessary to make some assumption on the mixing of intersecting streams. The dispersion in an individual channel is modeled by a diffusion-like random process. Full mixing is assumed to occur at intersections. Tracers are assumed to be nonreactive and it is further assumed that there is no diffusion into the rock matrix or into other stagnant waters.

For the purpose of computation, the model structure reduces to a 3-dimensional network of resistances and volumes. The length of the channels is obtained from the stochastically generated network of fracture planes, and the hydraulic properties of the channels in the fracture planes are obtained from an independent stochastic process. In this application of the model it has been assumed that the volume of a channel is determined by its conductivity, using the assumption of a circular channel in which there is Poiseuille flow. The latter assumption gives a direct relation between the radius of a tube and its hydraulic resistance.

With the fracture sizes and densities measured it was found in the simulations that, on average, every fracture was connected to 5 other fractures.

Interpretation of Hydraulic Information

The interpretation of the experiments was done in a stepwise manner. First, the local scale measurements were used to evaluate the stochastic properties of the fractures, then an intermediate scale was selected which was as large as could be handled practically in the computations. Finally a global scale was used to compare the results with the overall inflow to the tunnel section.

The geometric information was assumed to be correct and relevant. The fracture conductivities were assumed to be log-normally distributed and the distribution was assumed to be ergodic. The mean and standard deviation of the distribution was determined by generating 200 networks in each of which 2.5 m packer tests were simulated. The simulated tests were also found to be log-normally distributed. By fitting the simulations to the experiments the standard deviation of the PDF for the fractures was found to be 7.4 on the natural logarithm scale, or 3.2 on the 10-base logarithm scale. This is a very wide distribution and implies that the difference between the first 10% and the last 10% of the conductivities had an order of magnitude of 8. It was shown in simulations that almost all the flow occurs in a limited number of pathways and that the less-conductive fractures can be neglected in the calculations. It was found that the number of fractures was reduced by half in the simulations if those at the lower end of the distribution were left out.

These simulation results seem to explain the large-scale channeling obtained in the Stripa experiment (Neretnieks, 1987).

Because of computer limitations, it was impossible to simulate the whole region of interest with one single fracture network model. Instead, the notion was introduced that, at some intermediate scale, it might be possible to assign a block in the region with a value of the hydraulic conductivity and if need be with anisotropy properties. The punctual values of the intermediate scale could then be used to assess the properties of an equivalent continuum on the global scale even though some anisotropy was found on the intermediate scale.

Using the previously obtained values for the PDF of the fractures, 17 simulations of 10 m cubes were made. They showed that the medium appears to be isotropic on the global scale. For a porous medium with linear flow the global conductivity should lie between the geometric mean and the arithmetic mean of the intermediate scale values. Based on the simulated results the global conductivity can be expected to lie between $1.5 \cdot 10^{-8}$ and $2.3 \cdot 10^{-8}$ m/s. This is, then, essentially a prediction of the global conductivity using only geometric and small-scale hydraulic data.

The predicted global conductivity and that obtained from the global inflow measurements were compared. The experimental global conductivity was obtained from inflow and detailed hydraulic head measurements in the boreholes, using the porous media model. The analysis was complicated by the fact that the rock had been partially dewatered when the holes were drilled, which reduced the inflow by a factor of 2.6. The obtained average value of the global conductivity was found to be $1.8 \cdot 10^{-8}$ m/s. This is within the expected range.

One important question addressed was if the local scale injection test values could be used to estimate the global conductivity. It should be expected to fall between the geometric and arithmetic means, which were $3 \cdot 10^{-8}$ and $1.1 \cdot 10^{-6}$ m/s, respectively. This range is far too wide to be useful.

Interpretation of Tracer Information

Simulations of the tracer movement were done by the particle tracking method (Robinson, 1984). The crucial information on the volume of the flow channels was not available. In this application of the conceptual model, the volume of each channel was assumed to be directly related to the hydraulic conductivity by the relation of Poiseuille flow in a circular tube. The actual volume of all channels was assumed to be directly proportional to the Poiseuille volume. The proportionality factor, called a slowing-down factor, was subsequently found by calibration to be 32.

The model was used to study the sensitivity of some assumptions and parameters before proceeding to comparisons with the real tracer results. It was found that the dispersion in the individual channels contributed very little to the dispersion in the network. The slowing down factor has only the effect of scaling the time.

The calibration of the slowing-down factor was done by making 20 simulations for each of four tracer tests. All the resulting histograms of peak residence times were then scaled by the same slowing-down factor to obtain the best overall value. The histograms obtained and the experimental value are shown in Fig. 2.8.24. The scaling factor is 32. Other values having a factor of three higher or lower would also have been reasonable. Only four tests could be simulated because the fifth tracer required too large a simulation region.

It should be noted that the only additional information beyond the model calibrated by the hydraulic tests is the information on the volume of the channels. The dispersion data do not influence the results. The mean and standard deviation of the hydraulic conductivities obtained from the calibration of hydraulic data were used in these simulations.

The average radius of the pipes is 0.031 mm to attain the conductivity and it is 5.5 time larger, i.e., 0.171 mm (0.32 mm diameter), to account for the volume needed for the residence time. If the average flow channel were not a circular pipe but a ribbon with the same hydraulic conductivity and volume, the aperture and the width of the ribbon would be = 0.006 mm and

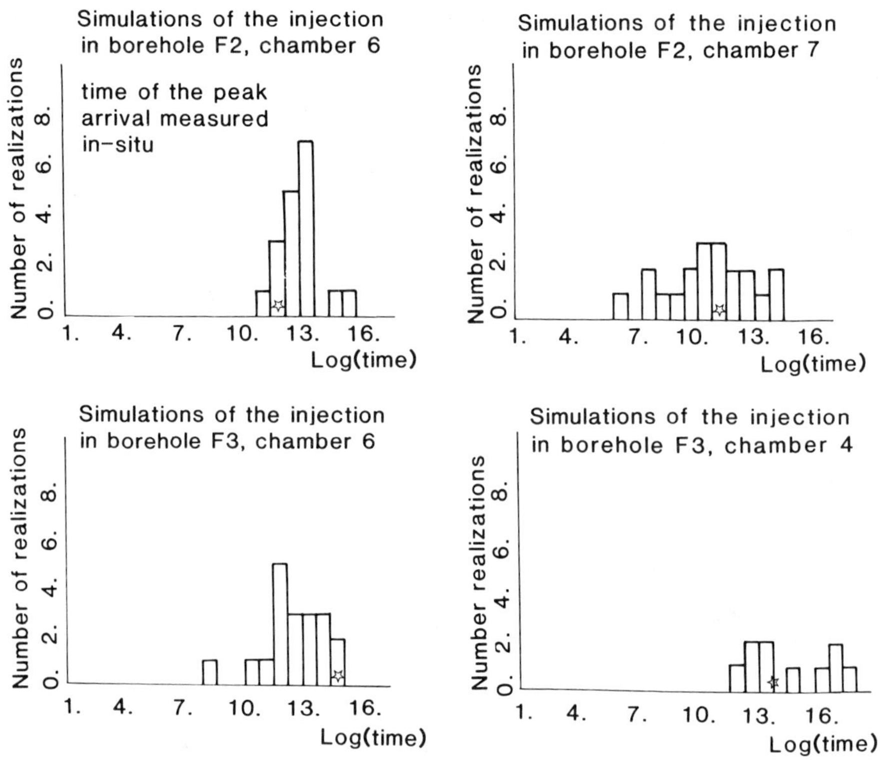

Figure 2.8.24. Histograms of simulated peak arrival times for 4 tracer paths and the experimentally obtained value. Scaling factor is 32.

15 mm respectively. Whatever the geometry, this implies that only a very small part, a few percent or less, of the fracture is in contact with the flowing water.

So far only the peak arrival times and not the residence time distribution nor the spatial distribution of the tracers have been addressed. The residence time distribution, RTD, is strongly influenced by the variation of the flow rates among the different channels, i.e., by the PDF of the conductivities as well as by the volumes that, in this application of the model, are obtained from the conductivity. It is expected that the larger the standard deviation σ in the PDF, the wider will be the RTD, and the more pronounced the channeling. The natural logarithm standard deviation of the radii obtained from the hydraulic calibrations was $\sigma_r = 1.85$. (This is equivalent to $\sigma = 3.2$ for logarithm of base 10 for the conductivities for circular tubes.) Simulations were also made with $\sigma_r = 0.1$, 1.0 and 1.85 to find the locations of the effluent points in the drift. For the larger value of σ_r, one location accounted for 60% of the recovery and also for $\sigma = 1.0$ the channeling was pronounced.

Further simulations were made to study the residence time distribution. This was done by determining the "breakthrough duration," which was arbitrarily defined as the time for 90% of the injected tracer pulse to arrive. For 15 linear flow simulations the breakthrough duration values were, on the average, about 50 to 100 times larger than the peak arrival times for a $\sigma_r = 1.85$.

This indicates a very large dispersion, but is nearly one order of magnitude larger than the values obtained in the experiments. The latter may, however, have been underestimated because they have been determined without having access to the whole tail of the breakthrough curves and because the overall recovery was so low. Further simulations of the tracer tests to the tunnel, where the flow is essentially radial, showed that the breakthrough duration values agreed with the experimental values if scaled by the same factor of 32 as the peak residence times. Because the variations in the simulated times typically span a range that has a factor of 500, the value of the slowing down factor could vary considerably.

The simulations of the recovery of the tracers spanned the same range as the experimental values; 2-70%. However, the comparison is complicated by the fact that the boundary conditions in the model are not quite comparable to those of the drift from the recovery point of view.

Discussion and Conclusions

The use of the stochastic network approach to the flow and tracer results in Fannay-Augeres has made it possible to use geometric information on fracture densities, orientations, and lengths together with small-scale measurements on hydraulic conductivity to predict global scale conductivity. The model also accounts for recent observations on strong channeling over large areas as well as for the observations that water does not flow evenly in fractures. It has been possible to predict some characteristic properties of the residence time distribution of tracers. The model also accounts for the observations in Fannay-Augeres and Stripa that tracers arrive preferentially in a few regions.

Several assumptions made in the model could not be validated with the data available. The assumption of the conducting channels being circular tubes has no impact on the hydraulic model but the coupled assumption that the volume of the conduits can be related to conductivity by the assumption of Poiseuille flow may have an impact on the transport of nonreacting tracers. Other reasonable and possible assumptions have not been tested in this study.

One important aspect of the model is that it proposes that only a small part of a fracture conducts water. The area of a fracture wetted by the mobile water is much smaller than the total area of the fracture surfaces. The simulations indicate that only at most a few percent of the fractures are open to flow. For reacting solutes this will have a very large impact because they will see much less rock to react with. This aspect of the model has not been tested yet.

NOTATION

a	Specific surface	m^2/m^3 water
C	Concentration	mol/m^3
C_p	Concentration in pores	mol/m^3
D_a	Apparent diffusion coefficient	m^2/s
D_e	Effective diffusion coefficient in matrix	m^2/s
D_p	Pore diffusion coefficient	m^2/s
D_L	Dispersion coefficient	m^2/s
D_s	Surface diffusion coefficient	m^2/s
D_w	Diffusion coefficient in water	m^2/s
F_f	Formation factor	-

K	Volumetric sorption coefficient	m^3/m^3
K_a	Surface sorption coefficient	m^3/m^2
K_d	Mass sorption coefficient	kg/m^3
K_p	Hydraulic conductivity	m/s
N	Flow rate of dissolved species	mol/s
Pe	Peclet number	-
R_d	Retardation factor	-
t	Time	s
t_w	Water residence time	s
u_f	Water velocity	m/s
W	Width of channel	m
x	Distance into matrix	m
z	Distance in flow direction	m

α	Rock capacity factor	-
δ_D	Constrictivity	-
δ	Fracture aperture	m
ε	Porosity	-
ε_p	Porosity of matrix	-
ε_t	Total porosity	-
$\overline{\eta}$	Penetration thickness	m
ρ_p	Density of rock	kg/m^3
τ	Tortuosity	-

REFERENCES

Abelin, H. Migration in a Single Fracture—An In Situ Experiment in a Natural Fracture, Ph.D. Thesis, Dep. Chemical Engineering, Royal Institute of Technology, Stockholm, Sweden, 1986.

Abelin, H., Neretnieks, I., Tunbrant, S., and Moreno, L. Final Report of the Migration in a Single Fracture - Experimental Results and Evaluation, Stripa Project Report 85-03, OECD/NEA, SKB, 1985.

Abelin, H., Birgersson, L., Gidlund, J., Moreno, L., Neretnieks, I., Widén, H., and Ågren, T. 3D Migration Experiment - Report 3, Part I, Performed Experiments, Results and Evaluation, Stripa Project Technical Report 87-21, Stockholm, Nov. 1987.

Abelin, H., Birgersson, L., Neretnieks, I., and Ågren, T. A Channeling Experiment to Study Flow and Transport in Natural Fractures, *Scientific Basis for Nuclear Waste Management XII*, Berlin, Nov. 1988, Proceedings, p. 661–668, 1989

Ahlbom, K., and Smellie, J.A.T. Characterization of Fracture Zone 2, Finnsjön Study-Site, SKB TR 89-12, 1989.

Andersson, J., and Dverstorp, B. Conditional Simulation of Fluid Flow in Three-Dimensional Networks of Discrete Fractures, *Water Resources Research*, 23, p. 1876–1886, 1987.

Atkinson, A., and Titchell, I. Diffusive Transport in Granite Studied Using Ionic Conductivity, UK Atomic Energy Authority Report AERE R 11827, 1985.

Bear, J. Hydrodynamic Dispersion, in Flow Through Porous Media, edited by R.J.M. de Wiest, Academic, New York, 1969.

Bird, R.B., Stewart, W.E., and Lightfoot, E.N. *Transport Phenomena*, Wiley, 1960.

Birgersson, L. Diffusion in the Matrix of Granitic Rock, Field Test in the Stripa Mine. M.S. Thesis, Dep. Chemical Engineering, Royal Institute of Technology, Stockholm, Sweden, May 1988.

Birgersson, L., and Neretnieks, I. Diffusion in the Matrix of Granitic Rock, Field Test in the Stripa Mine. Part 1, SKBF/KBS Technical Report 82-08, July 1982a.

Birgersson, L., and Neretnieks, I. Diffusion in the Matrix of Granitic Rock, Field Test in the Stripa Mine, *Scientific Basis for Nuclear Waste Management*, V, Berlin, June 1982, Elsevier, p. 519, 1982b.

Birgersson, L., and Neretnieks, I. Diffusion in the Matrix of Granitic Rock. Field Test in the Stripa Mine, Part 2, SKBF/KBS Technical Report 83-39, March 1983.

Birgersson, L., and Neretnieks, I. Diffusion in the Matrix of Granitic Rock. Field Test in the Stripa Mine, *Scientific Basis for Nuclear Waste Management*, VII, Boston, Nov. 1983, Elsevier, p. 247, 1984.

Birgersson, L., and Neretnieks, I. Diffusion in the Matrix of Granitic Rock. Field Test in the Stripa Mine, *Scientific Basis for Nuclear Waste Management*, Boston, Nov. 1987, Proceedings, p. 189–198, 1987.

Black, J., Holmes, D., and Brightman, M. Cross Hole Investigations - Hydrological Results and Interpretations, Stripa Project Report 87-17 OECD/NEA, SKB Nov. 1987.

Bourke, P. Channeling of Flow Through Fractures in Rock, *Proceedings of GEOVAL-87*, International Symposium, Stockholm, Sweden, April 7–9, p. 167–177, 1987.

Brace, W.F., Walsh, J.B., and Frangos, W.I. Permeability of Granite Under High Pressure, *J. Geophys. Res.*, 73, p. 2225, 1968.

Bradbury, M.H., Lever, D., and Kinsey, D. Aqueous Phase Diffusion in Crystalline Rock, *Scientific Basis for Nuclear Waste Management*, V, Berlin 1982, Elsevier, p. 569–578, 1982.

Bradbury, M.H., and Green, A. Measurements of Important Parameters Determining Aqueous Phase Diffusion Rates Through Crystalline Rock Matrices, *J. Hydrology*, 82, p. 39–55, 1985.

Bradbury, M.H., and Stephen, I.G. Diffusion and Permeability Based Sorption Measurements in Intact Rock Samples, *Materials Res. Soc. Symp. Proc.*, 50, p. 81–90, 1985.

Bradbury, M.H., and Green, A. Investigations into Factors Influencing Long Range Matrix Diffusion Rates and Pore Space Accessibility at Depth in Granite, *J. Hydrology*, 89, p. 123–139, 1986.

Cacas, M.C., de Marsily, G., Tillie, B., Barbreau, A., Durand, E., Feuga, B., and Peaudecerf, P. Modelling Fracture Flow With a Stochastic Discrete Fracture Network: Calibration and Validation - 1 The Flow Model, *Water Resources Research*, in print, 1990a.

Cacas, M.C., Ledoux, E., de Marsily, G., Barbreau, A., Calmels, P., Gaillard, B., and Margritta, R. Modelling Fracture Flow with a Stochastic Discrete Fracture Network: Calibration and Validation - 2 The Transport Model, *Water Resources Res.*, in print, 1990b.

Cathles, L.M., Spedden, H.R., and Malouf, E.E. A Tracer Technique to Measure the Diffusional Accessibility of Matrix Block Mineralization, paper presented at the 103rd annual meeting, Amer. Inst Mining Eng., Dallas Tex., 1974.

Dershowitz, W.S., Gordon, B.M., and Kafritsas, J.C. A New Three Dimensional Model for Flow in Fractured Rock, IAH Conference Jan. 7–12, 1985, Tucson Arizona, Volume XVII proceedings published by a committee of USA members of the International Association of Hydrologists, p. 449–462, 1985.

Gale, J., and Stråhle, A. Site Characterization and Validation Drift and Borehole Fracture Data. Stage 1, Stripa Project Internal Report 88-10, Sept 1988.

Garrels, R.M., Dreyer, R.M., and Howland, A.L. Diffusion of Ions Through Intergranular Space in Pore Water Saturated Rocks, *Bull. Geol. Soc. Am.*, 60, p. 1809–1828, 1949.

Gelhar, L.W. Applications of Stochastic Models to Solute Transport in Fractured Rocks, SKB Technical Report 87-05, 1987.

Gelhar, L.W., Mantoglou, A., Welty, C., and Rehfeldt, K.R. A Review of Field Scale Physical Solute Transport Processes in Saturated and Unsaturated Media, EPRI EA-4190S, Aug. 1985.

Gentier, S. Morphlologie et Comportement Hydromécanique d'une Fracture Naturelle dans un Granite sous Contrainte Normale. Etude Expérimentale et Théoretique, Thèse de Doctorat à l'Université d'Orléans, Mar. 1986.

Gilling, D., Jeffries, N.L., and Lineham, T.R. An Experimental Study of Solute Transport in Mudstones, Harwell Report AERE R12809, 1987.

Jefferies, N.L. Long Term Solute Diffusion in a Granite Block Immersed in Sea Water, CEC, Nuclear Science and Technology Report EUR 11627, 1988.

KBS-3, 1983, Final Storage of Spent Nuclear Fuel, Report by Swedish Nuclear Fuel Supply Co, SKBF, Stockholm, Sweden, May, 1983.

Lallemand-Barres, A., and Peaudecerf, P. Recherche des Relations Entre la Valeur de la Dispersivité Macroscopique d'un Milieu Aquifère, Ses Autres Caractéristiques et les Conditions de Mesure, Bull. BRGM (2) III, 4, 277, 1978.

Landström, O., Klockars, C-E., Holmberg, K-E., and Westerberg, S. In Situ Experiments on Nuclide Migration in Fractured Crystalline Rocks, KBS TR 110, 1978.

Lever, D.A., Bradbury, M.H., and Hemingway, S.J. The Effect of Dead End Porosity on Rock Matrix Diffusion, *J. Hydrology*, 80, p. 45–76, 1985.

Long, J.C.S., Endo, H.K., Karasaki, K., Pyrak, L., MacLean, P., and Witherspoon, P.A. Hydrological Behavior of Fracture Networks. Hydrogeology of Rocks of Low Permeability, IAH Conference, Jan. 7–12, 1985, Tucson, Arizona. Volume XVII, p. 44–468, 1985.

Long, J.C.S., and Billaux, D.M. From Field Data to Fracture Network Modelling: An Example Incorporating Spatial Structure, *Water Resources Research*, 23, p. 1201–1216, 1987.

Maloszewski, P., and Zuber, A. On the Theory of Tracer Experiments in Fissured Rock with a Porous Matrix, *J. Hydrol.*, 79, p. 333–358, 1985.

Moreno, L., Neretnieks, I., and Klockars, C-E. Evaluation of Some Tracer Tests in the Granitic Rock at Finnsjön, KBS TR 83-38, 1983.

Moreno, L., Neretnieks, I., and Eriksen, T. Analysis of Some Laboratory Tracer Runs in Natural Fissures, *Water Resources Res.*, 21, p. 951–958, 1985.

Moreno, L., and Neretnieks, I. Channeling and its Potential Consequences for Radionuclides Transport, presented at MRS Symposium at Boston, Nov. 30–Dec. 3, Proceedings, p. 169–178, 1987.

Moreno, L., Tsang, Y.W., Tsang, C.F., and Neretnieks, I. Flow and Solute Transport in a Single Fracture. A Two-Dimensional Statistical Model, Nuclear Fuel Safety Project, technical report TR 88-03, Stockholm, 1987.

Moreno, L., Tsang, Y.W., Tsang, C.F., Hale, F.V., and Neretnieks, I. Flow and Tracer Transport in a Single Fracture. A Stochastic Model and its Relation to Some Field Observations, *Water Resources Research*, 24, p. 2033–3048, 1988.

Moreno, L., and Neretnieks, I. Channeling in Fracture Zones and Its Potential Impact on the Transport of Radionuclides, Symposium on Scientific Basis for Nuclear Waste Management, Berlin Oct. 10–13, proceedings in print, 1988.

Moreno, L., Neretnieks, I., and Landström, O. An Evaluation of Some Tracer Tests Performed at Studsvik, KBS TR 89-09, 1989.

NAGRA Project Gewähr. Endlager Für Hochaktive Abfälle: Das System der Sicherheitsbarriären, Nagra, Baden, Switzerland, Jan. 1985.

Neretnieks, I. Diffusion in the Rock Matrix: An Important Factor in Radionuclide Retardation? *J. Geophys. Res.*, 85, p. 4379–4397, 1980.

Neretnieks, I. Age Dating of Groundwater in Fissured Rock. Influence of Water in Micropores, *Water Resources Res.*, 17, p. 421, 1981.

Neretnieks, I. A Note on Fracture Flow Mechanisms in the Ground, *Water Resources Res.*, 19, p. 364–370, 1983.

Neretnieks, I. Transport in Fractured Rocks. Paper at IAH 17th Congress, Tucson Arizona, Jan. 1985, *Memoires*, vol XVII, part 2, *Proceedings*, p. 301–318, 1985.

Neretnieks, I. Channeling in Crystalline Rocks. Its Possible Impact on Transport of Radionuclides from a Repository, Colloque International Impact de la Physico-Chimie sur l'Étude, la Conception et l'Optimisation des Procédes en Milieu Poreux Naturel, Nancy, 10–12 Juin 1987a.

Neretnieks, I. Channeling Effects in Flow and Transport in Fractured Rocks - Some Recent Observations and Models, Paper presented at GEOVAL symposium, Stockholm, *Proceedings*, p. 315–335, 1987b.

Neretnieks, I. The Swedish Repository for Low and Intermediate Level Reactor Waste- SFR. Radioactivity Release and Transport Calculations, Symposium on Scientific Basis for Nuclear Waste Management, Berlin Oct. 10–13, *Proceedings*, p. 537–544, 1988.

Neretnieks, I., Eriksen, T., and Tähtinen, P. Tracer Movement in a Single Fissure in Granitic Rock: Some Experimental Results and Their Interpretation, *Water Resources Res.*, 18, p. 849, 1982.

Neretnieks, I., and Rasmuson, A. An Approach to Modelling Radionuclide Migration in a Medium with Strongly Varying Velocity and Block Sizes Along the Flow Path, *Water Resources Res.*, 20, p. 1823–1836, 1984.

Neretnieks, I., Abelin, H., and Birgersson, L. Some Recent Observations of Channeling in Fractured Rocks. Its Potential Impact on Radionuclide Migration, DOE/AECL conference Sept. 15–17, 1987, San Francisco, *Proceedings*, p. 387–410, 1987.

Novakowski, K.S., Flavelle, P.A., Raven, K.G., and Cooper, E.L. Determination of Groundwater Flow Pathways in Fractured Plutonic Rock using a Radioactive Tracer, *Intern J. Appl. Radiat. Isot.*, 36, p. 399–404, 1985a.

Novakowski, K.S., Evans, G.V., Lever, D.A., and Raven, K.G. A Field Example of Measuring Hydrodynamic Dispersion in a Single Fracture, *Water Resources Research*, 21, p. 1165–1174, 1985b.

Pocos de Caldas Project: Progress Report March Through May 1987, SKB, Sweden, 1987.

Rasmuson, A., Analysis of Hydrodynamic Dispersion in Discrete Fracture Networks Using the Method of Moments, *Water Resources Res.*, 21, p. 1677–1683, 1985.

Rasmuson, A., and Neretnieks, I. Exact Solution of a Model for Diffusion in Particles and Longitudinal Dispersion in Packed Beds, *AIChE J.*, 26, p. 686, 1980.

Rasmuson, A., and Neretnieks, I. Surface Migration in Sorption Processes, SKBF/KBS Technical Report 83-37, March 1983.

Rasmuson, A., and Neretnieks, I. Radionuclide Transport in Fast Channels in Crystalline Rock, *Water Resources Res.*, 22, p. 1247–1256, 1986a.

Rasmuson, A., and Neretnieks, I. Radionuclide Migration in Strongly Fissured Zones - The Sensitivity to Some Assumptions and Parameters, *Water Resources Research*, 22, p. 559–569, 1986b.

Raven, K.G., Novakowski, K.S., and Lapcevic, P.A. Interpretation of Field Tracer Tests of a Single Fracture Using a Transient Solute Storage Model, *Water Resources Research*, 24, p. 2019–2032, 1986.

Robinson, P.C. Connectivity, Flow and Transport in Network Models of Fractured Media, Ph.D. Thesis, St. Catherine's College, Oxford University, Ref TP 1072, May 1984.

Robinson, P.C. Flow Modelling in Three Dimensional Fracture Networks, Report AERE R 11965 UK Atomic Energy Authority Harwell Theoretical Physics Division, May 1986.

Rouleau, A. Statistical Characterization and Numerical Simulation of a Fracture System—Application to Groundwater Flow in the Stripa Granite. Ph.D. Thesis, Univ. of Waterloo, Ontario, Canada, 1984.

Sauty, J.P. An Analysis of Hydrodispersive Transfer in Aquifers, *Water Resources Research*, 16, p. 145, 1980.

Skagius, K. Diffusion of Dissolved Species in the Matrix of Some Swedish Crystalline Rocks, Ph.D. Thesis, Dept. Chemical Engineering, Royal Institute of Technology, May 1986.

Skagius, K., Svedberg, G., and Neretnieks, I. A Study of Strontium and Cesium Sorption on Granite, *Nuclear Technology*, 59, p. 302–313, 1982.

Skagius, K., and Neretnieks, I. Porosities of and Diffusivities in Crystalline Rock and Fissure Coating Materials, *Scientific Basis for Nuclear Waste Management*, VII, Boston, Nov. 1983, Elsevier, p. 835, 1984.

Skagius, K., and Neretnieks, I. Porosities and Diffusivities of Some Nonsorbing Species in Crystalline Rocks, *Water Resources Res.*, 22, p. 389, 1986a.

Skagius, K., and Neretnieks, I. Diffusivity Measurements and Electrical Resistivity Measurements in Rock Samples Under Mechanical Stress, *Water Resources Research*, 22, p. 570, 1986b.

Skagius, K., and Neretnieks, I. Measurements of Cesium and Strontium Diffusion in Biotite Gneiss, *Water Resources Research*, 24, p. 75–84, 1988.

Smith, L., Schwartz, F., and Mase, C. Applications of Stochastic Methods for the Simulation of Solute Transport in Discrete and Continuum Models of Fractured Rock Systems, Proceedings of Conference on Geostatistical, Sensitivity and Uncertainty Methods for Ground-Water Flow and Radionuclide Transport Modelling, San Francisco, California, September 15–17, p. 425–440, 1987.

Tsang, Y.W., and Tsang, C.F. Channel Model of Flow Through Fractured Media, *Water Resources Res.*, 23, p. 467–479, 1987.

Tsang, Y.W., Tsang, C.F., Neretnieks, I., and Moreno, L. Flow and Tracer Transport in Fractured Media - A Variable Aperture Channel Model and its Properties, *Water Resources Research*, 24, p. 2049–2060, 1988.

Webster, D.S., Proctor, J.F., and Marine, I.W. Two Well Tracer Tests in Fractured Crystalline Rock, U.S. Geol. Surv. Water Supply paper 1544-I1, 1970.

Witherspoon, P.A., Wang, J.S.K., Iwai, K., and Gale, J.E. Validity of Cubic Law for Fluid Flow in a Deformable Rock Fracture, *Water Resources Res.*, 16, p. 1016–1024, 1980.

WP-Cave: Reference to a report under preparation.

KBS technical reports can be obtained from INIS Clearing House, International Atomic Energy Agency, P.O. Box 100, A-1400 Vienna, Austria.

3

SOLUTE TRANSPORT THROUGH FRACTURE NETWORKS

Leslie Smith
Department of Geological Sciences
University of British Columbia
Vancouver, British Columbia

Franklin W. Schwartz
Department of Geology
Ohio State University
Columbus, Ohio

3.1 INTRODUCTION

In the preceding chapters, examples have been presented that illustrate some basic features of joints and fractures, and observations of fluid flow and solute transport in fractured rocks. In this chapter we focus on the interplay between the geometric properties of a fracture network and the resulting character of solute transport. We then consider several methods for taking account of these effects in solute transport models applicable at the field scale. Basic concepts of hydrogeology make it clear that solute transport will be influenced by the geometry of a fracture network. That this is the case has been known since the early descriptions of hydrothermal ore deposits, which often suggest a complex pattern of mineralization related to a relic flow system. What is less clear is how various mass transfer processes, which act on a number of different scales, interact to determine transport patterns and solute concentrations, and how we can develop quantitative methods to describe transport in a rock mass where fractures provide the dominant pathways for solute migration.

Advective mass transfer of a nonreactive solute through a fracture network leads to spreading on several scales. These scales are:

i) microscopic scale—spreading occurs as a result of the velocity profile that exists between the two bounding walls of a fracture.

ii) scale of individual fracture planes—spreading occurs within individual fractures because streamlines take on a two-dimensional configuration with differing path lengths and advective velocities; reflecting both spatial variations in fracture aperture, and curvilinear pathlines between fracture intersections that act as line sources of fluid and those that act as line sinks for fluid leaving the fracture plane.

iii) macroscopic or network scale—with the mixing that occurs at fracture intersections, the effects of network geometry come into play, with mass moving along pathways through the network that are of differing length and variable fluid velocity.

The cumulative effect of these processes is to cause a plume to expand as it moves through the fracture network under the influence of the mean hydraulic gradient. By analogy to transport in porous media, the macroscale description of this spreading is termed dispersion. To this list of advective spreading mechanisms, it is necessary to add transverse diffusion within fractures, diffusion at fracture intersections, and diffusion of mass from the fractures into the intact rock matrix.

There are three approaches to simulate solute transport in fractured rocks; discrete network simulation, continuum approximations based on either porous medium equivalent assumptions or statistical representations of mass transfer, and hybrid models that combine elements of both discrete fracture models and continuum approximations. In the continuum approach, a geologic unit is considered to have a sufficiently high fracture density so that the structure of the network need not be described in detail, but rather the unit can be represented in terms of an equivalent porous medium. It follows that the continuum elements must be large enough to encompass a representative sample of the open, connected fractures and their variable influences on the fluid flow and transport behavior within the network. Coefficients must be formulated to characterize dispersive processes occurring within the representative elemental volume (REV) of the rock mass.

However, not all models developed for porous media are appropriate for fractured media. A case in point is the commonly adopted isotropic model of dispersion. Dispersion in fractured rocks is more complex, exhibiting anisotropic behavior. In this regard, three features can be noted:

i) the center of mass of a moving tracer may diverge from the flow direction defined by the mean hydraulic gradient,

ii) the major and minor principal axes of the plume may be rotated from the direction of the hydraulic gradient, and

iii) the direction of maximum spreading need not be greatest in the direction most closely aligned with the mean hydraulic gradient.

The form of the anisotropic dispersion tensor that describes this interplay of network geometry and hydraulic gradient has yet to be defined, except for some very simple geometries (see Sec. 3.7.2).

In media with relatively low fracture densities, or those where a significant proportion of the fractures are partially or fully closed by either mineral precipitates or due to the in situ stress, it is probable that mass can be transported a considerable distance without sampling a representative suite of fluid velocities in the network. Until that distance is reached, models based on the assumption of porous medium equivalency are not applicable. It is also conceivable that several scales of fracturing occur within the rock mass; with a small number of areally extensive fractures exerting a predominant influence on patterns of groundwater flow. All of these situations, where the assumption of a representative elemental volume is not justified, are termed sub-REV problems. A characteristic of these systems is the potential for solute transport to be significantly influenced by individual fractures and their interconnections with other fractures forming the network. A possible alternative in these instances is the discrete fracture modeling procedure. Instead of characterizing transport in terms of volume-averaged coefficients, one models the network on a fracture-by-fracture basis. Statistical concepts underlie the development of this approach, with probability distributions defining fracture frequency, location, dimension, and orientation (see, e.g., Long et al., 1985).

The discussion in this chapter focuses on the transport of non-reactive solutes at the network scale. We first review some experimental observations of solute transport, based on data collected in laboratory and field settings. Then, transport within single fracture planes, and mixing at fracture intersections are described, to set the stage for an examination of transport processes in networks of interconnected fractures. Far-field simulation will by necessity be based on some form of volume averaging. Two continuum approaches are described, one based on analytic calculation of the dispersion tensor, the other on a statistical procedure whereby mass moves in a way that mimics the effect of network geometry on spreading. A hybrid modeling approach is also outlined, where the objective is to permit an increase in the size of the domain that can be modeled, beyond that possible using a discrete network model, but without resorting to continuum approximations. This latter approach is suited to situations where open fractures with widely-varying length scales occur within the flow domain.

The purpose of this chapter is to examine relationships between network geometry, spreading patterns, and the consequent implications in formulating field-scale simulation models. We exclude from the discussion a number of processes that may potentially be key factors in certain field settings. These processes include: solute transport in a rock mass where the permeability of the intact rock is sufficiently high that advective mass transfer within the rock blocks becomes a significant component of the total mass flux; diffusive transfer of mass between the fracture network and the intact rock matrix; transient fluid flow; and coupled processes, such as a temporal variation in network permeability due to mineral dissolution, precipitation, or alteration of the wall rock, or changes in the in situ stress conditions.

3.2 FIELD OBSERVATIONS AND LABORATORY EXPERIMENTS

3.2.1 Field Observations

The published data base to characterize field-scale transport in fractured rock masses is limited. Interpretation of field experiments face the difficulty of having to simultaneously characterize network geometry, and to estimate values of coefficients in the equations used to define solute fluxes. Few data sets can be classified as coming from "controlled experiments." Invariably, key information is lacking, especially in regard to the detailed nature of pathways through the network. In a limited number of cases, a single fracture, forming a dominant pathway for tracer migration, has been isolated (e.g., Novakowski et al., 1985; Raven et al., 1988; Shapiro and Nicholas, 1989). The Stripa two-dimensional experiment described in Ch. 2 also falls in this category. Given a known geometry, it is possible to attempt a characterization of the hydraulic and dispersive properties of individual fractures.

Breakthrough curves that reflect transport through a network of interconnected fractures are generally interpreted in a more qualitative manner. Residence time distribution (RTD) curves provide an effective way of looking at breakthrough data (Robinson and Tester, 1984). The residence time distribution $E(t)$ is defined by the fraction of tracer with a residence time in the network between t and $t + \Delta t$. Numerical values of $E(t)$ can be calculated from the derivative of the cumulative breakthrough of mass recorded at a withdrawal well. Statistical properties of residence time distributions are reviewed by Rasmuson (1985). In the case of constant-rate withdrawal tests, it can be helpful to plot the residence time distribution as a function of the cumulative volume of fluid produced from the fracture network. Figure 3.2.1 is an example of a RTD curve, obtained during injection/withdrawal tests carried out as part of the Fenton Hill Hot Dry Rock experiments (Robinson and Tester, 1984). The distance separating injection and withdrawal points is on the order of 270 m. The residence time distribution reflects the summation of travel times along all pathways followed by the tracer. The long tail records pathways through more circuitous routes between the injection and withdrawal points.

Statistical parameters that quantify the residence time distribution include (Robinson and Tester, 1984):

modal volume — $\dot{V}$ the fluid volume corresponding to the maximum solute arrival rate on the RTD curve.

integral mean volume — total volume of flow through the network, calculated

$$\langle V \rangle = \int_0^\infty V / Q \, E(t) dV$$

where Q is the steady state, volumetric injection rate.

width at 1/2 peak height — as a relative measure of dispersion occurring during transport through the network.

For example, parameter estimates calculated from the data in Fig. 3.2.1 include a modal volume of 266 m^3, an integral mean volume of 2173 m^3, and a volumetric width at one half height of 479 m^3. The modal volume is an estimate of the volume of higher-permeability pathways providing a direct connection across the network, while the integral mean volume is an estimate of the total volume of fractures accepting fluid.

Under favorable circumstances, it may be possible to infer the large-scale character of network geometry from the RTD curve. In the simplest sense, a RTD curve with two distinct peaks suggests the presence of two preferred pathways across the network. However, a more complete analysis of the statistical moments of the RTD may be of limited value because of inaccuracies in specifying the tail of the distribution. Raven et al. (1988) illustrate the use of RTD curves in their analysis of transport through single fracture planes. A stochastic approach to the interpretation of arrival time data for a test configuration with multiple release points is given by Cacas et al. (1990).

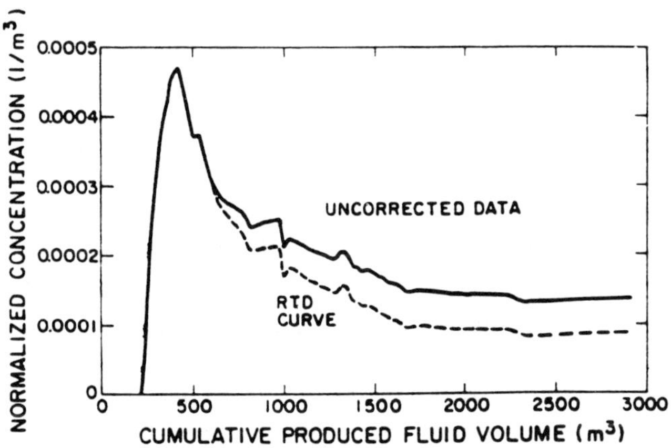

Figure 3.2.1. An example of a residence-time distribution curve, based on a ^{82}Br tracer test from the Fenton Hill, New Mexico hot dry rock experiments. The dashed curve has been corrected to account for the effects of fluid recirculation (adapted from Robinson and Tester, 1984).

3.2.2 Laboratory Experiments

Laboratory-scale models of a fracture network provide an alternative approach to study advective mass transfer. This approach has the advantage that the exact structure of the network geometry is known, and numerous sensors can be placed in the model to obtain data to define spreading patterns. The limitation of physical models is in their simplified representation of fracture geometry. Their primary use to date has been in the validation of numerical simulation models.

Laboratory experiments vary with the kind of material used to build the network model. If Plexiglas is used, tracer is effectively restricted to moving only within the fractures. Hull et al. (1987) describe an experiment in which two sets of orthogonal fractures, evenly spaced and continuous across the model domain, were cut into a Plexiglas sheet. The native solution for the flow system was KCl, with CsCl used as a tracer. These two solutions were chosen to have the same fluid density, but different electrical conductivities. Breakthrough of a pulse injection of tracer could be readily monitored at a number of electrode locations throughout the network. If a porous material is used to represent the rock mass, then mass transfer will also occur within the matrix. Hull and Clemo (1987) constructed a model of an irregular network of fractures within a porous polyethylene sheet. A complex plume developed as a result of the interaction between flow through the fractures and within the matrix. The response of the physical model suggests that without detailed knowledge of network geometry, apertures, and matrix properties, predictive simulation of the plume evolution would not be possible in a deterministic sense.

3.2.3 The Geologic Record

Another potential source of data on the effects of network geometry on spreading patterns exists in the geologic record. There are many descriptions of inferred transport patterns associated with mineral deposition in a fractured rock system. Much of this literature is descriptive and interpretations are not linked to rigorous hydrologic analyses, but with continuing advances in analytical capabilities, it is now possible to use isotopic and other geochemical data in an attempt to constrain the gross character of circulation patterns. However, when viewed at the local scale, it seems that many systems are too complex to be incorporated easily within the framework of existing modeling techniques. This apparent complexity occurs because mineralization reflects an integration over geologic time of changing hydraulic and geochemical conditions. An important research opportunity exists in examining how such observations could be used to best advantage in guiding the development of quantitative transport models.

3.3 TRANSPORT IN INDIVIDUAL FRACTURES

The earliest conceptual models of transport represented a fracture as a smooth, parallel plate. Arguments are made that point asperities, with no direct impact on the flow field, prop open the fracture. This model leads to the well-known cubic law to describe the hydraulic behavior of a fracture (Ch. 1). Dispersion is explained in terms of both the transverse velocity profile across the fracture (Taylor dispersion) and a small-scale surface roughness that locally perturbs the mean flow. Spreading is characterized by values for the longitudinal and transverse dispersivity.

Recent field observations, reviewed in Ch. 2, suggest that such a model may not be representative of in situ conditions, except perhaps for through-going, open fractures. In the majority of fractures, fluid flow may be restricted to discrete channels within the plane of each fracture. Channels occur because of the natural tendency of fluid to move towards and through the pathways of least resistance, that is, those regions of the fracture plane with larger apertures.

Channeling concepts are closely linked to perceived notions of the internal geometry of fractures. Three factors are important here:

i) undulations of the opposing fracture surfaces,

ii) partial or total closure of the fracture plane under the in situ stress conditions, and

iii) the presence of mineral precipitates clogging sections of the fracture plane.

The net effect of aperture variation is to focus the flow into a set of discrete, tortuous channels (Brown, 1987). Solutes moving through a network will disperse in a manner governed by the geometry of these channels. An extended discussion of channeling within a single fracture is presented in Ch. 5. In Sec. 3.6 of this chapter, a methodology is outlined for introducing aperture variation and channeling effects in a three-dimensional discrete fracture model.

3.4 MIXING AT FRACTURE INTERSECTIONS

Mass transfer at fracture intersections is a key factor in dispersion, for it is at the intersections that the influence of network geometry is brought into play. In this section, we review the approaches that have been adopted to model mass transfer at fracture intersections. For the purpose of discussion, we adopt a geometry corresponding to the intersection of two through-going fractures. However, the relationships described below will also apply to other types of intersections, such as when one fracture terminates against a second, through-going fracture.

The segments of a fracture network immediately upstream and downstream of a fracture intersection are referred to as the inflow and outflow branches (Fig. 3.4.1). The intersection itself is called a junction. There are two possible flow configurations at a fracture junction. A continuous junction occurs when the inflow branches are contiguous, as are the outflow branches. At a discontinuous junction, the sequence of inflow branches is interrupted by one or more outflow branches. In this latter case, fluid flow is converging at the intersection along branches that are on opposite sides of an intersection. Wilson and Witherspoon (1976) and Robinson and Gale (1990) describe experimental studies of flow through a continuous junction, while Hull and Koslow (1986) report on a laboratory experiment for a discontinuous junction.

Three models have been proposed to simulate the partitioning of mass among the fractures carrying water away from an intersection:

i) streamtube routing,

ii) streamtube routing with diffusion within fracture intersections, and

iii) complete mixing.

Models (i) and (iii) are approximations to the more complete analysis embodied in model (ii). All of these models are based on laminar flow. The possibility that turbulent mixing can occur at an

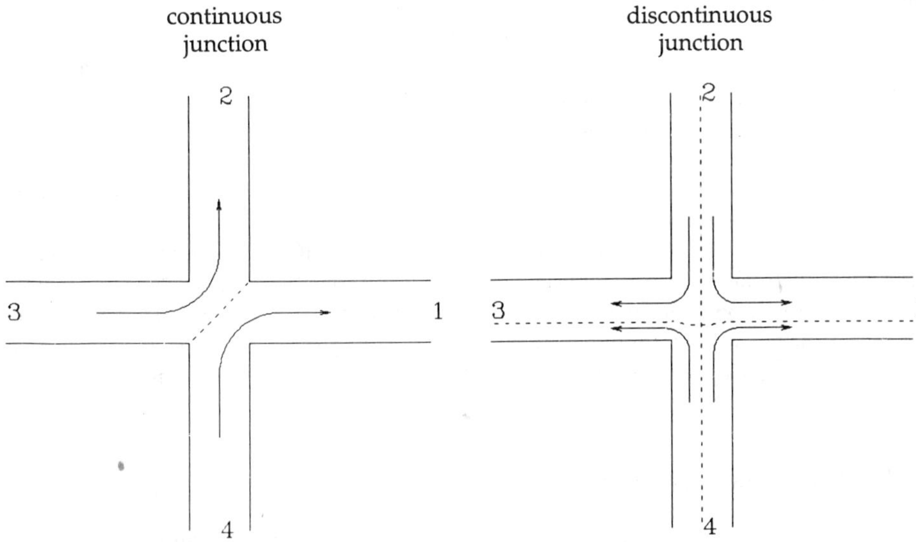

Figure 3.4.1. Schematic diagram of an intersection of two orthogonal fractures, showing examples of continuous and discontinuous junctions. Also shown is the numbering scheme used to label the inflow and outflow branches.

intersection, under natural gradient conditions, is not considered here. Furthermore, as Fig. 3.4.1 suggests, mass transfer at an intersection is viewed as a two-dimensional process. While some authors have speculated that mass may be preferentially transported by advection along the axis of an intersection, no experimental data exist to rigorously examine this possibility.

The majority of model studies at the network scale assume complete mixing at fracture intersections. For a laminar flow regime, it is argued that solute residence time within an intersection is sufficient for mass from an inflow branch to diffuse across all the streamlines within the intersection. Thus, outlet concentrations are uniform. Mass is transported away from the intersection in proportion to the discharge in each outflow branch. This model leads to the simplest algorithm to implement because the relationship between the inflow branch through which mass enters an intersection, and its outflow branch, can be expressed as a simple probabilistic function, without considering details of the flow patterns within an intersection. Similarly, if a local-scale advection dispersion equation is used to represent mass transfer within individual fractures, then it is possible to assign single-valued concentrations to nodal points located at fracture intersections.

Streamtube routing involves tracking streamlines through a fracture intersection to determine how fluid in the inflow branches is divided among the outflow branches. The outflow pattern, considered in relation to the inflow branch from which mass enters the junction, will determine how mass is distributed among several possible outflow branches.

3.4.1 Advective Mass Transfer at Fracture Intersections

A description of the advective transfer of mass through a continuous junction requires that streamlines do not cross. Using this rule, and knowing the flow rates into and out of an intersection, it is a straightforward bookkeeping operation to determine how mass is distributed to outlets, given its entry branch. To transfer mass by advection through a discontinuous junction, a more detailed analysis of the flow pattern is required. Philip (1988) has solved the boundary value problem that describes the microscale flow pattern at an intersection of two orthogonal fractures of equal aperture. The solution is given as a mapping of the streamfunction within the intersection. Figure 3.4.2 shows a typical result for a discontinuous junction. Once the streamfunction values are known within an intersection, advective mass transfer can be determined in a direct manner.

Hull and Koslow (1986) proposed a simpler approach to routing mass through a discontinuous junction. They based their calculation on distributing the inlet flow to the outlet branches in proportion to the two outlet flows. If the fluid flow in the branches is similar, then this calculation is reasonably accurate. However, Philip (1988) shows that if the fluxes are not of similar magnitude, then a calculation based on simple proportional routing can be in substantial error.

3.4.2 Approximations Underlying Mixing Models

Smith et al. (1992) describe a model study designed to evaluate mixing assumptions at fracture intersections. Streamtube routing models are based on the assumption that a negligible amount of mass is transferred by diffusion across the streamline that divides the incoming fluid into the

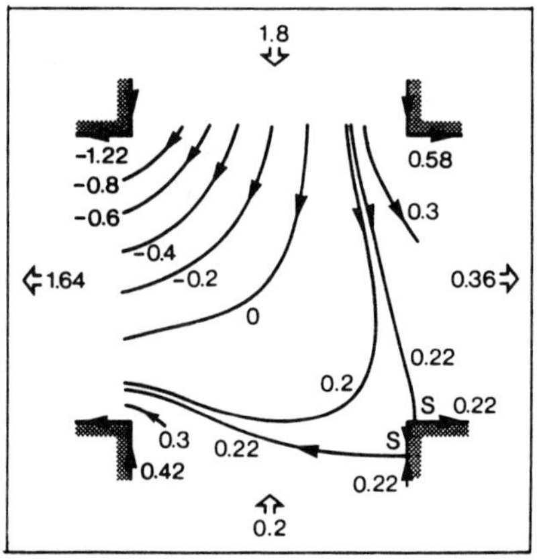

Figure 3.4.2. A plot of streamfunctions within a discontinuous junction. The numbers by the larger arrowheads are the inflow/outflow rates along the branches that form the intersection (adapted from Philip, 1988).

outflow branches. Complete mixing models are based on an assumption that the residence time of mass within a junction, and the magnitude of the diffusion coefficient, are sufficient to yield a uniform concentration at the outflow branches. The interplay of advection and diffusion within a fracture intersection can be characterized in terms of a Peclet number:

$$P = \frac{v\,r}{D}$$

where D is the diffusion coefficient, r a characteristic radius of the fracture intersection, and V the average fluid velocity within the inflow branches.

Consider a numerical experiment where a large number of particles are distributed uniformly along the entry to the intersection from branch four (see Fig. 3.4.1). If the inflow velocities are relatively high, then the effect of diffusion will be minor because of a limited residence time within the intersection. This flow regime has been examined in a laboratory model by Robinson and Gale (1990), which led them to conclude that no mixing occurs at fracture intersections. However, as fluid velocities decrease, residence time in the junction increases, and some fraction of the particles transfer across the dividing streamline by diffusion. This process can be simulated using a particle tracking technique. Fluid velocities can be determined from the analytic solutions for the streamfunctions within a fracture intersection, as given by Philip (1988).

Figure 3.4.3 shows how particles are distributed in outflow branches one and two in a case where the fluid flux in branches three and four is equal. The vertical axis is scaled according to the relative proportion of particles that enter the respective outflow branches. For Peclet numbers greater than approximately 0.1, a streamtube routing model appears to provide a reasonable approximation of the mass transfer process. In a range of Peclet numbers from 0.0001 to 0.1, the proportion of mass leaving the intersection via the two outflow branches depends upon both the advective transfer pattern, and the extent of diffusive transfer across the streamlines. For Peclet numbers less than 0.0001, diffusion is the dominant process controlling the particle distribution to

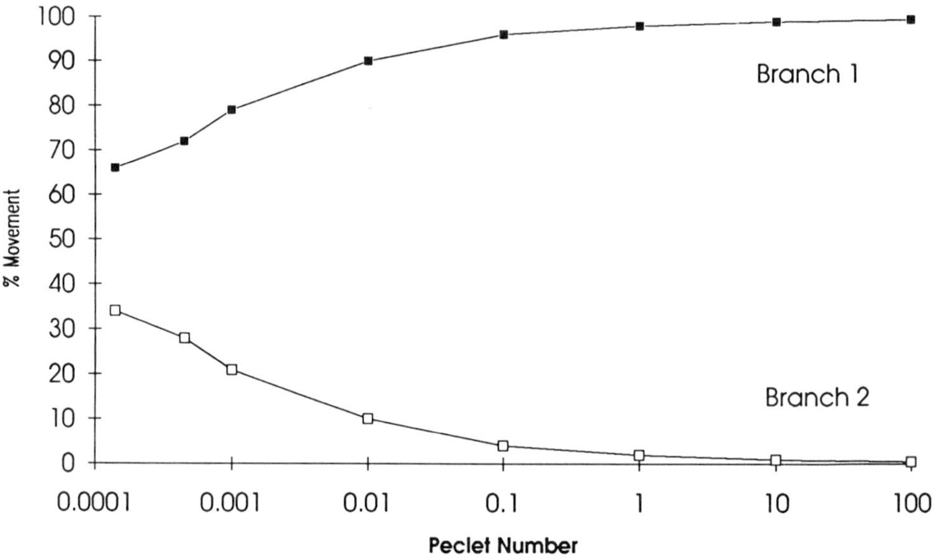

Figure 3.4.3. Distribution of particles in outflow branches one and two, as a function of the Peclet number. Particles enter the intersection from branch four. See Figure 3.4.1 for problem geometry. The distribution of particles in the outflow branches is independent of the magnitude of the diffraction coefficient and fracture aperture, when plotted as a function of the Peclet number (adapted from Smith et al. (1992).

the outflow branches. Note that complete mixing is not observed. For purposes of comparison, in a complete mixing model, 50% of the particles would, on average, exit from each of branches one and two. In this intersection, for Peclet numbers less than 0.0001, 66% of the particles exit through branch one, and 34% exit through branch two.

In an intersection where diffusive transfer comes into play, the microscale geometry of the intersection plays a key role in determining how mass is partitioned among the outflow branches. The pathlengths between the inflow and outflow branches are not all of the same length, so that, for example, particles that enter the intersection on a pathline close to the left-side wall of the fracture (Fig. 3.4.1) have a higher probability of leaving the intersection through branch two than branch one. This geometric effect modifies, in a predictable way, how mass enters the outflow branches, relative to that which would be calculated assuming complete mixing.

This example implies that the simple mixing relationships adopted in the majority of existing network models may not be adequate. Complete mixing (routing in proportion to the fluid flux in the outflow branches) is not observed. Due to the wide range of fluid velocities that could be expected to occur within a fracture network, a full spectrum of mixing ratios can be expected. Computationally-efficient techniques for modeling transport at the network scale, that account for both diffusion within the intersection, and the geometry of the intersection, need to be developed.

3.5 TWO-DIMENSIONAL DISCRETE FRACTURE MODELS

3.5.1 Introduction

Numerical studies of solute transport in discrete fracture networks are important for several reasons. They provide a method for examining the requirements underlying the representation of a discrete network of fractures by an equivalent porous medium. Because the network is modeled on a fracture-by-fracture basis, and no a priori assumptions are made about the form of the macroscopic dispersion process, they can also be used to examine the applicability of a conventional Fickian model of dispersion. Finally, they illustrate the sensitivity of spreading patterns to network geometry. For the most part, discrete network studies have been limited to two-dimensional geometries. Although the results of numerical calculations cannot be directly transferred to a field setting because of the restrictive two-dimensional nature of the models, the trends observed in comparing simulation results shed light on the nature of the transport process.

A number of researchers have examined solute transport at the network scale (Schwartz et al., 1983; Smith and Schwartz, 1984; Robinson, 1984; Endo et al., 1984; Shimo and Long, 1987; Robinson and Gale, 1990). Fractures are represented as linear features of finite length, with fracture midpoints located randomly within the rock mass. These studies extended pioneering work by Castillo et al. (1972) and Krizek et al. (1972) on deterministic, regularly-spaced networks of fractures of infinite length. By adopting a stochastic description of the network for fractures of finite length, the connectivity of the network takes on a dominant role in controlling spreading patterns. An example of a discrete network model, used in subsequent discussions, is shown in Fig. 3.5.1. This diagram shows the backbone of the fracture network, after removal of dead-end and isolated fracture segments. The network is composed of two orthogonal fracture sets, each with a scanline density of 7.2 fractures/m. Scanline density is defined by the average number of fractures observed per meter along a line trace.

An alternate model of a fracture network, based on a parallel set of fractures with mixing zones located along the flow path, was studied by Rasmuson (1985) using residence time distribution theory. This analysis extended calculations by Neretnieks (1983) for dispersion in a parallel set of fractures without mixing zones located between the inlet and outlet. In both studies, the focus was on an examination of the relationship between the probability distribution function for fracture aperture and the form of the breakthrough curve at the outlet.

One possible approach to modeling solute transport in networks like that shown in Fig. 3.5.1 combines a numerical solution for hydraulic head at fracture intersections with a particle tracking technique to determine the spatial distribution of mass within the network, and to estimate breakthrough at a downstream boundary (Schwartz et al., 1983; Smith and Schwartz, 1984; Robinson, 1984). It is commonly assumed that solute is transported between two fracture intersections at an average velocity given by:

$$v = \frac{\rho g b^2}{12\mu} \, dh/dl$$

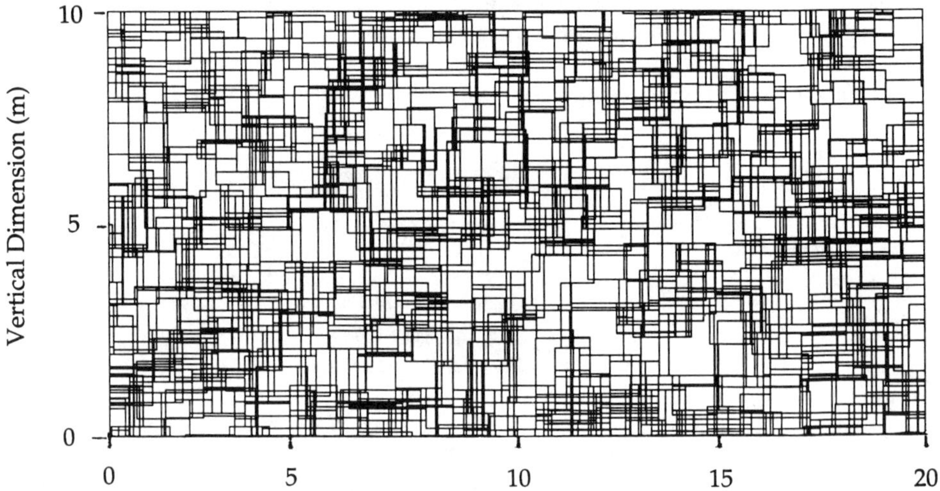

Figure 3.5.1. Example of one realization of a fracture network, showing the backbone of the active network. See text for details of the network geometry.

where μ is fluid viscosity, b fracture aperture, and dh/dl the local hydraulic gradient. The validity of this approximation rests on the requirement that transverse diffusion is rapid enough to create a uniform concentration across the fracture by the time mass reaches a downstream intersection. Except for the case of large-aperture fractures (greater than 1000 microns), and with residence times between intersections on the order of only seconds, this assumption appears reasonable (Hull et al., 1987).

To simulate advective mass transfer in the case of a continuous injection of mass, where particle tracking techniques are inefficient, Shimo and Long (1987) describe a front tracking method. Concentration values are transferred from one intersection to an adjacent intersection according to the travel time along the branch that connects them. At the intersection, a flux-averaged concentration is calculated on the basis of the incoming flux and concentration values in the inflow branches. Once the concentration value is calculated at one intersection, the data is used to calculate the concentration at the next downstream intersection. The method is continued to track concentration values across the entire network to the outflow boundary.

A different formulation of the transport problem bases calculations on the routing of streamtubes from the upstream to downstream side of the flow domain (Endo et al., 1984). Streamtubes are created in a sequential manner in a way that ensures that flowlines do not cross, and at an intersection, fluid satisfies concepts of proportional routing. For each streamtube, the advective travel time of the fluid can be calculated by summing the residence time along each branch of the streamtube between fracture intersections. In contrast to the models referenced above, this approach is purely advective, diffusive mass transfer is not considered; either within a fracture intersection, or along individual branches of the network.

3.5.2 Macroscopic Dispersion

Spreading patterns and transit times can differ significantly for realizations of fracture networks generated from the same geometric model. There will be regions within a network where the local connectivity of fractures is below average; these regions, of variable size, act as barriers to flow. Similarly, local regions within a network with higher than average connectivity have the potential to transmit a higher fluid flux than average, creating preferred pathways for mass transfer. Each realization will be unique in terms of the frequency, arrangement, and spatial extent of these regions with differing local connectivity. For this reason, comparisons relating network geometry to spreading patterns must be made in a statistical framework. Frequency distributions on various breakthrough fractions can be formed by repeating the transport calculations for a set of realizations in a Monte Carlo simulation.

Mass transfer in networks of the kind shown in Fig. 3.5.1 has the following characteristics:

i) the longitudinal and transverse distribution of mass can be quite variable among different realizations generated from the same geometric model,

ii) within any one realization, the distribution of mass on the local scale can be quite heterogeneous,

iii) the mass distribution has a consistent negative skew in the direction of the mean hydraulic gradient,

iv) a marked longitudinal dispersion can develop a short distance from the source,

v) the extent of spreading varies as a function of the angle between the direction of the hydraulic gradient and the attitude of the fracture sets,

vi) the directions of maximum and minimum spreading are less related to the direction of flow than to fracture geometry,

vii) factors that contribute to spreading in the major and minor principal directions change with the orientation of the fracture network relative to the mean hydraulic gradient.

The distribution of mass within the network shown in Fig. 3.5.1, calculated for a time 35 days after release from the source location, is plotted in the upper diagram of Fig. 3.5.2. The hydraulic gradient is parallel to the x axis. A particle tracking algorithm was used to obtain this result. In Figure 3.5.2, the dot size has been scaled to be proportional to the number of particles located within a square region of dimension 0.05 m.

The locally-heterogeneous form of the mass distribution reflects the limited number of pathways available for mass to pass through a fracture network. The negative skew develops because mass enters fractures aligned at right angles to the mean hydraulic gradient. In fractures of this set, fluid velocities are lower and thus mass lags behind the mean flow. In general,

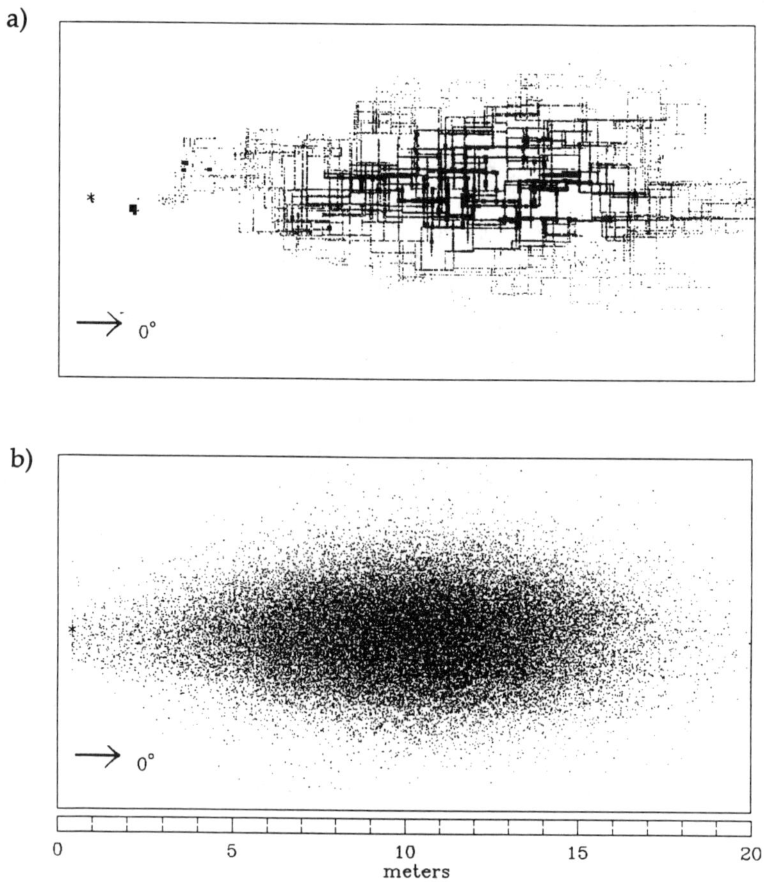

Figure 3.5.2. Simulation of the mass distribution within the fracture network shown in Figure 3.5.1. The mean and standard deviation in log aperture for both fracture sets is −4.50 and 0.20, respectively, for aperture measured in meters. Time since pulse injection is 35 days, with the source location denoted by "*". (a) Discrete network simulation. (b) Statistical continuum prediction, see Sec. 3.7.2 (adapted from Robertson, 1990).

significantly less time is required for the first 50% of the mass to leave the system, than is required for the last 50%, leading to a long tail on the breakthrough curve. Note however that if the hydraulic gradient is rotated so that it is aligned at 45° to both fracture sets, the velocity distributions in the two fracture sets become similar, the negative skew on the mass distribution decreases, and there is less variability among the set of realizations comprising a Monte Carlo simulation. If there is a positive correlation between fracture length and aperture (i.e., longer fractures tend to have a greater aperture), then the spatial distribution of mass is likely to be more regular, with less dispersion (Shimo and Long, 1987).

Simulation results also suggest that assumptions underlying Fickian models of dispersion are generally not applicable to networks of the kind illustrated in Fig. 3.5.1. An important consideration here is the occurrence of fractures in a limited number of sets with preferred orientations. Several features can be noted when the spatial distribution of mass is examined on a realization by realization basis. First, a Gaussian distribution of mass predicted by a Fickian model of dispersion is observed in only a narrow range of conditions. In the general case, the spatial distribution of mass is better represented by a skewed distribution. Second, a single dispersivity value, or simple dispersivity function, does not characterize the rate at which a plume grows in time. These model-based inferences are consistent with the observational data reviewed in Ch. 2.

Spatial moments provide an effective means of expressing the overall character of the mass distribution within a fracture network. The first moments (x_c, z_c) describe the center of mass of the plume, the second moments (s_{xx}, s_{zz}) the longitudinal and transverse variance about the mean position of the plume. For the mass distribution shown in the upper plot of Fig. 3.5.1, $x_c = 11.0$ m, $z_c = 0.5$ m, and $s_{xx} = 7.6$ m^2, $s_{zz} = 0.9$ m^2. In line with earlier comments, spatial moments can be expected to vary appreciably between realizations generated from the same fracture model, especially in cases where the scanline density is relatively low. However, by averaging over a set of realizations generated from the same geometric model, it is possible to observe a regularity in the spatial moments, that is, the ensemble mean behavior. The solid line in Fig. 3.5.3 shows the

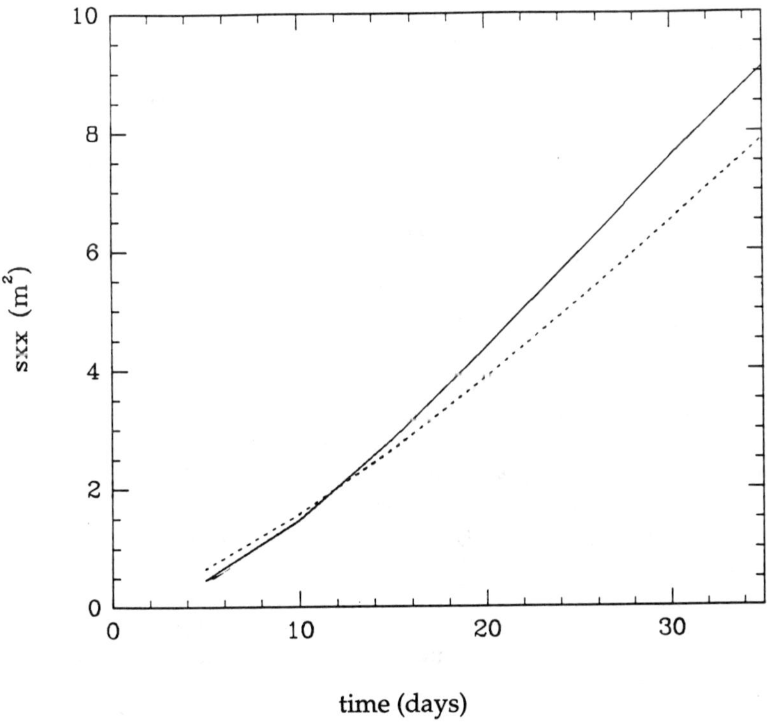

Figure 3.5.3. Temporal variation in the second spatial moment of the longitudinal mass distribution. The solid line is the result from a discrete fracture simulation model, formed as an average over 200 simulations. The dashed curve is the prediction using a statistical continuum model, see Sec. 3.7.2 (adapted from Robertson, 1990).

second spatial moment in the direction of mean flow, calculated as a function of travel time. Estimates are calculated by averaging over 200 discrete network simulations. The network shown in Fig. 3.5.1, and the mass distribution in Figure 3.5.2, are one realization from this set. An initial, non-linear rate of growth in the ensemble variance is observed. An analogous time dependence in the dispersion process has been documented for heterogeneous porous media (e.g., Dagan, 1984).

3.5.3 Influence of Network Geometry on Breakthrough

To explain how the geometry of a fracture network influences solute transport, it is helpful to view the mass transfer process relative to the directness of the hydraulic connection across the fracture network. Connectivity influences the mean velocity and variations in velocity within fractures of the same set. Figure 3.5.4 shows three fracture networks, each generated with different fracture densities. Fracture density is expressed in terms of the scanline density. In plots (a) and (b), the vertical and horizontal scanline density for both fracture sets is 2.0 and 4.0 fractures/m, respectively; in plot (c), the scanline density is 2.0 fractures/m along a vertical line trace, and 4.0 fractures/m along a horizontal line trace. The hydraulic gradient is aligned with the horizontal axis, and the upper and lower boundaries of the domain are assumed to be impermeable. For a scanline density of 2.0 fractures/m for both sets, the fracture network is near the percolation threshold. Just over one half of the realizations generated in a Monte Carlo simulation failed to develop connected pathways between the upstream and downstream boundaries. At a scanline density of 4.0 fractures/m, only 0.4% of the realizations generated did not form a connected network.

Differences in fracture geometry that have the effect of increasing the probability that more indirect or circuitous pathways exist through the network generally lead to an increase in the mean travel time to a downstream boundary, and to greater variability in travel time among a set of realizations generated from the same geometric model. Changes in fracture geometry that increase the probability that a relatively direct pathway exists across the network will have the opposite effect (Smith and Schwartz, 1984).

Examples that illustrate how differences in scanline density influence arrival times at a downstream boundary are given in Table 3.5.1. Included in this comparison are network geometries represented by the individual realizations shown in Fig. 3.5.4. Two suites of simulations are considered; in one all fractures are of equal aperture (32 microns), in the other the log aperture standard deviation is 0.2. A value of this magnitude leads to apertures that vary by approximately one order of magnitude. Mass enters the flow system through a single fracture located closest to the midpoint between the upper and lower impermeable boundaries. Inspection of Table 3.5.1 shows:

i) a reduction in the scanline density tends to delay the arrival of mass at the outflow boundary, while the variability in the arrival time of the various breakthrough fractions increases.

ii) a reduction in the number of fractures aligned in the direction of the hydraulic gradient (the vertical scanline density) leads to relatively longer arrival times, and greater macroscopic dispersion, when compared to the case where the number of fractures

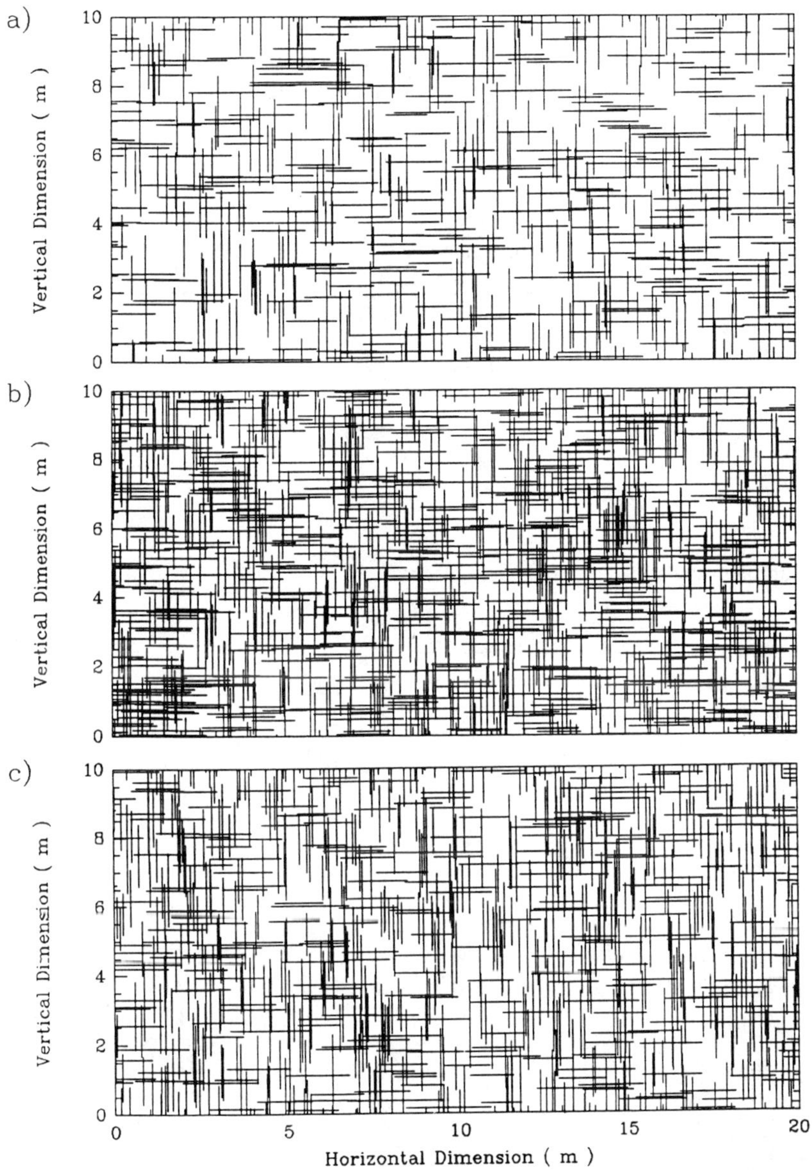

Figure 3.5.4. Examples of three discrete fracture networks constructed with differing scanline densities. (a) A scanline density for both sets of 2.0 fractures/m, (b) A scanline density for both sets of 4.0 fractures/m, (c) A scanline density of 2.0 fractures/m along a vertical line trace and 4.0 fractures/m along a horizontal line trace.

Table 3.5.1. Statistics on breakthrough times (days). Here, for example, T_{25} corresponds to the arrival time for 25% of the mass released at the upstream boundary to cross the outflow boundary. Results from Monte Carlo simulations based on 500 realizations with connectivity. Trials (a) through (e) are for networks with uniform apertures, in trials (f) through (h), the log aperture standard deviation is 0.20.

(i) Scanline density (fractures/m)

	Vertical scanline density	Horizontal scanline density
Trials (a), (f)	2.0	2.0
Trials (b), (g)	3.0	3.0
Trials (c), (h)	4.0	4.0
Trial (d)	4.0	2.0
Trial (e)	2.0	4.0

(ii) Breakthrough statistics

Trial		T_1	T_{25}	T_{50}	T_{75}	T_{90}
(a)	mean	128.5	174.1	205.0	253.7	335.0
	std.dev.	110.2	115.3	124.3	135.4	176.0
(b)	mean	68.2	94.8	107.8	125.4	148.7
	std.dev.	34.1	36.2	37.8	41.9	47.7
(c)	mean	53.5	73.4	82.1	93.4	107.8
	std.dev.	12.5	13.2	13.6	14.5	16.0
(d)	mean	74.7	98.1	110.3	128.2	151.5
	std.dev.	83.5	84.9	86.4	96.5	104.0
(e)	mean	93.6	134.2	154.8	184.1	222.3
	std.dev.	106.1	108.1	110.7	117.8	126.6
(f)	mean	189.9	265.7	324.2	419.1	576.1
	std.dev.	161.5	181.8	208.2	245.7	317.5
(g)	mean	71.8	106.9	126.1	154.4	192.9
	std.dev.	90.2	93.9	96.7	106.9	125.5
(h)	mean	42.2	66.1	77.5	92.7	112.7
	std.dev.	20.1	22.2	23.4	25.8	35.3

aligned orthogonal to the hydraulic gradient is reduced (the horizontal scanline density). The variability in arrival time is also greater when the connectivity provided by the vertical fractures assumes a more important role in transmitting mass through the network.

iii) macroscopic dispersion tends to be greater in networks with greater variability in fracture aperture, and the variability in travel time between realizations increases.

iv) in the fracture network that is relatively sparse (trials a and f), the plume moves, on average, at a slower rate in the network with variable fracture aperture, than in the corresponding uniform aperture case. However, in the well-connected network (trials c and h), the plume is moving, on average, at a higher rate in the network with variable fracture apertures. This behavior demonstrates the sensitive interplay between channelization at the network scale, and fracture density. It also supports the contention that generalized statements on relationships between fracture geometry and spreading behavior must be applied with caution in site-specific interpretations.

The influence of the fracture length distribution on transport can also be explained in terms of the connectivity of the network. Given similar length distributions for two fracture sets, the length distribution of the fracture set more closely aligned with the direction of the mean hydraulic gradient exerts a stronger influence on macroscopic dispersion than does that of the other fracture set. A reduction in the average length of fractures forming the set most closely aligned with the direction of the hydraulic gradient delays the arrival of mass at the downstream boundary, and increases the extent of macroscopic dispersion (Smith and Schwartz, 1984).

The results in Table 3.5.1 can also be interpreted in terms of the uncertainty in predicting breakthrough at the downstream boundary. The statistical distributions that define the network geometry (fracture location, length, orientation, and aperture) can be interpreted as a measure of the uncertainty in specifying the exact structure of a network. This uncertainty can then be propagated through the fluid flow and solute transport models to estimate the uncertainty in model output. For example, in case (h), the uncertainty in predicting initial breakthrough, measured in terms of one standard deviation on either side of the mean, is $\pm$ 20 days, for the average initial breakthrough time of 42.2 days. Uncertainties in predicting breakthrough are greater:

i) in networks with lower fracture densities, especially in cases near the percolation threshold,

ii) in networks where the variability in fracture aperture is higher, and

iii) in networks where the fracture frequency and length distributions promote relatively tortuous (indirect) connections across the network.

The results in Table 3.5.1 were developed for an unconditional simulation. Individual fractures, whose locations may be known from core or borehole mapping, are not preserved across the set of realizations. Andersson et al. (1984) describe a method of conditional simulation of fracture networks that preserves data on fracture locations. Using this approach, Andersson and Thunvik (1986) explored the extent to which this geometric constraint on the network could reduce the uncertainty in predicting solute breakthrough at a downstream boundary. As expected, the value of adding more core data depends upon the fracture length distribution. In their examples, additional geometric data from new cores contributed to only minor reductions in the uncertainty in breakthrough time. However, by conditioning the fracture network with a measurement of the bulk permeability of the network (via an adjustment in the fracture aperture distribution), much greater uncertainty reduction could be achieved. In effect, the uncertainty in breakthrough time is reduced by constraining the extent to which the mean advective velocity varies in the set of realizations forming a simulation.

3.5.4 Representation of a Fracture Network as a Continuum

Discrete network models can be used to investigate conditions under which it is possible to represent a fracture network with continuum elements, and thus treat the rock mass as an equivalent uniform medium. A minimum scale exists below which a fractured rock does not behave as a continuum. This minimum scale is related to, but not coincident with the percolation threshold. The continuum elements must be large enough to encompass a representative sample of the open, connected fractures, *and* their variable influences on the flow and transport behavior of the network. It is also necessary that the statistical parameters describing network geometry must be constant over a length scale that is larger than that of the continuum elements. It follows that the domain of interest must be larger than that defining the representative volume. The first two conditions are clearly not independent of one other.

The definition of what constitutes a representative sample of the fracture network is problem dependent. It depends upon both features of the network, and whether one is considering a description of fluid flux or solute transport. Long and Witherspoon (1985) have examined relationships between the connectivity of a network, and the size requirements for defining an REV for fluid flow. Solute transport is more sensitive to network geometry than is fluid flow. For networks of the kind shown in Fig. 3.5.1, the variability in the equivalent hydraulic conductivity among different realizations generated from the same geometric model is relatively small. There are a sufficient number of fractures within the domain so that the equivalent hydraulic conductivity is approximately independent of changes in the arrangement of the fractures forming the network. However, the variability in breakthrough times at the downstream boundary for this same network geometry remains appreciable. In placing the network shown in Fig. 3.5.1 in the context of a field setting, it is important to recognize that it was generated in a way that leads to nearly all fractures being part of the "active" network. While maps or borehole logs of fracture density might suggest much higher densities of fractures, field observations indicate that a significant proportion of the fractures in any given set may either be closed or sealed by mineral precipitates. It is the density of hydraulically-open fractures that is the critical factor in determining the scale of a continuum element.

The issue of continuum behavior for transport can also be examined using the concept of an hydraulic effective porosity (Endo et al., 1984). The hydraulic effective porosity is defined as the product of specific discharge and a mean travel time, divided by the linear length of travel. For each realization of a fracture network, values of the effective porosity can be calculated for a number of different orientations of the hydraulic gradient. For equivalent porous medium behavior, the hydraulic effective porosity must be independent of the direction of flow. On this basis, Endo et al. (1984) show for a fracture system with two sets of parallel, continuous fractures of constant aperture; that although the fracture system may behave as a continuum for fluid flux;

it may not behave as a continuum for transport. Because of computational limitations at the time, they were unable to simulate a system that was large enough to meet the REV requirement for a network system composed of discontinuous fractures (Endo, 1984).

Little is known of the distances over which solutes must travel to sample a representative suite of fluid velocities in a fractured rock mass. Some insight to this issue has been gained in simulation studies. However, compelling evidence can only be established in field settings. One of the great challenges in field experimentation is to provide evidence to aid in documenting characteristic values for this length scale.

3.6 THREE-DIMENSIONAL NETWORK MODELS

Two-dimensional studies provide insight to the factors influencing transport in fractured media. However, the three-dimensional structure of a network is required for these analyses to be more realistic. While several three-dimensional models for fluid flow are described in the literature (Long et al., 1985; Shapiro and Andersson, 1985; Elsworth, 1986; Liggett and Medina, 1988; Dverstrop and Andersson, 1989), little work has been carried out in modeling solute transport.

In this section, we outline an approach to examine features of transport in three-dimensional networks of discontinuous, rough-walled fractures. Figure 3.6.1 is a schematic of a simple three-dimensional network, presented here to guide the discussion. In a more general case, the network would be described stochastically, based on probability distributions for the length and width of individual fracture planes, fracture location, and fracture aperture. Unlike two-dimensional models, dead-end fractures must be retained. Differential values of hydraulic head along a fracture intersection result in circulation within dead-end segments. To solve for values of

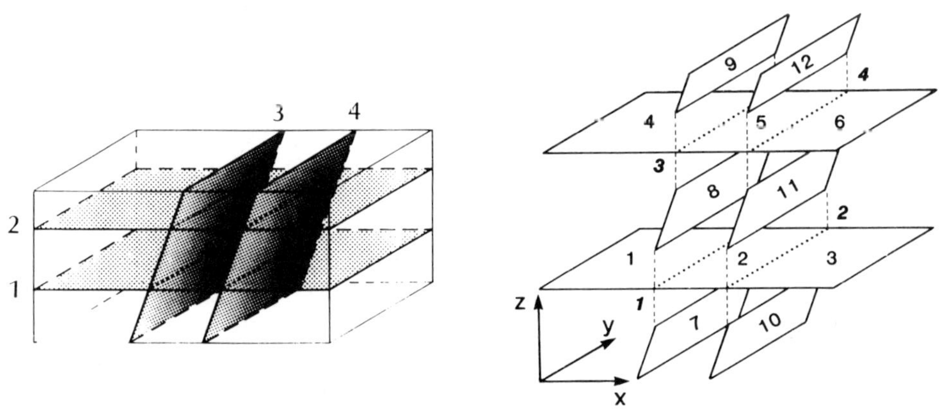

Figure 3.6.1. A schematic of a simple three-dimensional fracture network, showing the division into fracture quadrants. A finite element mesh is constructed within each fracture segment (numbered 1 through 12). Mass transport can be simulated using a particle tracking technique. From Smith et al. (1985).

hydraulic head within the network, a finite element mesh can be constructed within each fracture segment (Fig. 3.6.1), and the segments linked by common nodes at fracture intersections. Once the patterns of flow in the network are established, particle tracking is a feasible approach to simulate transport in networks constructed from many hundreds of fracture planes.

Earlier, in Sec. 3.3, we noted evidence for channeling of fluid flow and solute transfer due to aperture variability. Few in situ data exist to quantify this spatial variability. As an initial approximation, we adopt a geostatistical model where aperture variability is represented by a lognormal probability function, with an exponential function describing the spatial correlation structure. The length scale of the correlation can be defined by integral scales (λ_1, λ_2). In Ch. 5, examples of aperture variability, and the resulting channelized flowpaths across single fracture planes are presented. The degree to which higher-aperture regions on one fracture plane connect with those on other fractures, at the fracture intersections, will determine pathways and transport patterns on the network scale.

An example is presented to illustrate several features of transport through a three-dimensional network of rough-walled fractures (Smith et al., 1985). The flow domain is a cube with a dimension of 10 m on the side. Figure 3.6.2 is a plot of the trace of fractures on a vertical section through the rock mass. For this example, fractures are generated in a way that ensures they are continuous across the flow domain in the direction of the y axis. Fluid flow is established between the left and right faces by assigning constant values of hydraulic head to all fractures intersecting those faces. Impermeable boundaries are assigned to the remaining four faces. Superimposed on this natural gradient is the effect of an injection-withdrawal well pair, with $Q_i = Q_w$. Solute enters the flow system through the injection well (open circle), and exits at either the withdrawal well (hatched circle) or the downstream boundary.

Residence-time distributions are shown for three simulations in Fig. 3.6.3. The vertical axis is the fractional arrival rate, or the fraction of the total number of particles that reach the exit boundary per unit time. Panels to the left are residence time distributions recorded at the withdrawal well, while panels to the right are recorded at the downstream boundary. The uppermost plot is for a case where all fractures in the network are assigned a uniform aperture of 100 microns. The lower plots are for the same network geometry, but for two realizations with rough-walled fractures. To highlight dominant features in the RTD curve, a filter is applied to the breakthrough data to eliminate high frequencies introduced by using discrete timesteps and a finite number of particles.

The initial pulse reaching the withdrawal well marks the arrival of solute that moves within the fracture open to the well pair. This segment of the RTD curve carries information on the dispersive properties of that fracture. The tail of the RTD curve records the arrival of mass that either encounters lower fluid velocities near the bounding edges of the fracture plane, and/or follows a more circuitous route that includes pathways through the second fracture set, where fluid velocities are lower. Approximately 27% of the injected solute bypasses the well to move to the downstream boundary. In plot (b), the initial pulse at the well is narrower and has a higher peak than that observed in the parallel plate model. This response occurs if solute is channeled between the injection and withdrawal points. The opposite conditions are observed in plot (c). Here there is greater spreading within the plane of the fracture than in the simulation with a uniform aperture.

To draw general conclusions on the effects of three-dimensional network geometry on transport, it is necessary to carry out a Monte Carlo analysis as was done with the two-

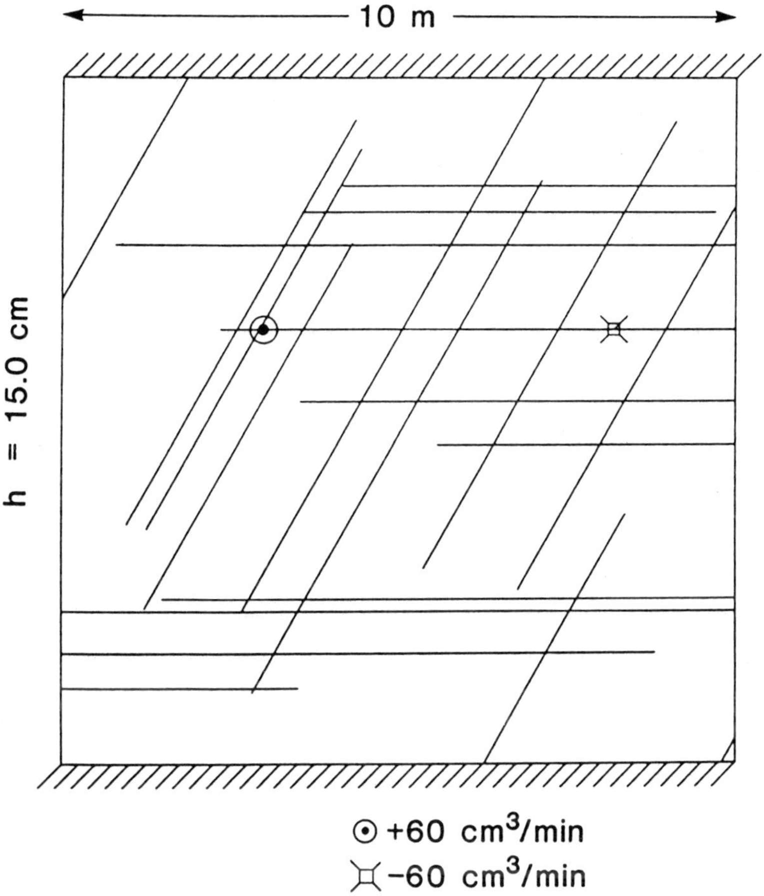

Figure 3.6.2. Realization of a three-dimensional fracture network, showing the trace of the network on the *xz* face of the flow domain. The projection of an injection-withdrawal well pair, located in the center of the fracture plane, is also shown.

dimensional network models. A difficulty here is the computational burden of solving for fluid flow within even just one network containing many hundreds of rough-walled fractures. Single realizations may contain from 10^5 to 10^6 nodes to ensure an accurate solution of the flow equation. The analysis of hundreds of realizations generated from the same geometric model likely exceeds available computer resources in most instances. The modeling approach described here should be viewed as a tool to examine transport processes (e.g., Smith et al., 1987). Other techniques, based either on simplified approximations of the network, or of fluid flow through the network, are needed to model site-specific field applications (e.g., Cacas et al., 1990).

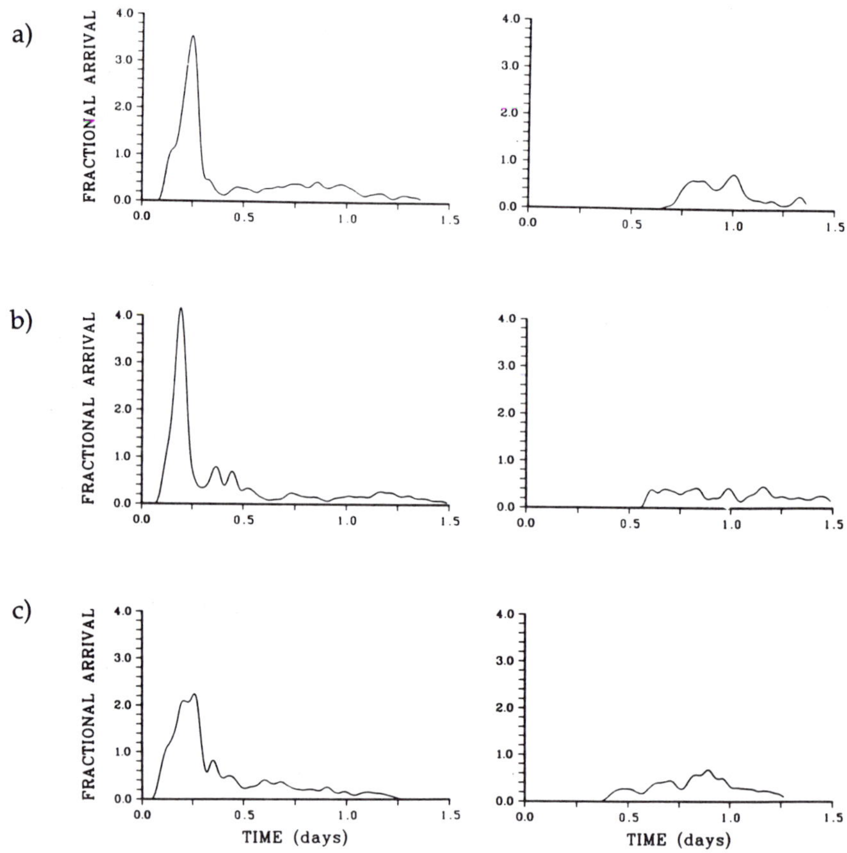

Figure 3.6.3. Residence-time distribution curves for the system shown in Fig. 3.6.2. Figures on the left are the RTD curves for the well, those on the right are formed at the downstream boundary. Plot (a) is for a network with uniform aperture, plots (b) and (c) are from two realizations incorporating aperture variation within the plane of each fracture. From Smith et al. (1985).

3.7 CONTINUUM MODELS OF SOLUTE TRANSPORT IN FRACTURE NETWORKS

Discrete network models have proven to be very useful for concept evaluation. However, their usefulness in a practical sense is limited. It is simply not possible to model a fracture network in its complete detail, except in a limited near-field simulation. The simulation of fluid injection and tracer tests at the Fannay-Augeres site that is described in Ch. 2, and in more detail by Cacas et al. (1990), is illustrative of the strength and limitations of discrete network simulation. At the field scale, some form of spatial averaging is necessary. How this spatial averaging should be carried out is at the crux of the debate on what modeling approaches are feasible for simulating mass transport in fractured geologic media. One alternative is to represent the rock mass as an equivalent porous medium. Assuming that the conditions necessary to adopt this approach are met, a methodology is then needed to identify coefficient values of the dispersion tensor, and the value of effective porosity. If a tracer test is carried out, it will in all probability be at the local scale. There are large uncertainties in attempting to extrapolate from a local scale test (a few tens of meters) to the field scale (hundreds of meters). At the field scale, the data most likely to be available are geologic observations from borehole logging, supplemented by single-borehole hydraulic tests, and, under favorable circumstances, identification of the major fracture zones that typically control the larger-scale flow patterns in crystalline rock. The issue of how best to use such data for far-field prediction of mass transfer is still unresolved.

In this section we first describe continuum approaches that are based on identifying transport coefficients for a representative elemental volume (REV) of the fracture network. Then we describe a new hybrid approach that combines features of discrete fracture simulation with volume averaging to model transport in fracture networks that may not meet the statistical homogeneity requirement of an REV.

3.7.1 Analytic Continuum Approaches

A continuum model described by de Josselin de Jong and Way (1972) simulates transport in a two-dimensional network of regularly-spaced fractures of infinite length, subject to a uniform hydraulic gradient. The network is replaced by an equivalent porous medium, with spreading characterized in terms of a dispersion tensor. The form of the dispersion tensor is related to features of the network. Consider a network consisting of two fracture sets (Fig. 3.7.1). The network geometry is defined with respect to the attitude of each fracture (θ_a,θ_b) in relation to the x-axis, and the distances between intersections (l_a,l_b). Flow is in response to a hydraulic gradient J acting at an angle α relative to the x-axis. Equations for calculating the components of the dispersion tensor, and the angle between the principal axes of the tensor and the direction of the hydraulic gradient, are reviewed in Schwartz and Smith (1988). Once determined, these coefficients would be used in a conventional form of the advection dispersion equation.

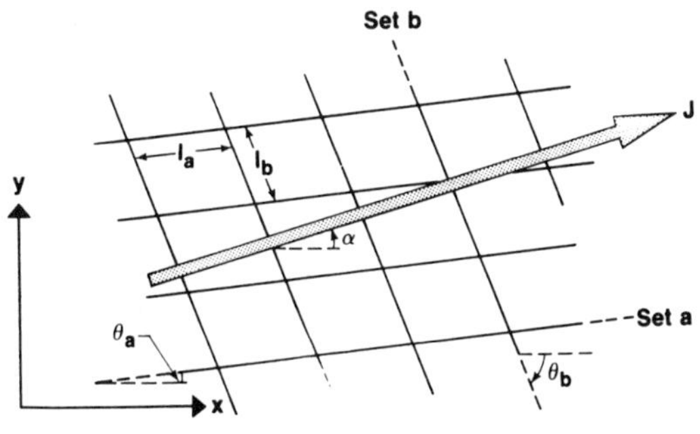

Figure 3.7.1. Geometry of a fracture network consisting of two sets of infinite fractures, for which the de Jong model is applicable. Fractures in each set are parallel and equidistant (adapted from Schwartz and Smith, 1988).

However, as noted by Schwartz and Smith (1988), in general the idealized de Jong model is not applicable to networks formed from sets of finite length, irregularly-spaced fractures.

In the de Jong model, the dispersion tensor is calculated knowing the geometric structure of the network, and the apertures of the fracture sets. However, in this and similarly-based models, the technical difficulty of translating a geometric model of a fracture network into a corresponding hydraulic model is an important issue, the significance of which should not be underestimated. Neuman (1987) discusses this problem at some length.

An alternate procedure for computing the dispersion tensor for a fractured rock mass has been proposed by Neuman and Depner (1988). The method combines single borehole and cross-hole hydraulic tests with stochastic transport theory as developed originally for anisotropic porous media. Their approach is based upon hydraulic conductivity estimates, rather than parameters describing network geometry. The key assumption is made that the large-scale hydraulic properties of the rock mass can be represented as an anisotropic porous medium. Their model characterizes the medium in terms of a mean, variance, and spatial covariance function for hydraulic conductivity. Single borehole packer tests are interpreted in terms of a locally isotropic continuum, which may be at a scale smaller than a representative elemental volume. If cross-hole tests indicate, at a given scale, that it is possible to define an effective hydraulic conductivity tensor for the fractured rock mass, then given that hydraulic conductivity tensor, stochastic flow theory provides an estimator for the principal directions and integral scales of the log hydraulic conductivity distribution. Knowing the principal directions and integral scales for log K, stochastic transport theory can be used to calculate estimates of an asymptotic Fickian dispersion tensor (e.g., Neuman et al., 1987). These dispersion coefficients describe the spreading of a plume that has traveled over a distance that exceeds tens of integral scales of the hydraulic conductivity field. Neuman and Depner (1988) illustrate this approach using data from the Oracle granite in Arizona. As the authors note, the calculations are founded on small variance approximations, a condition not met at the Oracle site, so the reliability of their dispersivity estimates is difficult to assess.

3.7.2 A Statistical Approach to Continuum Modeling

The analytic theory of de Jong and Way applies to a network geometry that is highly idealized (see Fig. 3.7.1), and not representative of many field settings. In this section we outline an approach that accounts for dispersion in a network of finite-length, randomly-located fractures, for a flow system with a nonuniform hydraulic gradient. The modeling concept is statistical, involving the application of particle tracking techniques to simulate transport within a continuum (Schwartz and Smith, 1988).

The modeling technique proceeds in two steps. First, a transport simulation is carried out using a discrete sub-domain model, adopting a particle tracking technique to collect statistics on particle motion. This network is a small sub-domain from a much larger continuum. Second, the statistics on particle motion are transferred to the larger-scale, continuum model. Predictions of solute travel times are developed with the continuum model. By characterizing particle motion in the sub-domain, the influence of fracture geometry can be incorporated in the larger-scale continuum model without an analytic calculation of a dispersion tensor. The method first proposed by Schwartz and Smith has recently been modified to provide better estimates of the second spatial moments of the mass distribution in fracture networks with variable aperture (Robertson, 1990).

To use statistics on particle motion from the discrete sub-domain model to predict transport patterns on a larger scale, the sub-domain itself must be large enough to respond as a REV with respect to the transport process. The size of the sub-domain model, for a given fracture geometry, is determined in a sensitivity analysis that examines variations in particle motion statistics as a function of domain size. The main difference between this simulation approach and particle tracking methods used in heterogeneous porous media (e.g., Smith and Schwartz, 1980) is that here particle motion is tailored to a particular fracture geometry by repeatedly sampling from a velocity distribution, rather than as a velocity calculation based on local values of hydraulic conductivity and hydraulic gradient.

The Discrete Sub-Domain Model

Statistics to describe transport in the continuum are collected by observing particle motion in a discrete network of fractures. The geometric properties of the fractures in the sub-domain are assumed to be identical to that of the larger-scale continuum. This assumption of statistical homogeneity is the key element linking the two models.

The sub-domain for the discrete network has a square shape (Fig. 3.7.2). By assigning appropriate values of hydraulic head to the corner nodes of the sub-domain model, it is possible to generate flow with a specific orientation relative to the attitude of the fracture sets. Typically, seven different orientations of flow are examined in a realization of the fracture network. This scheme can be used to collect statistics at 15° intervals over a 90° sector of flow. In this way, it is possible to account for anisotropic dispersion, as the particle motion changes depending upon the orientation of the mean gradient with respect to the fracture network.

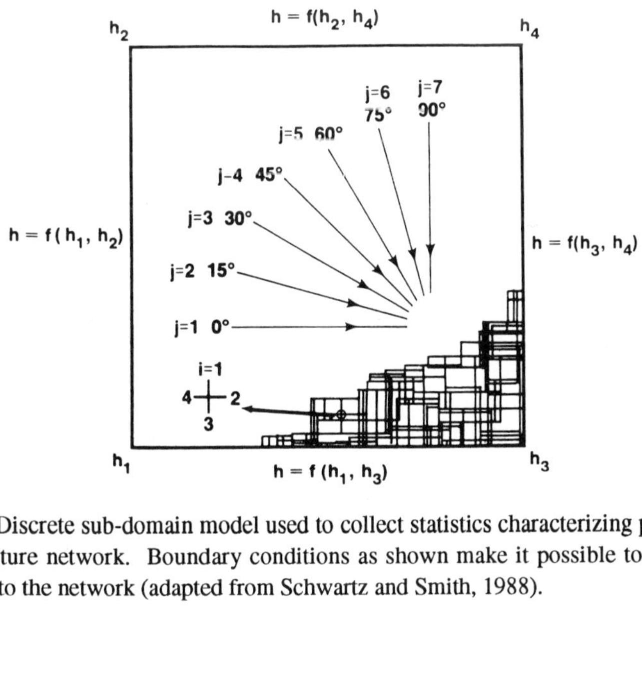

Figure 3.7.2. Discrete sub-domain model used to collect statistics characterizing particle motion within the fracture network. Boundary conditions as shown make it possible to orient flow at various angles to the network (adapted from Schwartz and Smith, 1988).

Statistics characterizing particle motion are collected as the reference particles move through the discrete network. A separate set of statistics is collected each time the orientation of the mean hydraulic gradient is changed relative to the orientation of the fracture sets. The following statistics are collected:

i) frequency distributions on lengths between fracture intersections for the two fracture sets (*ra* and *rb*),

ii) distributions characterizing the frequency of particles choosing one of four possible directions of motion for each of the seven different orientations of the network relative to the gradient (d_{ij}:i=1,4;j=1,7),

iii) frequency histograms on the negative logarithm of the fluid velocity as a function of particle direction (i) and orientation of the hydraulic gradient (j), and

iv) correlation coefficients for particle velocities, as a function of direction of motion.

A lag one correlation for movement in a given direction is defined as the correlation between the value of the particle velocity on the downgradient side of a fracture intersection, with respect to the value of particle velocity the last time it was moving down a fracture segment in the same direction. The lag two correlation coefficient extends back through two fracture segments in the same direction of motion.

Schwartz and Smith (1988) collected discrete sub-domain statistics from a single realization of the sub-domain model. We now use statistics on particle motion based on averages formed from multiple realizations of the discrete sub-domain. This additional step removes noise in the sub-domain statistics that occurs when the sub-domain model is close to the threshold of REV behavior for solute transport.

The Continuum Model

To simulate mass transport in the continuum, it is necessary to first determine how the magnitude and orientation of the hydraulic gradient varies through the region of flow. These data come from a solution of the steady state groundwater flow equation, assuming the domain is a homogeneous porous medium. The local hydraulic gradient is calculated from the resulting flownet.

The steps in mimicking particle motion in the continuum are as follows:

i) The direction that a particle located at point (x,y) will take during part of a time step Δt is determined by sampling the frequency distributions on particle direction (d_{ij}), given the orientation of the local hydraulic gradient in the vicinity of (x,y). Recall the sub-domain statistics are calculated for seven discrete orientations (0°, 15°,.., 90°). To refine the technique, the orientation of the hydraulic gradient at (x,y) can be used to develop a unique frequency distribution d_i. This latter distribution is formed by linear interpolation from d_{ij}. The choice of particle direction (i.e., 1, 2, 3, or 4) is determined by randomly sampling d_i.

ii) Having determined a direction for transport, the appropriate distribution for length (*ra* or *rb*) is sampled to determine how far the particle moves in the selected direction. This distance represents the location of the next decision point at which a possible change in the direction of motion could occur.

iii) By sampling the velocity histograms it is possible to determine the travel time to the next hypothetical fracture intersection. Particle velocities are sampled by first fitting three-parameter gamma distributions to the histograms obtained from the discrete sub-domain model. The correlation in particle velocity for each of the possible directions of motion is represented by sampling from second order autoregressive models, given the correlation structure estimated in the discrete sub-domain model.

These steps of moving the particle and accumulating values of travel time over each fracture segment are repeated until the particle completes the time step Δt, and another particle is transported. One time step is completed when all particles are moved. Note the motion of individual particles is statistically independent of one another.

Example Applications

The approach can be demonstrated by collecting sub-domain statistics from a discrete model at one scale, and then comparing a second discrete network simulation and a statistical continuum prediction in a larger-scale domain. This larger-scale domain provides data not unlike a field or laboratory experiment. The network used for this purpose was illustrated earlier in Fig. 3.5.1. The sub-domain used to collect particle statistics is 10 m by 10 m. This diagram illustrates the kind of open fracture density needed within the sub-domain to collect representative statistics on particle motion. Figure 3.5.2 shows a comparison of the mass distribution in the full-domain, discrete network with the statistical continuum prediction, for T = 35 days. Several features should be noted. First, the continuum nature of the simulation is obvious, the plume in the lower plot represents an "average prediction" of the mass distribution, the local character of the mass distribution in the discrete fracture model is not preserved. By taking fracture porosity into account, this plot could be reinterpreted as an estimate of ensemble mean concentrations. Second, a reasonable match is apparent between the discrete network simulation and the continuum prediction. The skew in the mass distribution is preserved. Comparison of the first and second moments of the mass distribution shows a reasonable agreement. Moments of the statistical continuum plume are $x_c = 9.6$ m, $z_c = 0.0$ m, $s_{xx} = 8.4$ m^2, $s_{zz} = 1.1$ m^2. Recall that the spatial moments of the discrete network plume are $x_c = 11.0$ m, $z_c = 0.5$ m, $s_{xx} = 7.6$ m^2, $s_{zz} = 0.9$ m^2. Because the full-domain discrete network simulation retains a sensitivity to the unique arrangement of fractures, it is expected that the first and second moments will not replicate the ensemble averages given by the continuum prediction. A comparison should instead be made on the basis of moment estimates from multiple realizations of the full-domain discrete simulation. Such a comparison is given in Fig. 3.5.3, which illustrates the temporal evolution of s_{xx}, calculated for both models. The match indicates that the statistics on particle motion collected in the discrete sub-domain model capture the major features of the influence of network geometry on the rates and patterns of spreading.

3.7.3 Dual Permeability Modeling

For some, and perhaps many fractured rock masses, the requirements for conventional continuum approaches may not be met. It might not be possible to identify a length scale for which representative transport coefficients can be defined in a meaningful way. Consider two examples. First, the density of fractures contributing to the active network may be large enough to exceed the percolation threshold for fluid flow, but insufficient to define a REV for transport at the scale of interest. Second, more than one statistical population of fractures forming the network could be present, each with differing length scales. The length scale of the REV must be of sufficient size to encompass a large number of the fractures with the greater length scale. Unfortunately, in these instances, computational limitations in the application of discrete network models do not provide us with a way to tackle this issue, except for near-field simulation.

Clemo and Smith (1989) describe a method with the potential to model far larger volumes of fractured rock than can be accommodated with the discrete approach, without resorting to continuum approximations. Their concept is based on delineating a hierarchy among the fractures forming a network. Those fractures that are deemed more important to the flow and transport process, called primary fractures, are modeled individually. The large majority of the fractures are modeled in groups using a lumped-parameter representation called network blocks. A schematic example of the concept is shown in Fig. 3.7.3. Together the combination of primary fractures and network blocks form a so-called dual permeability model. While smaller scale fractures submerged within network blocks are individually of less importance on a flux-weighted basis, their collective effect is on par with that of the primary fractures. This method is well-suited to modeling solute transport in fracture networks where fractures have a broad range of length scales.

Network blocks are an approximation to the effects of a group of adjacent fractures. They are not a continuum in the sense of an equivalent porous medium; the blocks are sub-REV in size. Each block is different, just as each discrete fracture is different. Network blocks are bounded by primary fractures, which define the geometry of the network blocks, and thus their hydraulic properties. This dual permeability concept is distinct from that of a dual porosity model (see Ch. 8). Dual permeability reflects the division of fracture permeability into two categories (primary fractures and network blocks), and it should not be thought of as two overlapping continua linked by a transfer function as in a dual porosity model.

A dual permeability model for solute transport contains two sub-models. The flow model determines the fluid flux in the network blocks and primary fractures. With the Galerkin finite element method, network blocks can be represented by two-dimensional rectangular elements, primary fractures by line elements. The hydraulic conductance of each network block is calculated knowing the hydraulic properties of the fractures that it replaces. In the transport model these fluid fluxes are used as the basis for tracking a large number of discrete particles through the rock mass. Entrance to a network block is governed by a probability function, based on the relative fluid fluxes in the bounding primary fractures and the cross flow into the network blocks. Mass transfer across network blocks is expressed in terms of stochastic residence time functions on travel time and on exit location to a down-gradient primary fracture. These probabilities are calculated from the statistical parameters of the smaller-scale fracture sets, the orientation of the local hydraulic gradient, and the location of the larger-scale primary fractures which define the geometry of the network blocks. An example illustrating how the residence

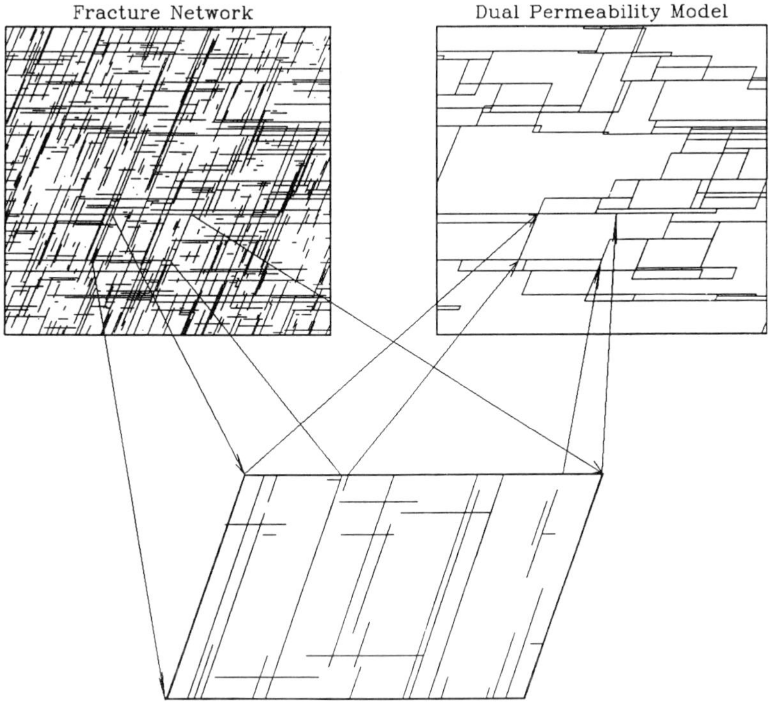

Figure 3.7.3. Creation of a dual permeability model for a fracture network. Primary fractures are retained as individual discrete fractures. Most fractures are submerged into network blocks, as shown for the single block that has been isolated in this schematic representation (adapted from Clemo and Smith, 1989).

time function is determined for a single network block is given in Fig. 3.7.4. The fluid flux within each fracture, as a fraction of the total flow through the network block, gives the probability of a particle traveling across the block in a given fracture. A particle that leaves a primary fracture travels through a network block in a single step.

Figure 3.7.5 illustrates the method. The network shown in plot (a) is one realization from a statistically-defined network composed of two orthogonal fracture sets. The horizontal fractures, while less abundant, have a greater length scale and higher mean aperture than the vertical fracture set (an average length of 6 meters and mean aperture of 50 microns, versus an average length of 1 meter and mean aperture of 10 microns for the vertically-oriented fractures). The vertical scan line density is two fractures per meter, the horizontal scan-line density is ten fractures per meter. The log aperture standard deviation is 0.10 for both fracture sets. The boundary conditions are defined by a uniform hydraulic head gradient of 0.01 at an angle 80° counterclockwise from the x direction. Plot (b) shows the flow distribution in the rock mass.

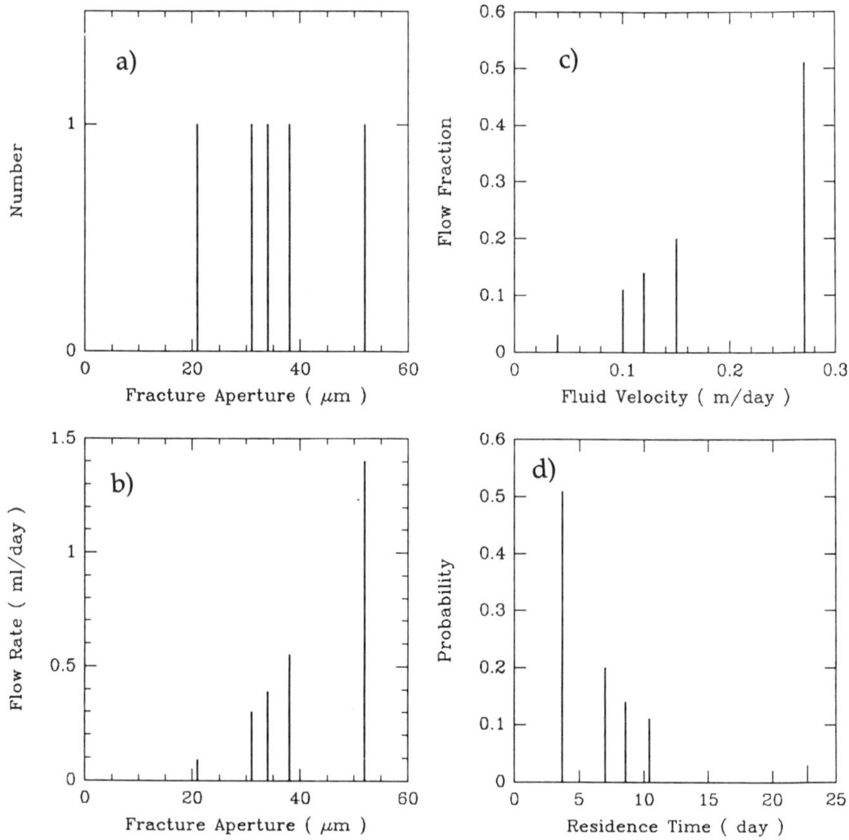

Figure 3.7.4. Example illustrating the creation of the residence time function for a single network block. (a) Distribution of fracture apertures within the network block. (b) Flow rate dependence on fracture aperture, determined by first solving the fluid flow equation at the network scale to determine the flow across the network blocks. (c) Fluid velocity distribution for the network block, computing knowing the flow rates and aperture values. (d) Residence time function determined from the fluid velocity distribution given in (c), and the geometry of the network block (adapted from Clemo and Smith, 1989).

Each fracture segment that is part of the active network is plotted with a line width proportional to the flow rate in the fracture. Fractures that carry less than 3% of the greatest flow rate are all plotted with a line width 3% of the maximum width. Plot (c) in Fig. 3.7.5 shows the flow distribution predicted by the dual permeability model. The dashed lines are the vertical boundaries used to define the network blocks. The solid line plotted in the center of most network blocks represents the total flow through the blocks. Fluid flow through the network blocks is only plotted if it is greater than the 3% threshold. Note that some of the network blocks will be assigned zero conductance if no fractures extend between the bounding primary fractures.

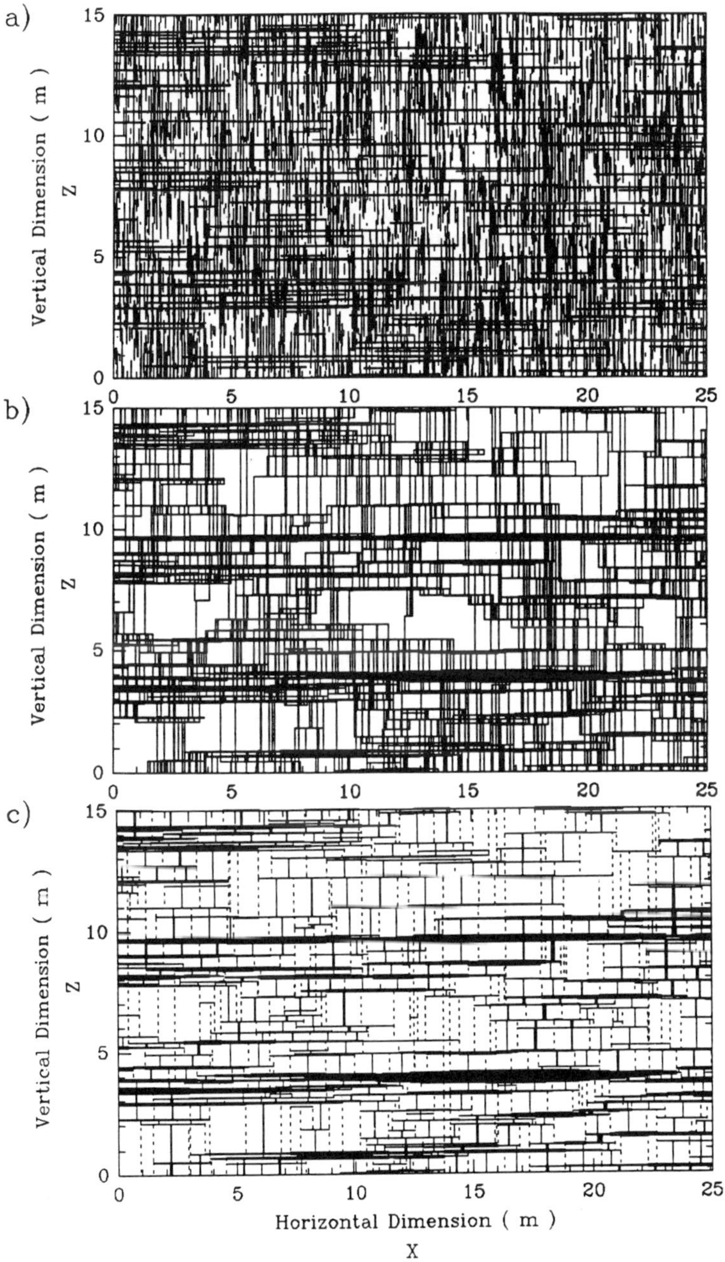

Figure 3.7.5. An example of a dual permeability model. (a) Fracture network. (b) Distribution of fluid flux within the network, where each fracture segment that is part of the active network is plotted with a line width that is proportional to the flow rate in the fracture. See text for details. (c) Fluid flux predicted by the dual permeability model, the dashed vertical lines identify boundaries of the network blocks (adapted from Clemo and Smith, 1989).

Heterogeneity of flow within the domain is evident in Fig. 3.7.5. Many of the horizontal fractures have relatively low flow rates. High flows are seen in large aperture fractures that are well connected to other horizontal fractures. Although individually the vertical fractures rarely have flow rates larger than 3% of the greatest flow, taken collectively they provide essential connectivity to the network.

Figure 3.7.6 compares the breakthrough curves at the downstream vertical boundary calculated using both a discrete network model, and the dual permeability model. Mass enters the fracture network as a pulse source located at coordinate position (0,2.8). The upper plot shows particle arrival integrated along the entire downstream vertical boundary, the lower plot the

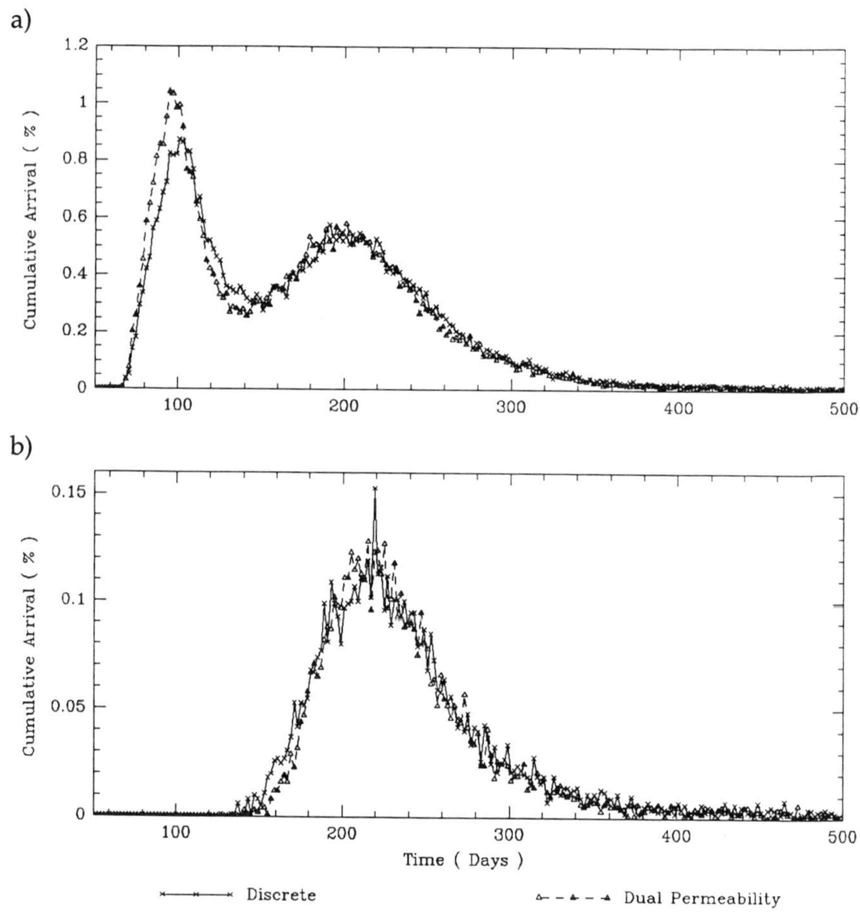

Figure 3.7.6. Comparison of breakthrough curves formed at the downstream vertical boundary, for the discrete fracture and dual permeability simulations. (a) Particle arrival integrated along the entire downstream vertical boundary. (b) Particle arrival at a specific fracture location crossing the downstream vertical boundary, $z = 9.6$ m (adapted from Clemo and Smith, 1989).

arrival time distribution for the fracture crossing the vertical downstream boundary at (25.0,9.6). The match is considered to be excellent. Considerably less computational effort was needed to obtain the dual permeability results. The flow component of the discrete network model required the solution of 7430 equations, the dual permeability model required the solution of 1392 equations. Particle tracking in the dual permeability model is also less computer intensive, although this is less of a factor than the efficiencies in determining the fluid fluxes. It is these efficiencies, coupled to the concept of sub-REV simulation and identification of primary fractures, that hold considerable promise in moving toward field-scale simulation.

Our most recent experience with the dual permeability model indicates that for fracture systems with a more complex geometry than that shown in Fig. 3.7.5, it is necessary to modify the techniques used to solve for fluid flow and mass transfer in the dual permeability model. A stiffness matrix is now built that includes midpoint nodes in the primary fractures to accept the volumetric flow from the network block which it bounds. A method similar in concept to domain decomposition is used to solve for the hydraulic head distribution. For each network block, probabilities on exit locations and arrival times are determined using the method of Shimo and Long (1987) to model mass transfer through all the possible pathways across a network block.

3.8 CLOSING COMMENTS

Even in the simplest case of a nonreactive solute in a steady state flow system, important questions are unanswered at the field scale as to the exact character of solute migration. Complexity arises from both the indeterminate structure of a fracture network, and the wide range of scales over which transport processes interact. Because fractures typically occur in a limited number of sets with preferred orientations, dispersion exhibits an anisotropic behavior. To make matters even more troublesome, it is not yet clear whether the conventional approach built on the concept of a representative elemental volume provides a generally suitable basis for formulating mass transport models.

Far-field simulation will by necessity be based on some form of spatial averaging. In favorable circumstances, it may be possible to recognize the length scale characteristic of the REV, and to estimate appropriate volume-averaged transport coefficients. This case requires that the scale of the domain of interest, and the scale on which the fracture network has statistically uniform properties, exceed the length scale required for solutes to sample a representative set of fluid velocities within the open fracture network. Given that this is the case, several approaches are possible. Conventional transport models adopting porous medium approximations are based on an advection-dispersion equation. The analytic form of the dispersion tensor, expressed in terms of the orientation of the hydraulic gradient relative to the orientation of the fracture sets, is known only for a highly-idealized geometry. Alternatively, for networks of randomly-located, finite-length fractures, it may be possible to derive values for an asymptotic dispersion tensor from single borehole and cross-hole hydraulic tests. This approach leaves unanswered the issue of how best to predict early-time behavior, for transport distances that are less than the scale of the REV.

Rather than formulating a transport model expressed in the form of a partial differential equation, there may be greater advantage to cast the problem in stochastic terms. One such approach was discussed in this chapter. Statistics on particle motion observed in a discrete sub-domain of the network can be used to incorporate the influence of fracture geometry in a larger-scale continuum simulation. In this way, the dispersive properties of a network can be related to its geometric characteristics, as expressed in the probability distributions on fracture frequency, location, length, and aperture. This is the type of data that may be most readily available in site-specific applications. These models preserve the statistical character of mass transport (i.e., the ensemble average), and do not replicate the movement that would be observed in a single realization of the network at comparable length scales.

For some fractured rocks, the requirements of conventional continuum models may not be met. No firm evidence exists whether it will be possible or not, at the field scale, to identify a length scale for which transport coefficients can be defined in a meaningful way. Stochastic techniques, such as the dual permeability formulation outlined in this chapter, appear of potential value in dealing with these sub-REV problems. The dual permeability method is a hybrid approach suited to rock masses where fractures occur on a range of length scales. The essential features of the dual permeability approach are the separation of fractures or fracture sets on a hierarchical basis, and the use of stochastic residence time functions to describe mass transfer between the larger-scale, discrete fractures within the network. These stochastic concepts provide a framework to tackle questions related to moving from observations at one scale, to transport predictions on a larger scale.

ACKNOWLEDGMENTS

Permission to reproduce unpublished results by Curt Naumann and Tom Clemo is gratefully acknowledged. Tom Clemo also provided assistance in updating results of our two-dimensional discrete network simulations.

REFERENCES

Andersson, J., Shapiro, A.M., and Bear, J. (1984). A stochastic model of a fractured rock conditioned by measured information. *Water Resour. Res.* 20(1), 79–88.

Andersson, J., and Thunvik, R. (1986). Predicting mass transport in discrete fracture networks with the aid of geometrical field data. *Water Resour. Res.* 22(13), 1941–1950.

Brown, S.R. (1987). Fluid flow through rock joints: The effect of surface roughness. *J. Geophys. Res.*, 92(B2), 1337–1347.

Cacas, M.C., Ledoux, E., de Marsily, G., Barbreau, A., Calmels, P., Gaillard, B., and Margritta, R. (1990). Modeling fracture flow with a discrete network model: calibration and validation 2. The transport model. *Water Resour. Res.*, 26.

Castillo, E., Krizeck, R.J., and Karadi, G.M. (1972). Comparison of dispersion characteristics in fissured rocks. *Proc. Symp. on Fundamentals of Transport Phenomena in Porous Media.* Guelph. 778–797.

Clemo, T.M. and Smith, L. (1989). Solute transport in fractured media: Dual permeability models. EOS, 70(43).

Dagan, G. (1987). Theory of solute transport by groundwater. *Ann. Rev. Fluid Mech.*, 19, 183–215.

de Josselin de Jong, G., and Way, C.A. (1972). Dispersion in fissured rock, Unpub. Rept., 30 pp. New Mexico Inst. Mining and Tech., Socorro.

Elsworth, D. (1986). A model to evaluate the transient hydraulic response of three-dimensional sparsely-fractured rock masses. *Water Resour. Res.* 22(13), 1809–1819.

Endo, H. (1984). Mechanical Transport in Two-Dimensional Networks of Fractures. Ph.D. Thesis. University of California, Berkeley. 203 pp.

Endo, H., Long, J.C.S., Wilson, C.R., and Witherspoon, P.A. (1984). A model for investigating mechanical transport in fracture networks. *Water Resour. Res.* 20(10), 1390 —1400.

Hull, L., and Clemo, T. (1987). Dual permeability model: laboratory and FRACSL validation studies. *Geothermal Injection Tech. Prog.* Idaho National Engineering Lab. Annual Progress Report.

Hull, L.C., and Koslow, K. (1986). Streamline routing through fracture junctions. *Water Resour. Res.* 22(12), 1731–1734.

Hull, L., Miller, J., and Clemo, T. (1987). Laboratory and simulation studies of solute transport in fracture networks. *Water Resour. Res.* 23(8), 1505–1513.

Krizek, R.J., Karadi, R.M., and Socias, E. (1972). Dispersion of a contaminant in fissured rock. *Proc. Inter. Soc. of Rock Mechanics Symp. on Perculation Through Fissured Rock,* Stuttgart.

Liggett. J.A., and Medina, D. (1988). Flow in three-dimensional fracture networks using a discrete approach. *Groundwater Flow and Quality Modeling*, E. Custodio et al. (eds.), 363–374.

Long, J., Gilmour, P., and Witherspoon, P.A. (1985). A model for steady state fluid flow in random, three-dimensional networks of disk-shaped fractures. *Water Resour. Res.* 21(8), 1105–1115.

Long, J. and Witherspoon, P.A. (1985). The relationship of the degree of interconnection to permeability in fracture networks. *Jour. Geophys. Res.*, 90(B4), 3087–3098.

Neuman, S.P. (1987) Stochastic continuum representation of fractured rock permeability as an alternative to the REV and fracture network concepts. *Proc. 28th U.S. Rock Mechanics Sympos.*, Balkema Press, 533–561.

Neuman, S.P., Winter, C.L., and Newman, C.M. (1987). Stochastic theory of field-scale Fickian dispersion in anisotropic porous media. *Water Resour. Res.*, 23(3), 453–466.

Neuman, S.P., and Depner, J.S. (1988). Use of variable-scale pressure test data to estimate the log hydraulic conductivity and dispersivity of fractured granites near Oracle, Arizona. *Jour. Hydrol.*, 102, 475–501.

Neretnieks, I. (1983). A note on fracture flow dispersion mechanisms in the ground. *Water Resour. Res.*, 19(2), 364–370.

Novakowski, K., Evans, G.V., Lever, D.A., and Raven, K.G. (1985). A field example of measuring hydrodynamic dispersion in a single fracture. *Water Resour. Res.* 21(8), 1165–1174.

Philip, J. (1988). The fluid mechanics of fracture and other junctions. *Water Resour. Res.* 24(2), 239–246.

Rasmuson, A. (1985). Analysis of hydrodynamic dispersion in discrete fracture networks using the method of moments. *Water Resour. Res.*, 21(11), 1677–1683.

Raven, K.G., Novakowski, K.S., and Lapcevic, P.A. (1988). Interpretation of field tracer tests of a single fracture using a transient solute storage model. *Water Resour. Res.*, 24(12), 2019–2032.

Robertson, M. (1990). A Statistical Continuum Approach For Mass Transport in Fractured Media. M.A.Sc. thesis, University of British Columbia, 141 pp.

Robinson, P. (1984). Connectivity, flow, and transport in network models of fractured media. Ph.D. Thesis, Oxford Univ., London.

Robinson, B.A. and Tester, J. (1984). Dispersed fluid flow in fractured reservoirs: An analysis of tracer determined residence time distributions. *J. Geophys. Res.* 89(B12), 10374–10384.

Robinson, J.W. and Gale, J.E. (1990). A laboratory and numerical investigation of solute transport in discontinuous fracture systems. *Ground Water*, 28(1), 25–36.

Schwartz, F.W., Smith, L., and Crowe, A.S. (1983). A stochastic analysis of macroscopic dispersion in fractured media. *Water Resour. Res.* 19(5), 1253–1265.

Schwartz, F.W. and Smith, L. (1988). A continuum approach for modeling mass transport in fractured media. *Water Resour, Res.* 24(8), 1360–1372.

Shapiro, A.M. and Andersson, J. (1985). Simulation of steady state flow in three-dimensional fracture networks using the boundary element method. *Adv. in Water Resour.* 8(3), 106–110.

Shapiro, A.M., and Nicholas, J.R. (1989). Assessing the validity of the channel model of fracture aperture under field conditions. *Water Resour. Res.* 25(5), 817–828.

Shimo, M. and Long, J.C.S. (1987). A numerical study of transport parameters in fracture networks. *Flow and Transport Through Unsaturated Fractured Rock.* Amer. Geophy. Union Geophys. Mono. 43, 121–131.

Smith, L., Berkowitz, B., and Naumann, C. (1992). An analysis of solute transport within fracture intersections. Manuscript in preparation.

Smith, L. and Schwartz, F.W. (1980) Mass transport 1. A stochastic analysis of macroscopic dispersion. *Water Resour. Res.*, 16(2), 303–313.

Smith, L., and Schwartz, F.W. (1984). An analysis of the influence of fracture geometry on mass transport in fractured media. *Water Resour. Res.* 20(9), 1241–1252.

Smith, L., Mase, C.W., and Schwartz, F.W. (1985). A stochastic model for transport in networks of planar fractures. *Proc. Inter. Assoc. Hydraul. Res. Symp. on the Stochastic Approach to Subsurface Flow*, Paris. 523–536.

Smith, L., Mase, C.W., and Schwartz, F.W. (1987). Estimation of fracture aperture using hydraulic and tracer tests. *Proc. 28th U.S. Symp. on Rock Mechanics*, Balkema Press, 453–463.

Way, S.C. and McKee, C.R. (1981). Restoration of in situ coal gasification sites from naturally occurring flow and dispersion. *In Situ*. 5(2), 77–101.

Wilson, C., and Witherspoon, P.A. (1976). Flow interference effects at fracture intersections. *Water Resour. Res.* 12(1), 102–104.

4

STOCHASTIC MODELS OF FRACTURE SYSTEMS AND THEIR USE IN FLOW AND TRANSPORT MODELING

Jean-Paul Chilès
Bureau de Recherches Géologiques et Minières
Orléans, France

Ghislain de Marsily
Université Pierre et Marie Curie
Paris, France

Fluid flow studies of fractured rocks require three-dimensional modeling of the fracture system. The complexity of reality does not allow a complete description of the actual field. It is therefore necessary to use stochastic models. The first task is the statistical study of the fracture population: identification of sets and distribution of orientation, size, and aperture. The second one is the choice of a network model: random planes, random discs, or models derived by mixing, clustering and regionalization of parameters. The third step is the study of the geometry of the single fracture: the shape of the fracture walls and of the channels. The fourth step is the derivation of the equations for fluid flow and solute transport in the network model. These various aspects are reviewed, including the conditioning of 3D simulations to the data. Emphasis is laid on the determination of the parameters by means of morphological, statistical, geostatistical and fractal tools that can be used with 1D information (boreholes) or 2D data (outcrops and drift walls); the use of hydraulic testing in boreholes and of tracer tests for determining flow and transport parameters is also presented. The link between geometry, connectivity, percolation theory, and flow is examined. Stationarity and ergodicity assumptions used for modeling stochastic fracture systems are discussed.

4.1 INTRODUCTION

The rock mechanics, flow and transport properties of hard rocks depend mainly on the characteristics of the fracture field, which ought to be considered separately from the matrix. Applications are increasingly complex, and some involve a complete description of the fracture field (e.g., Long, 1986). As fractures are not directly accessible in their entirety, but only at their intersections with boreholes, outcrops or drifts, one of the first steps in the study of a given site is modeling and simulation of the fracture field.

This modeling can be done at various scales, e.g., large faults, fractures, and microfissures. This chapter deals with the scale ranging from tens of centimeters to tens of meters. Large faults (up to a kilometer) are not found very often and are assumed to have been identified by means of a deterministic approach. In contrast, microfissures (millimetric or centimetric in scale) are considered as a property of the matrix, and their influence on flow is summarized through an equivalent continuous medium. Thus, our aim is modeling at the metric scale. Of course, the method is general and can easily be transposed to another scale (e.g., at the microfissure scale) in order to define an equivalent continuous medium.

Whatever the scale, two distinct features can be studied:

i) the fracture network: each element is considered as relatively simple (e.g., a planar fracture) and stress is laid on the imbrication of the various elements;

ii) the single fracture: generally, an open fracture is not a pair of parallel plates: there are voids and contact areas that depend on stress, and this has an enormous influence on conductivity (Feuga, 1986).

The main part of this chapter (sections 4.2 to 4.10) is focused on the fracture network. A 3D domain, even limited, can correspond to several thousand or over a million fractures. Few of these fractures intersect the boreholes, outcrops, or drift walls where the network is observed, which is the reason why deterministic modeling is not possible; hence stochastic modeling will be considered. The mathematical morphology approach is the reference, but simpler methods are usually used. Statistical tools describe the fracture population. Geostatistical tools and fractal methods are used to study the regionalized behavior of the network. The combination of these approaches leads to a realistic modeling and simulation of the 3D network. All these aspects of modeling are presented, with a final section devoted to the conditioning of a 3D simulation by observed 1D or 2D fracture intersections.

Section 4.11 is devoted to the single fracture. The approach can be global (definition of an average permeability) or local (2D modeling of both surfaces and void, using geostatistical or fractal methods). Both are presented. Section 4.12 briefly examines the link between fracture field, connectivity, and flow. The relationship between percolation theory and connectivity is introduced in section 4.13.

Section 4.14 describes how a discrete fracture flow model is built, calibrated, and used, and section 4.15 addresses the same issues for a transport model. Finally section 4.16 describes another stochastic fracture model based on the representation of a continuous equivalent medium.

In the study of the fracture network, fractures are considered as planar. This is, of course, a simplification, but one that does not usually modify the conclusions on connectivity or stability. According to the scale, they will be considered as either infinite or convex, so that fracture traces are either straight lines or segments.

4.2 MATHEMATICAL MORPHOLOGY APPROACH

Mathematical morphology (Matheron, 1975; Serra, 1982) provides a means of fully characterizing a fracture network. As we are considering stochastic models, let us denote by A the (random) set of all the fractures (fracture walls and void spaces). If we neglect the aperture of the fractures, this set will be fully determined by the knowledge of the functional moment $Q(B)$ for any (nonrandom) set B, where $Q(B)$ is defined by:

$$Q(B) = \text{Prob}\{B \cap A = \varnothing\} = \text{Prob}\ \{B \text{ does not intersect any fracture}\} \tag{4.2.1}$$

This characterization is the exact equivalence of the spatial law for random functions. The random set A is usually assumed to be stationary. $Q(B)$ is then unchanged by translation of B and can be estimated from the sole realization of the fracture network.

In practice it is not possible to consider any set B. Conrad and Jacquin (1973) used this approach to study a 2D network, with B being a segment, a disc, or a crown. The moments $Q(B)$ were measured with a texture analyzer. This is a very powerful method, which deserves to be rediscovered. Its implementation for 3D studies is not obvious and hence, other tools, less powerful but simpler, are usually used. Before presenting them, it is necessary to explain how data are collected.

4.3 TYPES OF SURVEY

There are three main types of survey:

4.3.1 Borehole Surveys

The measurements are usually made on oriented cores or deduced from resistivity logs. For each intersected fracture, one notes its location, orientation, existence of possible fill (nature, thickness), aperture, surface geometry, etc. Borehole measurements give information about the distribution of fracture orientation, the fracture density and its regionalization, the presence of fracture clusters. But only 2D surveys allow the evaluation of the fracture size, which is a major parameter in the study of the mechanical and hydraulic properties of the rock.

4.3.2 Scanline Surveys

This is an extension of the preceding technique to a 2D area. As for boreholes, only the fractures intersecting a line (usually a horizontal one), painted on an outcrop or on a batter, are collected. But each fracture is now represented by its trace, which can be measured (Fig. 4.3.1.a). This method is commonly used for pit batters and high bench walls that make it possible to observe long traces. The scanline is established about one meter above the bench level and, due to the presence of debris on the bench area, only the portion of trace occurring above the scanline is visible and measured (half-trace line sampling). If the batter is clean, it is of course better to also measure the length of the fracture below the line (full-trace sampling). In comparison with a borehole survey, two additional observations are required for each trace intersecting the line:

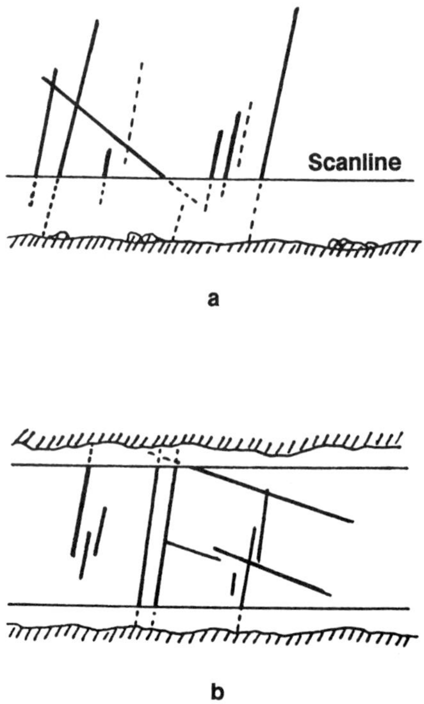

Figure 4.3.1. 2D survey of fracture traces: a) scanline survey; b) areal survey.

i) the coordinates (on the wall) of the apparent termination above the line (the half trace length will be deduced from this measurement and from the location of the intersection with the line); and

ii) whether or not the trace extends beyond the top of the batter.

Additional qualitative observations are noted, such as the termination type (in rock, or at the contact with another fracture) or the fracture relays. Vertical lines are also surveyed in order to collect data concerning subhorizontal fractures.

4.3.3 Areal Surveys

All traces intersecting a rectangle are collected (Fig. 4.3.1.b). The traces are censored to the rectangle, which means that if they extend beyond the rectangle, only their inner portion is considered. This method is recommended for drift or tunnel walls, but can be used for outcrops. As the height of the rectangle is limited to the wall height (commonly around two meters), the available information is reduced, and this type of survey makes it possible to retain all the available information. The same parameters are sampled as for a scanline survey, but the fracture is now represented by its full—possibly censored—trace; its location is characterized by the coordinates of it two endpoints within the surveyed surface. The visible length of the trace can be deduced from these data. For each endpoint it must also be noted if it is really a termination of the fracture trace, or if the trace extends beyond the surveyed surface. The rectangle must be carefully delimited, for it is tempting not to collect traces that are at the foot or at the top of the drift.

4.4 STATISTICAL STUDY OF THE FRACTURES

4.4.1 Fracture Sets

The orientation of fractures, though not regular, is not purely random. Usually, many of the fractures observed in a single outcrop are approximately parallel to one or several planes. These fractures, which have approximately the same orientation, constitute a fracture set. The existence of sets is due to the fact that the orientation of fractures is related to tectonic history. The first tectonic event has created two or more conjugate sets of fractures, whereas the last events have usually reactivated previous fractures. It is best to study the various sets separately, as individual sets are expected to be more homogeneous than combined sets (e.g., the size or the aperture can differ from one set to another). A first task is to identify the various sets. This may not be easy, in particular because of the geometric bias associated with the data. This problem is first introduced below together with the definition and calculation of the fracture density, and will be developed later.

4.4.2 Fracture Density

The 3D fracture density d_3 (of all the fractures or of a given set) is the average fractured surface per unit rock volume. Similarly, the 2D fracture density d_2 on a planar surface is the average fractured length (of traces) per unit surface area; the 1D fracture density d_1 along a straight line is the average number of fracture intersections per unit length. Both d_1 and d_2 depend on the orientation of the line or plane considered, except in the case of isotropy of the fracture network. Therefore, the most interesting parameter is d_3. It can be obtained from a survey with boreholes or scanlines, with the help of a weighting of the observed fractures. In order to determine the correct weights, it is sufficient to consider the surveyed straight line of length L as a cylinder with a small section s (as is the case for a borehole). If n fractures intersect the line, and if the ith fracture intersects it with an acute angle θ_i, then the fracture surface within the cylinder is $s/\sin\theta_i$. Hence we get:

$$d_3 = \frac{1}{Ls}\sum_{i=1}^{n}\frac{s}{\sin\theta_i} = \frac{1}{L}\sum_{i=1}^{n}\frac{1}{\sin\theta_i}$$

Similarly, for an areal survey, the plane can be considered as a very thin layer with thickness e. The surface, within the layer, of a fracture of trace length l_i within the surveyed rectangle, and acute angle θ_i with the plane, is $el_i/\sin\theta_i$. Hence, denoting the surveyed area as S, and the number of fractures intersecting the surveyed rectangle as n we get:

$$d_3 = \frac{1}{Ls}\sum_{i=1}^{n}\frac{s}{\sin\theta_i} = \frac{1}{L}\sum_{i=1}^{n}\frac{1}{\sin\theta_i}$$

(N.B.: Many authors also use the $1/\sin\theta_i$ weighting for an areal survey; this is only acceptable when one considers a set of infinite parallel fractures, for which l_i is constant, except at the edges.)

If all the fractures are parallel, the angle θ_i is constant and can be denoted as θ. The link between 1D and 3D fracture densities is then $d_1 = d_3 \sin\theta$. A similar relationship between 2D and 3D fracture densities exists. In all cases, the apparent density d_1 or d_2 is smaller than the true density d_3. The only exception is the case where the borehole, outcrop or wall is orthogonal to the fracture set.

In the case where the fracture orientations are purely random, the mean value of $\sin\theta$ between a fracture plane and a survey line is 1/2. It is $\pi/4$ between a fracture plane and a survey plane. The 1D, 2D, and 3D fracture densities are then linked by the relations:

$$d_1 = \frac{1}{2}d_3 \qquad d_2 = \frac{\pi}{4}d_3$$

4.4.3 Definition of the Orientation

a) Strike and Dip

The orientation of a fracture is usually expressed by its strike and dip. The strike is the trace of the intersection of the fracture with a horizontal plane; its direction can be specified by its azimuth, counted in degrees clockwise from the north. For example, the azimuth 300° corresponds to the direction N60W. The dip (or inclination, or plunge) is the magnitude of the angle between the fracture and a horizontal plane expressed in degrees.

If the strike is oriented so that the dip plunges on the right-hand side, the strike azimuth (a unique value between 0 and 360°) and the dip angle (between 0 and 90°) define the fracture orientation. If the strike is not oriented, and hence is defined up to 180°, it is necessary to also specify the approximate orientation of the dip (N, S, E, W) In the following, the strike is considered as oriented. There are also other systems: some authors use the dip direction instead of the strike (dip direction = strike direction + 90°), others express the angles in grades instead of degrees.

For calculations, it is more convenient to characterize the orientation by a unit vector orthogonal to the fracture plane (the fracture is translated so as to pass through the origin 0). This unit vector is defined either in polar coordinates or by its direction cosines.

b) Polar Coordinates

The polar coordinates of a unit vector OP are usually defined by the angle θ between the z axis and vector OP (colatitude), and by the angle ϕ between the x axis and the projection of OP on the xOy plane (longitude), counted anticlockwise (Fig. 4.4.1). For the unit vector orthogonal to a fracture plane and oriented upwards, the link between strike and dip and the polar coordinates, is the following:

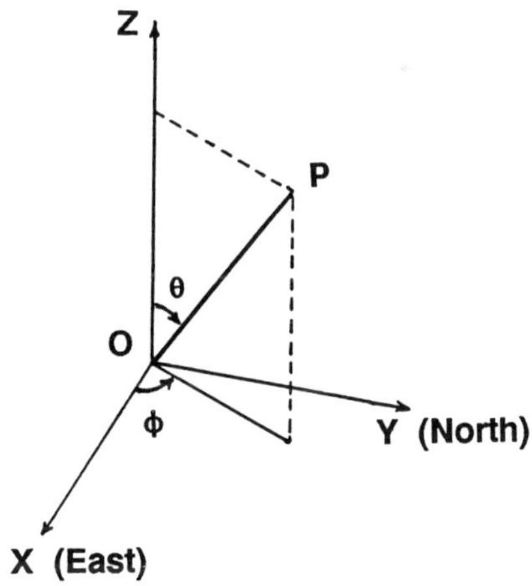

Figure 4.4.1. Definition of the polar coordinates of a unit vector orthogonal to the fracture plane.

$$\theta = \text{dip angle} \qquad \phi = 2\pi - \text{strike azimuth}$$

c) Direction Cosines

The direction cosines (u, v, w) of vector OP are the components of the vector. These components are equal to the cosines of the angles a, b, c, which the vector OP makes with the positive directions of the coordinate axes. The relationship between polar coordinates and direction cosines is the following:

$$u = \sin\theta \cos\phi \qquad v = \sin\theta \sin\phi \qquad w = \cos\theta \tag{4.4.1}$$

4.4.4 Projections and Polar Diagrams

Techniques to represent fracture orientations are well known. A comprehensive presentation can be found in Terzaghi (1965) with references to other authors. The main elements are presented in the following.

a) Poles and Axial Points

In order to visualize and to define the various fracture sets observed in an area, fracture orientations may be represented by points on the projection of a reference sphere with unit radius and center at the origin. One does not represent the actual fracture, but the direction of the normal to the fracture plane. More precisely, the reference sphere diameter that is orthogonal to the fracture plane intersects the reference sphere at two points, one in the lower hemisphere, the other in the upper hemisphere. These points are called the poles of the fracture (Fig. 4.4.2). Obviously,

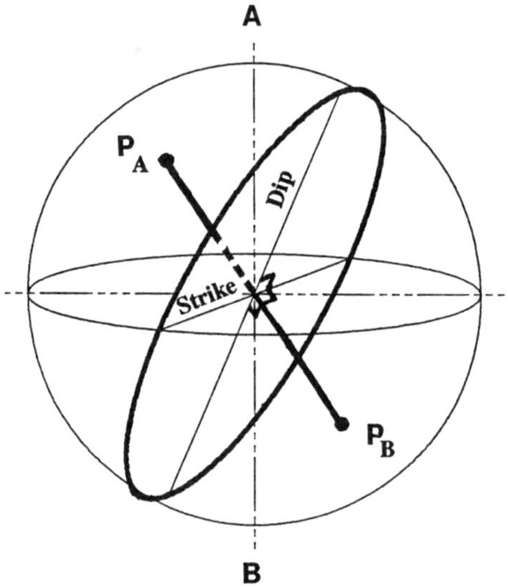

Figure 4.4.2. Upper pole P_A and lower pole P_B of a fracture on the reference sphere.

it is enough to retain only one of them. For the purpose of visualization, we will use the pole of the lower hemisphere, although the other one is sometimes used; the other pole will occasionally be used in subsequent sections.

Linear elements, such as the axis of a drill hole or a drift, may be represented by one of the points at which an identically oriented diameter of the sphere would meet the surface of the sphere. Such a point is called the axial point of the linear element to which it refers. It coincides with the pole of the fractures that are orthogonal to the linear elements.

The poles and axial points located on the surface of the reference sphere may be projected onto a plane by one or several methods. Those most commonly used are the Lambert equal-area and the stereographic projections.

b) Equal-Area Projection and Schmidt Diagram

The Lambert equal-area, or simply equal-area, projection is illustrated by Fig. 4.4.3.a. The projection plane is the plane tangent to the lower point B of the vertical diameter of the reference sphere. The vector joining B to the fracture lower pole P is rotated in its vertical plane in such a way as to become a vector BP' in the projection plane. In the 2D polar coordinates of the projection plane centered on B, the coordinates of P' are:

$$\text{polar distance: } \rho = 2\sin(\theta'/2) \qquad \text{polar angle: } \psi = \varepsilon'$$

$$\text{with } \theta' = \pi - \theta \qquad \phi' = \phi \pm \pi$$

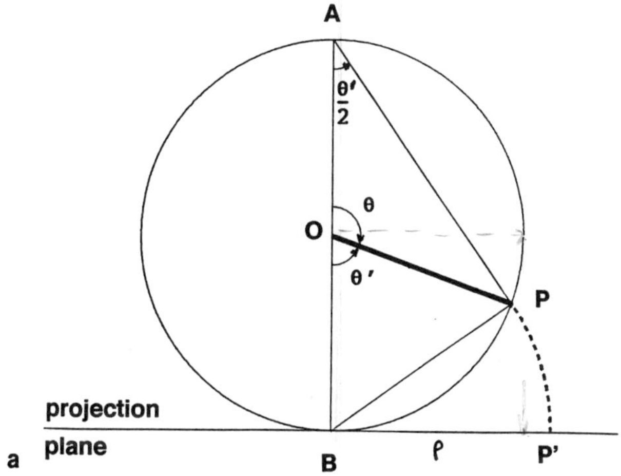

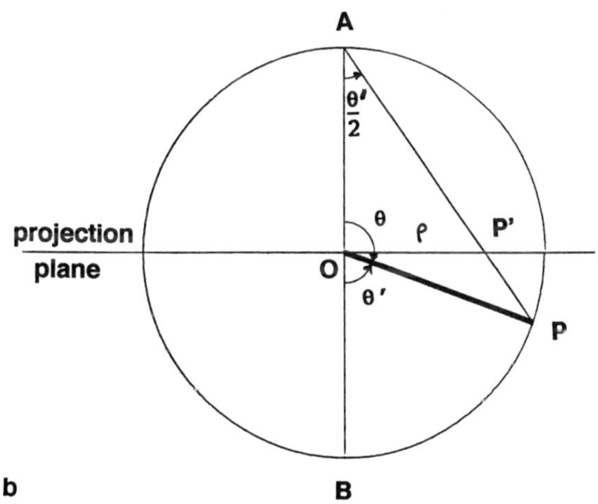

Figure 4.4.3. Pole projections: a) equal-area projection (in the vertical plane of the pole); b) stereographic projection (in the vertical plane of the pole).

In this projection, two equal areas in the projection represent two equal areas in the reference hemisphere. This feature is advantageous in the statistical study of the fracture density. A counterpart of this projection lies in the distortion, which is greatest at the circumference of the projection circle. Plotting the fracture poles according to this projection is called a Schmidt diagram. Poles near the center of the diagram correspond to subhorizontal fractures, and those near the circumference correspond to subvertical fractures. For a lower-hemisphere projection, the left-hand poles represent eastward dipping fractures, whereas right-hand poles represent westward dipping fractures (in case of an upper-hemisphere projection, the diagram would be transformed according to a symmetry around the center of the projection circle, and hence, the opposite applies).

c) Stereographic Projection and Wulff Diagram

The equal-angle stereographic, or simply stereographic projection, is illustrated in Fig. 4.4.3.b. The projection plane is now the horizontal xOy plane to which the center O of the sphere belongs. The projection P' of the fracture pole P is the intersection of the line joining the upper point A of the vertical diameter to the fracture pole P with the xOy plane. In the 2D polar coordinates of plane xOy, the coordinates of P' are:

polar distance: $\rho = tg(\theta'/2)$ polar angle: $\psi = \phi'$

An advantage of the stereographic projection is that it presents no distortion; a small circular area drawn anywhere on the reference sphere is represented by a very nearly circular area in the stereographic projection. However, an area of a given size on the reference sphere is represented in the stereographic projection by an area the size of which increases conspicuously with the distance from the center of the projection circle. Plotting the poles according to a stereographic projection produces a Wulff diagram. Its interpretation is similar to that of a Schmidt diagram. In our domain, the Schmidt diagram is usually preferred, as its equal-area property is very convenient for the statistical study of the fracture density and the delineation of fracture sets.

d) Contoured Polar Diagrams

In general, the results of a fracture survey are most conveniently represented by a contoured polar diagram, as illustrated in Fig. 4.4.4.a. A unit circle (small compared to the reference sphere) is moved systematically from position to position, across the reference sphere. In each position, the number of poles inside the unit circle is counted. This number, or its ratio to the total number of fractures, is recorded on the diagram at the projection of the center of the unit circle. From this new data set, the fracture density is interpolated and plotted. Diggle and Fisher (1985) present variants of this method with numerous references and a program.

e) Geometric Bias

When such a diagram is interpreted, it is necessary to account for the geometric bias resulting from the orientation of the surveyed boreholes or outcrops. Fractures that are normal to the surveyed line or surface occur more often than oblique fractures. This bias can be reduced by collecting data from boreholes, outcrops, or drifts with various orientations. On the contrary, if the orientation diagram corresponds to a single borehole or outcrop, it may be sufficient to represent the axial pole of the surveyed line or the pole of the outcrop and to read the diagram accordingly. In other cases, it is advisable to weight the data. The bias has the same origin as in the study of the 3D fracture density, and hence the weights are the same ($1/\sin\theta_i$ for a borehole or scanline survey, $1_i/\sin\theta_i$ for an areal survey). Figure 4.4.4 presents an example of the difference between a weighted and a nonweighted diagram, used by Courrioux and Jacquot (1984).

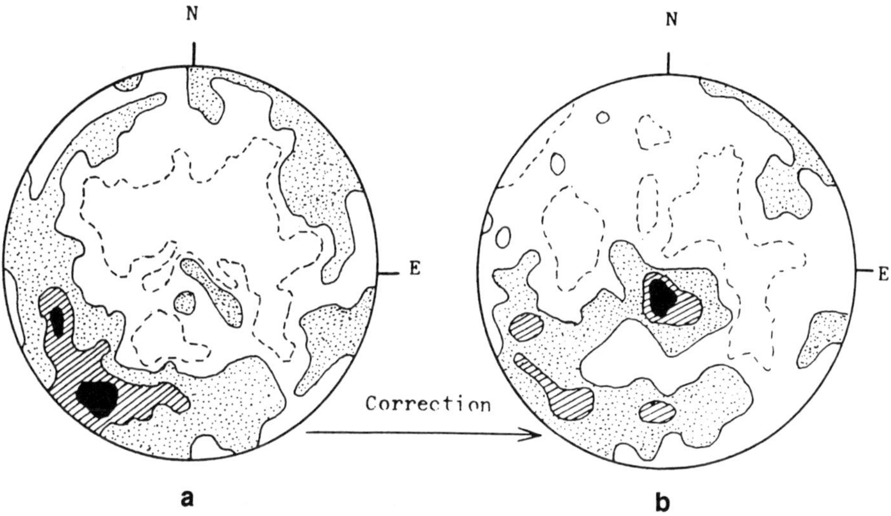

Figure 4.4.4. Contoured polar diagram: a) non-weighted; b) weighted. After Courrioux and Jacquot (1984).

This weighting is valid only under ideal conditions, with a strictly planar outcrop or drift wall, strictly planar fractures, and perfect angle measurements. As these conditions never occur, the corrective factor (inversely proportional to $\sin\theta_i$) has its own uncertainty, which is small if θ_i lies around 90°, but becomes infinite if θ_i equals 0°: as the surface of drift is not strictly planar, it is possible to observe fractures that have the same orientation as the mean drift wall orientation, even if this is impossible in theory. Thus, most authors recommend not using any fracture with θ_i less than 20 or 30 degrees. Otherwise the corrective factor must take account of the actual measurement conditions:

i) the non-planarity of outcrops and fractures can be accounted for in a similar way as the cut effect in petrographic diagrams (Courrioux and Jacquot, 1984). For example, the convex envelope of a fracture trace can be considered as an ellipse with a small eccentricity e (ratio small axis/great axis); in this case, $(\sin^2\theta_i + e^2 \cos^2\theta_i)^{1/2}$ replaces $\sin\theta_i$ in the expression of the corrective factor;

ii) the uncertainty concerning θ_i can be accounted for by randomization according to the measurement precision (see Yow, 1987).

Note that these remarks also apply to the estimation of the 3D density.

4.4.5 Identification of Fracture Sets

The Schmidt or Wulff diagram shows the existence of different sets. Several cluster-analysis methods have been developed to automatically define the various sets (e.g., Bailey, 1975; Schaeben, 1984). However, these methods do not take into account the geometric bias nor the measurement errors, and it is therefore sufficient to define the various sets from a visual inspection of the diagram. Moreover, structural geologists can usually relate fractures to tectonic history, which further helps in the definition of sets.

4.4.6 Modeling of the Orientation Distribution

The statistical modeling of directional data has been studied for a long time. See for example Mardia (1972) and his references to previous authors. The elements necessary for our studies are presented later. The most commonly used models for spherical orientations concerning a fracture set are the truncated Fisher, the Bingham, and the uniform distributions. The expression of these distributions is simpler when the coordinate system, linked with the reference sphere, is changed in such a way that the "new" north pole coincides with the mean direction of the given fracture set.

Change of Reference System

Denote (θ_0,ϕ_0) as the mean orientation in the reference system (Ox,Oy,Oz) with spherical coordinates (θ,ϕ). The change from this system to a reference system (Ox', Oy', Oz') with spherical coordinates (θ',ϕ'), such that the mean orientation coincides with the new north pole axis $\theta' = 0$, can be obtained by a single rotation, of angle θ_0, which leads Oz to Oz'. In this case (θ',ϕ') is defined from (θ,ϕ) by:

$$\begin{aligned}
\sin\theta'\cos\phi' &= \cos\theta_0 \quad \cos(\phi-\phi_0)\sin\theta - \sin\theta_0\cos\theta \\
\sin\theta'\sin\phi' &= \qquad\quad \sin(\phi-\phi_0)\sin\theta \\
\cos\theta' &= \sin\theta_0 \quad \cos(\phi-\phi_0)\sin\theta + \cos\theta_0\cos\theta
\end{aligned}$$

The "new" great circle with longitude $\phi' = 0$ is the "old" great circle with longitude $\phi = \phi_0$.

For the reverse transformation, a rotation of angle $\theta'_0 = -\theta_0$ around the Oy' axis would lead the Oz' axis to the "old" north pole axis, and give θ, $\phi - \phi_0$. Hence (θ,ϕ) is defined from (θ',ϕ') by:

$$\begin{aligned}
\sin\theta\cos(\phi-\phi_0) &= \cos\theta_0 \quad \cos\phi'\sin\theta' + \sin\theta_0\cos\theta' \\
\sin\theta\sin(\phi-\phi_0) &= \qquad\quad \sin\phi'\sin\theta' \\
\cos\theta &= -\sin\theta_0 \quad \cos\phi'\sin\theta' + \cos\theta_0\cos\theta'
\end{aligned}$$

Estimation of the Mean Direction

There are sophisticated methods for estimating the mean direction (θ_0,ϕ_0), e.g., Mancktelow (1981). Because of the uncertainties associated with the definition of the "direction" of a fracture and with the measurement itself, and due to the geometric bias (usually not taken into account in the sophisticated methods), it is sufficient to identify the mean direction with the direction of the weighted sum of all the unit vectors orthogonal to the fractures of the set (use relations 4.4.1 to compute the components of the unit vectors). In principle, each unit vector must be weighted according to the geometric bias corrective factor. In the study of undirected orientations, the choice of the direction of each unit vector must be consistent: starting from an approximate value of (θ_0, ϕ_0), each (θ_i, ϕ_i) is chosen such that both associated unit vectors form an acute angle (if this is not the case, take π-θ_i and $\phi_i \pm \pi$).

Uniform Distribution

If the directions are uniformly distributed on the upper hemisphere, the probability density function (pdf) associated with a surface element *ds* is $ds/(2\pi)$. Expressed in spherical coordinates, as the surface associated with $d\theta$ and $d\phi$ is $ds = \sin\theta \, d\theta \, d\phi$, the bivariate pdf of the polar coordinates θ, ϕ referred to the reference sphere is:

$$f(\theta,\phi) = \frac{1}{2\pi}\sin\theta \qquad 0<\theta<\pi/2 \qquad 0<\phi<2\pi$$

θ and ϕ are independently distributed, ϕ is uniformly distributed on $[0,2\pi]$, and $\cos\theta$ uniformly distributed on $[0,1]$.

Fisher Distribution

This is an isotropic nonuniform distribution. When referred to new axes Ox', Oy', Oz', such that Oz' is the pole of the mean direction, the bivariate pdf of the polar coordinates (θ', ϕ') referred to the reference sphere is:

$$f(\theta',\phi') = \frac{1}{2\pi}\frac{k\sin\theta'}{e^k - e^{-k}}\exp(k\cos\theta') \qquad 0<\theta'<\pi \qquad 0<\phi'<2\pi$$

k is a concentration parameter: for $k = 0$, the mass is uniformly distributed; for large k, the mass is concentrated on a small portion of the sphere around the pole of the mean direction.

In our case, undirected orientations are considered, and the Fisher distribution must be truncated to $\theta' = \pi/2$ (it is then known as the Arnold distribution). In this case, the multiplying factor $k/(e^k - e^{-k})$ must be replaced by $k/(e^k - 1)$. It can be truncated to a lower value θ'_0, with the corresponding multiplying factor $k/(\exp(k) - \exp(k\cos\theta'_0))$

From the expression of the pdf, it follows that θ' and ϕ' are independent random variables: ϕ has a uniform distribution on $[0,2\pi]$, and $T = 1 - \cos\theta'$ has a truncated exponential distribution $ke^{-kt}/(1 - e^{-kt_0})$, where $t_0 = 1 - \cos\theta'_0$ is the point of truncation (usually $\theta'_0 = \pi/2$, and hence $t_0 =$ 1). As k is usually high, the exponential tail probability beyond the truncation point, e^{-kt_0}, is negligible, and the distribution of T can be regarded as exponential, with mean k. This provides an easy estimation of k, and enables the user to check whether the bivariate distribution of (θ,ϕ) can be considered as truncated Fisher (see Lewis and Fisher, 1982).

Bingham Distribution

Expressed in the polar coordinates (θ',ϕ'), it has the bivariate pdf:

$$f(\theta',\phi') = \frac{1}{2\pi}\frac{\sin\theta'}{M_k}\exp(k\cos^2\theta') \qquad 0<\theta'<\pi/2 \qquad 0<\phi'<2\pi$$

with

$$M_k = \int_0^1 e^{kt^2}\, dt$$

More general forms exist of the Bingham distribution, which can model anisotropic distributions (see Mardia, 1972).

Gaussian Distribution in a Projection Plane

It is also possible to project (θ', ϕ') in the plane tangent to the north pole (upper hemisphere projection) by using an equal-area projection, a stereographic projection or simply the intersection of the direction (θ', ϕ') with this tangent plane (projection from the sphere center). The projection has in this case the following 2D polar coordinates:

polar distance:	equal-area projection:	$\rho = 2\sin(\theta'/2)$
	stereographic projection:	$\rho = tg(\theta'/2)$
	projection from center:	$\rho = tg\theta'$
polar angle:		$\psi = \phi'$

and the Ox axis in the projection plane is the great circle with longitude $\phi' = 0$.

The advantage of the last projection lies in that the half-hemisphere is represented by the whole plane. So, as suggested by Grossman (1985), it is possible to model the orientation with a bivariate Gaussian distribution, which is easy, even in case of anisotropy.

4.4.7 Trace Length and Fracture Size

If fractures cannot be considered as infinite at the scale of the study, the distribution of their size largely determines the frequency of intersections and hence the mechanical and hydraulic behavior of the rock mass (existence of distinct blocks, flow, etc.). For a fixed 3D fracturation density, there is a better connectivity with a few large fractures than with many small ones. In practice, it is not possible to observe the exact size of a fracture. A 2D survey, however, provides the trace-length distribution, which is linked with the fracture-size distribution, but the experimental histogram of trace length is affected by several biases that have to be corrected.

Biases of the Experimental Trace-Length Distribution

As pointed out by several authors (see Baecher and Lanney, 1978), there are three main causes of bias:

i) **Censoring.** When one or both terminations of a trace are not observable, as is often the case on drift walls for subvertical fractures, the length recorded is a censored length, being shorter than the true length; hence, a reduction of the histogram dispersion.

ii) **Truncation.** Traces that are shorter than some cut-off length are not recorded, either because they are considered of secondary importance or because it is difficult to discriminate between natural fractures and fractures due to mining when they are small; hence, a truncation of the histogram.

iii) **Size.** First consider a horizontal scanline survey of fractures with the same orientation; if the level of the scanline is randomized, it clearly appears that the probability of intersecting a given trace is proportional to its length. In the calculation of the experimental histogram, each trace appears as weighted by its length: it is a histogram in measure and not in number (see e.g., Serra, 1982). In the case of an areal survey, the weighting is more complex, but the phenomenon remains.

Derivation of the True Trace-Length Distribution

These biases are well-known, and methods for deriving the true (in number) mean or the underlying distribution have been developed (Cruden, 1977; Baecher and Lanney, 1978; Pahl, 1981; Laslett, 1982, who presents most methods). Considering an areal survey, both Pahl and Laslett separate the traces into three classes according to the number of visible terminations (0, 1, or 2). Pahl proposes a distribution-free estimation method of the mean trace length, whereas Laslett presents a parametric method (maximum likelihood) for the estimation of the trace length distribution. Pahl's method can be easily extended to the estimation of the distribution (Massoud, 1987); here this last method is presented.

Consider a set of traces with the same apparent (2D) orientation (if the orientation varies, the result will simply have to be randomized), and consider the areal survey of a drift wall within a rectangle of length L and height H. More precisely, in order to avoid edge effects, the survey area is a parallelogram of height H, whose lateral sides are parallel to the considered traces (Fig. 4.4.5). Note:

θ	angle between the base-line and the traces
τ_2	density of trace centers within the rectangle plane
$F(l)$	cumulative distribution function (cdf) of trace length (in number)
$f(l)$	probability density function (pdf) associated with $F(l)$
N_0	number of traces with 0 visible termination (all with length $a = H/\sin\theta$)
$N_1(dl)$	histogram of censored length of traces with 1 visible termination
$N_2(dl)$	histogram of length of traces with 2 visible terminations

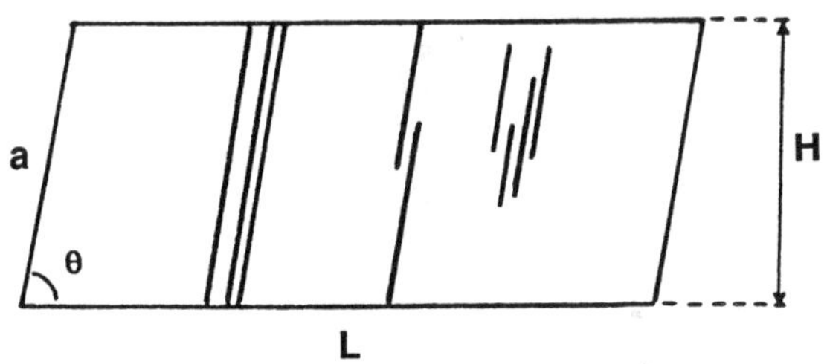

Figure 4.4.5. Derivation of the true trace-length distribution in the case of an areal survey; refer to text for explanations.

It is easily proved, using geometric probability arguments, that if measurements are made up to the limits of the parallelogram, these histograms have the following expectation:

$$E[N_0] = \tau' \int_a^{\infty} [1 - F(l')]\,dl' \qquad (l = a)$$

$$E[N_1(dl)] = 2\tau'[1 - F(l)]\,dl \qquad (l < a) \qquad (4.4.2)$$

$$E[N_2(dl)] = \tau'(a - l)\, f(l)\,dl \qquad (l < a)$$

where $\tau' = \tau_2\, L \sin\theta$ and $a = H/\sin\theta$.

These results take account of the various types of bias. If τ_2 is known, the histograms of N_1 and N_2 lead to two different estimations of the trace-length distribution for $l < a$, and N_0 makes it possible to fit the behavior of the distribution for $l > a$. In practice, the density τ_2 of trace centers is not exactly known (the actual center of the censored traces can be located outside the surveyed rectangle), and one simultaneously determines density τ_2 and distribution $F(l)$ by trial and error, in such a way as to get a good fit of the three theoretical histograms on the three experimental ones.

In the case of a scanline survey, the method is the same, except that only N_0 and N_1 are available. If H is the height of the bench above the scanline, the mathematical expectation of the experimental histograms becomes:

$$E[N_0] = \tau' \int_a^{\infty} [1 - F(l')]\,dl' \qquad (l = a) \qquad (4.4.3)$$

$$E[N_1(dl)] = \tau'[1 - F(l)]\,dl \qquad (l < a)$$

As explained below, the method is distribution-independent, but, in practice, one tries several classical models such as the exponential and the lognormal distributions.

c) Derivation of the Fracture-Size Distribution

Knowledge of the trace-length distribution is not enough to determine the fracture-size distribution, unless assumptions are made regarding the shape of fractures. A particularly simple case is the one where fractures can be considered as discs. The 3D parameters (3D density of disc centers τ, pdf $h(D)$ of disc diameter) and the 2D parameters (2D density of trace centers τ_2, pdf $f(l)$ of trace length) are then linked by simple relations that were developed for fractures by Warburton (1980) and are standard results in the field of stereology (Stoyan et al., 1987, Ch. 11):

$$\tau_2 = D\,\tau$$

$$f(l) = \frac{1}{D} \int_l^{\infty} \frac{h(D)}{\sqrt{D^2 - l^2}}\, dD \qquad (4.4.4)$$

where D is the mean disc diameter

$$D = \int_0^{\infty} D\, h(D)\, dD$$

The relation (4.4.4) between pdf's can be integrated to get the cdf $F(l)$, which is more useful:

$$F(l) = 1 - \frac{1}{D} \int_{l}^{\infty} \sqrt{D^2 - l^2}\, h(D)\, dD \qquad (4.4.5)$$

The density of disc centers τ should not be confused with the 3D fracturation density d_3. They are linked by the relationship $d_3 = \tau S$, where S is the average disc surface, i.e.,

$$S = \pi / 4 \left(D^2 + \sigma_D^2\right)$$

where σ_D is the disc diameter standard deviation.

From relation (4.4.5), it is known that, even if the disc-diameter distribution is not lognormal, the trace-length distribution tends to look lognormal. The inversion of relation (4.4.5) to derive the cdf $H(D)$ from the cdf $F(l)$ is possible in theory, but unstable in practice; it is an ill-posed problem, as pointed out by O'Sullivan (1986) in a very similar case. Therefore in practice one simply tries to fit an exponential or lognormal distribution. This step can be regrouped with the previous step (derivation of the true trace length distribution): combining relations (4.4.2) or (4.4.3) and (4.4.5), one chooses a disc-diameter distribution (type, mean D, standard deviation σ_D if necessary) and a 3D density of disc centers τ, such that the theoretical histograms of N_0, N_1, and possibly N_2, fit the corresponding theoretical ones.

4.4.8 Aperture and Thickness

Aperture is an important parameter for flow calculations, but it is difficult to handle, for two main reasons:

i) aperture is not constant: there are voids and contact areas, and flow between two parallel plates separated by the mean aperture has nothing in common with the actual flow;

ii) because of stress release, the measured aperture has nothing in common with an in-situ aperture.

Thus, the study of the apparent aperture is unsatisfactory. Indirect methods are used, e.g., water injection tests (Snow, 1970; see also Ch. 9). In some cases, however, the fracture is filled with calcite or another material, and the aperture is replaced by a thickness, which is not affected by stress release and is commonly found to have a lognormal distribution.

4.4.9 Correlations between Parameters

All parameters can be correlated, but, as the fractures have been separated into several sets, the correlations are usually low. An important exception is the thickness, which is usually found to be strongly positively correlated with the fracture size (see Loiseau, 1987, who studied fractures in gneiss and found correlation coefficients around 0.90 between thickness and trace length).

4.5 BASIC NETWORK MODELS

All models can be deterministic or stochastic. The first models were deterministic, as illustrated by the orthogonal model by Snow (1965), which is defined by three orthogonal directions of equidistant planes. After this, models with purely random parameters were used, such as the Poisson flat process (Priest and Hudson, 1976), which presents the advantage of displaying many "good" properties. Stochastic models are considered here, because they are better adapted to usual applications. Their transposition to deterministic models is easy.

From a stochastic point of view, fracture networks are realizations of random sets. The basic models can be found among the random-set models, mainly developed by Matérn (1960), Matheron (1967, 1975), Miles (1972), and many others. Most models are reviewed by Serra (1982). Fractures are peculiar sets, because their third dimension is negligible with regard to the other two, so that they are considered here as planar. Therefore, the basic models of fracture networks are rather simple.

They will be described succinctly. A comprehensive description from a rock mechanics point of view can be found in Dershowitz (1984). A first presentation can be found in Chilès (1989a,b).

All the models described in this section share common assumptions:

i) All fractures are planar (assumption applicable to the whole chapter).

ii) All fracture locations are equally probable.

iii) All fracture orientations are independent of fracture locations.

However, the orientation is not necessarily uniformly distributed. This is even exceptional if directional sets are studied.

4.5.1 Infinite Fractures: Random Planes

In some cases, the fractures are long enough when compared to the studied domain, so that they can be considered as infinite. The simplest stochastic model is the Poisson flat model, defined by a single parameter λ and the following two properties:

i) if we consider the planes, whose poles lie within an elementary solid angle $d\omega$ centered on direction ω, their intersections with the line of direction ω that cuts across the origin of the coordinate system form a 1D Poisson process with density $\lambda\, d\omega$;

ii) the 1D Poisson processes induced along different directions are independent.

This model is obviously isotropic. A nonuniform distribution of fracture orientations can be achieved by replacing the constant density λ by a function $\lambda(\omega)$ of the orientation. Therefore, a simulation of Poisson flats within a ball of radius R can be achieved in two steps (Fig. 4.5.1):

i) simulation of a 1D Poisson process of density $\lambda_0 = \int \lambda(\omega)\, d\omega$;

ii) for each point of the Poisson process, selection of a random orientation according to the pdf $\lambda(\omega)/\lambda_0$; the corresponding random plane is the plane orthogonal to this orientation, whose algebraic distance to the origin is equal to the abscissa of the point of the 1D Poisson process.

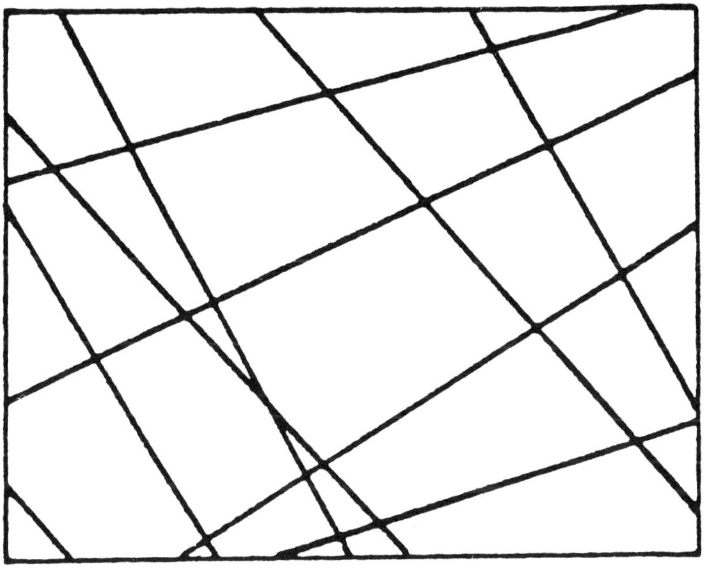

Figure 4.5.1. Section of a 3D model of random planes (Poisson flat process). After Chilès (1989a,b).

The simplicity of this model is very gratifying: ease of parameter fitting, availability of numerous properties of this kind of network through geometric probability theory (Kendall and Moran, 1963; Miles, 1969; Santalo, 1976). It is sometimes used in hydrological applications (see Andersson et al., 1984). The counterpart of this simplicity is lack of realism in most cases. Misuse of this model produces a too-systematic block network and an excessively connected fracture network.

4.5.2 Finite Fractures: Tessellation Models

Usually fractures are studied at a scale where they cannot be considered as infinite. Tessellation models give finite fractures. These models are built in two steps:

i) simulation of a set of jointed polygons, by partitioning of space or of fracture planes;

ii) selection, among these polygons, of those that will be considered as fractures.

The second step is achieved by randomly tossing for each polygon if it is a fracture (probability p) or intact rock (probability 1-p). The tosses are independent of each other. The probability p, or persistence, can be a function of the polygon orientation, in order to simulate anisotropic networks.

The first step can be achieved in several ways. The most commonly used in the present case are the following (Fig. 4.5.2):

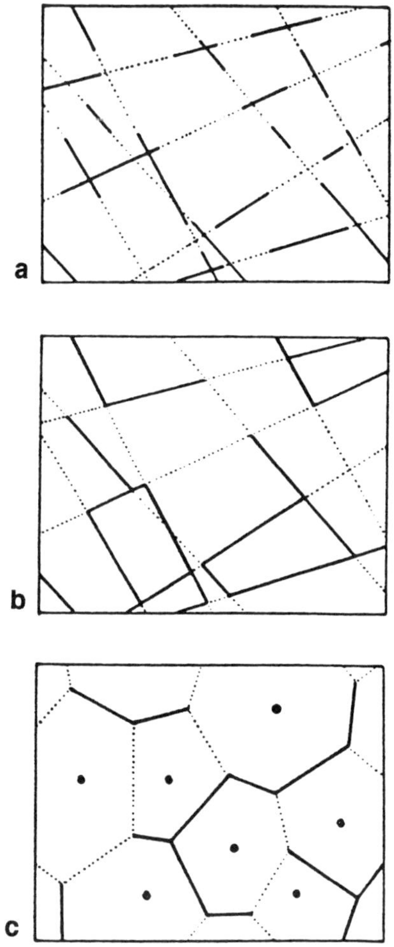

Figure 4.5.2. Section of 3D tessellation models: a) Veneziano model; b) Dershowitz model; c) Voronoï polyhedra. After Chilès (1989a,b).

a) tessellation of Poisson flats by Poisson lines: each Poisson flat is subdivided into Poisson polygons by an independent Poisson line process, which can be simulated with the same method as the Poisson flat process; this model was introduced for rock mechanics studies by Veneziano (1978);

b) tessellation of space by Poisson flats: a network of random planes already subdivides space into polyhedra, whose faces are polygons; fractures are chosen among these polygons. This model has been studied by Dershowitz (1984);

c) tessellation of space by Voronoï polyhedra: each point of a 3D Poisson point process (i.e., random points) is considered as the seed of a grain. Each seed grows at constant rate in all directions. Growth is stunted where two grains come into contact. This process subdivides the 3D space into polyhedra whose faces form a network of jointed polygons. Voronoï polyhedra and similar models (see Serra, 1982, Dershowitz and Einstein, 1988) enable us to model metallic grains, but are less suitable for rock fractures, with a few exceptions such as basalt pipes.

Each model corresponds to a different network type: Voronoï polyhedra produce noncoplanar fractures, whereas the other two models produce coplanar fractures. Fractures commonly stop when encountering another fracture in the Dershowitz model, whereas this never occurs in the Veneziano model.

The required fracture density and size are obtained by calibrating the Poisson density (of planes, lines, or points) and the persistence p. For example, Dershowitz reduces the polygon size by adding planes of zero persistence: no polygon of these planes is considered as a fracture, but these planes subdivide the polygons of the other planes. Persistence p has a strong influence on connectivity. In the first two models, where polygons are coplanar, each fracture is formed by one or several jointed polygons, so that the fracture size is a function of p.

4.5.3 Finite Fractures: Boolean Model of Random Discs

A Boolean model consists in the setting-up of a volumetric element at each point of a 3D Poisson process. This element can be the same all the time (for example, a ball of fixed radius r), or can be random (for example, a ball of radius r_i is set up at point x_i, where the r_i's are independent random variables with the same distribution). The definition and the general properties of Boolean models are due to Matheron (1967, 1975). For fracture networks, this model is used with a disc as the volumetric element, and is known as the random-disc model (Fig. 4.5.3).

It was introduced for rock mechanics applications by Baecher et al. (1977). Each fracture is assumed to be a circular (or elliptical) disc, and is defined by its central location, diameter and orientation. For this reason the model is defined by the following properties:

- the disc centers constitute a 3D Poisson point process;
- the disc diameters are independent and have the same distribution;
- the disc orientations are independent and have the same distribution;
- diameter and orientation are independent.

The last property may seem a limitation for the applications, but in practice a specific model is used for each fracture set. So, the fracture size can differ from one set to the other, whereas it is acceptable to consider that, within a set, size is independent of orientation. This model also has a limited number of parameters. They are directly related to the shape of the fractures.

The intersection of the network with a plane is a 2D model of segments centered on the points of a 2D Poisson process. The parameters of the 3D and 2D models (Poisson process densities, disc diameter and trace-length distributions) are related by simple stereological relationships (see Sec. 4.4.7). The 3D model parameters are hence easily deduced from 2D measurements.

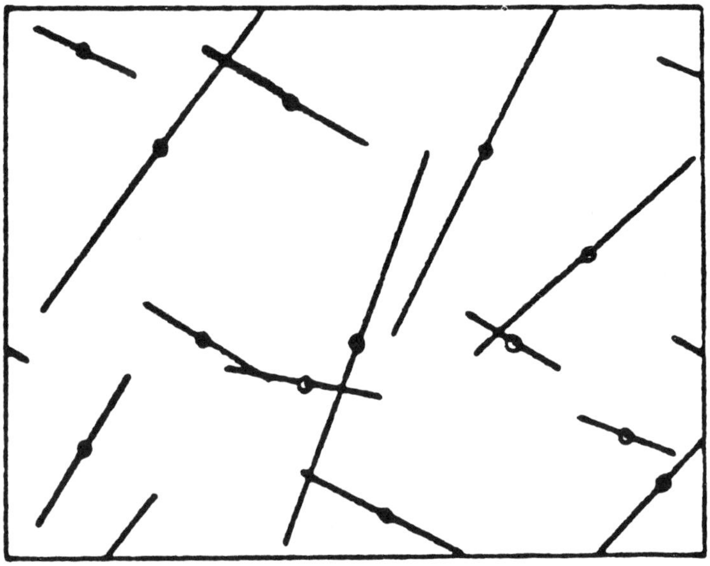

Figure 4.5.3. Section of a 3D random disc model. After Chilès (1989a,b).

In this model, blocks are only formed when fracture density or disc size is sufficiently high to cause intersections between triples of discs rather than pairs. Blocks are not shaped as convex polyhedra. Thus, this model seems suitable for rock masses that do not exhibit distinct rock blocks.

4.6 FRACTAL APPROACH

The fractal theory of Mandelbrot (1983) is now widely used in the earth sciences. Applications to fracture-network modeling are, however, not numerous, in spite of the kinship of the terms "fracture" and "fractal." The main parameter of fractal models is the fractal dimension.

4.6.1 Fractal Dimension – Box Dimension

The exact definition of fractal dimension, or Hausdorff dimension, does not provide a simple means of measuring it in a fracture network, but several alternative definitions exist that yield the same value as the Hausdorff dimension in the case of self-similar fractals. For fracture networks, the tool that is usually used is the box-counting dimension, or simply box dimension, initially called similarity dimension, as defined by Mandelbrot (1985).

To define this box dimension, suppose that we are studying a 2D pattern (a network of traces in our case). We can cover the study domain with a lattice made of square boxes of side r, and count the number $N(r)$ of boxes that intersect the pattern (Fig. 4.6.1). If the pattern is a rectangle of area S and r is small enough, we have approximately $N(r) = S/r^2$. Likewise, if the pattern is reduced to a segment of length L parallel to one side of the boxes, we have $N(r) = L/r$. In both cases, the Euclidean dimension (2 or 1), is D so that $N(r)$ has the form $N(r) = A/r^D$, or equivalently:

$$\log N(r) = \alpha - D \log r$$

D, the box dimension, is the absolute value of the log-log plot slope of N versus r. This relation is generalized to more complex patterns, where D can be a non-integer value.

4.6.2 Scale Dependence

In practice, one must be very careful when analysing a log-log plot. The mechanical use of the method always provides a graph, even if the phenomenon is neither fractal nor self-similar. Of course, in this case the graph is not necessarily linear, but it may look approximately linear at the scale of the study. In the case of a fracture survey, the graph cannot be linear: for large r values, every box is intersected by one or several traces, leading to a global box dimension of 2; in the opposite case of very low r values, the local box dimension is 1, for the traces have been collected as segments. Of course this local dimension is obtained by using the data beyond their precision, but this shows that it is possible to get any box dimension value between 1 and 2, unless the scale, or more precisely the inner cut-off, is specified. This may explain why very dissimilar results have been presented: for example, Barton and Larsen (1985) found a dimension between 1.12 and 1.16 for three tuff outcrops; Jacquin and Adler (1987) gave a dimension of 1.8 to 1.9 to several fracture sets; Thomas (1987), with another method, concluded that the fractal behavior of several calcareous or granitic outcrops had a dimension slightly above 1. There are several ways to overcome these difficulties:

i) characterize the network by its whole box-dimension graph (the derivative of the log-log plot) instead of a single dimensional parameter (Chilès, 1988).

ii) specify the inner cut-off for which the local dimension is computed; Barton (1992) proposed to use as box cut-off the truncation length of the collected fractures; with such a

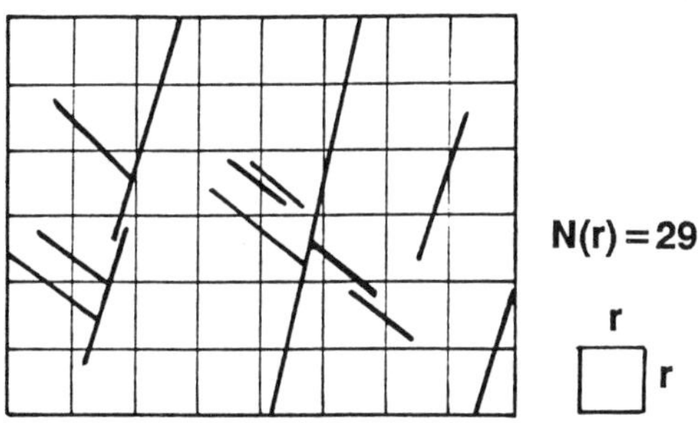

Figure 4.6.1. Definition of the box dimension. After Chilès (1989a).

definition, and also refinements in the computation of the box-dimension graph, he found a dimension between 1.6 and 1.8 for 15 different data sets, including the three tuff outcrops studied by Barton and Larsen (1985); the box-dimension graph of Chilès (1988) gives a dimension of 1.6 with Barton's definition of the inner cut-off.

iii) transpose to fracture networks other dimensional parameters, such as those developed for lattice models (see, e.g., Feder, 1988, Ch. 7).

4.6.3 Fractal Models

Even if the phenomenon is considered as fractal, the fractal dimension D itself yields no information on the model: D is simply the parameter of a model that must be chosen on the basis of other considerations. This is easy to understand if we compare the box dimension definition and the functional moment $Q(B)$ defined by Eq. (4.2.1) in the mathematical morphology approach: the knowledge of the box dimension graph is exactly equivalent to the knowledge of $Q(B)$ when B is a square box. This is not sufficient to fully characterize the network, as we know that it would require $Q(B)$ for any set B.

An exception is the 1D case (the fracture network is reduced to a network of points) when successive spacings can be considered as independent. This is why the few attempts to simulate a fracture network with a fractal model start from a 1D model: Thomas (1977) simulated infinite planar fractures from a fractal simulation of intersections with a sampling line. Chilès (1988) simulated finite fractures of fixed orientation and random length in a 2D case; the method takes account of the whole box dimension graph (Fig. 4.6.2). Barton (1992) combined fractal models (Cantor dust) and statistical characteristics (distribution of length and orientation) to provide a 2D simulation.

4.7 GEOSTATISTICAL APPROACH

The above basic models provide networks of independent fractures (random planes, random discs). These models are a good starting point for the study of actual fields. However, reality is often more complex, combining variable density, relays of fractures, and clusters. The geostatistical approach (Matheron, 1965) enables one to question, to validate, or, if necessary, to

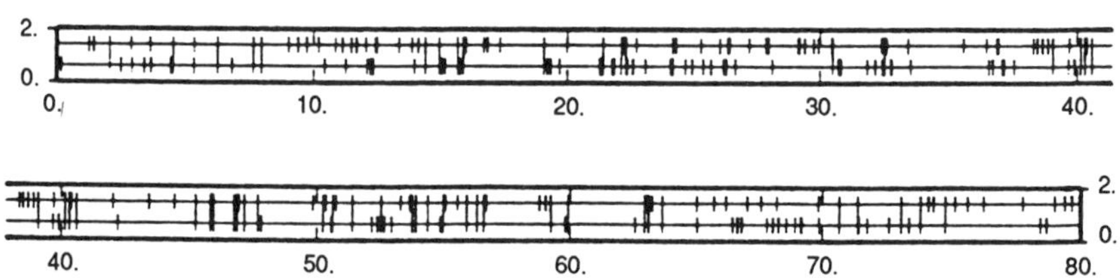

Figure 4.6.2. Simulation of a set of subvertical fracture traces on a 80-m by 2-m drift wall with a model reproducing a given box dimension graph. Only the intersections with two lines and the continuity are simulated. The distance between these two lines is equal to the mean trace length. The extension of the traces is conventional. After Chilès (1988).

weaken the assumptions of independence and of constant density that are the basis of these models. Among the first works on the subject see Miller (1979) and La Pointe (1980). For more recent developments refer to Chilès (1988, 1989a,b) and Massoud (1987).

Remember that the main geostatistical tool is the variogram. Considering a (random) function $Z(x)$, the variogram is the ordinary function of distance (or more precisely vector) h defined by

$$\gamma(h) = \text{mean} \frac{1}{2} [Z(x+h) - Z(x)]^2$$

It summarizes the dissimilarity of Z between two points at a distance h apart. The variogram is linked with the covariance function

$$C(h) = \text{mean} [Z(x) - m] [Z(x+h) - m]$$

where m is the mean value of Z, by the relation

$$\gamma(h) = C(0) - C(h).$$

4.7.1 Spacing

The intersection of any of the basic models with a line provides a 1D Poisson process. Successive spacings are independent and follow an exponential distribution (so that their relative variance equals 1). Instead of simply computing the mean spacing, it is worth verifying if the actual data satisfy these properties. This can be achieved by computing the experimental histogram and the variogram referenced to the fracture sequence number introduced by Miller (1979): the spacing values are not assigned to points on the sampling line, but to successive integer abscissae: the first observed spacing is assigned to the abscissa $x = 1$, the second one to $x = 2$, and so on. The h distance of variogram $\gamma(h)$ is then measured in number of fractures. For a Poisson process this variogram shows a pure nugget effect with sill equal to the square of the mean spacing.

Massoud (1987) for a granite massif, and Loiseau (1987) for gneiss, obtained flat variograms (pure nugget effect), but with a sill two to four times too high to correspond to a Poisson process. The histograms were obviously not exponential. Observation of fracture traces along drift walls shows that this is due to clustering. Fractures are grouped in clusters causing short spacing between fractures from the same cluster and long spacing when the two fractures belong to two distinct clusters. The same phenomenon was already observed by Snow (1970).

4.7.2 Fracture Trace Density

Another property of the Poisson process is that the number of points in disjointed segments (or volumes in space) are independent variables with a Poisson distribution. If we subdivide a surveying line into equal segments of length b and if, for distances $h = b, 2b, 3b, \ldots$, we compute the variogram of the number of intersections per segment, this variogram will also show a pure nugget effect. Its sill is equal to $d_1 b$, where d_1 is the density of intersections with the surveying line. If a 2D survey is available, the surveyed zone can be subdivided into rectangles, and the same property holds for the variogram of the number of trace centers per rectangle; this is always true for the random disc model, and approximately exact if the rectangle is large enough for the other models.

In the Fannay-Augères granite formation, most variograms computed by Massoud (1987) from 2-m-high drift wall surveys present:

- a nugget effect, corresponding to the random part of the fracture location;
- a first range of several meters, which may correspond to the extension of fracture clusters (increased by the regularization length b);
- longer-range components (30 to 300 meters), which can only be accounted for by a regionalized variation of the fracture density.

Loiseau (1987) found similar results in the Cézallier orthogneiss. La Pointe (1980) obtained ranges of about 80 m in dolomite.

4.7.3 Fracture Parameters

Other fracture parameters (orientation, size, thickness) can also be studied. The pole orientation within a set has been studied by many authors. La Pointe (1980) found an 80-m range in dolomite, Miller (1979) a 20-m range in granite, whereas Barla et al. (1987) in limestone and Massoud (1987) in granite, found ranges of several meters. In the latter case, this can be understood as the fact that fractures belonging to a cluster tend to have the same orientation, whereas cluster orientations are less correlated.

4.8 MORE GENERAL NETWORK MODELS

In order to take account of the complexity of reality, several simple generalizations of the basic models can be used.

4.8.1 Hierarchical Models

In their 2D study, Conrad and Jacquin (1973) observed long fractures and short ones that usually ended when encountering a long fracture, which led them to use a hierarchical model:

i) a primary network of infinite Poisson lines, that subdivides the plane into polygons;

ii) within each polygon, an independent process of random segments (the 2D version of the 3D random disc model); the segments are of course censored if they encounter the edge of the polygon (Fig. 4.8.1).

Another model is presented by Gervais et al. (1992). Such models are very useful for networks where a fracture set is stopped by another one.

4.8.2 Cluster Models

Models with fracture clusters are very simple. As all the basic models are based on Poisson point or flat processes, one has only to replace the Poisson process by a procedure in two steps:

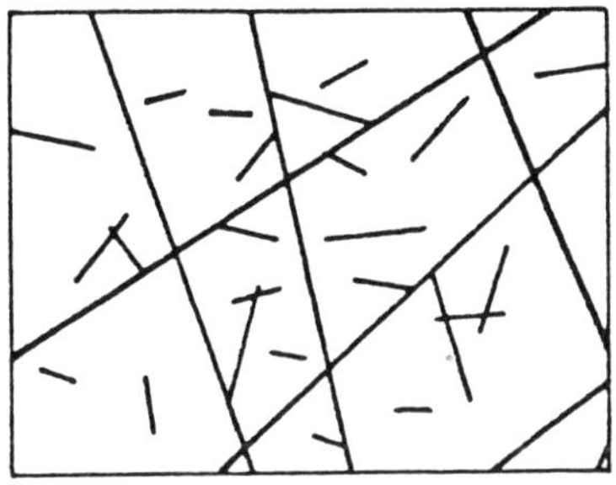

Figure 4.8.1. 2D hierarchical model of Conrad and Jacquin. After Chilès (1989a,b).

i) a Poisson process of lower density, which generates seeds (points or planes);

ii) around each seed, the generation of a cluster of independent points or planes (Fig. 4.8.2)

This kind of model has two new parameters: the mean number of elements of a cluster (the distribution of this number is difficult to determine, and thus a Poisson distribution is usually assumed) and the dispersion of the elements around the seed (for similar reasons, a Gaussian distribution is usually assumed). The density of seeds is taken as the ratio of the density of elements to the mean number of elements per cluster.

4.8.3 Regionalization of Density

If the fracture density is not constant, there are two ways to introduce a regionalization into the model, depending on the type of model:

i) for models based on Poisson points (Boolean model, Voronoï model), the density λ of disc centers can be replaced by a stationary random function $\lambda(x)$ with variogram $\gamma(h)$;

ii) for models based on Poisson flats, persistence p can be considered as a stationary random function (the randomization of the density of planes would introduce less flexibility).

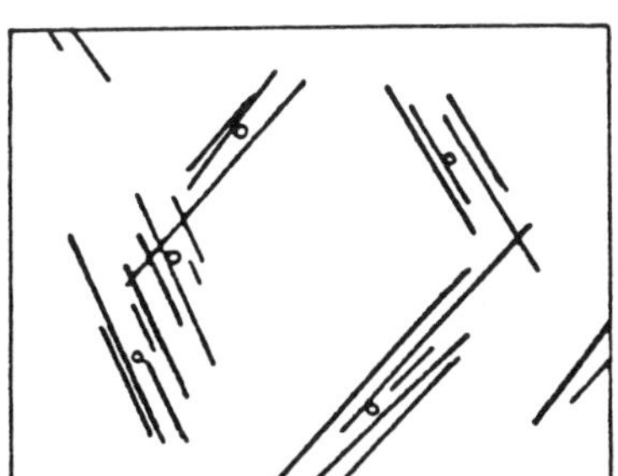

Figure 4.8.2. Section of a 3D model of disc clusters centered on Poisson seeds. After Chilès (1989a,b).

In this case, a parameter (e.g., the fracture density) is replaced by a distribution plus a variogram.

4.8.4 Parameter Fitting

These models are easy to simulate: regionalization can be achieved by a classical geostatistical simulation of the parameter (see e.g., Matheron, 1973; Mantoglou, 1987); mixed models and cluster models simply add one more step. The real problem is fitting the various additional parameters. In some cases, it is not more difficult than for the other parameters: for the 2D mixed models, Conrad used the functional moment $Q(B)$; for a model of Poisson twin flats, Matheron (1971) gave the spacing distribution of the twin planes from the distribution of the spacing of twin points that can be measured along a borehole, but usually the parameter fitting is more indirect. It can be carried out without major problems when the parameters of the model remain close to observable quantities. From this point of view, models of random planes or random discs offer much flexibility, as illustrated below for the latter.

4.8.5 Example of Disc Clusters with Regionalized Density

This is a rather complex model, as it generalizes the random disc model in two ways. Its parameters are the following:

- the cluster centers or seeds locally form a Poisson process of density $\lambda(x)$; this density is a stationary random function with mean λ and covariance $C(h)$;

- each cluster includes n discs; n is random, independent of all other clusters, and distributed according to a Poisson distribution with mean θ;

- the disc centers spread independently around the seeds according to a distribution with pdf $f(x)$ (usually an isotropic Gaussian distribution);

- all fractures of a cluster show the same orientation, but the cluster orientations are independent and have the same distribution;

- all disc diameters are independent and follow the same distribution, with pdf $h(D)$.

Orientation distribution, diameter distribution and average density of disc centers $\lambda\theta$ are determined for each studied fracture set as in the basic random disc model. Apart from complications due to a 3D modeling from 1D or 2D data, there are two difficulties in obtaining the other parameters:

- several clusters may overlap and it is not necessarily clear to which seed a fracture belongs; in this case, splitting the overall density $\lambda\theta$ into λ and θ is not easily done;

- the number of disc centers within a volume of size V centered at point x is not equal to the local density $\lambda(x)$ multiplied by θV, but to a realization of a random variable with mathematical expectation $\lambda(x)\ \theta V$; in this case the variogram of $\lambda(x)$ is not directly accessible.

When all parameters are fixed, it is possible to define the theoretical value of the variograms of fracture spacing and trace center number, which allows us to fit the remaining parameters by a trial and error method. The main theoretical results (Chilès, 1988, 1989a,b) are given below. Some demonstrations can be found in Massoud (1987).

The theoretical spacing variogram (or simply the relative spacing variance) is studied with a 1D Monte Carlo technique. This requires first the calculation of the distribution of the induced 1D clusters from the distribution of 3D clusters (not all fractures of a cluster intersect a straight line). If a fracture set orthogonal to the surveying line is considered, it is easy to demonstrate that the simultaneous 1D density of 1D clusters of m points that are part of a 3D cluster of n discs ($n \geq m$) is

$$\lambda_{mn} = \lambda \binom{m}{n} p_n \iint P(x,y)^m \, [1 - P(x,y)]^{n-m} \, dx\,dy \qquad (0 < m \leq n) \qquad (4.8.1)$$

We consider here a constant cluster density λ. The three coordinates x, y, z are explicit. The surveying line is the Oz axis; $p_n = e^{-\theta}\, \theta^n/n!$ is the marginal distribution of n; $P\,(x, y)$ is the probability for a disc of a seed located at point (x, y, z) to intersect the Oz axis. If the discs all have the same diameter D, one has:

$$P(x,y,D) = \iint C_D(-x,-y,0) \; f_1(u,v)\,du\,dv$$

but: $C_D = (-x,-y,0)$ shall be considered as an index

$f_1(u,v)\,du\,dv$ shall be in normal position:

$$P(x,y,D) = \iint_{C_D(-x,-y,0)} f_1(u,v)\,du\,dv$$

where $f_1\,(u, v)$ is the marginal pdf linked with pdf $f(u, v, w)$ of dispersion of disc centers around the seeds, where $C_D\,(-x, -y, 0)$ is the disc of diameter D centered on point $(-x, -y, 0)$.

Randomizing D, one gets:

$$P(x,y) = \int P(x,y,D)\,h(D)\,dD$$

The marginal density of 1D clusters of n points is then

$$\lambda'_m = \sum_{n \geq m} \lambda_{mn}$$

Figure 4.8.3 presents the relative standard deviation of fracture spacing as a function of the mean number of discs θ for several values of the dispersion (the pdf f is assumed to be isotropic Gaussian, and is characterized by its standard deviation). This makes a good evaluation of θ possible if the dispersion of a cluster is not too high. If the fractures are not normal to the surveying line, but intersect it at an acute angle β that remains close to $\pi/2$, these results remain

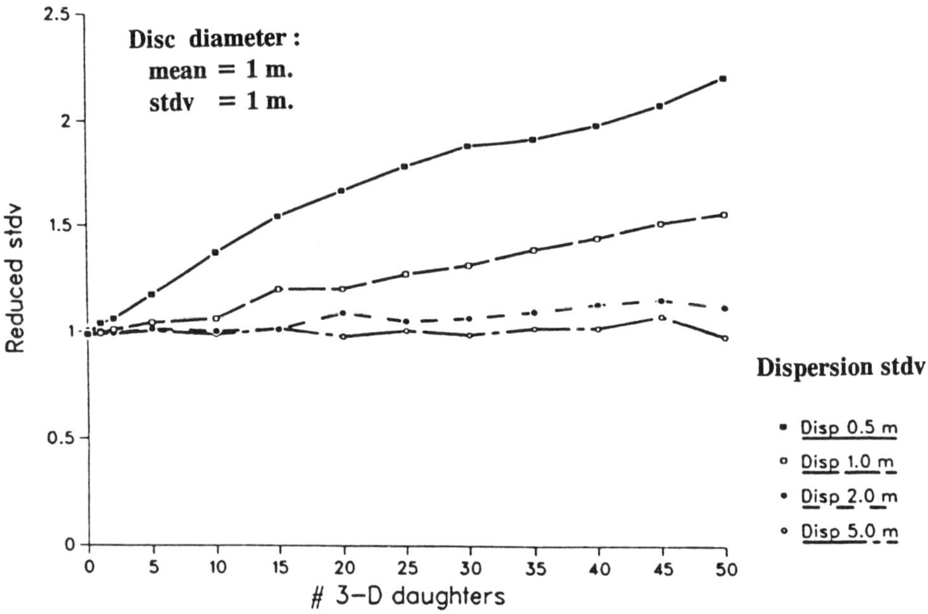

Figure 4.8.3. Theoretical relative standard deviation of fracture spacing for a model of disc clusters, as a function of the average number of discs per 3D cluster, for various dispersions of the disc centers. The disc diameters are lognormal with mean = 1 m and standard deviation = 1 m. After Chilès (1989a,b).

valid in a first approximation, provided that $P(x, y, D)$ is replaced by $P(x, y, D \sin \beta)$ in the expression of $P(x, y)$. If the density λ is not constant, the experimental relative variance must be computed locally.

The derivation of the theoretical variogram of the trace center number is explained in Massoud (1987) and Billaux et al. (1989). Figure 4.8.4 presents this variogram for several sets of parameters. If θ is known, a good evaluation of the dispersion of the disc centers and of the variogram $\gamma(h)$ of the density $\lambda(x)$ can be made.

The detailed application of this method to actual data collected in the Fanay granitic site is described in Massoud (1987) and Billaux et al. (1989).

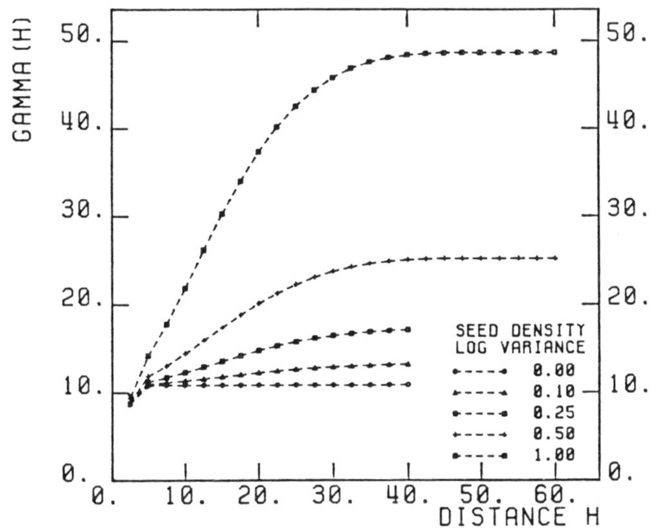

Figure 4.8.4. Theoretical variogram of trace center number for a regionalized model of disc clusters, for various variances of the seed density. The seed density is lognormal; its logarithm has a cubic variogram with a range of 50 m. The average seed density is 0.05 par m^3, and the average number of discs per 3D cluster is 20. The disc diameters are lognormal with mean = 1 m and standard deviation = 1 m. After Chilès (1989a,b).

4.9 CONDITIONING TO THE DATA

The tools examined above allow the fitting of all the parameters of the chosen 3D model from 1D or 2D data, and the simulation of a 3D fracture network that has the same statistical properties as the actual network. However, it would be even better if the simulated network also coincided with the actual network at the location of the 1D or 2D data.

4.9.1 Conditioning to Fracture Density

Seismic tomography methods are improving. They can at least give the relative variations of the fracture density between two boreholes, and define the precision of the method. These new data can be used to condition the simulation of fracture density in a model with regionalized density. The conditional simulation of random functions is well-known in geostatistics (e.g., Delhomme, 1978, 1979).

4.9.2 Conditioning to Observed Fractures

Conditioning to observed fractures can be easily made for models directly based on Poisson processes because all points, lines, or flats of a Poisson process are independent of each other. Thus the general outline of a conditioning is as follows (Fig. 4.9.1):

1. produce a nonconditional simulation;
2. reject all the fractures that intersect the surveyed lines or surfaces, and retain all those that do not intersect;
3. add the actual fractures that have been surveyed.

The simplest case is the Poisson flat model because fractures are infinite. This model was used by Andersson et al. (1984). For a random disc model (see e.g. Andersson and Dverstorp, 1987), step 3 introduces a slight complication because the extension of the fracture must be determined from an intersection or a trace; in other words, we need the joint distribution of disc diameter and disc center location conditioned by observation (point or trace length). This results from geometric probability calculations. Such a simulation, based on actual data, is presented by Chilès et al. (1992).

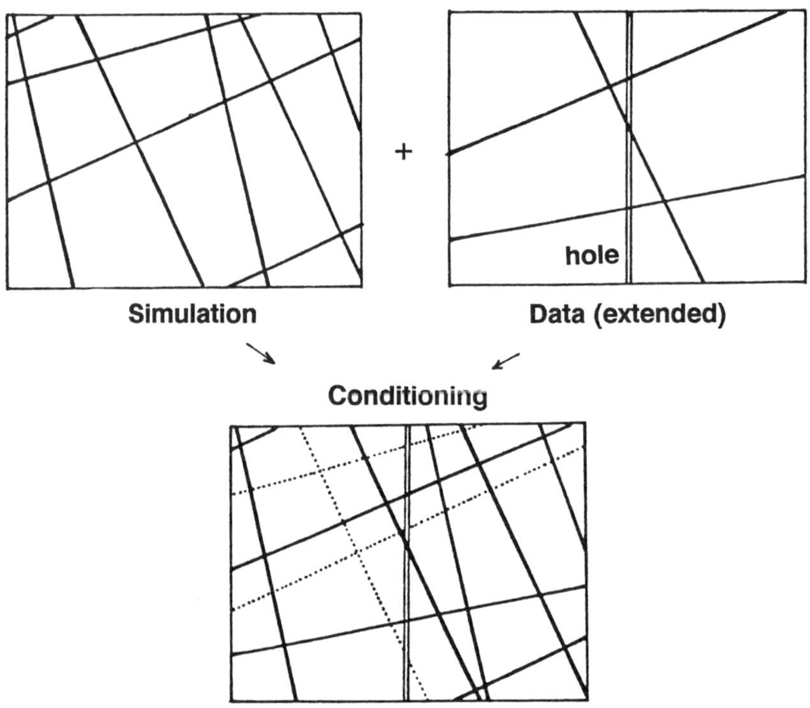

Figure 4.9.1. Conditioning of a simulation of random planes to borehole data. After Chilès (1989a,b).

With a disc cluster model, the same approach can be used (Chilès, 1989a,b), but, because independence is now at the cluster level, the rejection defined in step 2 must be applied at this level: we must reject all the fractures of the cluster as soon as one of them intersects the surveyed lines or surfaces. Conversely, in step 3, we must add to the surveyed fractures a simulation of the other fractures belonging to the same clusters (Fig. 4.9.2).

To achieve this, we need the distribution of the total number n of discs of a cluster conditioned by the number m of visible discs of the cluster. In the case of borehole data, this conditional distribution derives directly from the joint distribution of m, n, x, y that is the basis of expression (4.8.1) of λ_{mn}. The location of disc centers and their diameters are simulated from their a priori distributions by means of an acceptance/rejection technique. Thus, the theoretical tools are available. The main constraint is practical: the user must be able to define from the data which fractures belong to the same cluster; but this can usually be done with a good confidence.

4.10 CORRELATION BETWEEN SETS

The various directional fracture sets are usually studied independently but in fact, these sets are often found at correlated locations. Negative correlations (competition effect: recent tectonic

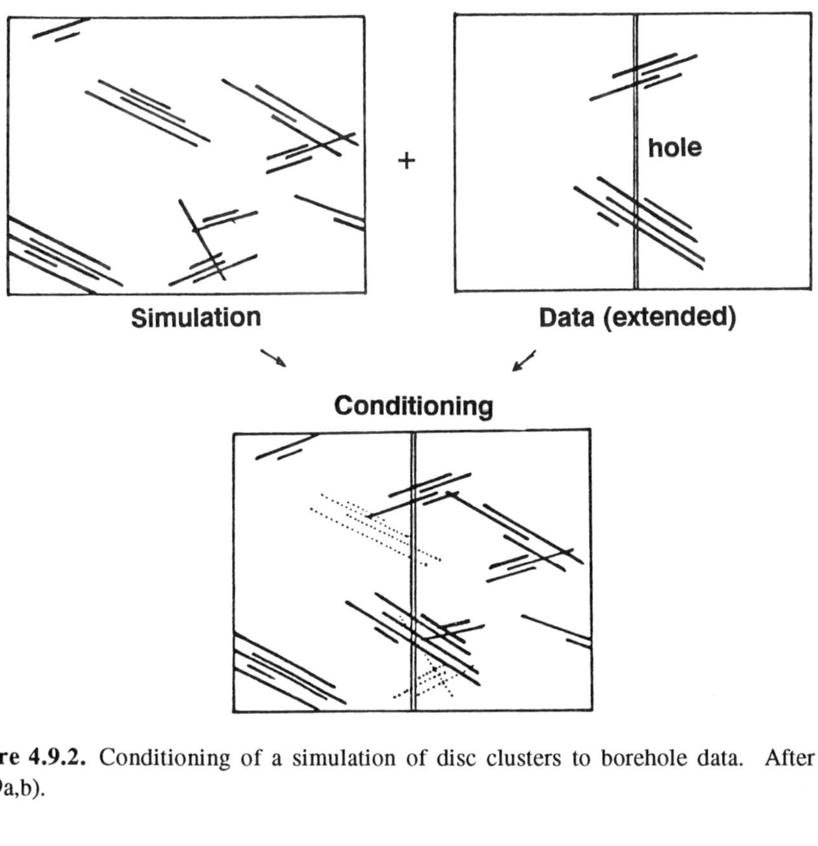

Figure 4.9.2. Conditioning of a simulation of disc clusters to borehole data. After Chilès (1989a,b).

episodes open new fractures in previously nonfractured areas only) as well as positive correlations (conjugate sets, zones of weakness in rock; e.g., Loiseau, 1987) can be observed. It is important to take these correlations into account because of their impact on connectivity. This can be done, for example, by means of correlated fracture densities.

4.11 THE SINGLE FRACTURE

If the fractures are considered as discs or polygons of constant aperture, it is possible to perform 2D flow calculations on the whole fracture surface, but actual fractures have nothing in common with perfect pairs of parallel plates. It is therefore necessary to use models that take account of the actual shape of fractures. Two approaches are possible:

i) A global approach to the single fracture: the flow and transport properties of a fracture are summarized by a few parameters. These parameters, or their distributions, are fitted by means of hydraulic tests.

ii) A local approach to the single fracture: the topography of the fracture walls and the induced aperture are modeled by (random) functions. This can be the basis for a representation of the void space by channels or conductive pipes, from which flow and transport calculations are performed.

4.11.1 Global Approach to the Single Fracture

Cacas (1989) and Cacas et al. (1990a) characterized each fracture by a mean hydraulic head. Flow between secant fractures is summarized by a 1D hydraulic link where flow rate is proportional to the head difference between the two fractures. The ratio of flow rate to head gradient is defined as the integrated hydraulic conductivity. This parameter represents the average resistance to flow between the two fractures and integrates the entire flow geometry: surface roughness, flow tortuosity, and channeling. The distribution of the integrated conductivity specifies the hydraulic model. A lognormal distribution is usually used, and its parameters are determined by trial and error from hydraulic tests in order for the model properties to correspond to the observed ones. The network of linear links replaces the network of discs. Flow and head are computed by using Kirchhoff's law. (See Sec. 4.14.)

Cacas (1989) and Cacas et al. (1990b) studied the particle transport with a particle-following method. Two parameters summarize the properties of the fractures: i) the retardation coefficient, related to the channel shape; and ii) the microscopic dispersivity, linked with dispersion phenomena within the fracture planes. (See Sec. 4.15.)

4.11.2 Geostatistical Modeling of the Single Fracture

The precise shape of a fracture can be studied at the laboratory scale on a core orthogonal to the average fracture plane. Measurements can be made by sampling profiles along both fracture walls (Gentier, 1986), or by analyzing the image of a translucent cast of the void (Billaux, 1990). The latter method offers the advantage of a complete description of the aperture, but gives no information concerning the shape of the fracture walls. The translucent cast data are sufficient for flow studies. For transport and mechanical studies, however, it is necessary to also consider the fracture wall profiles.

A fractal approach, combined with spectral analysis, was used by Mandelbrot et al. (1984) in the study of a metal fracture surface, considered as a fractional Brownian surface. Brown and Scholz (1985) used a fractal approach for the study of natural rock surfaces. A geostatistical approach seems preferable for two reasons: i) it includes a wider class of phenomena than the fractional Brownian model, which simply corresponds to intrinsic random functions with an h^{α} variogram; and ii) it enables the user to analyze simultaneously the shape of the fracture and the relationship between walls and aperture.

Gentier (1986) presents the variography of an actual fracture surface. Gentier et al. (1991) simultaneously analyze the topography and the gradient in order to ensure a good modeling of the roughness. But let us examine a very simple example of the usefulness of a simultaneous study of walls and aperture.

Denoting the 2D coordinates as x in the average fracture plane and the axis orthogonal to this plane as z, the elevations of the lower and upper fracture walls are represented by two functions $Z_1(x)$ and $Z_2(x)$. The aperture is then the function $Z_3(x) = Z_2(x) - Z_1(x)$. The walls Z_1 and Z_2 are often similar. Due to all these relations, Z_1, Z_2 and Z_3 are not independent.

Suppose that the two walls are identical, up to a displacement. Then one has:

$$Z_1(x) = Z(x) \qquad Z_2(x) = b + Z(x + a) \qquad Z_3(x) = b + Z(x + a) - Z(x)$$

where b is the average aperture (vertical displacement), and a is the translation (horizontal displacement). Denoting $C(h)$ the covariance of both walls and $\gamma(h) = C(0) - C(h)$ their variogram, such a displacement is easily detected on the wall cross-covariance, as $C_{12}(h) = C(h + a)$ is translated from $C(h)$, with its maximum for $h = a$ instead of $h = 0$. This displacement can also be detected on the aperture variogram $\gamma_{33}(h)$, as it is easy to see that:

$$\gamma_{33}(h) = 2\,\gamma(a) + 2\,\gamma(h) - \gamma(h - a) - \gamma(h + a)$$

When $h >> a$, $\gamma(h)$ can usually be considered as linear between $h - a$ and $h + a$, so that $\gamma_{33}(h)$ is at its sill $2\gamma(a)$. Figure 4.11.1 represents the aperture variogram $\gamma_{33}(h)$ linked with a linear wall variogram $\gamma(h)$. Along the translation direction, it has a range equal to the translation a. Its behavior below this range is also linear, but with twice the slope of the wall variogram. Along other directions, the behavior is similar, but with a smoother transition to the sill. This kind of variogram can be found on actual fractures. More complex phenomena can be studied in a similar fashion (Chilès and Gentier, 1992). The aperture variogram is more characteristic of the movement that created the fracture than of the topography of the fracture walls, so that its study is very useful from the point of view of the structural geology.

4.11.3 Channeling of the Single Fracture

When the value of the aperture is known at any point — from a void cast or from an interpolation of profile data — it is theoretically possible to calculate flow by assigning to any point a transmissivity as a function of the aperture and then using a finite differences model. This method would be very cumbersome and, in fact, is not suitable since the aperture is much smaller under stress than that measured on the void cast. Gentier (1986) and other authors have observed

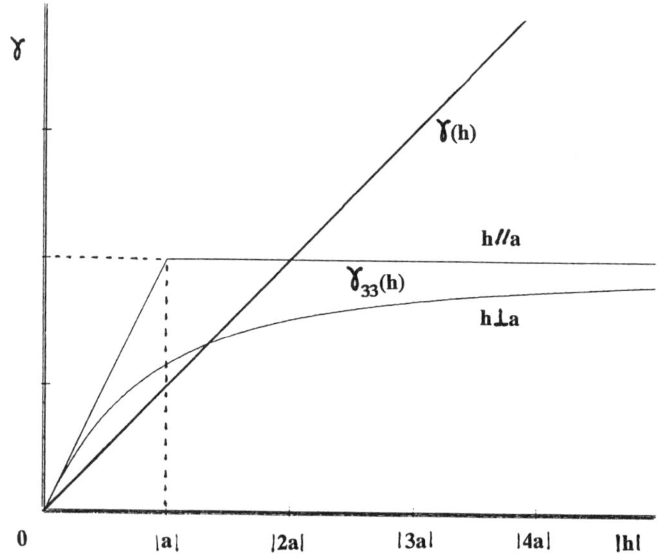

Figure 4.11.1. Aperture variogram linked with a linear variogram for the fracture walls, and a translation *a* along the *Ox* axis.

that, under stress, the water flow is usually along channels. Hence the idea, from the flow-studies point of view, of representing each fracture by a network of channels or conducting pipes. The approach can be global, in the sense that each fracture is replaced by a 2D network of Poisson lines or of Boolean random segments with a priori given parameters (Billaux, 1989). However, a model based on the observation of actual fractures is much better. This approach has been used by Billaux (1990) on a void cast. Figure 4.11.2 presents the successive main steps of the treatment: i) threshold at a given aperture value, visualizing the contact and void areas under normal stress (Fig. 4.11.2.a); ii) morphological skeletonization, followed by a suppression of dead ends and small loops, reducing the void areas to a network of channels (Fig. 4.11.2.b); and iii) reduction to a network of nodes and linear links (Fig. 4.11.2.c). The skeletonization is performed in such a way as to associate with each skeleton point the area of the corresponding channel cross-section. The flow calculation within the fracture is computed on this network. In this example, the diameter of the core is 11 cm. It is of course not useful, for a study at the 100-m scale, to model the precise fracture shape at the millimetric scale, and these results have to be transposed at least to the scale of the whole fracture.

4.12 FROM FRACTURE SYSTEM GEOMETRY TO CONNECTIVITY AND FLOW

If every fracture or fracture intersection is associated with a hydraulic parameter or if every fracture is replaced by a network of channels with given parameters, it is possible to compute the water flow through the network (see Cacas, 1989; Cacas et al. 1990a; Billaux, 1989). However,

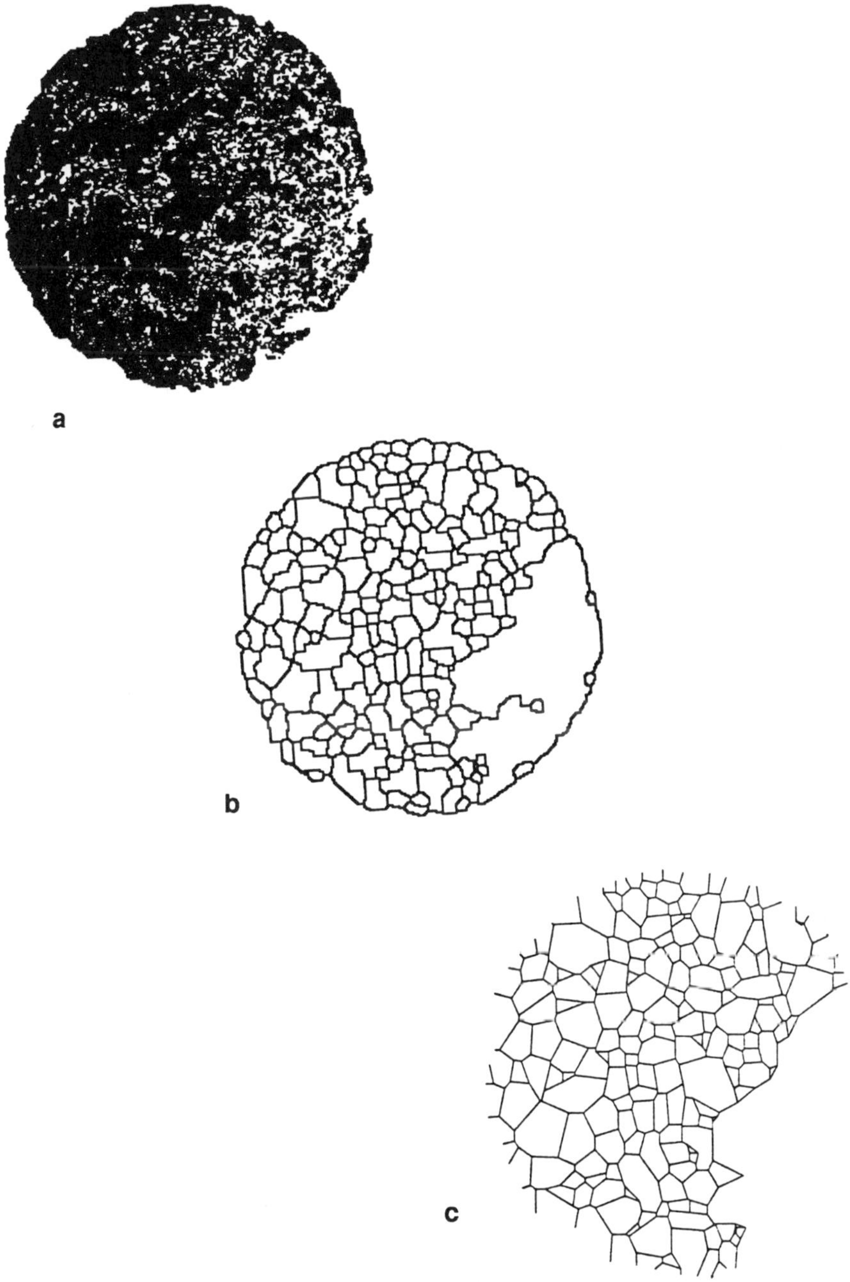

a b c

Figure 4.11.2. Representation of a single fracture by linear channels: a) binary image at a given threshold, visualizing the contact and void areas under normal stress; b) skeleton (after suppression of dead ends and small loops); c) network of linear channels. After Billaux (1990).

as actual fracture systems are complex, this requires long calculations. But flow is related to connectivity: if the fractures are not connected or form small separate clusters, no flow is possible whereas, if the fractures are highly connected, the system behaves like a continuous medium. As it is simpler to compute connectivity than flow, the connectivity can be studied first, in order to avoid flow calculations when they are not necessary.

Billaux (1989) presents a numerical example of hydraulic tests, injections with constant flow at the central hole of a simulated channel network, and prescribed head at the limits. Two curves are computed for each of six simulated networks: i) the connectivity curve, defined as the number of points of the channel network that can be reached from the injection point as a function of the distance to the injection point; and ii) the hydraulic head at the injection point as a function of time. The study shows that the shape of the head curve is very well explained by the shape of the connectivity curve. This is true not only in the extreme cases of very low or very high connectivity but also in the intermediate cases. A very interesting example is presented in Fig. 4.12.1. It corresponds to a network that seems to be near the percolation threshold. Figures 4.12.1a and 4.12.1b display the two steps of the network simulation: i) simulation of random discs; and ii) simulation of channels (intersections of discs, plus random segments on each disc). Figures 4.12.1c and 4.12.1d present the connectivity and head curves, which are rather regular and express a fairly good connectivity. This field is, however, far from being perfectly connected in three dimensions. It is easy to see that in the case of a 3D continuous medium that is perfectly connected connectivity varies as the square of the distance. This would lead to a slope of 2 on a log-log plot such as in Fig. 4.12.1c. Similarly, for a 2D continuous medium, this slope would be equal to 1. The corresponding lines are shown in Fig. 4.12.1c. The actual slope is around 0.66 (the first part of the curve should not be taken into consideration, as it is too dependent on the precise location of the injection hole). Thus, the actual network is less connected than a 2D continuous medium. If the notion of dimension of a continuous medium is extended to noninteger values, the analogy with the 2D and 3D cases shows that the slope of the log-log plot is equal to the dimension, minus 1. In the present case, the network can be considered as an equivalent continuous medium of dimension 1.66. The lack of connections produces a deficit in dimension in comparison with the Euclidean dimension of the whole space (3 in this example). This result can be compared with fractal concepts. Barker (1988) has derived the analytical expression of flow in the case of a medium with a noninteger dimension. Hence, the characterization of the connectivity of a network should give access to a simple model of its hydraulic behavior. A numerical example is given by Billaux (1990).

Another approach was explored by Hestir and Long (1990). They determined how a 2D network of random segments can be considered as equivalent to a regular lattice of bonds. This makes it possible to apply to random networks the results available for lattice networks. In particular, the correlation length can be derived from the percolation theory, which leads to a theory for the size of the representative elementary volume.

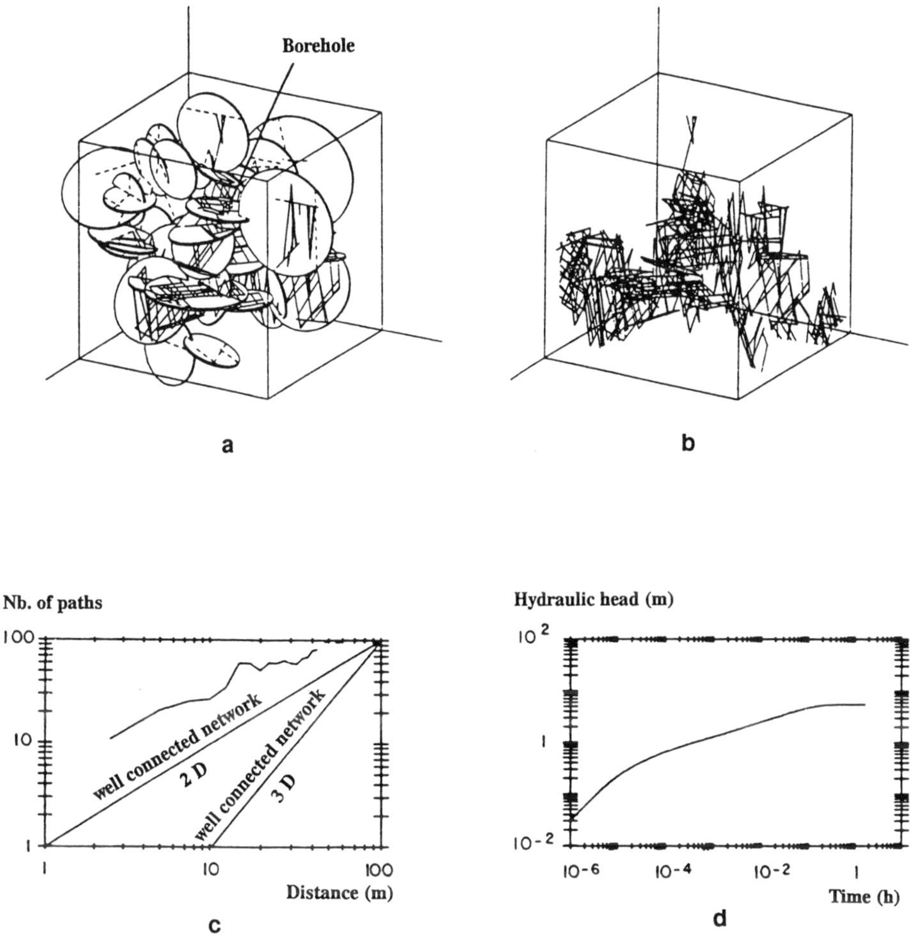

Figure 4.12.1. Connectivity and flow: a) network of random discs; b) network of channels; c) connectivity curve; d) head curve. After Billaux (1989).

4.13 PERCOLATION THEORY AND CONNECTIVITY

4.13.1 Definition of the Percolation Threshold

As stated in Sec. 4.12, a rock mass made of an impervious matrix and with a network of open fractures will only conduct water at a large scale if the fractures intersect; if one imagines that the density and/or length of the fractures in a network is progressively decreased, the connectivity of the network will also decrease; at a given density/length, the fractures, on average, no longer connect, and the rock mass becomes globally impervious. It may, however, happen that a small number of fractures in the network are still connected, in which case the network may be locally pervious: one may, for instance, inject water into one borehole and recover it in another one, provided that both boreholes belong to the same local set of connected fractures. If they belong to different local sets of connected fractures, no water can flow from one borehole to the next. It seems that such behavior can be observed in reality in some crystalline rocks (Marsily, 1985).

This problem of connectivity has already been addressed in other branches of physics, e.g., clogging of a filter, breaking of an electric circuit, and semi-conductor physics. A general mathematical theory for treating such issues, the "percolation theory," was proposed in 1957 by Broadbent and Hammersley (e.g., see also Shante and Kirkpatrick, 1971; Kirkpatrick, 1975; Clerc et al, 1983). In a network of increasing connectivity (starting from a low density/short length of the fractures without global connectivity), the local sets of connected fractures are called "finite clusters," and the sudden birth of a global permeability is said to happen at the "percolation threshold." At this threshold, one of the finite clusters becomes infinite: the medium is globally pervious; but local finite clusters may still exist together with the infinite one, at least when the network remains above, but close to, the percolation threshold.

Theoretical work as well as major numerical experiments have been carried out to determine, in general, the value of the percolation threshold in 2 or 3 dimensions as a function of the properties of the network as well as the conductive properties of the network in the vicinity of the percolation threshold (e.g., see Clerc et al., 1983).

It is therefore of interest to draw on these existing results for fracture networks. A first concept that emerges is the existence of a threshold in fracture conductivity; until recently, hydrogeologists considered fracture flow only as a scale problem: it was impossible to describe locally the flow in fractured media with continuum mechanics, but if a large enough "representative elementary volume" could be defined, any fractured medium ought, at that scale, to obey the flow equations of an (anisotropic) continuous medium. The only question was "how large is large enough?" and how to determine the large-scale hydraulic conductivity tensor (we will address these issues in Sec. 4.15). Percolation theory tells us that, before we attempt to answer such questions, we have to determine if the network is at all connected, i.e. whether it is above the percolation threshold. If not, the concept of "representative elementary volume" is useless and the global permeability is zero.

One must, however, strongly emphasize here that when one deals with fracture flow, the fractures that should be considered in any analysis are those that are pervious, at least locally (if channeling occurs): any crack or fracture that is totally sealed by, for example, calcite, silica, or clay, or is completely closed by the stress field plays no role or a very small one in fracture flow. The fractures that are observed in purely geometrical descriptions of the medium, e.g., from the observation of outcrops, or of drift walls or cores, are not necessarily those that conduct water. In fact it has been shown (e.g., see Ch. 2) that often less than 20% of the total number of fractures are water-conducting. It is very difficult in practice to determine which fractures are water-conducting; it is therefore common to initially consider all the fractures of the medium, whether sealed or not, and then to randomly suppress a certain number (e.g., 50% or 80%) in order to represent the remaining water-conducting network. The number of fractures to eliminate is then a parameter of the model that can be calibrated (perhaps fracture set by fracture set) or, possibly, estimated from observation of the ratio of "dry" fractures to "wet" fractures on outcrops. We will show in Sec. 4.14.2 and 4.14.3 how this calibration can be done, and how a logical order of fracture aperture or conductivity can be used to select the fractures to eliminate from the initial network.

A second useful procedure is to try to determine the value of the percolation threshold from some properties of the medium. Intuitively, one expects these properties to be the density and the length of the fractures or the average number of intersections of any given fracture with its neighbors. It can be shown that this is indeed the case. It is, however, necessary to first make sure that random stochastic fracture networks are a class of objects similar to the other networks studied in physics, such as electrical networks. This has been done, for example by Vere-Jones (1977), Chelidze (1982), Dienes (1982), Robinson (1984), Engelman et al. (1983), Robinson (1984), Guyon et al. (1984), and Wilke et al. (1985), and applies in principle to purely random Boolean sets of discs (i.e., not for discs structured in directional sets of fractures); it gives, however, only approximate values for the percolation threshold. One finds (Robinson, 1984; Charlaix et al., 1984; de Marsily, 1985) that:

i) in 2D (network of lines of random orientation and random finite length) Robinson showed that the average number of intersections of any fracture with all other fractures of the network must lie between 3.2 and 3.8 at the percolation threshold. Alternatively, the dimensionless number p

$$p = d \times r^2$$

must be on the order of 1.5 at the percolation threshold, where d is the density of the network (number of lines per unit area) and r the average length of the lines. If the orientation of the lines is not purely random, i.e., structured into families, the threshold is slightly larger than 1.5.

ii) in 3D (which is the only realistic case, the 2D case is only of academic interest) Charlaix et al. proposed that for a random network of fracture planes, the dimensionless number p

$$p = d \times (\text{average area of the fractures}) \times (\text{average half perimeter})$$

must lie between 1.5 and 3 at the percolation threshold, where d is again the density (number of fractures per unit volume of rock). If the fractures are circular discs, the average area of the fractures and the half perimeter are functions of the radius; the percolation threshold is then obtained when

$$p = d \times (\text{average of the square of radius}) \times (\text{average of radius})$$

lies between 0.15 and 0.3; if all the fractures have the same constant radius r, then this product is just

$$p = d \times r^3$$

4.13.2 Properties of the Network in the Vicinity of the Percolation Threshold

Apart from the definition and estimation of the percolation threshold, the percolation theory is also able to describe some fundamental properties of networks in the vicinity of this threshold. The power of the theory lies in the fact that the properties of the network at a large scale are not related to its local geometry or configuration but to general properties of connectivity, which are independent of the actual network of interest and can be addressed generally; "universal constants" can thus be determined. For instance :

i) below the percolation threshold, the "correlation length," i.e., the average size of a finite cluster, is given by :

$$X(p) = A\ \varepsilon^{-\nu}$$

ii) above percolation threshold, the conductivity is given by:

$$K(p) = B\ \varepsilon^{\tau}$$

where A and B are constants (that depend on the particular network of interest), $\varepsilon = p - pc$ is the "distance" of the network to the percolation threshold, p is the dimensionless number defined in Sec. 4.13.1, and pc is the value of this dimensionless number at the percolation threshold. ν and τ are "universal constants" that are approximately independent of the network. Clerc et al. (1983) provide the following values :

	2D	3D
ν	1.33	0.89
τ	1.00	1.75

Such results may be useful for sensitivity studies, e.g., to determine how the hydraulic conductivity of a fracture network would vary if the density or size of the fractures were varied. They can be used even without building a stochastic model of the fracture network. They are, however, only approximate.

4.14 USE OF STOCHASTIC FRACTURE NETWORK MODELS FOR FLOW PROBLEMS

In sections 4.5 to 4.10, it has been shown how geometric stochastic discrete fracture network models can be built in three dimensions from geometric information on the fracture system; in sections 4.12 and 4.13, the connectivity of this fracture network was discussed, as it is of overwhelming importance in understanding and simulating fracture flow. In section 4.12, it was also shown that it is difficult to actually measure the apertures of the fractures in the field, and therefore that this parameter has to be determined differently. In this section, we will review how the geometric fracture model can be built into a flow model and how this model can be used.

4.14.1 Building the Flow Model

Most discrete fracture flow models have, so far, assumed that the flow in the domain is laminar, i.e., a linear relationship can be found between flow and hydraulic head gradients. Only Bruel (1990) extended that approach, by an iterative technique, to the case where the opening of the fracture is a function of the pressure within it, and coupled a mechanical discrete network model to a flow model to calculate the change in opening due to the pressure field. The flow is still assumed laminar, but the flow equations are no longer linear, because the hydraulic conductivity is a function of the aperture of the fractures. For a presentation of nonlaminar flow in fractures, see for example Louis (1974), or de Marsily (1986).

So far, four approaches have been used to solve the linear flow equations in the fracture network.

i) Analytical expression. In this first approach, the flow in each fracture plane is calculated assuming that the whole surface of each fracture is the flow domain. If the fracture is a smooth parallel plane model of constant aperture, it is possible to develop an analytical expression for the flow in each fracture plane (Long, 1983). Using the image theory in a bounded circular geometry, Long developed a steady-state expression that relates the head along a line in the fracture plane to the flow within the fracture, given the geometry of that line and the aperture of the fracture. Of course, for flow to occur in steady state in a fracture plane, at least two lines must lie along each fracture plane: these lines represent the intersections of the fracture planes with other fractures (Fig. 4.14.1). The solution is approximate in that the head along each line is assumed constant over its length, whereas, in reality, it might vary with position. This head therefore represents an average head along each intersection and is common to two intersecting fractures. More than two intersections for any given fracture plane can of course be considered.

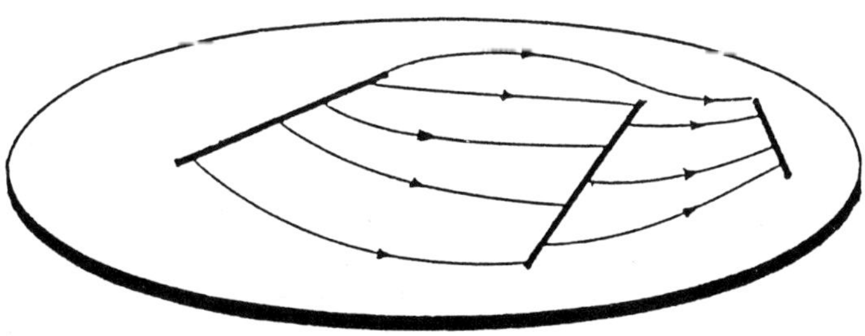

Figure 4.14.1. Flow between two parallel plates in a disc-shaped fracture with three intersections. After Cacas (1989).

For each fracture plane the analytical solutions provide the values of the flow entering or leaving that fracture plane at a given intersection, in the form of a linear function of the values of the heads along each intersection of that particular plane. Assembling these local flow models to form the global network model is then trivial, at least in principle. In steady state, at any given intersection of two fractures, the algebraic flow entering or leaving one fracture of the intersection must equal the negative of that calculated for the second fracture. This gives as many linear equations as there are intersections with just as many unknowns, which are the values of the heads at these intersections. This is known as a Kirchhoff problem in electricity or in pipe-network analysis, and it is simply the application of the principle of mass balance, assuming that the water compressibility can be neglected. Alternatively, the flow at the intersections can be the unknowns, and the heads, calculated as a linear function of the flows at the intersection, are made equal for each of the two intersecting planes.

The boundary conditions can be a prescribed head at any given intersection (e.g., when a fracture outcrops), thus reducing by one the number of equations and unknowns; a no-flow condition is simply represented by not allowing flow to occur through a given intersection, i.e., removing it from the given fracture plane. Source terms can be added if necessary at any intersection.

Once the heads at each intersection have been calculated by solving the linear system of equations, the flow can be calculated anywhere in the system by using the relevant analytical expression (which, of course, varies from plane to plane given the particular size and geometry of the intersections of that plane). Reciprocally, if the flows are calculated first, then the heads can be calculated next.

ii) Numerical solution. An alternative is to use a numerical solution for the flow problem in each fracture plane (e.g., see Smith et al., 1985; Andersson and Dverstorp, 1987; Stratford et al., 1990). This will allow the representation of a variable aperture in each fracture plane or of a noncircular geometry of the fractures or even of the partial filling of fractures with permeable sediments or alteration products. A finite element mesh is first generated in each plane. Each intersection is thus discretized with a certain number of nodes; such nodes are common to two intersecting fractures. The linear system of equations is then built by saying that, in steady state, the heads of two intersecting fracture planes are identical at the nodes, and that the algebraic flows calculated along each discretized segment of the intersections are also identical. Of course, the size of the linear system of equations becomes enormous, as there may be hundreds of unknown heads in each fracture plane, depending on the discretization; the final linear system is the sum of all the usual finite element equations in each plane, plus the Kirchhoff relations at the intersections.

There are, however, simplifications that can make the system tractable. One assumes again that the head is constant along each intersection; all nodes on an intersection in the finite element mesh are given the same value. Given that the flow equations are linear, it is then possible to solve the flow problem independently for each fracture plane: one determines numerically, once and for all, the coefficients that, for a given plane, relate the algebraic flow in each intersection of that plane to the constant heads on these same intersections (one can use the principle of superposition to do so). One then writes the same Kirchhoff equations as for the analytical method (i) above. It is also possible to solve numerically for each fracture plane successively even when the heads are kept different along the nodes of an intersection: this only increases the size of the resulting Kirchhoff problem.

An alternative is to use the boundary element method to solve the flow equation in each fracture plane (Andersson and Dverstorp, 1987). Each fracture must, however, have constant properties over its area.

It would be conceivable to extend the numerical approach to transient situations, but this has not yet been done.

iii) Channel model. Cacas (1989) and Cacas et al. (1990a) assumed that in each fracture plane flow did not occur over the whole surface of the fracture, but only in a limited number of "channels." A series of such "channels" are assumed to connect the intersections in a fracture plane to each other. To simplify the problem even further, these authors also assumed (Fig. 4.14.2) that:

a) Each intersection is connected to a fictitious "node" at the center of each circular fracture by a single equivalent straight "channel" starting from the center of the intersection; no assumption is made on the exact geometry of the real channel.

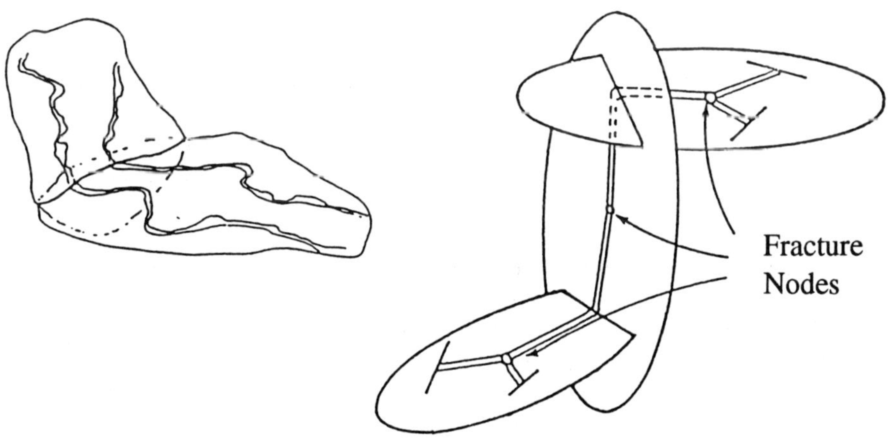

Figure 4.14.2. Channeling in a fracture plane and equivalent "channel model." After Cacas (1989).

b) The property of each equivalent channel is uniquely defined by an equivalent "hydraulic conductivity" K (in cubic meters per second) relating the flow through the channel to the hydraulic gradient within the channel (the difference of heads at the two ends of the channel divided by its length).

c) The head at the fictitious central "node" of each fracture becomes the unknown; note that the number of unknowns is then drastically reduced compared to the previous approach, since there are far fewer fractures than fracture intersections.

d) Between two central nodes of intersecting fractures, two channels are thus in series (one in each fracture plane). The flow through the channels is then given by the harmonic mean of their equivalent "hydraulic conductivity" K_i, weighted by their length:

$$Q = \frac{\Delta H}{L_1 / K_1 + L_2 / K_2}$$

where L_i is the length of each channel and ΔH the head difference between the two central nodes.

e) The equivalent "hydraulic conductivity" K_i for a given fracture plane is assumed identical for all the channels of that plane; K_i can, however, vary from plane to plane.

The mass-balance Kirchhoff relation in steady state is then easily written, with the heads at the central nodes as unknowns, by stating that the algebraic sum of all the flows calculated at each node is zero, again giving as many linear equations as unknowns. This approach could quite easily be extended to transient state, by giving to each central node a "storage coefficient" representing the change in mass in the fracture as a function of the change in head (for compressibility of the water and mechanical opening or closing of the fracture, see, for example, Bruel, 1990 for the mechanical coupling).

In the two previous approaches (i) and (ii), it is conceivable to estimate independently the aperture of the fractures (or its distribution) and to provide that information to the model. On the contrary, the "equivalent hydraulic conductivity" of the channel model cannot easily be anticipated, as it also includes some unknown geometric properties of the channels (e.g., tortuosity and spatial variability), and can thus only be calibrated from local flow experiments. This will be described in section 4.14.2.

One other property of the above channel model is that it might slightly overestimate the connectivity of a network model: as each fracture has a single "equivalent hydraulic conductivity," all the fractures that intersect it will be connected with each other; it could happen, in practice, that only a few fracture intersections within a fracture plane are interconnected, the others being sealed. It might be possible, however, to include such an effect by removing the concerned intersections from the system.

Still, the channel model is numerically the most efficient: it can include the largest number of fractures for a given computational effort (measured in terms of both CPU time and memory).

iv) Multi-channel model. This approach has been suggested, for example, by Billaux (1990). In this case, each fracture plane is said to include a prescribed number of fixed "channels," forming, for example, a network of lines in the plane. Flow can only occur along these lines. They can be generated randomly, as a 2D network of fractures, and they can randomly be given a directional hydraulic conductivity. Tsang and Tsang (1987) used another approach (a random fracture aperture model) to generate the flow channels in the plane.

In each such planar network, the calculation of the flow is again a Kirchhoff problem, the nodes of which are the intersections of the lines, and the unknowns of which are the heads at these intersections. The intersection between two fracture planes is considered as a particular channel belonging simultaneously to the two intersecting planes, thus forming the link in the global Kirchhoff model with the local Kirchhoff models in each plane. These intersection lines are generally given a larger hydraulic conductivity than the other channels, fracture intersections being in general assumed to be preferential pathways.

This model is very attractive but, again, very demanding in terms of computer resources.

4.14.2 Simplification of the Flow Model

Once a flow model is built, it can be "trimmed" and simplified.

i) First, all fractures that form an isolated cluster not linked to the infinite cluster can be removed, as they do not contribute to the flow (see a definition of the clusters in Sec. 4.13.1). Second, all fractures that have only one connection can also be removed, since no water can flow through them (a minimum of two connections is, of course, required). Different search algorithms are needed to carry out this screening.

ii) Then, another series of fractures may be removed from the model, even if they belong to the infinite cluster. For instance, it can be assumed that, let us say, only 50% of all the fractures are water-conducting; after having generated the network, one can randomly remove 50% of the generated fractures (this could also be done by arbitrarily reducing the density of the fractures during the generation process). A more common practice is to remove fractures according to their aperture or equivalent hydraulic conductivity. Suppose that in the fracture network, each fracture has been given an aperture, or an equivalent hydraulic conductivity, and let us order the fractures by decreasing value of their aperture or hydraulic conductivity (in other words, fracture number 1 has the largest value, and the last fracture the lowest). Let us imagine that the fracture network is now built by locating in space first fracture number 1, then fracture number 2, etc. Until the number of fractures in the domain is large enough for the network to be at the percolation threshold, the fracture medium is globally impervious. However, after the percolation threshold has been reached, adding new fractures with decreasing hydraulic conductivity contributes, after a while, very little to the global permeability of the medium, since the "pervious" fractures have already been located. In practice, one observes that it is not necessary to go very much beyond the percolation threshold, especially if the distribution of the apertures or hydraulic conductivities is large; this would not be true, however, if all fractures had the same properties. One can therefore, in general, reduce the size of the numerical problem by eliminating the less pervious fractures by a procedure opposite to the one just outlined above. Tests must, however, be made to determine how many fractures can be disregarded. One must then repeat step (i).

4.14.3 Stationarity, Ergodicity, Realizations and Averaging

Before going into the calibration of fracture network models, we need to discuss what they actually represent. Let us assume that a "block" of fractured rock (e.g., a cube or a sphere) is to be modeled with a discrete fracture network. As explained in section 4.5, such models are realizations of random sets. Once the statistics of the fracture geometry of the medium to be modeled have been determined, as well as the statistics of the apertures or "equivalent hydraulic conductivities" (as will be explained in section 4.14.4), it is possible to simulate (in the Monte

Carlo sense) an infinite number of equivalent fracture networks. This is done by selecting random numbers and sampling all the properties of a given network from their probability distributions (location, orientation, length, aperture, etc.). If these properties are not independent, correlated sampling must be used (e.g., the length of a fracture and its aperture, if they are correlated).

A single network, with all its fractures located in space and given all their properties, is called a "realization" of the stochastic model. The statistics on which the model is based refer in general to a real rock mass where the observations have been made. What are the relations between a realization and the real medium? They can never be identical, even if conditioning has been used (as described in section 4.9); stochastic models are built precisely because the real medium is unknown (apart from its outcrops), and therefore only the statistics of the description of the medium can be of use when representing what has not been observed.

An underlying assumption is that the statistics of the properties of the medium that have not been observed are identical to those that have. For instance, if a rock mass is analyzed through the examination of the wall of a 100 m-long drift excavated through it, and if the rock block that is to be modeled is a cylinder centered on that drift with a diameter of 100 m and a length of 100 m, the underlying assumption is that the statistics of the fracture properties that have been observed on the wall apply, in fact, to the whole cylinder, which has not actually been observed. This assumption is called "stationarity" and it is the one always used in practice.

One realization of the stochastic model of the rock cylinder will therefore only display one possible arrangement of the fractures in the cylinder; this arrangement, if it were used for determining the statistics of the fracture properties, should, by definition, produce the same statistics as those observed (within the sampling fluctuations, which are a function of the sample size, here the number of fractures in the cylinder). Nothing more can be said about the representativeness of one single realization of the real medium. If the realization has been conditioned on the observations on the drift wall, the fractures on the walls in the model and in the real medium will, by definition, be identical, but this involves such a small part of the rock volume that its global hydraulic behavior is largely unaffected, e.g., as shown by Andersson and Dverstorp (1987). One is therefore faced with the problem that a large number of (equally likely) realizations of the fractured medium need to be considered against a single real rock mass that is to be modeled. This leads us to use averaging.

Let us, however, discuss for a moment the issue of stationarity. One must understand that this is a question of scale. Let us take, for instance, the density of fractures, suppose that we measure the density of fractures along a drift in a mine and find a density d_1 over the first 100 m of the drift; on a later section, a density d_2 different from d_1 is found. Is the medium nonstationary? Yes and no! If our objective is to model a rock block around section 1, we are better off using density d_1 in the stochastic model and assuming that it very well represents the density in the vicinity of section 1; we would similarly use density d_2 in another model in the vicinity of section 2. The medium can be seen as nonstationary but, at a larger scale, if we want to represent a very large rock block including several such sections with different densities, can we simply use an average of the measured local densities? If the variations in the local densities are small enough to be compatible with the normal sampling fluctuations of a pure random model, the medium can be considered, in fact, as globally stationary with a density equal to the average experimental one. If the local densities differ too much, a model with a regionalized density such as the one described in Sec. 4.8.3 can be used; in such a model, the density still remains stationary but displays spatial variations; an application is presented by Long and Billaux

(1987) and Billaux (1990). We would expect from this geostatistical approach a model that is representative of the rock both at the local and global scales, which the pure random model could not do. In some circumstances, such as for modeling an increasing fracture density near the surface, due to decompression, a truly nonstationary model would be needed.

Let us go back to averaging. We cannot compare the real system with one realization, and this is all the more true if there are local property fluctuations that the model neglects because of the stationarity assumption. Can we then compare averages? In the real medium, we can do two kinds of averaging. We can first average over space. We can, for instance, design hydraulic tests that influence a large volume of rock and produce a single average property (e.g. an equivalent continuous medium hydraulic conductivity); this can be done, for example, by long duration pumping tests with distant observation wells. This is the approach used when one is trying to define "representative elementary volumes" in fractured rocks (with the restrictions already discussed in Sec. 4.13.1). We can, however, use other kinds of averaging. We can, for instance, measure locally the property of the medium, e.g., with local injection tests in short sections of boreholes. We can then analyze statistically the distribution of the results of these tests, and, incidentally, not just limit this analysis to the average.

Now what can we do with the stochastic model? We can actually do very similar things. We can design hydraulic tests to be performed numerically on one realization of the model and interpret them in terms of an equivalent hydraulic conductivity. Examples can be found, for instance, in Long (1983) and Long et al. (1982). The hydraulic test was used just to prescribe two different heads on two sides of a cube of rock made of discrete fractures and to measure the flow through the cube in a numerical experiment. If the size of the cube is progressively increased, and if the concept of "representative elementary volume" (REV) applies, one should reach a size beyond which the equivalent hydraulic conductivity no longer fluctuates significantly when the size is varied. This size would be that of the REV of the model. The property determined in this way is a space average of the local model properties obtained from one single realization. This approach has not been very successful so far, because either the concept of REV does not apply to the studied fracture networks or it is impossible, because of computer limitations, to reach the size beyond which the fluctuations are considered small enough to be negligible (a cubic block of 100 m in size may contain millions of fractures that cannot, as yet, be handled by any existing computer). If this approach of averaging could be used, one can anticipate that the same average value would be obtained for any realization of the medium, whereas this is certainly not true for small blocks below the REV size: each realization produces a different average value.

However, we can also perform the other type of averaging on the stochastic model: we can design a small-scale hydraulic test, e.g., with a small cube with two different heads prescribed on two sides, and carry out many such tests on many realizations of the model. As stated before, each realization will give a different equivalent hydraulic conductivity, as the size of the cube is below REV size. Now we can also do the statistics of these tests, among which we can calculate the average.

By doing this, we replace the statistics of the small-scale tests made on the real medium (at different locations in space) with the statistics of the small-scale numerical experiments made on the model (on different realizations). We are allowed to do this if for the fracture network we use an ergodic model, i.e., a model such that all realizations have the same global average density, which is, in practice, the average experimental density. If this average density is not well-known, it is more interesting to study the behavior of the network for some possible values of the density (especially if the network is near the percolation threshold) rather than to indistinctly mix all the possibilities into a nonergodic model.

The ergodic assumption is thus commonly made, and ensemble statistics of realizations of stochastic fracture network models are indeed compared with spatial statistics of measurements; this is, in fact, one of the ways of calibrating stochastic models, as will be explained in Sec. 4.14.4.

If the ergodic assumption can be used, it is then also possible to determine the large-scale equivalent hydraulic conductivity of the network by making averages of small-scale numerical experiments: the properties at the REV size of the model can be inferred from running many numerical experiments at the small scale, much below that of the REV. It is, however, necessary to assume that the small-scale numerical experiments have carried with them all the major features of the fractured medium properties: at the very large scale, for instance, the role of a major fracture could never be correctly accounted for by the small-scale experiments.

A question arises, however: which type of averaging of the small-scale numerical experiments should be used? Most authors, so far, have intuitively used the arithmetic average, which may be incorrect. If the property of interest is the hydraulic conductivity, it is well known concerning heterogeneous media that averaging hydraulic conductivities is not trivial (e.g., see Matheron 1967). This author has shown, in particular, that the true average always lies between the harmonic and the arithmetic mean. Only in two dimensions, for parallel flow conditions, and if the distribution of the hydraulic conductivity is log-normal and isotropic, is the geometric mean the exact average. If we now consider small block realizations of network models as having "local" properties, and the large-scale equivalent medium as constituted by the small blocks placed side by side, we can use these results and propose that the geometric mean of the realization is probably a better estimate of average model properties, and that harmonic and arithmetic means are the lower and upper brackets, respectively. See, for example, Cacas et al. (1990a).

4.14.4 Calibration of the Flow Model

The issue is here to determine the aperture of the fractures, or the "equivalent hydraulic conductivity" of the channels, in the model. We have indeed seen that these quantities cannot, in general, be determined directly from observation of the fractures on outcrops or cores. It is, however, always possible to assume a priori a value for such parameters (e.g., defined as constant, constant per set of fractures, or random with a given pdf) and to use the stochastic network model in a "what if?" manner: What if the fractured medium happened to have these properties? This can be acceptable for generic studies of fractured media, but it is inadequate for the study of a given real fractured medium.

One would therefore like to infer the hydraulic properties of the fractures from real measurements. So far, injection tests in short sections of boreholes have mostly been used for this purpose, but on one occasion water dripping from the ceiling of a drift in a mine was taken as basic information. Three methods have been proposed.

i) One method is to interpret the injection test as if all the fractures in the tested section are orthogonal to the axis of the borehole, of infinite extension, and have the same aperture; the interpretation of the results of the test can then be given in terms of the equivalent aperture of the fractures with the parallel plate assumption, if the total number of fractures in the section is known (e.g., from the coring or borehole televiewer). Another model is to assume that the apertures of the fractures of a given section are related to the ones observed on cores; Long and Billaux (1987) assumed, for instance, that the cubes of the real apertures are proportional to observed ones and fit the proportionality coefficient. The shorter the section the better, until, in the end, one single fracture per section is tested. By interpreting several such tests, an idea of the probability distribution function of the apertures can also be obtained. Other methods of interpreting injection tests can also be used, the point being that many tests must be performed and interpreted in order to get a value of the pdf of the aperture, possibly set by set of fractures, as the aperture may be very different according to the orientation if the stress field is very anisotropic.

ii) Another method, proposed by Cacas et al. (1990a), is to use the ergodic assumption outlined in Sec. 4.14.3. A large number of injection tests performed in the real medium are compared with simulated tests on a large number of realizations of the fracture network model; initial assumptions are made on the type of distribution of the apertures (or in this case "equivalent hydraulic conductivity" of channels) and on the parameters of these distributions. By trial and error, the calculated results are compared with the measured ones, the parameters or types of pdf's are changed until the results are considered close enough; the model is then said to be calibrated. In their work, Cacas et al. had 180 injection tests over 2.5-m long sections, and they performed 208 simulations of similar tests on 208 realizations of a spherical fracture network (10 m in diameter) with a central borehole where water was injected at a prescribed head, while the outer surface of the sphere was assumed at constant head. Figure 4.14.3 shows the histograms of the measured and calculated flows in the 2.5-m sections when calibration was obtained. Note that the calibration is not just made on the mean, but on the comparison of the histograms. For their case, they selected a lognormal distribution for the equivalent "hydraulic conductivity" of their channels, which was thus defined by two unknown parameters, a mean and a variance. These two parameters were calibrated by trial and error. They suggested that more parameters could have been calibrated, if, for instance, a different distribution of the hydraulic conductivity of the channels had been selected for each directional set of fractures.

iii) Dverstorp and Andersson (1989) did not use injection tests but observations of the water dripping from the ceiling in a drift in the Stripa mine, collected in 350 plastic sheets glued to the ceiling (see Ch. 2). Assuming the hydraulic gradient to be unity and all fractures reaching a given sheet to have the same aperture, they derived the fracture transmissivity (product of the equivalent hydraulic conductivity by the aperture) and its distribution.

Once the model has been calibrated, i.e., the unknown distributions of the hydraulic conductivities or apertures have been determined, it can be used for simulations. Cacas et al. (1990a) also proposed to validate the model by a change of scale: they used the calibrated model to calculate the large-scale (REV) hydraulic conductivity of the stochastic fracture network, using the averaging method explained in Sec. 4.14.3 and showed that it compared very well with a large-scale hydraulic conductivity determined by a large-scale flow experiment in the real fractured system.

4.14.5 Results Given by Stochastic Fracture Network Flow Models

As presented in section 4.14.3, stochastic fracture network models, once calibrated, can be used to calculate large-scale equivalent hydraulic conductivities of a continuum. They are especially well-suited for determining the anisotropy of this conductivity as exemplified by Long et al.

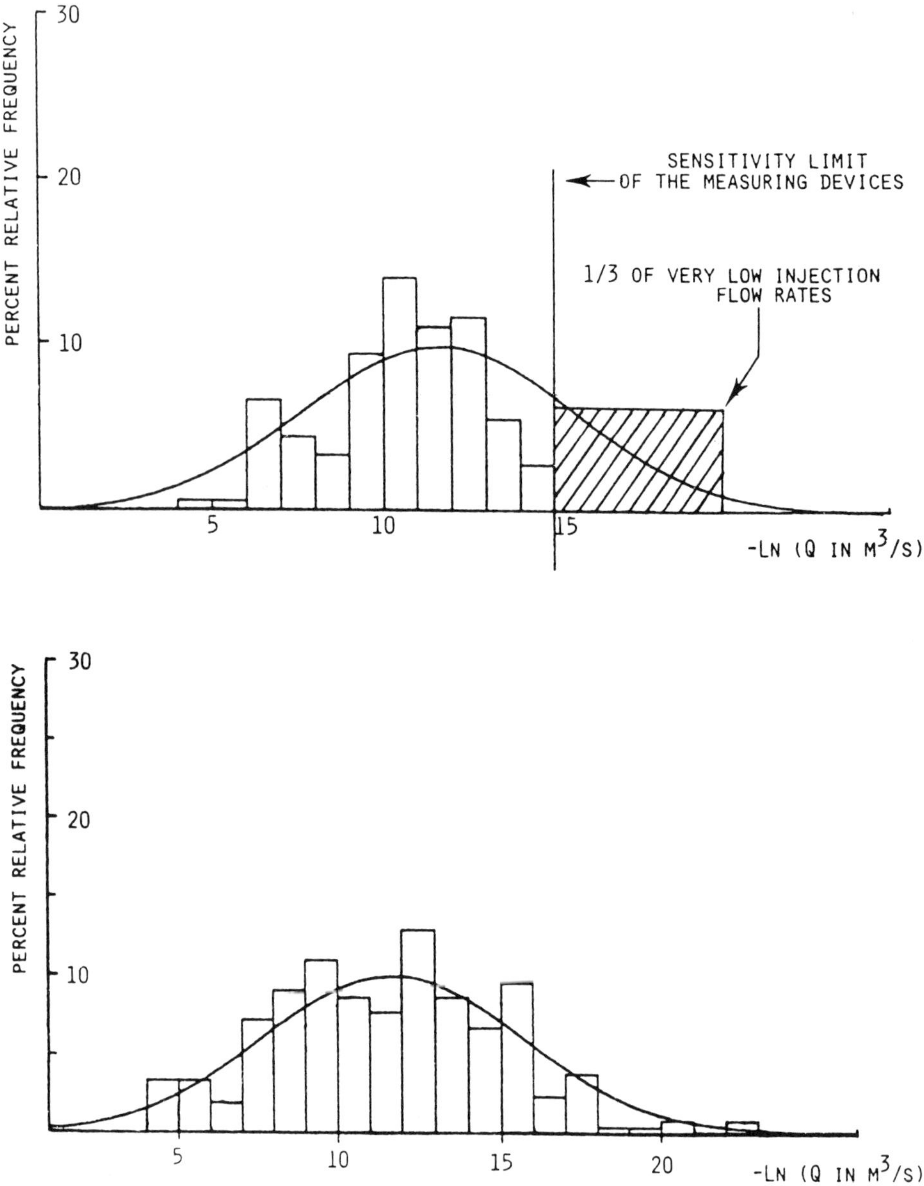

Figure 4.14.3. Histograms of measured and calculated flows in borehole injection tests. After Cacas et al. (1990a).

(1982) and Cacas et al. (1990a). In the numerical flow experiment in a cube, the orientation of the cube can be rotated continuously, by "cutting" the cube for the flow experiment within a larger fixed realization (Fig. 4.14.4); a directional hydraulic conductivity is then calculated along the orientation of the cube. The anisotropy must, however, be determined by averaging the directional hydraulic conductivity of a good number of realizations.

Stochastic fracture network models can also be used directly to solve flow problems but they can only handle problems of a limited size (because of computer limitations, the dimension of a block can only be on the order of tens of meters, if all the water-conducting fractures are to be represented). Examples are the injection of water into deep hot dry rock wells for hydraulic fracturing (Bruel, 1990), and calculation of the variability of the flux of water coming into a deposition hole in a granitic nuclear waste repository (Dverstorp, 1990). In the latter case the interest of the stochastic model is to show, with many realizations, the distribution of the flow rates in the deposition holes and not just the mean: the corrosion rates of the canisters is indeed a function of this flow rate, and for safety studies the distribution is sometimes more important than the mean.

It would also be possible to predict fracture flow around the walls of a dam; this would similarly show the uncertainty of the predictions if a large number of realizations of the fracture network were run with the same boundary conditions and, possibly, conditioned by the same set of observations on the walls of the valley. A coupled mechanical-hydraulic model (Bruel, 1990) should be used to take into account the potential build-up of underpressures below the dam and its potential consequences for the dam stability.

It would be possible to represent a larger flow domain (e.g., in kilometers) only if the number of fractures introduced into the model were drastically reduced. This could be done, for example, by keeping only the largest fractures; it is not clear, however, if the largest fractures alone will create a percolating network or if some of the smaller fractures must be introduced in order to raise it above the percolation threshold. This topic deserves further research.

4.15 USE OF STOCHASTIC FRACTURE NETWORK MODELS FOR TRANSPORT PROBLEMS

Most of what has been explained in general for flow models can be extended to transport models. We will therefore only present the specifics of discrete fracture transport models.

4.15.1 Building the Transport Model

The method most widely used to represent transport in a stochastic fracture network model is the "particle following" method (Robinson, 1984; Cacas et al., 1990b; and Nordqvist et al., 1992). It is, of course, first necessary to build a flow model; the transport is then calculated for each realization of the model. We describe below the method as applied to a channel model (or along lines in 2D).

In its original form, the particle following method consists in following discrete particles added to the flow, whose trajectories are determined from the flow pattern. The main assumptions are the following: (i) the particles are moved by advection only (longitudinal dispersion is disregarded); (ii) chemical reactions and sorption are not taken into account; and (iii) diffusion both in the fluid and in the rock matrix is neglected.

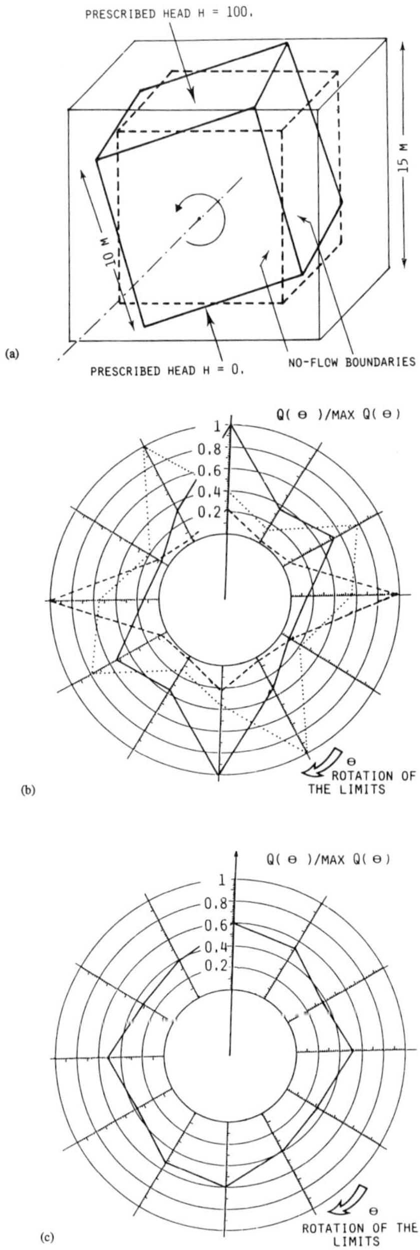

Figure 4.14.4. Determination of the anisotropy of a fracture model, after Cacas et al. (1990a): a) selection of the orientation of a fracture network cube in a realization for a numerical experiment; b) directional hydraulic conductivity of three realizations; c) average directional hydraulic conductivity of 17 realizations.

The particle trajectories are drawn at random according to the following rules: (i) The particle trajectory is divided into elementary steps corresponding to the displacement between two adjacent connections. An average travel time is computed at each step. It is inferred from the flow rate and the volume of the channel through which the particle travels. The total residence time of a particle along its path is given by the summation of the elementary travel times computed at each step from the entrance to the exit points. (ii) When the particle arrives at a new connection, the next channel through which it is going to move is drawn at random. The assumption of a perfect mixing of the different fluxes entering a connection is made (Fig. 4.15.1a). Therefore the probability that the particle will follow a given channel is proportional to the ratio of its flow rate to the total flux that goes through the connection.

a

b

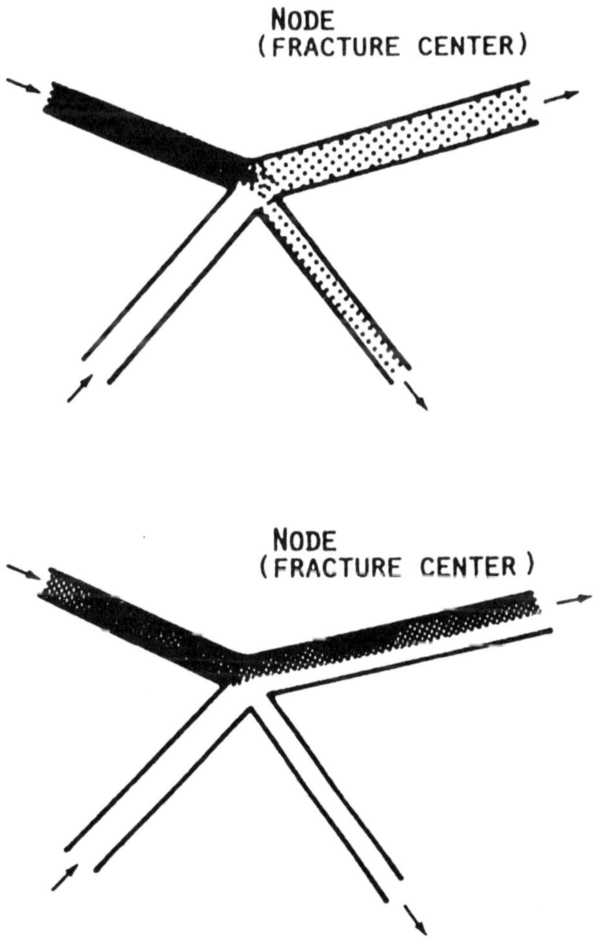

Figure 4.15.1. Mixing assumptions at fracture intersections, after Cacas et al. (1990b): a) perfect mixing; b) no mixing.

The assumption of complete mixing at each fracture node needs to be discussed (e.g., see Novakowski et al., 1985, and Philip, 1988). In practice, perfect mixing at the nodes means that all the outflows of a given fracture of the model have the same tracer concentration. In the assumption of laminar flow, the streamlines define stream tubes with a constant flow rate (Endo et al., 1984), as shown in Fig. 4.15.1b. In the case of a dense and well-connected network of channels inside the fracture surface, the complete mixing hypothesis tends to be realistic because of the spreading of the stream tubes over all the different paths open to the flow, thus creating an intense microscopic dispersion. Conversely, if the channel network is sparse and not connected, the tracer moves in a few preferred paths and no mixing occurs.

The particle-following method could easily take into account biased mixing or even no mixing in the fractures; but no field experiments have yet been carried out that have yielded quantitative results concerning the tracer mixing in channelized fracture flow.

In the channel model, since the exact geometry of the channels is not known, the volume of the channel used to calculate the travel time when the flow rate is known is a parameter of the model; Cacas et al. (1990b) calculated this volume by first assuming that each channel was a straight cylindrical pipe; the radius of this pipe was calculated assuming that the hydraulic resistance of the channel, as calibrated in the flow model, was that of laminar flow in a smooth cylinder: this gives the smallest possible volume of each channel. They then had to multiply this volume by a constant number, identical for all channels, to obtain calibration with tracer test data, without changing the hydraulic resistance of the channels. This multiplying factor was one of the two parameters of the transport model. After the fitting, this number turned out to be on the order of 32 for a fractured granite (see Sec. 4.15.2). This can easily be explained since a "channel" in a fracture plane is far from being a straight cylindrical pipe: its hydraulic diameter varies along the path, and the sections with a small diameter control the hydraulic conductivity, whereas the sections with a large diameter control the total volume of the channel. The hydraulic calibration can estimate the hydraulic diameter of the small sections, but only information obtained through tracer tests can provide the means necessary for the calibration of the diameter of the large sections. The calibrated value of 32 for the multiplying factor thus shows that the channels have an equivalent mean diameter approximately 6 times larger than their hydraulic conductivity would imply. The second parameter of the model is an extension of the original particle-following method to include, approximately, some longitudinal microscopic hydrodynamic dispersion. This is done by randomly modifying the average residence time at each step. The velocity in each channel is perturbed by the addition of a random number normally distributed with zero mean and standard deviation sigma given by:

$$\sigma = \frac{1}{T_m}\sqrt{2U\alpha T_m}$$

where T_m is the average residence time (in s), α is the microdispersivity coefficient in the fracture plane (in m), and U is the average velocity in the channel (in m/s). The dispersion created in this way is an approximation of the 1-D Fickian model of dispersion. α is thus the second parameter of the transport model.

To represent transport, several tens of thousands of particles are injected successively into the model at the entrance (a point, a line, or a surface) and their exit points and transit times are calculated. A histogram of this transit time at any exit point can then be calculated and represents the breakthrough curve for a slug injection of tracer at the entrance point. Other injection patterns can be calculated by convolution. No concentration within the domain can be calculated, however.

This model is computationally very efficient.

When the flow model is made of finite element meshes for each fracture plane, it is possible to solve the usual convection-dispersion equation in each fracture with mass transfer from fracture to fracture at the intersections. This becomes computationally very demanding but would provide an in situ concentration. Stratford et al. (1990) improved this approach by calculating response functions for each fracture plane for a step tracer injection and then assembling these response functions in series along the flow path by convolution.

4.15.2 Calibration of the Transport Model

In any fracture network transport model, a number of parameters must be calibrated from comparison with tracer experiments. These parameters essentially represent porosity and microscopic dispersion, as, for example, in the Cacas et al. (1990b) model, where only two parameters were defined, representing average properties for the whole model. These parameters must be derived from a comparison of the model outputs with experiments. However, there is one difficulty that frequently crops up here: generally only a few tracer tests are available and furthermore, they more often than not represent transport over different distances of the fractured medium. For a given transport distance, one therefore has just one experiment to compare with the model results and, by definition, a stochastic fracture network model will provide as many results as realizations.

This contrasts with the calibration of a flow model, where in general many injection tests with the same geometry can be compared with a large number of realizations of the stochastic medium. Cacas et al. (1990b) proposed a way out of this dilemma. In their case, four tracer tests were available, on four independent pathways with different lengths; they assumed that all four tests were characterizing the same fractured medium on which a stochastic flow model had already been calibrated; thus they had two parameters to calibrate on these four tests. They developed four different submodels, each representing one of the four tracer tests, with its particular geometry, distance, flow conditions, etc. They then characterized each test by a few global numbers, e.g., time of arrival of the peak concentration, the total breakthrough duration (for a slug injection of tracer), or the time of first arrival. Several realizations of each tracer test were then made, from which a histogram of the corresponding numbers for each test could be established (they used only time of peak arrival and total breakthrough duration).

For a single test, it is impossible to compare one such number with a histogram: there are no valid reasons why the experimental value should fall around the mean of the histogram, nor why it should be on the right or on the left of the distribution. However, when four tests are compared simultaneously, it is justified to require that all four tests should not simultaneously be on the same side of their histograms: if this were the case, one could logically assume that the model was biased.

It thus becomes possible to calibrate the model by modifying its parameters until the four tests are reasonably well distributed over their respective histograms. This is shown in Fig. 4.15.2 a,b. The four calculated histograms of the time of peak arrival (a) or (open circles) breakthrough duration (b) are shown, one for each tracer test; a star also shows the single measured value for the corresponding test; it is easy to see that the model is approximately calibrated, since for test

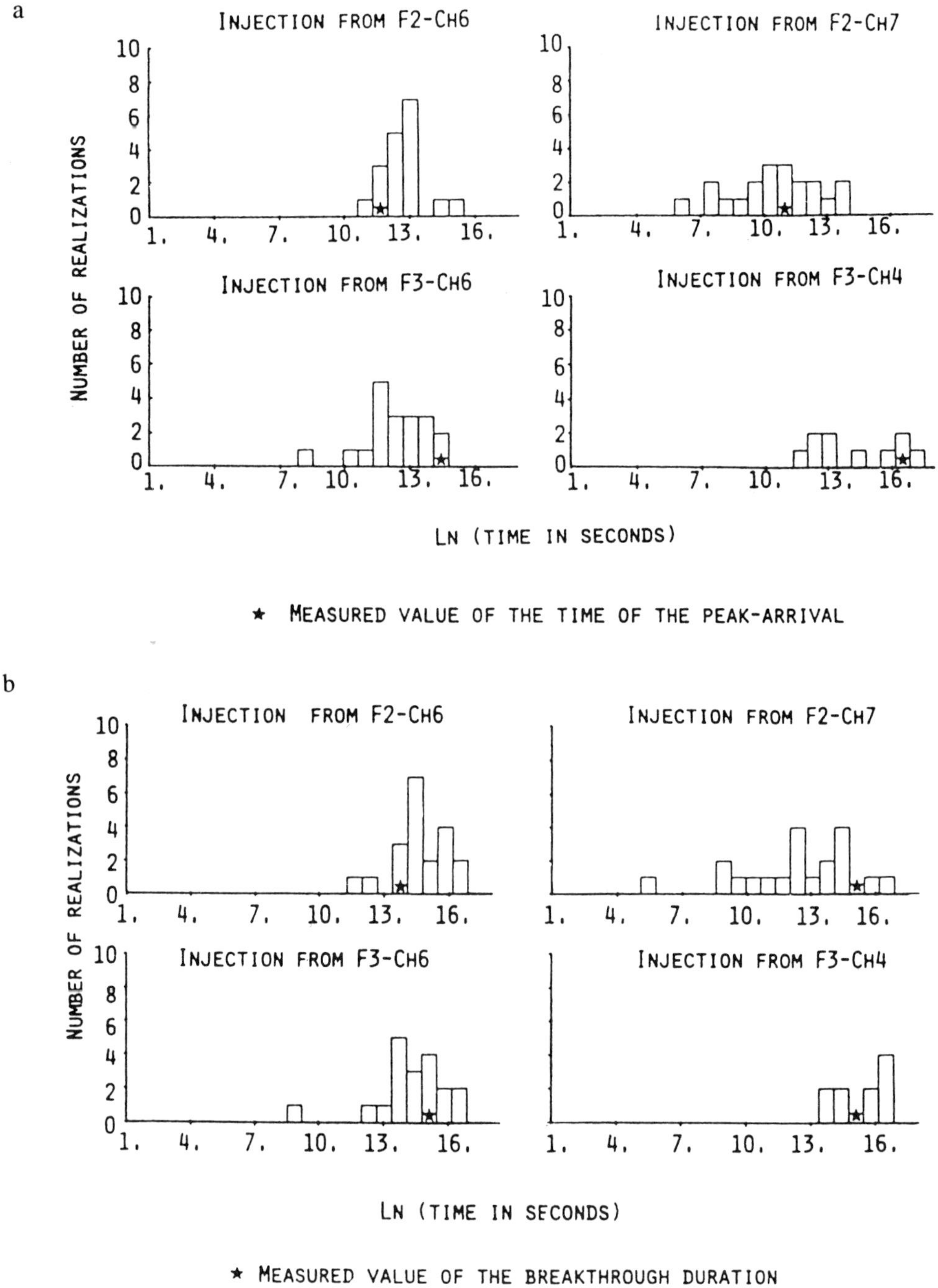

Figure 4.15.2. Calibration of a stochastic transport model, after Cacas et al. (1990b): a) comparison of observed and calculated peak arrival times; b) comparison of observed and calculated breakthrough durations.

F2-CH6 the star is on the left side of the histogram, for F2-CH7, it is on the right side, for F3-CH6, on the right side, and for F3-CH4, it is close to the middle, in Fig. 4.15.2a, and the same argument also applies to Fig. 4.15.2b. It was with this comparison that the first multiplying parameter of 32 quoted in Sec. 4.15.1 was obtained.

The second parameter of the Cacas et al. (1990b) model, representing longitudinal microscopic dispersion, was calibrated by comparing the "smoothness" of the calculated responses with the actual breakthrough curves measured in the field; Fig. 4.15.3 is a sensitivity study of the response of the model to the α parameter of the longitudinal microscopic dispersion, as explained in section 4.15.1. From this sensitivity study, a value of α of 0.8 m was selected. Note that this microscopic longitudinal dispersion introduced in the particle-following algorithm only provides a small part of the global dispersion; most of the global longitudinal dispersion and all of the transverse dispersion is given by the connections between the different channels in the fracture network. No additional parameter can modify the transverse dispersion, at least in the Cacas et al. (1990b) model, which is only due to the branching-out of the different fractures in the geometric network. Tracer tests are thus a way to validate the geometry of the fracture network model, by comparing observed and calculated transverse dispersion.

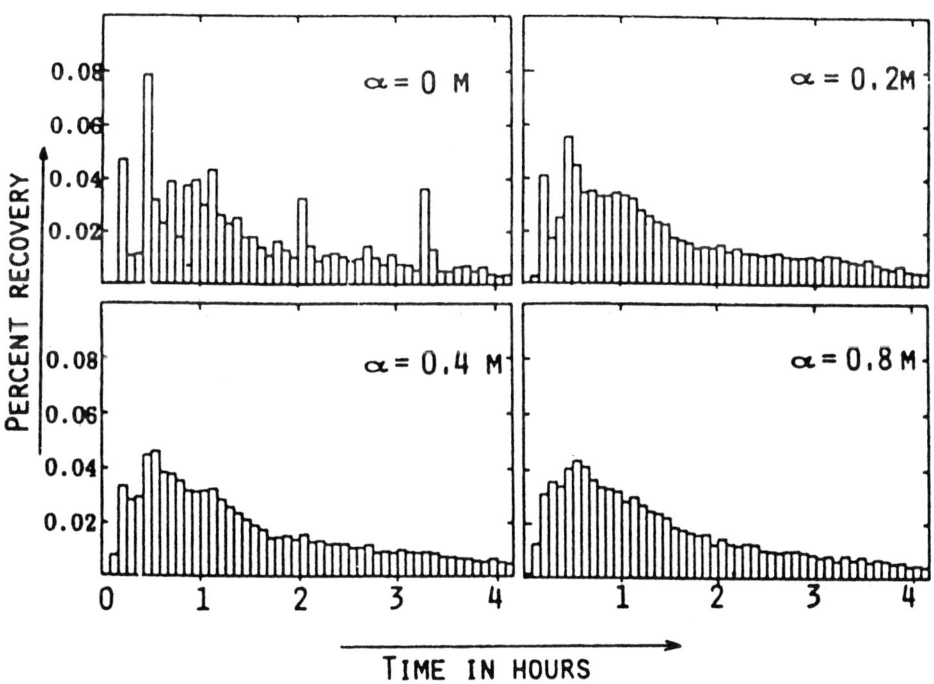

Figure 4.15.3. Study of the sensitivity of the channel transport model to the microdispersion parameter α. After Cacas et al. (1990b).

To actually determine the real transverse dispersion of the fracture system from the observations it is necessary to observe the "spread" of the tracer in a direction orthogonal to the main flow direction; in a two-well tracer test, this is impossible; it can, however, be done in an underground test, when a tracer is injected into a borehole and recovered at the walls of a drift, as in Cacas et al. (1990b) or Nordqvist (1992). In such a case, the observed "spread" of the tracer in the drift is compared, for each tracer test, with the calculated spread for an ensemble of realizations.

4.15.3 Representation of Matrix Diffusion

To represent matrix diffusion in a discrete fracture network model (or to represent sorption or chemical reactions), a method of representing transport other than the particle-following algorithm must be used. When the flow model is made of finite element meshes for each fracture plane and the usual convection-dispersion equation is solved in each fracture, then matrix diffusion (or sorption and chemical reactions) can be introduced as in any numerical model. For the channel model, concentrations must be calculated at each node of the model by first solving an advection equation in each channel of the model and then a mass balance equation at each node. Perfect mixing of all the fluxes coming from each channel connected to a node is assumed. Matrix diffusion (or sorption and chemical reactions) can be included in this mass balance equation assuming that at each node a given area of the fracture is available for contact with the fluid. Preliminary tests of this method can be found in Cacas (1989) or Bruel (1990) for a thermal problem, where heat conduction in the unfractured rock takes the place of matrix diffusion.

4.15.4 Results Given by a Stochastic Fracture Network Transport Model

Like the flow models, the discrete fracture network transport models are severely limited by the size at which they can represent transport. Thus their use is presently limited to the interpretation of tracer tests or to the prediction of transport over small distances, e.g., tens or hundreds of meters at most. Their advantage over more classical continuous models is that they can help to predict the probability of existence of a "short-circuit" in a given rock mass made of connected fractures. When the flow and transport properties of the discrete model have been calibrated, one can indeed run transport calculations over a large number of realizations on the distance of interest and determine the percentage of the results that show "rapid" transit, defined as a given fraction of the mean travel time. This percentage can then be seen as the probability that, in the real medium, a "short-circuit" may exist over the given travel path. Any other statistics of the travel time can be determined just as well.

4.16 OTHER STOCHASTIC FRACTURE MODELS

Neuman (1988) has suggested another method of stochastically representing a fractured medium. It is based on continuous equivalent medium representation on a regular grid (finite differences or finite elements); each grid block being given an equivalent hydraulic conductivity value.

The generation of several possible such hydraulic conductivity fields is made in three steps: (i) a large number of injection tests performed in the medium over small injection chambers are interpreted using one of the usual analytical expressions describing such tests, e.g., the Dupuit or

Lugeon formulas; (ii) the variogram of these local values is determined with a geostatistical analysis; and (iii) conditional simulations of the hydraulic conductivity are generated using the procedure described, for example, by Delhomme (1979).

As in discrete fracture network models, the method needs a large number of realizations of the fractured medium in order to analyze its behavior, not just a single one; the author also claims that the size of the grid blocks on which the conditional simulations are done can be on a smaller scale than the representative elementary volume of the fracture network, since the domain investigated by the injection tests can also be below the size of the REV. The geostatistical simulations are a means to generate the below-REV spatial variability of the medium.

One of the problems of this approach is, however, that it assumes that the analytical expressions used to interpret the injection tests provide representative local values, which may not always be true. Cacas (1989) has tried to determine to what extent such local radial tests are representative of the true local hydraulic conductivity of the medium. Using a theoretical 3D discrete fracture network model representing a cube, she simulated radial injection tests in such cubes and determined numerically, for 155 realizations, the local hydraulic conductivity K using the classical formula

$$K = \frac{Q}{L \Delta H}$$

(Louis, 1974), where Q is the calculated injection flow rate, L is the length of the injection chamber, and ΔH is the prescribed increase in hydraulic head in the chamber. This formula assumes that the radius of influence R of an injection test is

$$R = r_0 e^{2\pi}$$

where r_0 is the radius of the injection chamber.

She then determined the local hydraulic conductivity of the same theoretical fractured medium by running parallel flow experiments in a cube on the same realizations, as explained in Sec. 4.14.5 and Fig. 4.14.4. She observed that, whatever the variance of the hydraulic conductivity of the channels, the distribution of the hydraulic conductivity given by the radial tests had a larger variance than that of the parallel flow experiment in a cube, and furthermore, that the mean of the former was about 5 times larger than that of the latter.

These results should be taken with some caution since they apply to a theoretical model and not to a real medium. They show, however, that using classical analytical expressions to interpret injection tests in fractured media may not always be justified. Using discrete fracture network models to make such interpretations, as explained in Sec. 4.14.4, may be a better approach.

4.17 CONCLUSION

Stochastic methods can realistically model and simulate the geometry of actual fracture fields. Much care must be taken during the statistical study of the fractures because of the numerous bias sources that affect the data. Network models of variable sophistication are available. The parameters of the 3D models can be deduced from 1D and 2D surveying. Simulations visualize the possible 3D geometry of reality and can be conditioned to the actual data.

Several approaches can be used to study the variability at the scale of the single fracture. The reduction of fracture networks to channel networks opens new ways, such as the comparison with media of noninteger dimension or the analogy with regular lattice networks.

There is obviously a need for further development. In the field of geometric modeling, it would be useful to develop models that take into account the chronology of the brittle deformation that generated the fractures as well as the mechanism of crack propagation (e.g., Granier, 1985; Meakin, 1990). This may involve refinements in the survey and modeling techniques, for example in order to describe the points where some fractures are stopped by others (Gervais et al., 1992). Concerning the shape of the single fracture and its effect on flow, the above-mentioned approaches have to be investigated in more depth. This should lead to new developments concerning the general hydraulic behavior of fractured fields and its link with connectivity.

Concerning flow and transport modeling, the assumptions of channeling in the fracture plane for various rock types and history should be further investigated; the degree of mixing at channel or fracture intersections should also be experimentally examined. More efficient computer codes capable of handling a larger number of fractures and thus of representing a larger rock volume should be developed. The linking of discrete fracture network models with more usual continuous models should also be considered in order to have local discrete representations of fractured media imbedded in discretized continuous models.

REFERENCES

Andersson, J., and Dverstorp, B. Conditional simulations of fluid flow in three-dimensional networks of discrete fractures, *Water Resour. Res.*, vol. 23(10), p. 1876–1886, 1987.

Andersson, J., Shapiro, A.M., and Bear, J. A stochastic model of fractured rock conditioned by measured information, *Water Resour. Res.*, vol. 20(1), p. 79–88, 1984.

Baecher, G.B, and Lanney, N.A. Trace length biases in joint surveys, *Proceedings,* 19th US Symp. Rock Mechanics, AIME, vol. 1, p. 56–65, 1978.

Baecher, G.B., Lanney, N.A., Einstein, H.H. Statistical description of rock properties and sampling, *Proceedings,* 18th US Symp. Rock Mechanics, AIME, p. 5C1:1–8, 1977.

Bailey, A.I. A method of analyzing polymodal distributions in orientation data, *Math. Geol.,* vol. 7(4), p. 285–293, 1975.

Barker, J. A generalized radial flow model for hydraulic tests in fractured rock, *Water Resour. Res.*, vol. 24(10), p. 1796–1804, 1988.

Barla, G., Scavia, C., Antonellis, M., and Guarascio, M. Characterization of rock mass by geostatistical analysis at the Masua mine, *Proceedings,* 6th Int. Congr. Rock Mechanics, ISRM, Montreal, vol. 2, p. 777–786, 1987.

Barton, C.C. Fractal analysis of the scaling and spatial clustering of fractures in rock. In: "Fractals and their application to geology," *Proceedings*, 1988 GSA Annual Meeting Symposium on Fractals, GSA, in press, 1992.

Barton, C.C., and Larsen, E. Fractal geometry of two-dimensional fracture networks at Yucca Mountain, southwestern Nevada, *Proceedings*, Int. Symp. on Fundamentals of Rock Joints, Bjorkliden, p. 77–84, 1985.

Billaux, D. Influence de la connectivité d'un réseau de fractures sur sa réponse à un essai de pompage, *Rock at Great Depth-Rock Mechanics and Rock Physics at Great Depth*, Proceedings of an International Symposium, Pau, 28-31.08.89, V. Maury and D. Fourmaintraux (eds.), 1620 p., 3 vols., Hfl.295, A.A. Balkema, P.O. Box 1675, Rotterdam, p. 473–480, 1989.

Billaux, D. Hydrogéologie des milieux fracturés. Géométrie, connectivité et comportement hydraulique, Ph.D. thesis, Ecole des Mines, Paris, 277 p., 1990.

Billaux, D., Chilès, J.P., Hestir, K., and Long, J. Three-dimensional statistical modeling of a fracture rock mass – an example from the Fanay-Augères mine, *Int. J. Rock Mech. Min. Sci. and Geomech. Abstr.*, vol. 26(3/4), p. 281–299, 1989.

Broadbent, S.R., and Hammersley, J.M. Proc. Camb. Phil. Soc., Vol. 53, p. 629, 1957.

Brown, S.R., and Scholz, C.H. Broad bandwidth study of the topography of natural rock surfaces, *J. Geophys. Res.*, vol. 90(12), p. 575–582, 1985.

Bruel, D. Exploitation de la chaleur des roches chaudes et sèches. Etude des phénomènes hydrauliques, mécaniques et thermiques au moyen d'un modèle à fractures discrètes, Ph.D. thesis, Ecole des Mines de Paris, Fontainebleau, 161 p., 1990.

Cacas, M.C., Ledoux, E., de Marsily, G., Tillie, B., Barbreau, A., Durand, E., Feuga, B., and Peaudecerf, P. Modeling fracture flow with a stochastic discrete fracture network: 1. The flow model, *Water Resour. Res.*, vol. 26(3), p. 479–489, 1990a.

Cacas, M.C., Ledoux, E., de Marsily, G., Barbreau, A., Calmels, P., Gaillard, B., and Margritta, R. Modeling fracture flow with a stochastic discrete fracture network: 2. The transport model, *Water Resour. Res.*, vol. 26(3), p. 491–500, 1990b.

Cacas, M.C. Développement d'un modèle tridimensionnel stochastique discret pour la simulation de l'écoulement et des transports de masse et de chaleur en milieu fracturé, Ph.D. thesis, Ecole des Mines de Paris, Fontainebleau, 1989.

Charlaix, E., Guyon, E., and Rivier, N. A criterion for percolation threshold in a random array of plates, *Solid State Communication*, vol. 20(11), p. 999–1002, 1984.

Chelidze, T.L. Percolation and fracture, *Physics of Earth and Planetary Interiors*, vol. 28, p. 93–101, 1982.

Chilès, J.P. Fractal and geostatistical methods for modeling a fracture network, *Math. Geol.*, vol. 20(6), p. 631–654, 1988.

Chilès, J.P. Three-dimensional geometric modelling of a fracture network, *Geostatistical Sensitivity, and Uncertainty Methods for Ground-Water Flow and Radionuclide Transport Modeling*, B.E. Buxton (ed.), Battelle Press, Columbus, Ohio, p. 361-385, 1989a.

Chilès, J.P. Modélisation géostatistique de réseaux de fractures, *Geostatistics*, M. Armstrong (ed.), Kluwer Academic Publ., Dordrecht, Netherlands, Vol. 1, p. 57–76, 1989b.

Chilès, J.P., and Gentier, S. Geostatistical modeling of a single fracture. *Proceedings,* 4th Int. Geostatistics Congress, Troia, Portugal, 1992, to be published.

Chilès, J.P., Guérin, F., and Billaux, D. 3D stochastic simulation of fracture network and flow at Stripa conditioned on observed fractures and calibrated on measured flow rates, *Rock Mechanics*, J.R. Tillerson and W.R. Wawersik (eds.), Balkema, Rotterdam, Netherlands, p. 533-542, 1992.

Clerc, J.P., Giraud, G., Roussenq, J., Blanc, R., Carton, J.P., Guyon, E., Ottavi, H., and Stauffer, D. La percolation: modèles, simulations analogiques et numériques, *Annales de Physique*, vol. 8(1), p. 5–105, 1983.

Conrad, F., and Jacquin, C. Représentation d'un réseau bidimensionnel de fractures par un modèle probabiliste. Application au calcul des grandeurs géométriques des blocs matriciels, *J. of IFP*, vol. 28(6), p. 843–890, 1973.

Courrioux, G., and Jacquot, T. Etude de l'effet de coupe dans l'interprétation des diagrammes de pétrofabrique: application au granite de Beauvoir (Echassières, Massif Central Français). *Comptes Rendus*, Académie des Sciences, Paris, vol. 299, série II(9), p. 549–554, 1984.

Cruden, D.M. Describing the size of discontinuities, *Int. Rock. Mech. Min. Sci. and Geomech. Abstr.*, vol. 14, p. 133–137, 1977.

de Marsily, G. Flow and transport in fractured rocks. Connectivity and scale effect. In: Hydrogeology of Rocks of Low Permeability, *Memoires Int. Assoc. Hydrogeol.*, vol. 17, p. 267–277, 1985.

de Marsily, G. *Quantitative Hydrogeology, Groundwater Hydrology for Engineers*, Academic Press, New York, 440 p., 1986.

Delhomme, J.P. Kriging in the hydrosciences, *Advances in Water Resources*, vol. 1(5), p. 251–266, 1978.

Delhomme, J.P. Spatial variability and uncertainty in groundwater flow parameters: a geostatistical approach, *Water Resour. Res.*, vol. 15(2), p. 269–280, 1979.

Dershowitz, W.S. Rock joint systems, Ph.D. thesis, MIT, Cambridge, Mass., 764 p., 1984.

Dershowitz, W.S., and Einstein, H.H. Characterizing rock joint geometry with joint system models, *Rock Mechanics and Rock Engineering*, vol. 21, p. 21–51, 1988.

Dienes, J.K., 1982 Permeability, percolation and statistical crack mechanism. Issues in Rock Mechanics, Proc. 22nd Symp. on Rock Mechanics, Berkeley, Univ. of Calif. Aug. 1987.

Diggle, P.J., and Fisher, N.I. Sphere: a contouring program for spherical data, *Computer and Geosciences*, vol. 11(6), p. 725–766, 1985.

Dverstorp, B., and Andersson, J. Application of discrete fracture network concept with field data: possibilities of model calibration and validation, *Water Resour. Res.*, vol. 25(3), p. 540–550, 1989.

Dverstorp, B. Report to SKI, Royal Institute of Technology, Stockholm, 1990.

Dverstorp B., and Andersson, J. Estimation of effective transport parameters for predictive use in fractured rock, (1991), *Water Resour. Res.*, to appear,

Endo, H.K., Long, J.C.S., Wilson, C.R., and Witherspoon, P.A. A model for investigating mechanical transport in fracture networks, *Water Resour. Res.*, vol. 20(10), p. 1390–1400, 1984.

Engelman, R., Gur, Y., and Jaeger, Z. Fluid flow through a crack network in rocks, *J. Applied Mechanics*, vol. 50, p. 707–711, 1983.

Feder, J. *Fractals*, Plenum Publ. Corp., New York, 283 p., 1988.

Feuga, B. Hydrogeologic modeling of fractured rock, *Jornadas sobre modelos matematicos aplicables al almacenamiento de residuos radioactivos*, Esc. Sup. de Engenieros de Minas, Madrid., 1986.

Gentier, S. Morphologie et comportement hydromécanique d'une fracture naturelle dans un granite sous contrainte normale, Ph.D. thesis, Univ. of Orléans, Doc. BRGM no. 134, 597 p., 1986.

Gentier, S., Chilès, J.P., and Riss, J. Analyse et simulation d'une surface de fracture rocheuse. *Cahiers de Géostatistique*, Fasc. 1, ENS des Mines de Paris, p. 77–87, 1991.

Gervais, F., Chilès, J.P., and Gentier, S. Geostatistical analysis and hierarchical modeling of fracture network in a stratified rock mass, *Proceedings*, *Fractured and Jointed Rock Masses*, Lake Tahoe, California, to be published by A.A. Balkema, Rotterdam, 1992.

Granier, T. Origin, damping and pattern of development of faults in granite, *Tectonics*, vol. 4(7), p. 721–737, 1985.

Grossman, N.F. The bivariate normal distribution on the tangent plane at the mean attitude, *Proceedings,* Int. Symp. on Fundamentals of Rock Joints, Bjorkliden, Sweden, p. 3–11, 1985.

Guyon, E., Hulin, J.P., and Lenormand, R. Applications de la percolation à la physique des milieux poreux, *Les Annales des Mines*, vol. 5-6, p. 17-40, 1984.

Hestir, K., and Long, J. Analytical expression for the permeability of random two-dimensional Poisson fracture networks based on regular lattice percolation and equivalent media theories, J. Geoph. Res., vol. 95, B13, p. 21565–21581, 1990.

Jacquin, C.G., and Adler, P.M. Fractal porous media. II. Geometry of porous geological structures, *Transport in Porous Media*, 2, p. 571–596, 1987.

Kendall, M.G., and Moran, P.A. *Geometrical probability*. Hafner, New York, 1963.

Kirkpatrick, S. Percolation and conduction. *Rev. of Modern Physics*, vol. 45(4), p. 574–588, 1975.

La Pointe, P.R. Analysis of spatial variation in rock mass properties through geostatistics, *Proceedings,* 21st US Symp. on Rock Mechanics, Rolla, Missouri, p. 570–580, 1980.

Laslett, G.M. Censoring and edge effect in areal and line transect sampling of rock joint traces, *Math. Geol.*, vol. 14(2), p. 125–140, 1982.

Lewis, T., and Fisher, N.I. Graphical methods for investigating the fit of a Fisher distribution to spherical data, *Geophys. J.R. Astr. Soc. 69*, p. 1–13, 1982.

Loiseau, P., 1987. Etude structurale et géostatistique des gneiss de la région du Cézallier: modélisation tridimensionnelle de réseaux de fractures; application à l'écoulement des fluides, Ph.D. thesis, University of Orléans, Doc. BRGM no 162, 1988, 200 p.

Long, J.C.S. Investigation of equivalent porous media permeability in networks of discontinuous fractures, Ph.D. thesis, Lawrence Berkeley Laboratory, Univ. of California, Berkeley, CA, 1983.

Long, J.C.S. Modeling of fluid flow and transport in fracture networks, *Jornadas sobre modelos matematicos aplicables al almacenamiento de residuos radioactivos*, Esc. Sup. de Engenieros de Minas, Madrid, 1986.

Long, J.C.S., Remner, J.S., Wilson, C.R., and Witherspoon, P.A. Porous media equivalents for networks of discontinuous fractures, *Water Resour. Res.*, vol. 18(3), p. 645–658, 1982.

Long, J.C.S., and Billaux, D. From field data to fracture network modeling: An example incorporating spatial structure, *Water Resour. Res.*, vol. 23(7), p. 1201–1216, 1987.

Louis, C. Introduction à l'hydraulique des roches, *Bull. Bur. Rech. Géol. Min.*, Ser. 2, Sec. III, vol. 4, 1974.

Mancktelow, N.S. A least-square method for determining the best-fit point maximum, great circle and small circle to nondirectional orientation data, *Math. Geol.*, vol. 13(6), p. 507–521, 1981.

Mandelbrot, B.B. *The Fractal Geometry of Nature*, Freeman, New York, 468 p., 1983.

Mandelbrot, B.B. Self-affine fractals and fractal dimension, *Physica Scripta*, vol. 32, p. 257–260, 1985.

Mandelbrot, B.B., Passoja, D.E., and Paullay, A.J. Fractal character of fracture surfaces of metals, *Nature*, vol. 308, p. 721–722 , 1984.

Mantoglou, A. Digital simulation of multivariate two- and three-dimensional stochastic processes with a spectral turning bands method, *Math. Geol.* vol. 19(2), p. 129–149, 1987.

Mardia, K.V. *Statistics of Directional Data.* Academic Press, London, 357 p., 1972.

Massoud, H., 1987. Modélisation de la petite fracturation par les techniques de la géostatistique, Ph.D. thesis, Ecole des Mines, Paris, Doc. BRGM no. 155, 1988, 210 p.

Matérn, B. Spatial variation. Stochastic models and their application to some problems in forest surveys and other sampling investigations, *Meddelande fran Statens Skogsforskningsinstitut*, vol. 49(5), Almaenna Foerlaget, Stockholm, 144 p., 1960.

Matheron, G. *Les variables régionalisées et leur estimation. Une application de la théorie des fonctions aléatoires aux Sciences de la Nature*, Masson, Paris, 306 p., 1965.

Matheron, G. *Eléments pour une théorie des milieux poreux*, Masson, Paris, 1967.

Matheron, G. Les polyèdres poissoniens isotropes, *Actes du 3e Colloque Européen sur la fragmentation*, Cannes, II-9, p. 509–534, 1971.

Matheron, G. *Random Sets and Integral Geometry*, Wiley, New York, 1975.

Meakin, P. Simple models for crack growth. In: Crystal Properties and Preparation on "Elastic and Plastic Properties of Crystals," Trans. Tech Publications, Aedermannsdorf, Switzerland, 1990.

Miles, R.E. Poisson planes in Euclidean space. Part I: A finite number of random uniform flats. *Advances in Applied Probability*, no 1, p. 211–237, 1969.

Miles, R.E. The random division of space. *Special Supplement to Advances in Applied Probability*, p. 243–266, 1972.

Miller, S.M. Geostatistical analysis for evaluating spatial dependence of fracture set characteristics. *Proceedings,* 16th A PCOM Symp., SME-AIME, New York, p. 537–545, 1979.

Neuman, S.P. A proposed conceptual framework and methodology for investigating flow and transport in Swedish crystalline rocks, a position paper submitted to SKB. Ref. SKB 88-37, Stockholm, Sweden, 1988.

Nordqvist, A.W., Tsang, Y.W., Tsang, C.F., Dverstorp, B., Andersson, J. A variable aperture fracture network model for flow and transport in fractured rocks, *Water Resour. Res.*, vol. 28(6), p. 1703–1713, 1992.

Novakowski, K.S., Gordon, V.E., David, A.L., and Kenneth, G.R. A field example of measuring hydrodynamic dispersion in a single fracture, *Water Resour. Res.,* vol. 21(8) p. 1165–1174, 1985.

O'Sullivan, F. A statistical perspective on ill-posed inverse problems, *Stat. Sci.*, vol. 1(4), p. 502–527, 1986.

Pahl, P.J. Estimating the mean length of discontinuity traces, *Int. J. Rock Mech., Min. Sci. and Geomech. Abstr.* vol. 18, p. 221–228, 1981.

Philip, J.R. The fluid mechanics of fractures and other junctions, *Water Resour. Res.*, vol. 24(2) p. 239 – , 1988.

Priest, S.D., and Hudson, J.A. Discontinuity spacings in rock, *Int. J. Rock Mech., Min. Sci. and Geomech. Abstr.*, vol. 13, p. 135–148, 1976.

Robinson, P.C. Connectivity of fracture systems. A percolation theory approach, Ph.D. thesis, St. Catherine's College, Oxford Univ., UK, 1984.

Rouleau, A. Statistical characterization and numerical simulation of a fracture system. Application to a groundwater flow in the Stripa granite, Ph.D. thesis, Univ. of Waterloo, Ontario, Canada, 1984.

Santalo, L.A. *Integral Geometry and Geometric Probability*, Addison-Wesley, Reading, Mass., 434 p., 1976.

Schaeben, H. A new cluster algorithm for orientation data, *Math. Geol.*, vol. 16(2), p. 139–153, 1984.

Serra, J. *Image Analysis and Mathematical Morphology.* Academic Press, London, 628 p., 1982.

Shante, V.K.S., and Kirkpatrick, S. An introduction to percolation theory, *Adv. Phys.*, vol. 20, p. 325–357, 1971.

Smith, L., Maze, C.W., and Schwartz, F.W. A stochastic model for transport in networks of planar fractures, *Int. Symp. on the Stochastic Approach to Subsurface Flow*, Int. Ass. of Hydraulic Res., Montvillargenne, France, G. de Marsily, ed., 1985.

Snow, D.T. A parallel plate model of fractured permeable media, Ph.D. thesis, Univ. of California, Berkeley, 331 p., 1965.

Snow, D.T. The frequency and apertures of fractures in rock, *Int. J. Rock Mech. Min. Sci.*, vol. 7, p. 23–40, 1970.

Stoyan, D., Kendall, W.S., and Mecke, J. *Stochastic Geometry and Its Applications*, Wiley, New York, 1987.

Stratford, R.G., Herbert, A., Jackson, C.P. A parameter study of the influence of aperture variation on fracture flow, and consequences on fracture network, Proc. ISRM Rock Joint Conference, Balkema, Amsterdam, 1990.

Terzaghi, R.D. Sources of error in joint surveys, *Geotechnique*, vol. 15(3), p. 287–303, 1965.

Thomas, A. Structure fractale de l'architecture de champs de fractures en milieu rocheux. *Comptes Rendus*, Acad. Sci., Paris, vol . 304, ser. II(4), p. 181–186, 1987.

Tsang, Y.W., and Tsang, C.F. Channel model of flow through fractured media, *Water Resour. Res.*, vol. 23(3), p. 467–479, 1987.

Veneziano, D. Probabilistic model of joints in rock, unpub. manuscr., MIT, Cambridge, Mass., 1978.

Vere-Jones, D. Statistical theories of crack propagation, *Math. Geol.*, vol. 9, 445, 1977.

Warburton, P.M. A stereological interpretation of joint tracer data, *Int. Rock Mech. Min. Sci. and Geomech. Abstr.* vol. 17, p. 181–190, 1980.

Wilke, S., Guyon, E., and de Marsily, G. Water penetration through fractured rocks : test of a tridimensional percolation description, *Math. Geol.*, vol. 17(1), p. 17–27, 1985.

Yow, J.L. Jr. Blind zones in the acquisition of discontinuity orientation data, *Int. J. Rock Mech. Min. Sci. and Geomech. Abstr.*, vol. 24(5), p. 317–318, 1987.

5

TRACER TRANSPORT IN FRACTURE SYSTEMS

Chin-Fu Tsang
Earth Sciences Division
Lawrence Berkeley Laboratory
Berkeley, California

5.1 INTRODUCTION

The theory of tracer transport in a water-saturated homogeneous porous medium is well developed. Typically, tracers introduced into the porous medium are carried by the water flow, so that at a point downstream the tracers arrive after a delay in time, and are dispersed because of the multiple flow paths imposed by the tortuous porous structure of the medium. The equation that describes such transport is called the advective-dispersive equation. However, this equation cannot be applied directly in the study of a medium that is heterogeneous rather than homogeneous. Heterogeneous media have been actively studied by a number of authors, such as Gelhar, Dagan, Neuman and coworkers (e.g., Dagan, 1979; Gelhar and Axness, 1983, Neuman et al., 1985). Many of the approaches used by these authors involve statistical approaches, by which the medium's heterogeneity is defined, for example, in terms of a probability distribution of the hydraulic conductivity. However, for a medium containing multiple fractures, many of the established methods may not be applicable. New methods must then be developed, and this is an area of active current research.

It may be useful to point out that the transport of a tracer through a fractured medium is quite different from thermal transfer or pressure transmission through the system. Thermal transfer is a strong diffusive process involving both the fluid in the fractures and the solid components of the medium; therefore, it averages out formation heterogeneity very effectively. Pressure transmission, while not as strong a diffusive process, can still transmit between the fluid, the fracture, the pores and the matrix of the medium and can average out small-scale medium heterogeneity to a large degree. Tracer transport, however, is subjected to the very weak process of molecular diffusion and is restricted to movement within the fracture network. The result is that tracer transport is extremely sensitive to medium heterogeneity, i.e., to the specific presence and distribution of the fractures.

The purpose of the present chapter is threefold. Firstly, a number of simple calculations and sample results are presented to provide insights into the transport of tracers through fracture systems, from small transport distances where the single fracture effects are important, to larger distances where the fracture network plays the main role. Secondly, the implications of these results on experimental observations of tracer transport in fractured rocks are pointed out. Thirdly, an example of an analysis of actual field data of tracer transport in fractures is described to illustrate the special features as revealed in the calculations.

5.2 VARIABLE APERTURE NATURE OF ROCK FRACTURES

Until seven or eight years ago, most studies of rock fractures assumed that a fracture could be represented by a pair of parallel plates with a constant aperture between them. However, this has been found to be inadequate in describing fluid flow and transport through the fracture. A field experiment involving solute transport in single fractures in the Stripa mine in Sweden (Abelin et al., 1985; Neretnieks, 1985) showed that the fluid flow was very unevenly distributed along the fracture plane and that large areas did not carry water. The flow paths, or channels, made up only 5%-20% of the fracture area. The nonsorbing tracers injected were also unevenly transported through these channels. Furthermore, Abelin et al. (1985) showed that the equivalent fracture aperture derived from the constant head permeability measurements, assuming that the volumetric flow is proportional to the cube of the fracture aperture, was much smaller than that derived from the tracer migration experiments, which measured the residence volume of the tracer solution. The equivalent aperture values derived from these two different measurements would be identical if the parallel plate description of the fracture were valid.

In the tracer migration experiment carried out in granite at Cornwall, England (Heath, 1985) two parallel boreholes were drilled to intersect a number of fractures. It was found that the flow connection between the two boreholes varied considerably among the fractures. Another experiment carried out in a single fracture at the same site (Bourke, 1987) demonstrated that flow in a single fracture took place in a limited number of flow paths. In this experiment five parallel holes were drilled in the fracture plane. Each of the holes was pressurized in turn. The flow connections with the adjacent holes were measured by pressure response measurements in packered sections in these holes. The results showed the flow took place in only a few flow channels and that the channels occupied a total area of about 10-20% of the fracture plane.

Direct measurements of fracture aperture values in laboratory core samples (Gentier, 1986; Gale, 1987) revealed that the aperture, b, has variable values, which may be represented by a log normal distribution, $n(b)$, characterized by two parameters: the mean log aperture, $\log b_o$, and a standard deviation in $\log b_o$, σ.

$$n(b)\mathrm{d}b = \frac{1}{\sqrt{2\pi}(\ln 10)\sigma b} \exp\left(-\frac{(\log b - \log b_o)^2}{2\sigma^2}\right) \mathrm{d}b \qquad (5.2.1)$$

In addition to the aperture density distribution parameters, the spatial correlation length λ describing the spatial distribution of these aperture values over the fracture plane is also needed to define the fracture.

Given the parameters b_o, σ and λ we can generate a realization of the single fracture with variable apertures by geostatistical methods. For example, the numerical code COVAR

(Williams and El-Kadi, 1986) is used to generate different aperture values in a fracture plane that has been divided into square meshes. COVAR uses the matrix decomposition method to generate the log-normally distributed values of fracture aperture b, which are first transformed to a normal distribution Y.

$$Y = \log_{10} b \tag{5.2.2}$$

The values of Y are estimated from

$$Y = L\,\varepsilon + \nu \tag{5.2.3}$$

in which ν is the mean of Y, ε is a vector in $N[0,1]$, and L is defined in terms of the covariant matrix

$$A = L\,L^T\ . \tag{5.2.4}$$

Equation (5.2.3) represents a generated process because the mean is given by

$$E[Y] = LE[\varepsilon] + \nu = \nu \tag{5.2.5}$$

and the covariance is given by

$$E[(Y-\nu)(Y-\nu)^T] = LE[\varepsilon\varepsilon^T]\,L^T = LL^T = A \tag{5.2.6}$$

in which E stands for the expected value. We use the exponential form of the covariance function

$$A = \sigma^2 \exp(-\alpha|r|) \tag{5.2.7}$$

in which σ^2 is the variance of Y, r is the separation lag, and α is the autocorrelation parameter, which has a dimension of inverse length. The exponential form of (5.2.7) indicates that quantities within a distance of the order of $2/\alpha$ will be correlated, and thus we may define the correlation λ to be $2/\alpha$. The form of (5.2.7) indicates that the covariance chosen is isotropic. An anisotropic form of the covariance function may also be chosen.

Figures 5.2.1 and 5.2.2 show eight realizations of statistically generated apertures with identical mean and variance of the log normal aperture density distribution, with the mean $\nu = \log_{10} b_o$ equal to 1.7 and the standard deviation in $\log_{10} b$, σ, equal to 0.43. These parameter values are consistent with the aperture data derived from the fracture surface roughness profiles of granitic samples 12 cm in diameter (Gentier, 1986). The square regions with the dimension $L_x = L_y = 1$ in these figures represent, therefore, a single fracture with spatially correlated variable aperture as a flow region. Figure 5.2.1 differs from Fig. 5.2.2 only in the spatial correlation length λ of the apertures. The correlation length is expressed in terms of a fraction of the linear extent of the generated flow region. The spatial variations of the apertures in Fig. 5.2.1 correspond to a correlation length λ of 0.1 of the linear dimension of the fracture flow region and those in Fig. 5.2.2 correspond to a correlation length λ of 0.4 of the linear dimension. The variation of the apertures is represented by the different shading in these figures. The darker the shading, the smaller the aperture. The shadings are given in five levels. The dividing values of the logarithm of the aperture for the five levels are $\nu - 0.85\sigma$, $\nu - 0.25\sigma$, $\nu + 0.25\sigma$, and $\nu + 0.85\sigma$. It is apparent that the spatial correlations of the apertures are different in Figs. 5.2.1 and 5.2.2.

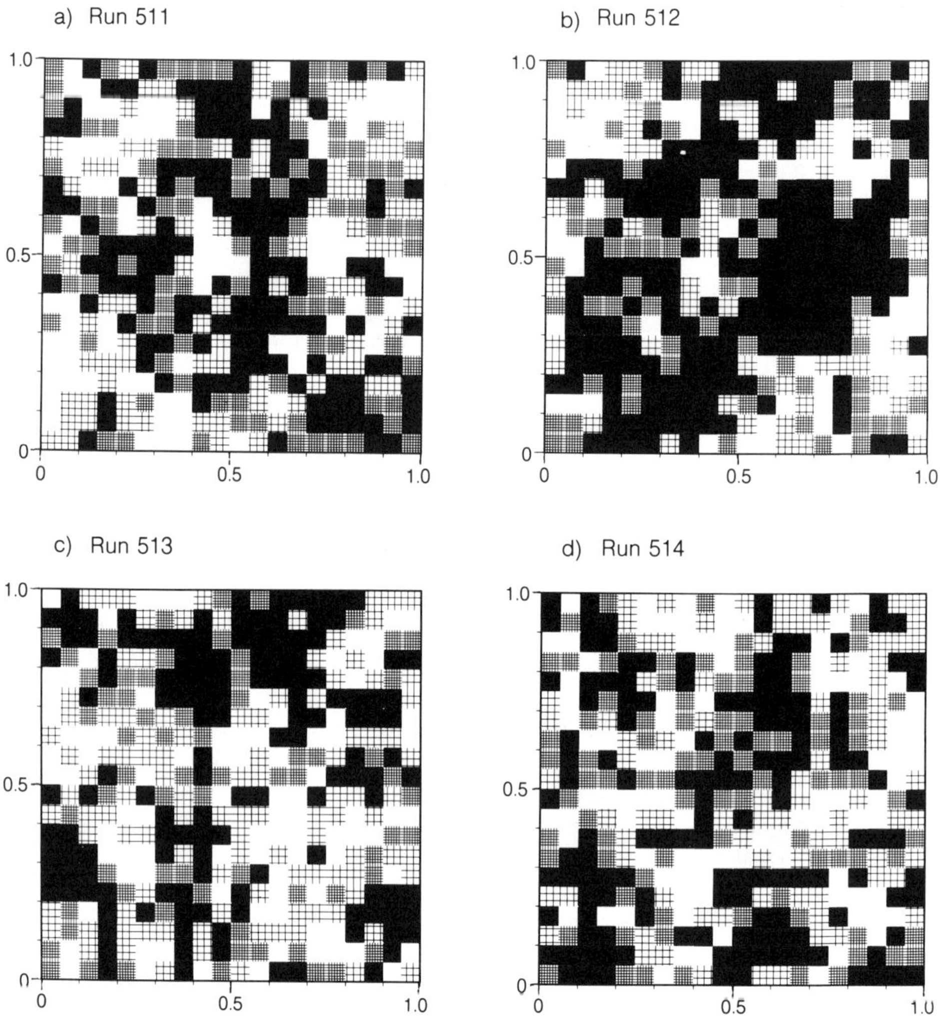

Figure 5.2.1. Statistically generated apertures with a spatial correlation length of 0.1 in the plane of a single fracture of linear dimension 1.0. The shadings are given in five levels from small aperture values (dark) to large ones (white). The dividing values correspond to the logarithm of apertures given by $\nu - 0.85\sigma$, $\nu - 0.25\sigma$, $\nu + 0.25\sigma$ and $\nu + 0.850\sigma$.

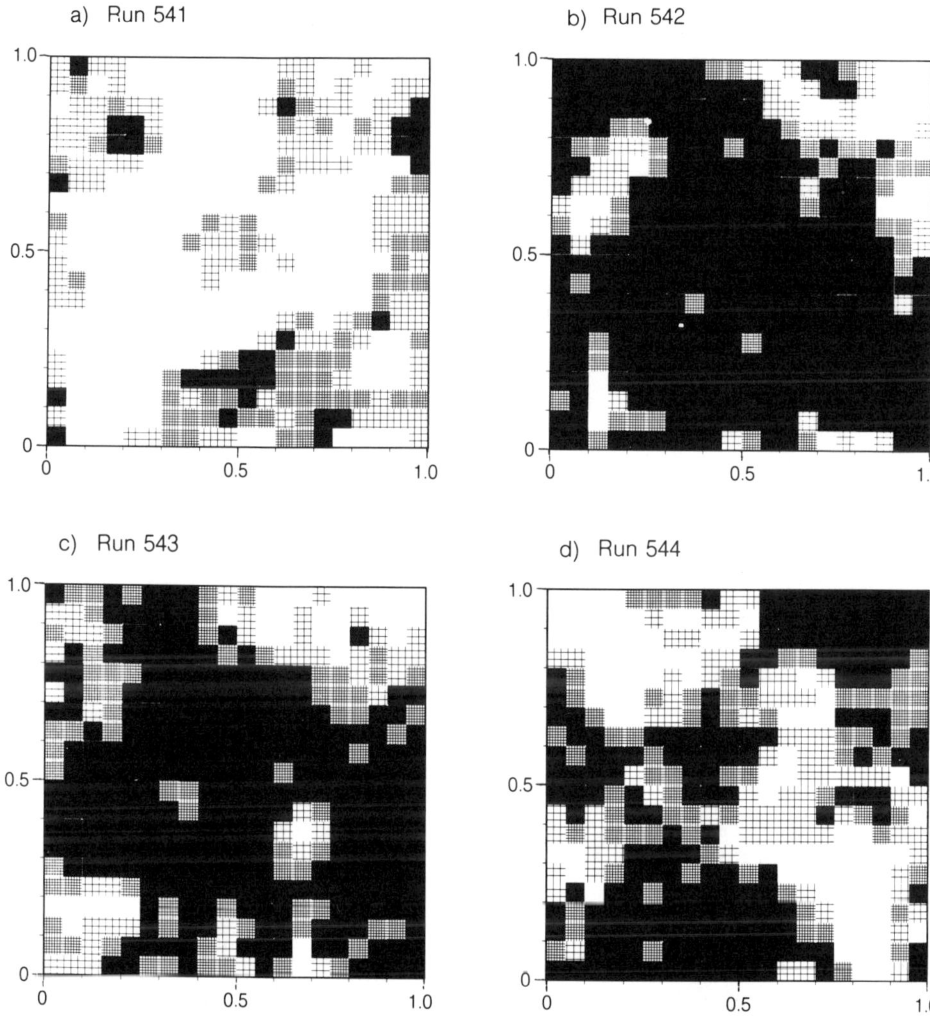

Figure 5.2.2. Statistically generated apertures with a spatial correlation length of 0.4 in the plane of a single fracture of linear dimension 1.0. The shadings are given in five levels from small aperture values (dark) to large ones (white). The dividing values correspond to the logarithm of apertures given by $v - 0.85\sigma$, $v - 0.25\sigma$, $v + 0.25\sigma$ and $v + 0.850\sigma$.

5.3 FLOW AND TRANSPORT IN SINGLE FRACTURES

Fluid flow and tracer transport through these variable-aperture fractures may be straightforwardly calculated for a constant pressure boundary condition (Moreno et al., 1988; Tsang et al., 1988; Tsang and Tsang, 1987, 1989). In Fig. 5.3.1a, the left boundary is kept at a higher potential, P_1, and the right boundary at a lower potential, P_2, with no-flow conditions imposed on the upper and lower boundaries. The steady laminar velocity of a viscous incompressible fluid through a pair of parallel, smooth plate walls separated by distance b satisfies the Darcy equation

$$v = -\frac{1}{12\mu} b^2 \nabla P \tag{5.3.1}$$

where v is the Darcy velocity, μ is the dynamic fluid viscosity, and P is the fluid pressure. Then, the volumetric fluid flow is

$$Q = -\frac{1}{12\mu} b^3 W \Delta P / L \tag{5.3.2}$$

where the pressure drop is ΔP over the length of L, and W is the extent of the parallel plate normal to the pressure gradient. Now (5.3.2) may be applied to each of the squares enclosed by the gridlines shown in Fig. 5.3.1a. Figure 5.3.1b shows a schematic diagram of two adjacent nodes with apertures b_i and b_j, respectively. If the volumetric flow rate from node i to node j is Q_{ij} we can write down the expression for the pressure drop from node i to node j

$$P_i - P_j = Q_{ij}\left[12\mu \frac{\Delta x}{\Delta y} \cdot \frac{1}{2}\left(\frac{1}{b_i^3} + \frac{1}{b_j^3}\right)\right] = Q_{ij} R_{ij} \tag{5.3.3}$$

where R_{ij} is the resistance to flow between nodes i and j as illustrated in Fig. 5.3.1c. The mass balance at each node i may be written as

$$\sum_j Q_{ij} = \sum_j \frac{P_i - P_j}{R_{ij}} = 0 \tag{5.3.4}$$

Except for the nodes at the left and right boundaries of the fracture region, the pressure at each node is an unknown to be solved. The system of equations (5.3.4) with the pressures as unknowns is solved using a sparse matrix solver. The solutions of the system of equations yield the pressure at each node, and flow between adjacent nodes is then calculated using (5.3.3).

After the steady state fluid flow rates are obtained the solute transport through the fracture is calculated using a particle tracking technique. A particle tracking technique was used by Smith and Schwartz (1980, 1984) to study dispersion in a heterogeneous porous medium. It was also used by Schwartz et al. (1983) and Robinson (1984) to address dispersion in an idealized fractured medium consisting of two sets of orthogonal fractures. We shall apply this technique in studying solute transport and its dispersion through a single fracture with variable apertures.

A large number of particles are introduced at the fracture inlet line (i.e., the left boundary of the square flow region as shown in Fig. 5.3.1a). Particles coming to an intersection are then distributed in the outlet branches according to a no-mixing streamline model. The residence time

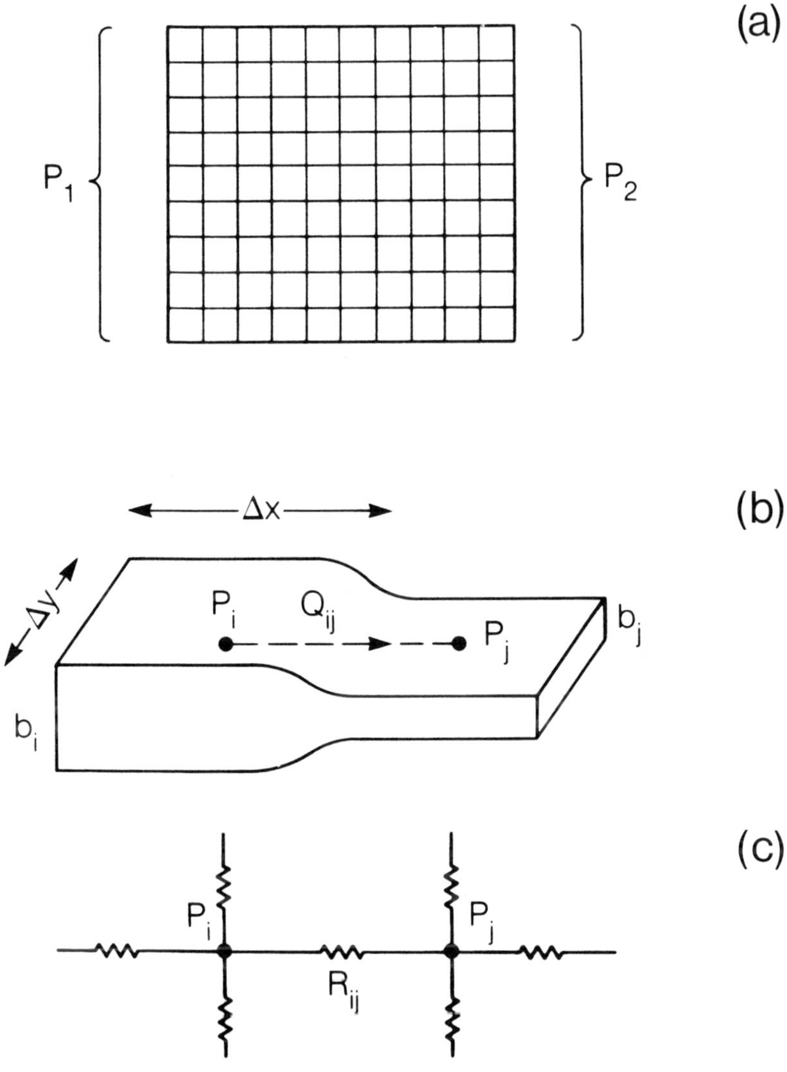

$$R_{ij} = 6\mu \frac{\Delta x}{\Delta y}\left(\frac{1}{b_i^3} + \frac{1}{b_j^3}\right)$$

Figure 5.3.1. (a) Schematic diagram for flow through a single fracture with different aperture values assigned to areas bounded by grid lines. (b) Schematic diagram for two adjacent nodes of different apertures: b_i and b_j and the fluid flow Q_{ij} between them. (c) Electric analog of fluid flow between adjacent nodes.

for particles within each square element is determined from the flow rate through this element and the volume involved:

$$t_i = \frac{b_i \Delta x \Delta y}{\frac{1}{2} \sum_j |Q_{ij}|} \quad . \tag{5.3.5}$$

The denominator in (5.3.5) is the amount of flow that passes through the nodal element i per unit of time. Summing the residence times t_i traversed by the particle over the entire path from inlet to outlet gives the total residence time of the particle. The results show that the mean residence time obtained by tracking all of the particles is within a few percent of the value calculated by dividing the total fracture volume by the total volumetric flow rate through the fracture. In this chapter we focus on the effect of the different residence times along the different pathways as the chief source of the overall dispersion of the solute in the fracture. The effects of molecular diffusion, matrix diffusion, or local dispersion within each channel are not included in the calculation.

Results of these calculations are shown in Figs. 5.3.2 and 5.3.3. They correspond to the flow rates in the fractures with aperture variations as shown in Figs. 5.2.1 and 5.2.2, respectively. Thus, the aperture variations in Fig. 5.2.1a give rise to the flow rate distribution displayed in Fig. 5.3.2a, and those in Fig. 5.2.2a give rise to flow distribution in Fig. 5.3.3a, and so forth. The flow rates between nodes vary over several orders of magnitude. The large range of values arises from the fact that a local resistance to fluid flow varies as the inverse of the local apertures raised to the third power and that the distribution of apertures assumed for the fracture plane takes on a wide range of values. The volumetric flow rates are plotted in Figs. 5.3.2 and 5.3.3 where, for display purposes, the thickness of the lines joining nodes is chosen to vary as the square root of the flow rate. Thus, the thicker the lines, the larger the flow rate. The plots in Figs. 5.3.2 and 5.3.3 show the following features: First, they all display preferred paths or channels of large volumetric flow rates. These channels are formed because of the variation of the apertures within the single fracture plane. Second, different statistical realizations and different spatial correlation of the variable apertures give rise to different flow patterns. Figures 5.3.2a through 5.3.2d display flow paths for apertures with spatial correlation length $\lambda = 0.1$, and Figs. 5.3.3a through 5.3.3d are for cases with $\lambda = 0.4$.

Solute transport phenomena are investigated by tracking the particles advected through the fractures. All the nodes on the left boundary (Fig. 5.3.1a) are connected at a higher constant pressure P_1, and all the nodes on the right boundary are connected at a lower constant pressure P_2. The input particles are distributed in proportion to the flow rates at different nodes along the inlet left boundary. A plot of the number of particles collected at all the outlets on the right boundary at different arrival times constitutes the tracer breakthrough curves. Calculations are carried out for the total number of input particles ranging from several hundreds to 10,000. The effect of the number of particles on the breakthrough curves is investigated, and it is found that calculations using 1,000 particles are more than adequate because they yield breakthrough curves that have few spurious artifacts, as a result of the finite number of particles employed.

Perhaps it is instructive to see how the flow field in the fracture plane varies as a function of σ, the standard deviation of the aperture values found in the fracture plane. For this purpose, a realization of the fracture is selected, and the flow patterns are calculated corresponding to σ = 0.1, 0.2, 0.43 and 0.6. The different values of σ for the same realization can be effected by scaling b values over the fracture plane. It can be straightforwardly shown that, given the lognormal distribution equation (5.2.1), if we scale all b values,

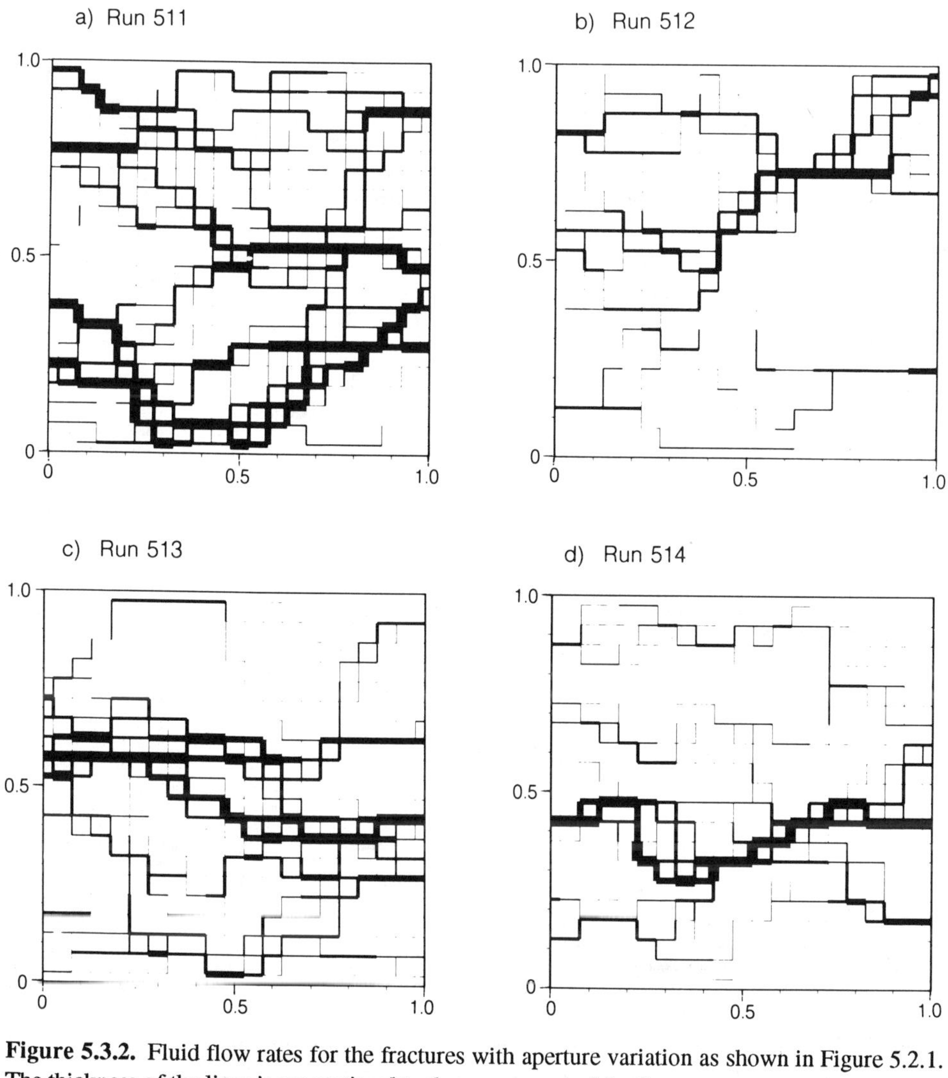

Figure 5.3.2. Fluid flow rates for the fractures with aperture variation as shown in Figure 5.2.1. The thickness of the lines is proportional to the square root of the flow rate.

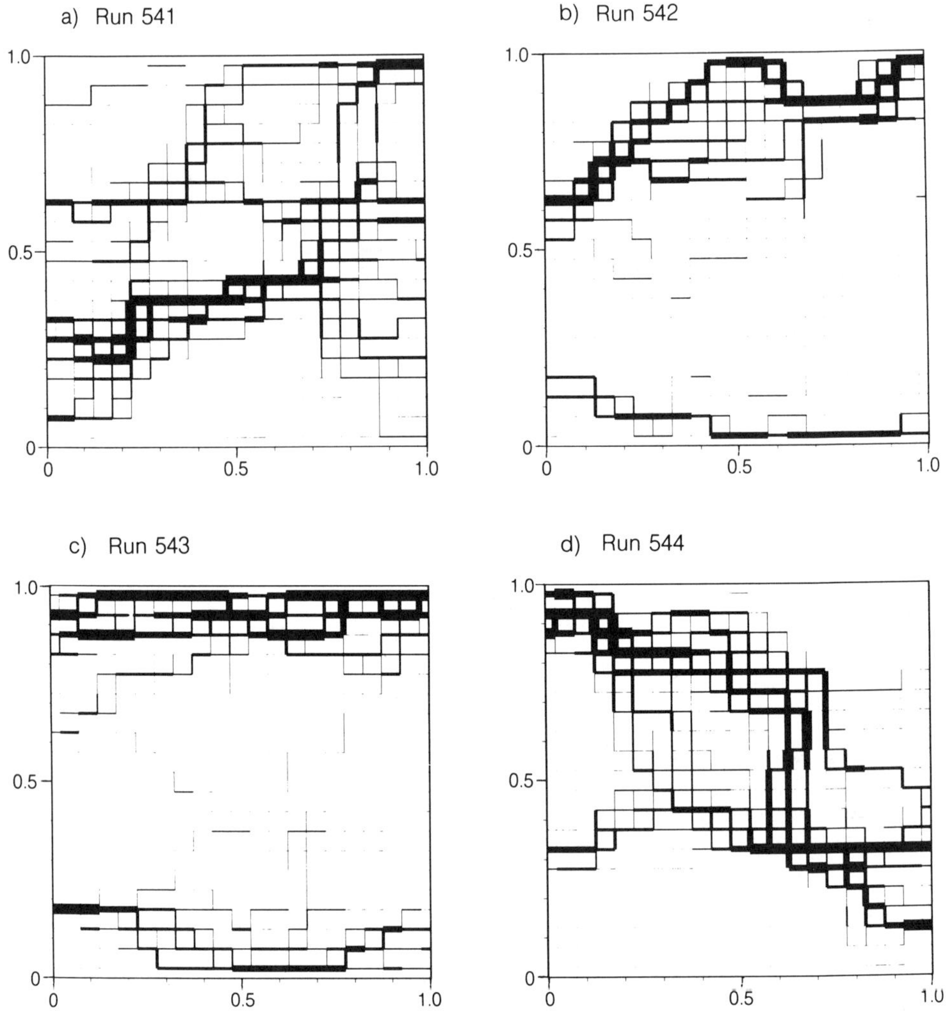

Figure 5.3.3. Fluid flow rates for the fractures with aperture variation as shown in Figure 5.2.2. The thickness of the lines is proportional to the square root of the flow rate.

$$b \rightarrow b^s \ ,$$

then the distribution is equivalent to the case where

$$\sigma \rightarrow \sigma/s \ .$$

This procedure is used for studying flow patterns for different σ values.

The results are shown in Fig. 5.3.4. It is apparent from this figure that for the case of $\sigma = 0.1$, the flow in the plane is quite uniform, whereas the channelization of flow in preferred flow paths becomes very pronounced for $\sigma = 0.6$. If one were to look at the distribution of the exit flow rate on the right boundary of the flow planes shown in Fig. 5.3.4, the result is shown in Fig. 5.3.5 where the exit flow rate at different points along the exit line corresponding to Fig. 5.3.4 are plotted. In this figure, for $\sigma = 0.1$ the exit flow is quite uniform along the outlet line, but this becomes more and more uneven as σ increases. Experimentally, the observation for flow distribution in a fracture (e.g., Abelin et al., 1985) resembles more the case of $\sigma = 0.6$ than $\sigma = 0.1$.

In principle, given the basic statistical parameters b_o, σ, and λ, one can generate a large number of realizations of single fractures with variable apertures. For each of the fractures, one can release a large number of particles on the left inlet boundary (Fig. 5.3.1a). These particles are all released at the same time, which corresponds to the case of a pulse injection of tracer. They traverse the fracture flow plane and emerge on the right side. If the number of particles is plotted as a function of arrival time, one obtains a tracer breakthrough curve corresponding to a pulse tracer injection. An example of the results for four different realizations for the same input statistical parameters is shown in Fig. 5.3.6. The horizontal axis is the arrival time normalized to the peak arrival time of all the particles and the vertical axis is the particle numbers normalized to the total number of particles introduced at the inlet boundary. It is seen in this figure that some cases display a multiply peaked breakthrough curve. This is a characteristic result for variable aperture fractures, where because of the range of apertures in the fracture, there is more than one major flow path in the fracture. In some cases, however, one of the flow paths dominates and there is only one major peak in the breakthrough curve. In many of these cases there is also a long tail in the tracer breakthrough curves.

In summary, for flow and transport through variable aperture single fractures, we expect that the transport is highly channelized in the fracture plane. In addition we expect the breakthrough curves for pulse tracer injection to display multiple peaks in general and the curves to have a long tail in many cases. All these features are very different from those of tracer transport through homogeneous porous media.

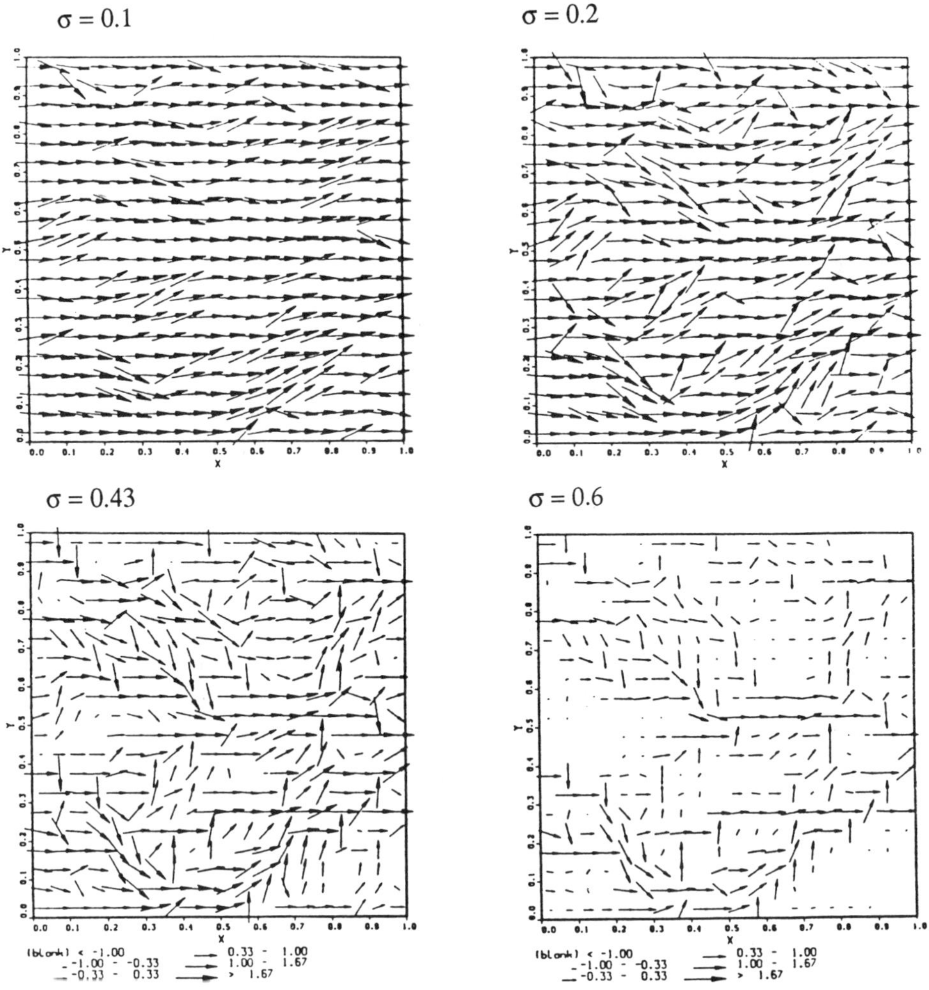

Figure 5.3.4. Flow through a variable-aperture fracture for four values of $\sigma = 0.1, 0.2, 0.43$ and 0.6. Flow rates are indicated by lengths of arrows in the figure.

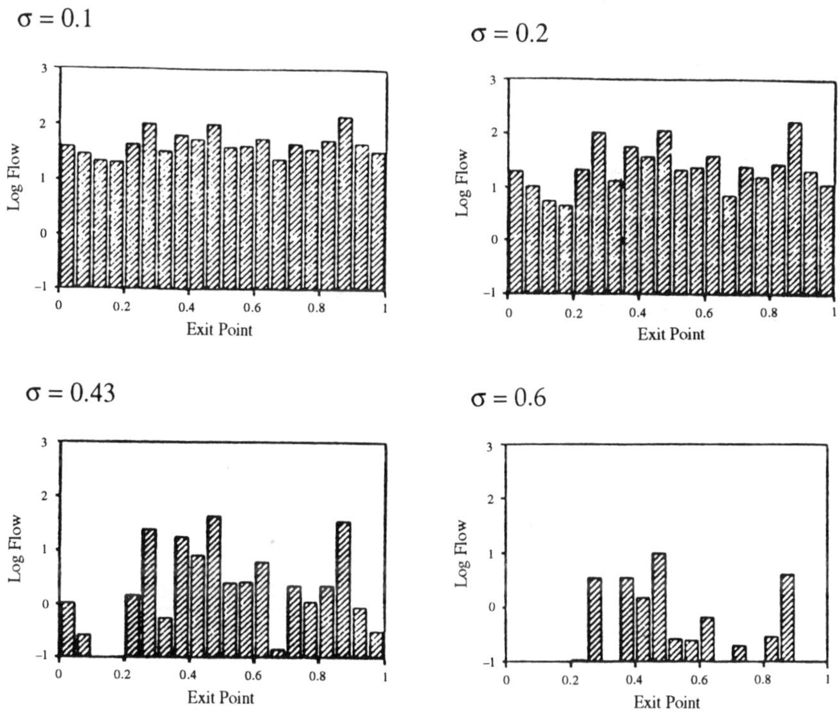

Figure 5.3.5. Exit flow rates at different points along the exit line corresponding to Figure 5.3.4, for $\sigma = 0.1, 0.2, 0.43$ and 0.6

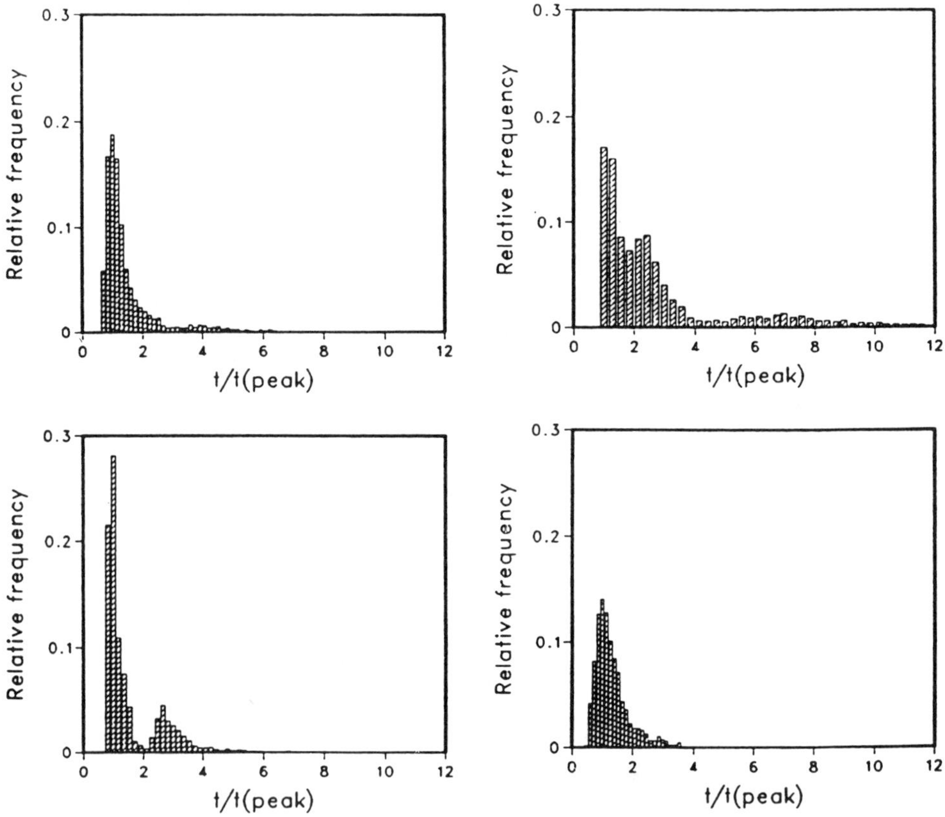

Figure 5.3.6. Tracer breakthrough curves (in terms of relative particle arrival frequency) displaying possible residence times (normalized to the time value of the highest peak) and their probabilities for four realizations of a variable aperture fracture, assuming a pulse injection of tracers.

5.4 FLOW AND TRANSPORT THROUGH THREE-DIMENSIONAL VARIABLE APERTURE FRACTURE NETWORKS

In Sec. 5.3 flow and transport in single fractures were discussed and demonstrated. In this section we will consider the case of a three-dimensional system where the rock is characterized by a large number of fractures forming a fracture network in the three-dimensional space. The problem of flow and transfer in fracture networks has been studied by Long et al. (1982), Endo et al. (1984), Long and Billaux (1987), Dverstorp and Andersson (1989), Dverstorp et al. (1992), Cacas et al. (1990a,b), and others. Nearly all these studies of fracture networks assumed that each fracture is represented by a pair of parallel plates (or a pipe) with constant aperture. Here we shall describe the more recent work of Nordqvist et al. (1992), who incorporates the variable aperture features of single fractures into the fracture network model of Dverstorp and Andersson (1989).

Following Andersson and Dverstorp (1987), one generates a fracture network consisting of plane circular disks randomly and independently distributed in the three-dimensional space. To accomplish this, the fracture centers are first located in a three-dimensional box by assuming a Poisson spatial generation process with density η fracture centers per cubic meter of rock. Then fractures in the form of disks with radius r and orientation (θ,ϕ) are constructed. When generating the radius r and the orientation (θ,ϕ) of the fractures, different density distributions may be used. Thus, the properties that characterize each fracture's position in space, radius and orientation are stochastic variables. Figure 5.4.1 shows an example of a fracture network generated with this model. The volume within which the fractures are generated is larger than the flow domain over which simulations of flow and transport will be made. By generating fractures in a larger volume, fractures intersected by the domain boundaries whose centers are outside the flow domain may contribute to the fluid flow. Thus the bias due to the absence of these contributions is prevented.

The incorporation of variable aperture fractures into the fracture network model is illustrated in Fig. 5.4.2. Parallel to the construction of the three-dimensional fracture network, a library of realizations of single fractures is also generated. These fractures are labeled as fractures 1 through n, generated with the input parameters b_o and σ. Each one of these single fractures has a

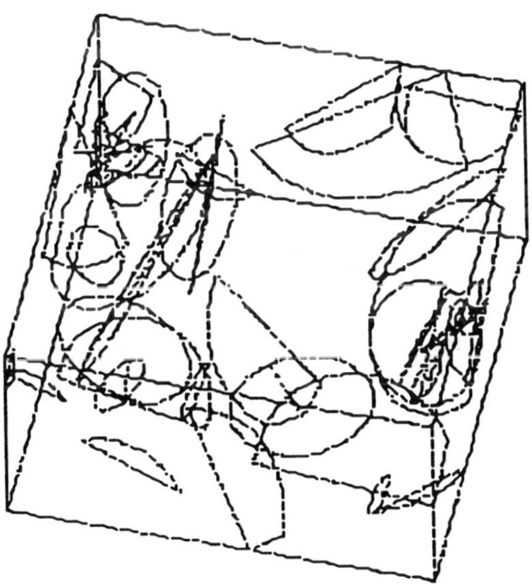

Figure 5.4.1. Statistically generated network of fractures. Model by Andersson and Dverstorp [1987].

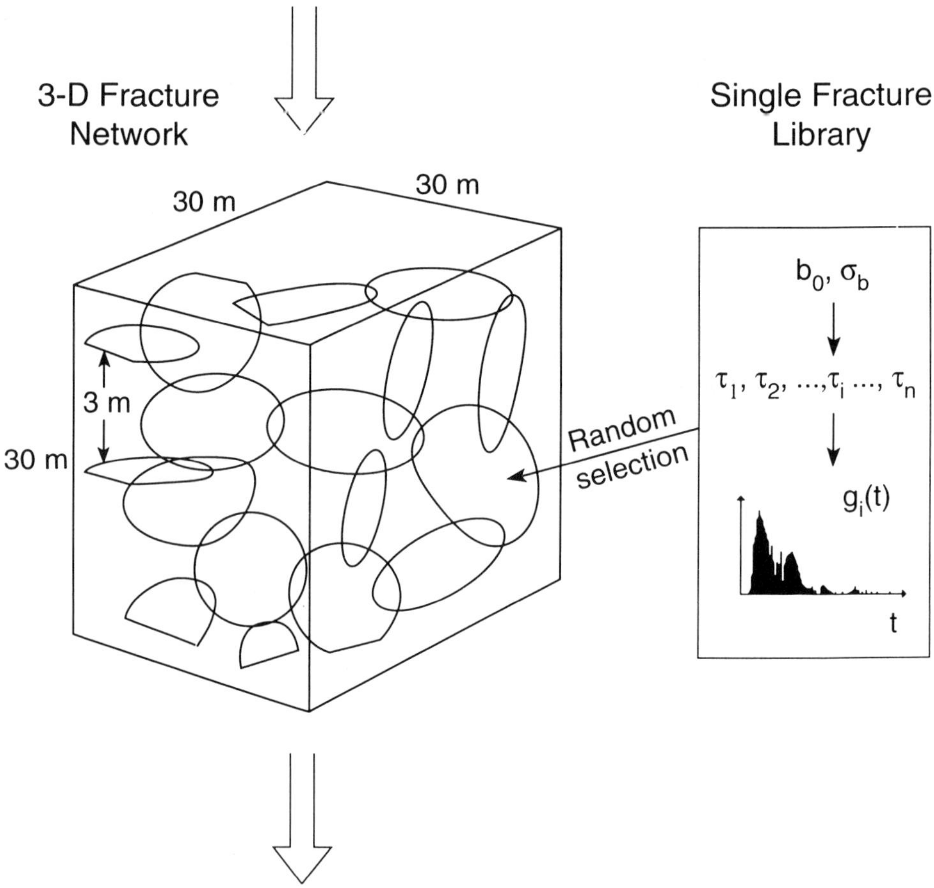

Figure 5.4.2. An illustration of the basic concept of how the variable aperture fractures are introduced into the fracture network model. b_o and σ are the two parameters in the lognormal aperture distribution. The τ_i's are the effective transmissivities of the fractures in the library. The $g_i(t)$ plots are graphical illustrations of the residence time spectra for the fractures in the library.

permeability value and an associated tracer breakthrough curve for a pulse injection (e.g., see Fig. 5.3.6). The single fractures are randomly selected to be associated with the different fractures in the network so that each network fracture corresponds to one realization within the library of single fractures. Several network fractures may, by chance, correspond to the same library fracture. The transmissivity values τ_i of the library fractures are then scaled to the network geometry to give the transmissivity of the fractures in the network. The fluid flow through the entire network is calculated on the basis of these fracture transmissivities, using the method of Andersson and Dverstorp (1987). Once the flow in each network fracture is obtained, the fracture residence time spectra $g_1(T)$ through $g_n(t)$ of the library (Fig. 5.4.2) are used to generate the residence times of transport through the entire fracture network.

For the cases considered here, the boundary conditions imposed on the flow domain of the three-dimensional fracture network are: no-flow along the four sides parallel to the flow direction, and constant hydraulic head values assigned to the two end faces. With the values of transmissivity given for all the variable aperture fractures, a system of linear equations is obtained, the solution of which yields the hydraulic heads at all fracture centers and all intersections. From these, all the flow rates in the network are computed. The inclusion of the fracture centers in the flow and transport calculation is an approximation that simplifies the solution of fluid flow considerably. One important assumption used in this model is that the fracture intersections are equipotential lines. This is a reasonable assumption based on field observations by Abelin et al. (1987), who found good correlation between large flow rates and fracture intersections. This means that a very large permeability value is present at fracture intersections and, hence, an approximate equipotential condition may be assumed there.

To perform tracer transport calculations for the network of variable aperture fractures, a particle tracking technique is used. As a particle traverses each fracture a residence time is chosen, in a random selection process, from the single fracture residence time spectrum weighted by the function $g_k(t)$ for that fracture.

At the center of each fracture in the network a particle has more than one direction to choose from. For this choice, which determines the new fracture into which the particle will continue in its travel, the following prescription is used: Firstly, the particles are not allowed to enter fractures conducting water towards the fracture center; secondly, among the rest of the directions, the probability of a particle choosing a certain fracture is proportional to the portion of the flow in that direction.

The particles are sampled on the exit boundary of the flow domain. The total travel time for each particle then is equal to the sum of residence times through the single fractures that was calculated during the particle transport. Breakthrough curves for solute transport through the fracture network are thus obtained by calculating total transport times for a large number of particles and displaying the result as frequency plots; in other words, the number of particles arriving in a given time interval is plotted versus the arrival time.

The following case gives a typical result for the three-dimensional variable aperture fracture network. In the three-dimensional rock mass, four classes or sets of fractures are assumed to be present. The classes differ from each other in two ways. First, the fracture density parameter in the Poisson process is not the same in the four classes. Second, the parameters in the distribution that was chosen to generate the fracture orientations are also different. The choice of orientation parameters divide the fractures into two subvertical sets; one subhorizontal set and one randomly oriented set. The fracture radius is modeled using an exponential distribution with a mean of 2.3 m in all four classes. Parameters are indicated in Table 5.4.1.

Table 5.4.1. Fracture network parameters [Dverstorp et al., 1992].

	Distribution	Parameters
Fracture radius, r	Exponential	$\mu_r = 2.3$ m
Fracture center density, η	Poisson space process	$\eta_i = 0.0018$ to 0.018 m^{-3} $i = 1,\ldots,4$ (no. of fracture set)
Fracture orientation, (θ,ϕ)	Fisher distribution fitted to each of three roughly orthogonal (one sub-horizontal and two subvertical) fracture sets and one random orientation set.	

It is important to make sure that the number of particles is large enough so that the characteristics of the tracer breakthrough curves do not change significantly when the number is increased. On the other hand, restrictions imposed by computer time and capability limit the number of particles to be used. After a number of test runs with particle numbers up to 100,000, it appears that 20,000 particles are sufficient.

Figure 5.4.3 shows typical tracer breakthrough curves for three realizations of the three-dimensional fracture network. Input data for all three realizations are identical, i.e., $b_o = 1.7$, $\sigma = 0.43$ and the parameters of Table 5.4.1. In general, the breakthrough curves from the variable aperture fracture network model have more than one peak. This is apparent in curves I and II in the figure. A probable interpretation of this is that each peak corresponds to particles transported along a different major route through the network of fractures. Thus, the fact that there is more than one major route available for transport through the fracture network will lead to this type of multiple peak dispersion. The width of the peaks indicates that the particles have been subjected to dispersion on a smaller scale. This second type of dispersion may be due to a large number of possible flow paths, within each fracture, that originated from the variable aperture nature of the single fractures. The two different scales of dispersion that this model produces have found support in results from the analysis of experiments on tracer transport through a fractured granitic formation (see next section).

Another interesting feature of the breakthrough curves obtained with the variable aperture fracture network model is the presence of peaks at a transport time considerably shorter than the mean value. An example of this is the peak at about 50 hours in curve I of Fig. 5.4.3, whereas the mean value of all the transport times is about 156 hours. This behavior has also been reported from the analysis of a number of experiments (Dverstorp et al., 1992). The early peaks correspond to comparatively fast pathways through the three-dimensional flow domain. The presence of such fast pathways illustrates so-called extreme channeling, by which the particles travel through the fracture medium at a relatively high velocity, experiencing only a low degree of dispersion.

The variable aperture fracture network model described above is able to provide insight into the very complicated phenomena that influence flow and transport through fractured rocks. The essential physics from the fracture aperture scale to the much larger fracture network scale is

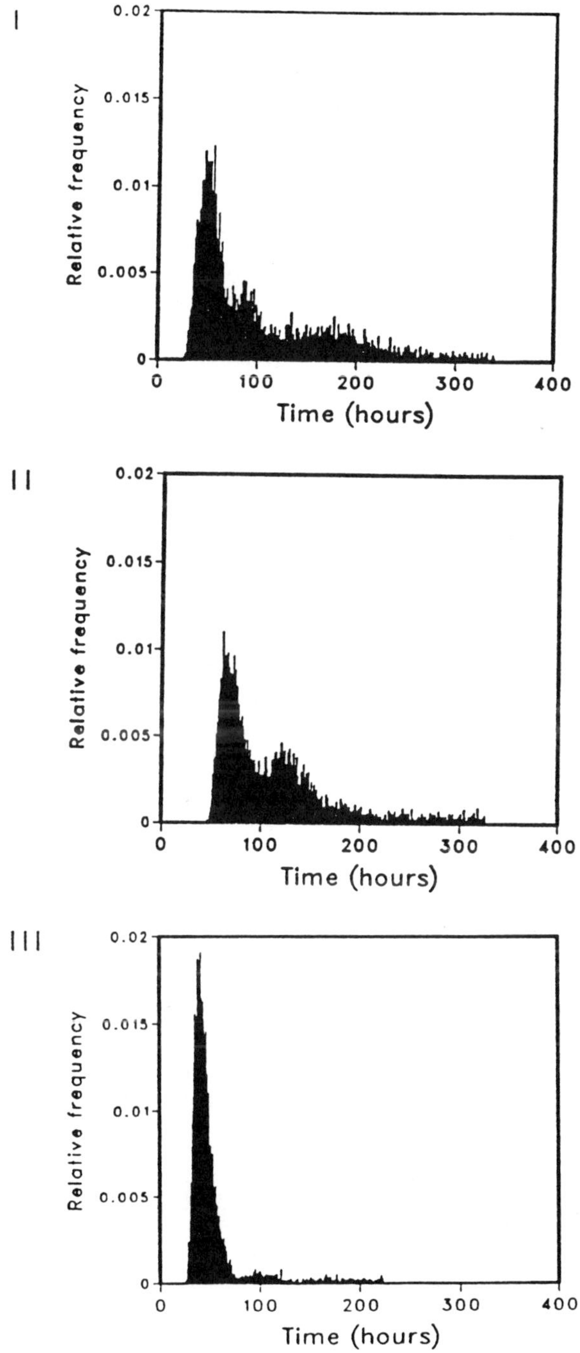

Figure 5.4.3. Breakthrough curves for 15 meter transport with the variable-aperture fracture network model. The curves were obtained from three different realizations of fracture networks.

included in a consistent and yet computationally feasible way. Results from the initial application of the model display dispersion on two different scales. It is of significant interest to investigate the transition between the two scales and, in general, to study the scale dependence of transport in fractured rocks.

5.5 ANALYSIS OF A TRACER TRANSPORT EXPERIMENT IN FRACTURED ROCKS

In this section we shall present as an illustration the analysis of data from a tracer experiment in fractured rock. As we will demonstrate, the results of this analysis confirm flow channeling and the multiple peak character of tracer transport in fractured rocks as shown by the calculations described above.

The experiment chosen for this illustration is known as the Stripa 3D experiment, whose details are given by Neretnieks (1987) and Abelin et al. (1987), as well as in Chapter 2 of this book. The experiment was carried out in the Stripa mine in two drifts which form a cross as shown in Fig. 5.5.1a. The longer drift is 75 m long and the shorter intersecting drift is 25 m long. Each drift is 2.5 m in width and 3 m in maximum height. Three vertical boreholes are drilled into the ceiling of the drifts, and into these boreholes nine packers of sections, each 1 m long, were identified as having large enough local hydraulic transmissivity to be suitable for tracer injections. Nine different tracers were used and injected in these sections in the three boreholes, as shown in Fig. 5.5.1b. These tracers were collected in over 300 plastic sheets of 1×2 m each, covering the ceilings and sides of the drifts. The flow rates at the collection sheets were approximately constant in time, indicating that steady state flow conditions prevailed. However, the majority of these sheets did not receive significant flow, and tracer tended to emerge at discrete regions of the drift (see Fig. 5.5.2).

Only five of the nine tracers were found in significant quantities in the different plastic collection sheets during the first 30 months of the experiment. These were eosin B, eosin Y, uranin, elbenyl, and iodine. The apparent linear transport distances range from 10 m to 45 m. Figure 5.5.3 shows injection flow rate as a function of time. Most of the tracer breakthrough into the sheets were localized to a small number of sheets positioned around the X location at 25–36 m, where X is measured along the axis of the longer drift (see Fig. 5.5.2).

What follows is our conceptual model of the Stripa 3D experiment. Because the underground drift is maintained at atmospheric pressure, it is a major sink for the water that saturates the rock around it. The large range of aperture values in the fractures of the rock gives rise to flow channeling so that the majority of flow takes place in only a few channels or groups of flow paths of least flow resistance, which occupy only a small fraction of the total 3-dimensional flow region (Fig. 5.5.4). Tracer is injected at a particular location in the flow field. The resultant solution moves downstream along these preferred flow paths towards a drift and emerges in a limited number of plastic collection sheets. From the experimental measurements we do not have knowledge of the actual flow path connecting the tracer injection and the particular collection sheets of tracer emergence. However, from the time-dependence of tracer collection it will be shown below that we can identify channels or groups of flow paths which have comparable residence times.

At Stripa the injected tracers become diluted because the tracer injection flow rate q is much smaller than the total exit flow rate at the drift. Also, while the injection flow rate is strongly

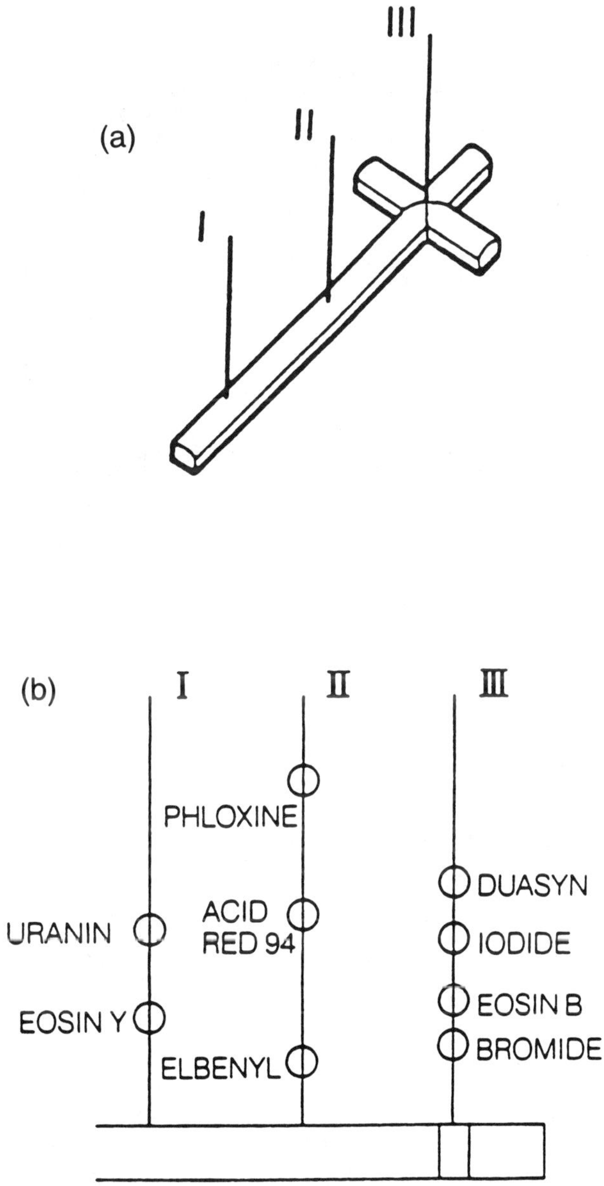

Figure 5.5.1. (a) The layout of the Stripa 3D test site with three vertical injection holes. (b) The location of tracer injection zones in the three vertical injection holes. Adapted from Abelin et al., 1987.

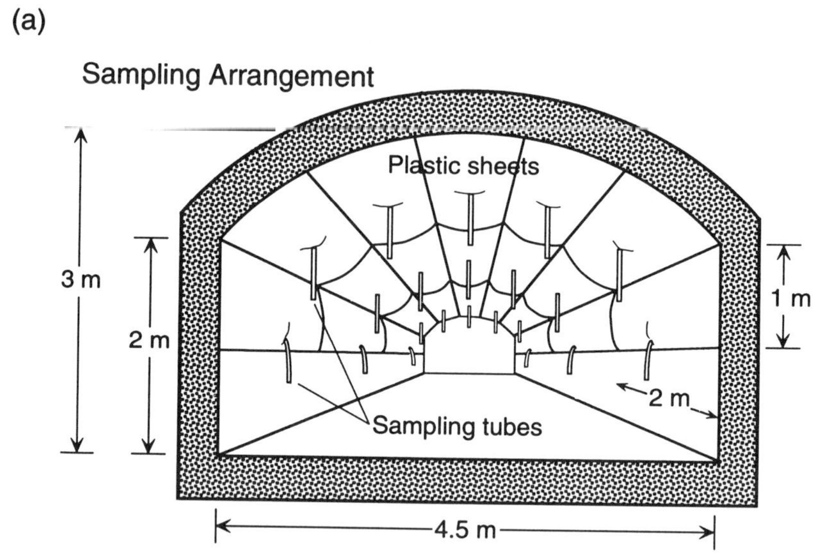

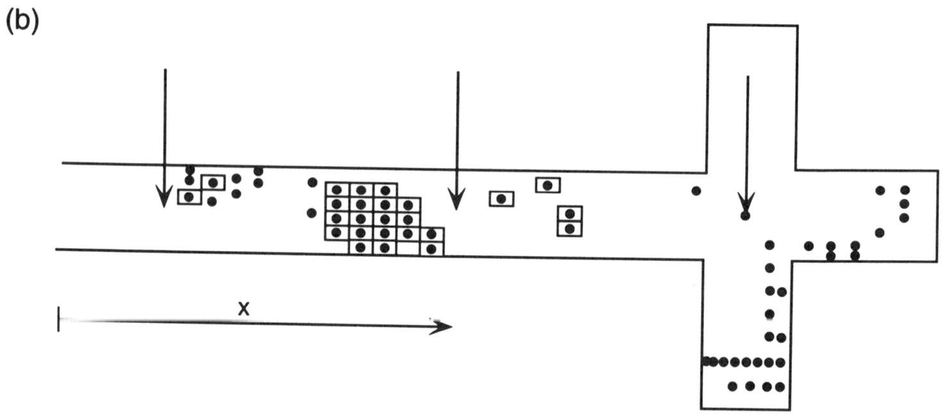

Figure 5.5.2. (a) The Stripa 3D sampling arrangement, showing placement of plastic sheets for tracer collection. (b) A diagram showing tracer distribution in the Stripa 3D test site. The arrows indicate the positions of the injection holes, dots indicate sheets with significant water flow, and squares indicate sheets where tracers were collected. Adapted from Abelin et al., 1987.

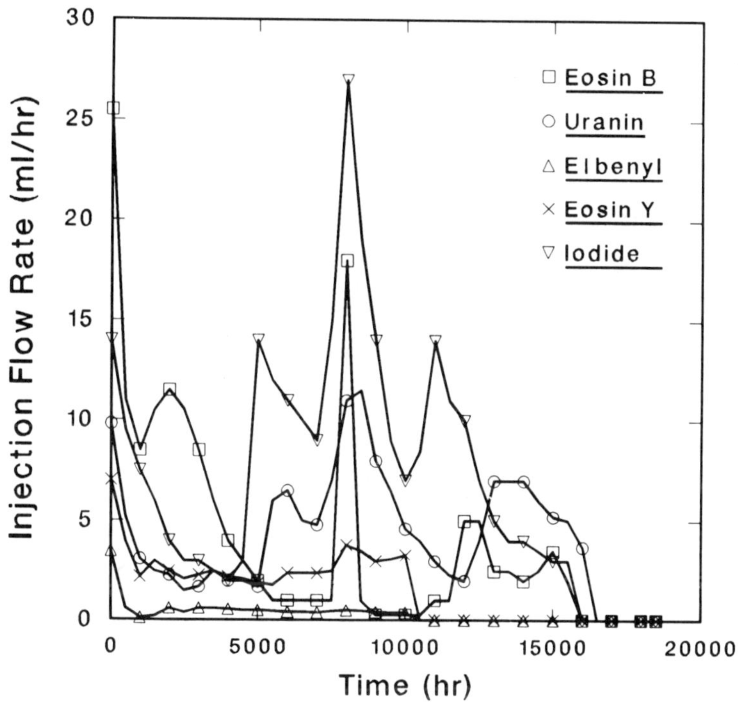

Figure 5.5.3. Injection flow rates in ml/hr for the five tracers; points are average flow rates over 500-hour intervals.

varying, the collection flow rate is approximately constant with time. Because of the unknown dilution effect of the channeled flow, the absolute value of the concentration of breakthrough curves is not being investigated. Actually, this is not needed to obtain the relevant parameters such as mean and standard deviation of the aperture values along the flow paths that govern and characterize the effect of transport in the fracture medium. In the analysis below the rate of tracer mass accumulation is used as the observed quantity, rather than the more commonly used tracer concentration.

In this conceptual picture, then, one has a flow channel with emergent flow rate Q at a collection point, while tracer is injected with a given mass per unit time at an upstream location. The tracer mass injected per unit time is proportional to injection flow rate q if the injection concentration is constant, as is the case in this experiment. Assuming that the one-dimensional effective dispersive transport equation holds in the one-dimensional channels, well-known analytical solutions are available for the analysis of tracer transport data. The main complication with the present data set is a strong temporal variation of the tracer injection flow rate (Fig. 5.5.3).

To handle the strong temporal variation of the injection flow rate, a deconvolution method based on Toeplitz matrices was developed, the details of which are given in Tsang et al. (1991). The method essentially reduces the variable injection flow rate and the observed tracer breakthrough curve to a breakthrough curve that would be expected if the tracer injection were a

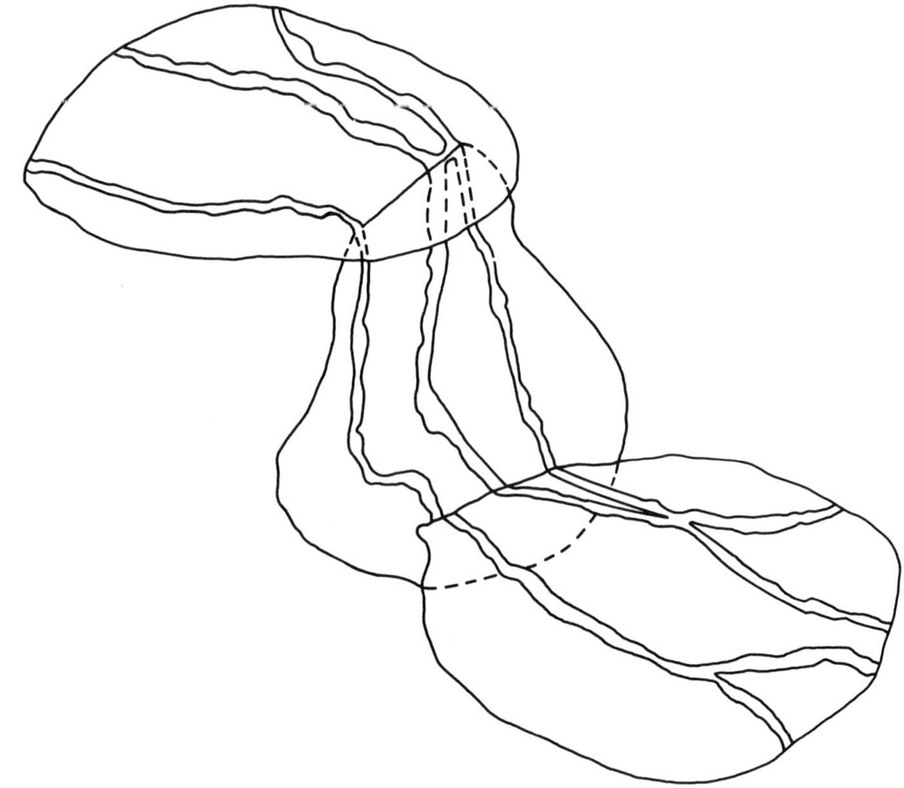

Figure 5.5.4. Conceptual model of flow channeling through three fractures in series.

delta function pulse input. When this procedure is applied to the five sets of tracer data under study, the results shown in Figs. 5.5.5, 5.5.6, 5.5.7, 5.5.8, and 5.5.9 are obtained. These figures clearly reveal the multiple-peak channeling nature of tracer transport in the Stripa 3D data. Over the range of distances and observation time periods, eosin B apparently has four channels with very different mean resident times, while the other tracers have two channels each. Note that in this context a channel is defined as a group of nearby flow paths with comparable mean velocities.

To quantitatively characterize these multiple peaks we choose to fit the solution of the one-dimensional advective dispersive equation, with delta function tracer injection, to each of the peaks in Figs. 5.5.5–5.5.9. That is, the data in these figures are fitted to

$$a(t) = \sum_{K} \frac{R_K}{\sqrt{4\pi D_K t}} e^{\frac{-(x - v_K t)^2}{4 D_K t}} \tag{5.5.1}$$

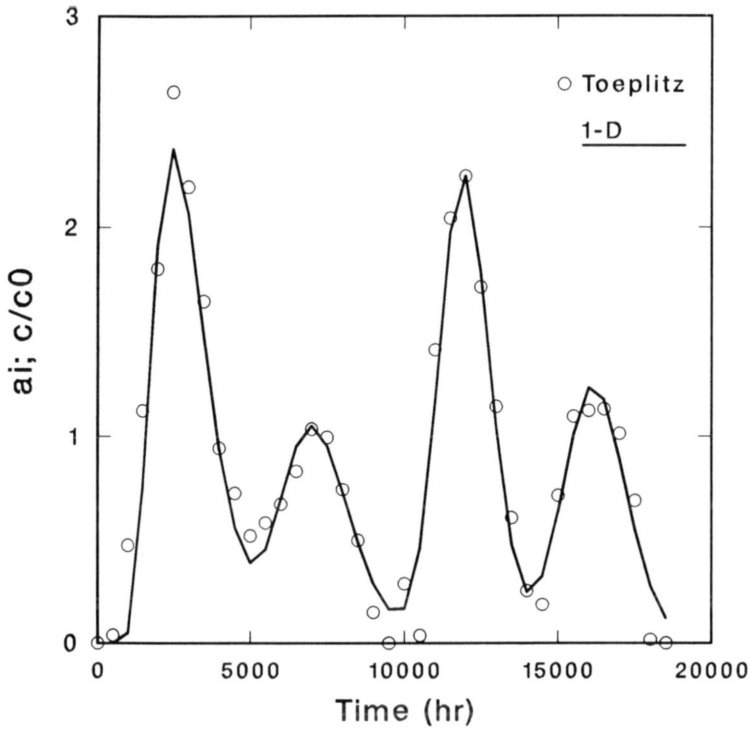

Figure 5.5.5. Results of Toeplitz analysis method for eosin B. The circles are the results of the Toeplitz analysis, and the solid line is the fit by Eq. 5.5.1, with the goodness of fit given by the correlation coefficient $r = 0.972$.

where κ designates different channels and x is taken as a linear geometric distance from the injection point to the weighted center of the tracer collection area. The results are also shown as the continuous line in Figs. 5.5.5–5.5.9. The goodness of the fit, especially for eosin B (correlation coefficient $r = 0.972$) and elbenyl ($r = 0.989$) indicates that transport through these channels may indeed be adequately represented by the one-dimensional advective dispersive equation.

The analysis of tracer breakthrough data as described above may also afford us a way to estimate the mean aperture $\bar{b}$ of the flow regime responsible for transport. This may be done based on the experimental flow rate Q to the 1×2 m plastic collection sheets and the calculated velocity v from the first peak or first channel. At a collection sheet on the drift ceiling the observed incoming flow rate is fed by flows containing different tracers, with velocity v_i. If the average width of the flow channel covered by the collection sheet is w and the mean fracture aperture is $\bar{b}$, then

$$Q = \sum_i \bar{b}\, w v_i = \bar{b}\, w \sum_i v_i = A \sum_i v_i \tag{5.5.2}$$

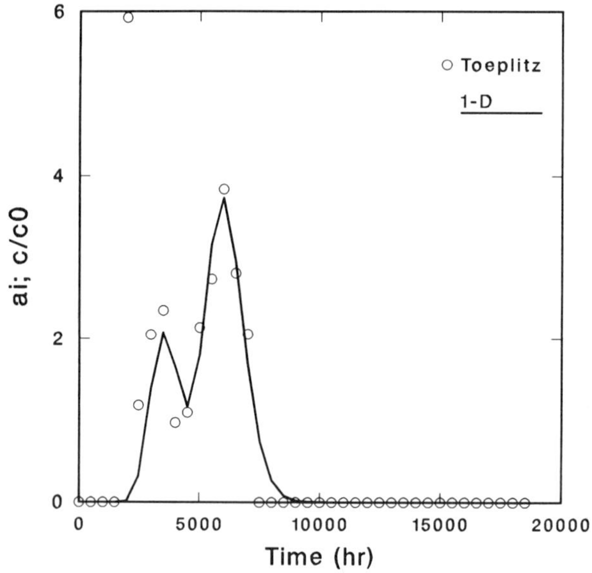

Figure 5.5.6. Results of Toeplitz analysis method for uranin, and their fit by Eq. (5.5.1). Note that a Toeplitz solution point at 2,000 hrs probably represents an early peak, but there is not sufficient data to determine its characteristics. Without this point, $r = 0.964$.

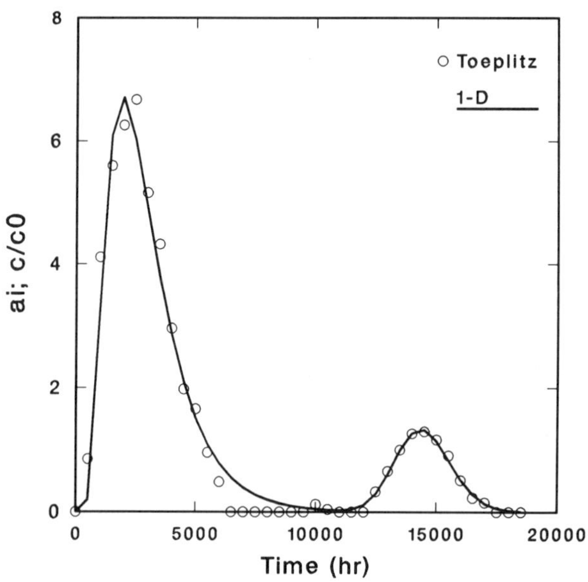

Figure 5.5.7. Results of Toeplitz analysis method for elbenyl, and their fit by Eq. 5.5.1. Goodness of fit $r = 0.989$.

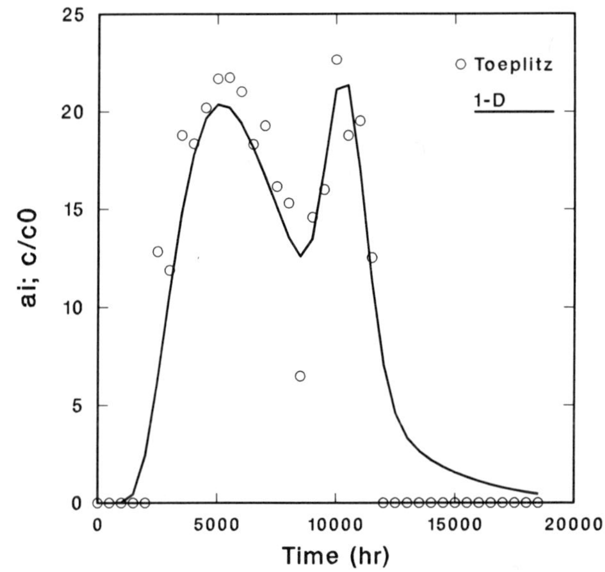

Figure 5.5.8. Results of Toeplitz analysis method for eosin Y, and their fit by Eq. 5.5.1. Goodness of fit $r = 0.965$.

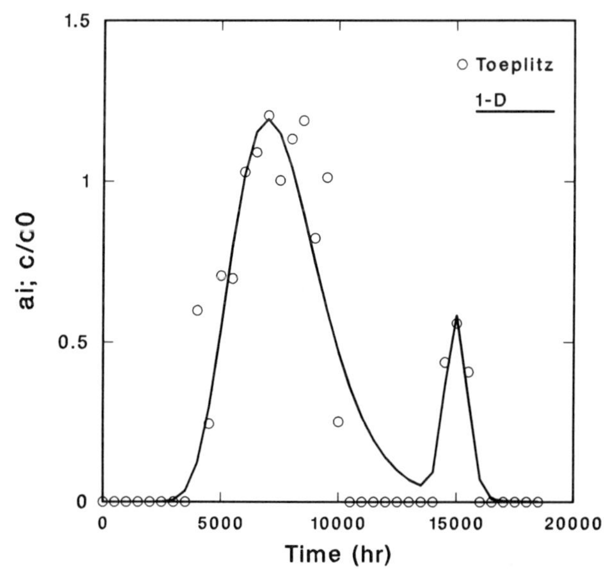

Figure 5.5.9. Results of Toeplitz analysis method for iodide, and their fit by Eq. 5.5.1. Goodness of fit $r = 0.936$.

where A is the cross-sectional area of the conducting fracture covered by the collection sheet and Σv_i is the sum of the mean velocities of the different tracers that arrive at the sheet. From Abelin et al. (1987), the inflow rate Q averaged over those sheets with large flow rates is about 20 ml/hr. The average width of the flow channel w is not determined within the present data set. For each of the 1 × 2 m collection sheets with large flow, we assume a fracture width w of 2 m. Further, from the fits of Figs. 5.5.5 through 5.5.9 to Eq. 5.5.1, Σv_i = 59 mm/hr. Combining all numbers it is shown that $A = Q/\Sigma v_i = 0.3 \times 10^{-3}$ m^2 and $\bar{b} = A/w = 150$ μm.

5.6 CONCLUDING REMARKS

As was indicated in Sec. 5.1, tracer transport in fracture systems is still an area of active current research. The problem has not yet been solved and predictions for tracer breakthrough curves for arbitrary field conditions are not yet possible. However, in this chapter we have endeavored to provide some insight into tracer transport in the fracture system from the single fracture scale to the fractured network scale. By analyzing one major set of field data we have confirmed some of the characteristics of tracer transport as revealed by theoretical or numerical considerations.

ACKNOWLEDGMENTS

The work presented in this chapter was based on various papers and reports done in close cooperation with Y.W. Tsang and F.V. Hale of Lawrence Berkeley Laboratory, University of California, Berkeley, and with L. Moreno, I. Neretnieks, and W. Norqvist of the Royal Institute of Technology, Stockholm. Their continued cooperation and input is gratefully acknowledged. I also appreciate the reviews and comments on these papers and reports by G.S. Bodvarsson, C. Carnahan, C. Doughty, T.N. Narasimhan, and J.S.Y. Wang. The chapter is prepared under the auspices of the Director of Office of Energy Research, Office of Basic Energy Sciences, Engineering and Geosciences Division of the U.S. Department of Energy under Contract No. DE-AC03-76SF00098.

REFERENCES

Abelin H., Birgersson, L., Gidlund, J., Moreno, L., Neretnieks, I., Widen, H., and Agren, T. 3-D Migration experiment, Report 3, Part 1, Performed Experiments, Results and evaluation, Stripa Project Technical Report 87-21, Svensk Karnbransleforsorjning, Stockholm, 1987.

Abelin, H., Neretnieks, I., Tunbrant, S., and Moreno, L. Final Report of the Migration in a single fracture: Experimental results and evaluation, Stripa Project Sven. Karnbransleforsorjning Tech. Rep. 85-03, Nuclear Fuel Safety Project, Stockholm, Sweden, May 1985.

Andersson, J., and Dverstorp, B. Conditional simulations of fluid flow in three-dimensional networks of discrete fractures, *Water Resources Research,* 23(10), 1876–1886, 1987.

Bourke, P.J. Channeling of flow through fractures in rock, Proceedings of GEOVAL - 87 International Symposium, Stockholm, Sweden, April 7–9, 1987.

Cacas, M.C., Ledoux, E., de Marsily, G., Tillie, B., Barbreau, A., Durand, E., Fuega, B., and Peaudecerf, P. Modeling fracture flow with a stochastic discrete fracture network: Calibration and Validation, 1. The flow model, *Water Resour. Res.,* 26(3), 479489, 1990a.

Cacas, M.C., Ledoux, E., de Marsily, G., Barbreau, A., Calmels, P., Gaillard, B., and Margritta, R. Modeling fracture flow with a stochastic discrete fracture network: Calibration and Validation, 2. The transport model, *Water Resour. Res.,* 26(3), 491–500, 1990b.

Dagan, G. Models of groundwater flow in statistically homogeneous porous formations, *Water Resour. Res.,* 15(1), p. 47–63, 1979.

Dverstorp, B., and Andersson, J. Application of the discrete fracture network concept with field data: Possibilities of model calibration and validation, *Water Resour. Res.,* 25(3), 540–550, 1989.

Dverstorp, B., Andersson, J., and Nordqvist, W. Discrete fracture network interpretation of field tracer migration in sparsely fractured rock, *Water Resour. Res.,* 28(9), 2327–2343, 1992.

Endo, H.K., Long, J.C.S., Wilson, C.R., and Witherspoon, P.A. A model for investigating mechanical transport in fracture networks, *Water Resour. Res.,* 20(10), p. 1390–1400, 1984.

Gale, J.E. Comparison of coupled fracture deformation and fluid flow models with direct measurements of fracture pore structure and stress-flow properties, *Proceedings of the 28th U S. Symposium of Rock Mechanics,* Tucson, Arizona p. 1213–1222, 1987.

Gelhar, L.W., and Axness, C.L. Three-dimensional stochastic analysis of macrodispersion in aquifers, *Water Resour. Res.,* 19(1), p. 161–180, 1983.

Gentier, S. Morphologie et comportement Hydromécanique d'une Fracture Naturelle dans un Granite Sous Contrainte Normale, Doctoral Thesis, Université d'Orléans, France, 1986.

Heath, M.J. Solute migration experiments in fractured granite, South West England., in Design and instrumentation of in situ experiments in underground laboratories for radioactive waste disposal, Proceedings of a joint CEC-NEA workshop, Brussels, 15–17 May 1984, pp 191–200. Published by Balkema, Netherlands, 1985.

Long, J.C.S., and Billaux, D.M. From field data to fracture network modeling: An example incorporating spatial structure, *Water Resour. Res,* 23(7), 1201–1216, 1987.

Long, J.C.S., Remer, J.S., Wilson, C.R., and Witherspoon, P.A. Porous media equivalents for networks of discontinuous fractures, *Water Resour. Res,* 18(3), 645–658, 1982.

Moreno, L., Tsang, Y.W., Tsang, C.F., Hale, F.V., and Neretnieks, I. Flow and tracer transport in a single fracture: A stochastic model and its relation to some field observations, *Water Resour. Res.,* 24(12), 2033–2048, 1988.

Neretnieks, I. Channeling effects in flow and transport in fractured rocks — Some recent observations and models, Paper presented at GEOVAL symposium, Swedish Nuclear Power Inspectorate (SKI), Stockholm, Proceedings, p. 315–335, 1987.

Neretnieks, I. Transport in fractured rocks, Proceedings, Memoires of the 17th International Congress of International Association of Hydrologists, Vol. XVII, 301318, International Association of Hydrologists, Tucson, Arizona, 1985.

Neuman, S.P., Simpson, E.S., Hsieh, P.A., Jones, J.W., and Winter, C.L. Statistical analysis of hydraulic test data from fractured crystalline rock near Oracle, Arizona, Proceedings, Memoires of the 17th International Congress of International Association of Hydrologists, Vol. XVII, 289–301, International Association of Hydrologists, Tucson, Arizona, 1985.

Nordqvist, W., Tsang, Y.W., Tsang, C.F., Dverstorp, B., and Andersson, J. A Variable-aperture fracture network model for flow and transport in fractured rocks, *Water Resources Research,* 28(6), p. 1703–1713, 1992.

Robinson, P.C. Connectivity, flow and transport in network models of fractured media, Ph.D. Thesis, Oxford University, Oxford, 1984.

Schwartz, F.W., Smith, L., and Crowe, A.S. A stochastic analysis of macroscopic dispersion in fractured media, *Water Resour. Res.,* 19(5), p. 1253–1265, 1983.

Smith, L., and Schwartz, F.W. An analysis of the influence of fracture geometry on mass transport in fractured media, *Water Resour. Res.*, 20(9), p. 1241–1252, 1984.

Smith, L., and Schwartz, F.W. Mass transport, 1, A stochastic analysis of macroscopic dispersion, *Water Resour. Res.*, 16(2), p. 303–313, 1980.

Tsang, C.F., Tsang, Y.W., and Hale, F.V. Tracer transport in fractures: Analysis of field data based on a variable-aperture channel model, *Water Resour. Res.,* 27 (12), 3095–3106, 1991.

Tsang, Y.W., and Tsang, C.F. Channels model of flow through fractured media, *Water Resour. Res.*, 23(3), p. 467–479, 1987.

Tsang, Y.W., and Tsang, C.F. Flow Channeling in a single fracture as a two-dimensional strongly heterogeneous permeable medium, *Water Resour. Res.,* 25(9), 2076–2080, 1989.

Tsang, Y.W., Tsang, C.F., Neretnieks, I., and Moreno, L. Flow and tracer transport in fractured media: A variable aperture channel model and its properties, *Water Resour. Res.,* 24(12), 2049–2060, 1988.

Williams, S.A., and El-Kadi, A.I. COVAR - A computer program for generating two-dimensional fields of autocorrelated parameters by matrix decomposition, International Groundwater Modelling Center, Holcomb Research Institute, Butler University, Indianapolis, IN, 1986.

6

MULTIPHASE FLOW IN FRACTURED PETROLEUM RESERVOIRS

H. Kazemi and J. R. Gilman
Marathon Oil Company
Littleton, Colorado

6.1 INTRODUCTION

In this chapter, we mainly deal with multiphase flow in naturally fractured hydrocarbon reservoirs. The mathematical principles involved very closely resemble those for flow in other fractured porous media such as geothermal formations, underground waste disposal rocks, aquifers and coal beds.

Folding, faulting, and subsidence of sediments over geologic time cause fracturing. The more brittle the rock, the more intensely it fractures. Carbonate reservoirs are typically more fractured than sandstone reservoirs for this reason. In general, the two rock surfaces that bound a fracture are rough. The degree of roughness may be a large fraction of the fracture gap. Solution enlargement of the pores near fractures is possible, as is partial or full sealing of fractures by mineral precipitation. Properties of rock in regions of contact between blocks may also differ from bulk rock because of large compressive stresses. Compressive stress in regions of contact is greater than the average stress in the surrounding rock. Therefore, the nature of fractures is reservoir-specific, depending on mineralogy, tectonic stress history, and diagenesis.

Natural fractures affect all phases of petroleum reservoir life from the accumulation of oil to the techniques used to manage oil production. The existence of fractures in oil reservoirs was known as early as the 1860's. However, in the last thirty or forty years a significant interest in the effect of fractures on oil production has developed. This interest was sparked by the discovery of the giant fields in the Middle East and the Spraberry trend of West Texas. The interest aroused by these discoveries made the industry more aware of the presence and effects of fractures in other reservoirs.

Naturally fractured reservoirs come in a wide variety of rock mineralogy (carbonate, diatomite, granite, schist, sandstone, shale and coal), porosity, and permeability. Carbonates include limestone, dolomite, and chalk. Fractured limestones are prevalent in the giant and prolific fields of the Middle East. Fractured dolomites are exemplified by the San Andres formation in many West Texas fields, and fractured chalks are found in Texas (Austin Chalk),

North Sea (Ekofisk), and other parts of the world. Limestones and dolomites have moderately low matrix porosities and permeabilities, while chalks can possess high matrix porosities (30 to 50%) but sometimes very low permeabilities (0.01 to 10 md). Limestone and dolomites have strong rock frames. High-porosity chalks and diatomites have much weaker rock frames, which lead to subsidence and collapse of the wellbore as a result of production and reservoir depletion.

Diatomites are siliceous remains of diatoms and usually have large matrix porosities and very low permeabilities (South Belridge, CA). The rock frame can be weak, as in chalks.

Fractured granite (igneous) and schist (volcanic) have essentially no permeability and no matrix porosity aside from a small matrix porosity in granite, caused by microfractures or microfissures in the rock, or a larger porosity resulting from weathering of the feldspar or plagioclase minerals in the rock.

Sandstones and shales can be fractured too. In fractured sandstone, the rock matrix has a very low permeability (Spraberry, Texas), while fractured shales (Devonian Shale, Eastern U.S.) have virtually no matrix porosity.

Interconnected open fractures normally contribute little pore volume to a reservoir, but do act as highly conductive flow channels throughout the reservoir. The main portion of the reservoir pore volume is attributed to the broken-up pieces of the reservoir rock known as matrix blocks, which have much smaller permeabilities than do fractures. The presence of interconnected open fractures significantly affects the choice of mechanisms used for oil recovery from the matrix blocks. Viscous displacement of oil by water or gas in the matrix is often not very significant because of the small pressure gradients across the individual matrix blocks. Solution gas drive, gravity drainage, and water imbibition are the most prevalent recovery mechanisms.

Gravity drainage is an effective means for recovering oil from fractured reservoirs that have a large vertical relief. The buoyancy of oil in the matrix relative to the surrounding gas or water can cause these fluids to invade the matrix blocks and displace oil. Imbibition of water from fractures to matrix blocks also recovers oil. The rate and extent of water imbibition depend on wettability of the rock, rock permeability, relative permeability, and the size of matrix blocks.

Naturally fractured reservoirs can often be classified as "dual-porosity." In a dual-porosity idealization, one porosity represents the matrix blocks and the second represents that of the fractures and vugs. In dual-porosity oil reservoirs, fractures provide the main path for fluid flow from the reservoir. Oil from the matrix blocks flows into the fracture space, which carries the oil to the wellbore. In water-wet reservoirs, water that comes in contact with the oil zone imbibes into the matrix blocks to displace oil. Combinations of large flow rates, low matrix permeability, and weak imbibition result in water fingering through the fractures into the wellbore. Once fingering of water occurs, the water-oil ratio increases to large values. In oil reservoirs containing free gas, gas fingering and high gas-oil ratios can also occur.

Many papers on single- and multiphase flow in naturally fractured porous media have appeared in the literature. For single-phase flow, Barenblatt et al. (1960) established the mathematical foundation of flow behavior in dual-porosity rocks. Later, Warren and Root (1963) developed a radial model for well transient testing purposes. Kazemi (1969) extended Warren and Root's model to more complex situations in two dimensions. Virtually all the work in single-phase flow has been directed toward transient testing applications (Odeh, 1965; Kazemi, et al., 1969; Najurieta, 1979; Abdassah and Ershaghi, 1986).

For two-phase flow, the pioneering work of Birks (1955) and Mattax and Kyte (1962), are notable. These early papers gave insight into the mechanism of oil displacement from matrix blocks in a fractured porous rock. Birks provided two formulations to calculate oil displacement by water. The first method used a capillary model, and the second a relative-permeability model. Both formulations make a dynamic balance between viscous, capillary and gravity forces to calculate cumulative oil with time. Mattax and Kyte focused on oil recovery by water imbibition from matrix blocks in water-drive reservoirs. From experimental imbibition data they developed an empirical oil-recovery correlation in which "the time required to recover a given fraction of oil from a matrix block is proportional to the square of the distance between fractures."

Other multiphase fracture flow work includes the theoretical works of Barenblatt (1964) and Braester (1971, 1972). Barenblatt considered the reservoir as two overlapping continua – matrix and fractures. He formulated equations describing immiscible two-phase flow in each continuum that use conservation of mass. Flow between the matrix and the fractures was accounted for by source functions. These source functions use Darcy's law over a mean path between the matrix and the adjacent fractures. Braester uses a formulation similar to Barenblatt's.

Several papers (Mattax and Kyte, 1962; Parsons and Chaney, 1966; Iffly et al., 1972; Mannon and Chilingar, 1972; Kleppe and Morse, 1974; Kazemi and Merrill, 1979; Hamon and Vidal, 1986; Beckner et al., 1988; Cuiec, et al., 1990; Keijzer and de Vries 1990) on experimental work have been concerned primarily with imbibition aspects of the flow mechanism in the matrix rather than the total flow problem in the fracture-matrix system. For instance, Mattax and Kyte (1962) developed equations for scaling up imbibition experiments. Parsons and Chaney (1966) used these equations in a study of imbibition in carbonate rocks. Iffly et al. (1972), in addition to the experimental work, used a one-dimensional, two-phase, mathematical model to match the experimental oil recoveries from the matrix. They treated flow in the fractures as a boundary condition for flow in the matrix. This approach is usually suitable for studying laboratory experiments. However, the method is not suitable for field applications where the reservoir geometry is far more complex than the laboratory cores. Kleppe and Morse (1974) used a similar mathematical model in two dimensions to match imbibition experiments.

Experimental results have been extended to large-scale field applications through dual-porosity/dual-permeability finite-difference simulation (Kazemi et al., 1976; Rossen, 1977; deSwaan, 1978; Thomas et al., 1983; Gilman and Kazemi, 1983; Saidi, 1983; Litvak, 1985; Quandalle and Sabatheir, 1987; Sonier, et al., 1988; Gilman and Kazemi, 1988). Capillary-gravity interaction between fracture and matrix is not rigorous in these simulators. Litvak (1985), and Sonier et al. (1988) simulated gravity effects by segregating phases for both fracture and matrix in each computational grid for each timestep. This was done by simple material balance without accounting for the time-dependent nature of gravity segregation as affected by capillary pressure and the vertical transmissibility of the matrix. Pruess and Narasimhan (1985) and Gilman (1986) reported a grid refinement scheme to calculate more accurately fluid movement and heat transfer in the matrix as a function of time.

The "dual-permeability" concept reported by Hill and Thomas (1985) allows both matrix-to-matrix and fracture-to-fracture flow between gridblocks. Dual-permeability models require much greater computing time than the dual-porosity models. Gilman and Kazemi (1988) used the dual-permeability idealization to develop an efficient algorithm to account accurately for gravity effects both in the fracture and the matrix. They also accounted for viscous displacement in matrix blocks caused by pressure gradient in the fracture network. Quandalle and Sabatheir (1987) reported a similar viscous displacement calculation method.

Recent work has shown that transfer functions using pseudocapillary pressure curves can be used (Rossen and Shen, 1987; Dean and Lo, 1988; Coats, 1989; Kazemi, et al., 1989) to enable coarse-grid, dual-porosity models to mimic fine-grid results.

Steam injection has been used to recover heavy oil from naturally fractured reservoirs (Sahuquet and Ferrier, 1982; Dreher, et al. 1986). Reis (1990) has reviewed the recovery mechanisms for steam injection in naturally fractured reservoirs and van Wunnik and Wit (1990) have explored the possibility of improving gravity drainage by injecting steam into the gas cap of a naturally fractured reservoir. Chen, et al. (1987) and Lee and Tan (1987) described dual-porosity simulators for this process.

Other applications of the dual-porosity concept include flow in fractured aquifers (Duguid and Lee, 1977; Shapiro, 1985), waste-disposal storage in fractured rocks (Sinnock, et al., 1987), and production of methane from coal beds (Cervik, 1973; Ancell and Lambert, 1980; King et al., 1986). In coal beds, gas is stored in the matrix pores (micropores) in an adsorbed state. Gas flows from the matrix to the fractures by diffusion. The matrix has large storage capacity, but does not contribute much to the flow capacity of the reservoir. The fractures (cleats) in the coal bed have uniform spacing with distances on the order of several inches. These fractures provide the bulk of the reservoir flow capacity.

In the following sections we will develop several geometrical idealizations of naturally fractured reservoirs, and will provide mathematical equations for describing flow in such reservoirs. These formulations represent the state-of-the-art foundation of the commercial and private simulators available in the petroleum industry. The reservoir engineering, geological, and petrophysical background for the physics involved are detailed in books by Reiss (1980), Aguilera (1980) van Golf-Racht (1982), Nelson (1985) and Saidi (1987).

6.2 IDEALIZATIONS OF PETROLEUM RESERVOIRS

Reservoir rocks are highly heterogeneous. Depending on the degree of heterogeneity, the reservoir is classified as "single-porosity", "dual-porosity", "triple-porosity" or "dual-permeability". In this section we discuss these various idealizations.

6.2.1 Single-Porosity

Nonfractured porous rocks are generally classified as "single-porosity." In such systems, porosity is usually described by a normal distribution function in each geologic layer. Permeability is similarly assumed to fit a log-normal distribution function in each geologic unit. The pore system has a large surface area on the order of 1 m^2/g of rock. In multiphase flow, relative permeabilities are used to calculate the relative flow of each phase. The relative shape of relative permeabilities are strongly influenced by the wettability of the rock. Flow takes place as a result of viscous, gravity, capillary, and diffusive forces. The physics of flow in single-porosity systems has been extensively reported in various petroleum and hydrology texts.

6.2.2 Flow in Fractures

In fractured rocks, individual fractures are often represented by a parallel plate idealization. The permeability, k_f, of an individual fracture with smooth surface is related to fracture width, w_f, as:

$$k_f = 84 \times 10^6 \, w_f^{\,2} \tag{6.2.1}$$

for k_f in md and w_f in mm. For fractures with rough surfaces and widths less than 0.6 mm, experimental work (Jones, et al., 1988) has shown that the right hand side of (6.2.1) must be multiplied by 96/C, where $96 \leq C \leq 400$. C = 96 represents a smooth fracture while C = 400 is for a fracture with a relative roughness of about 0.6. Relative roughness is the ratio of the roughness height and fracture width.

Interconnected fractures are assumed to form a high-conductivity, three-dimensional network. When describing fractured reservoirs, we generally refer to the interconnected fracture network as "fracture."

For an orthogonal fracture network, porosity, ϕ_f, is approximated by

$$\phi_f = 3.28 \times 10^{-3} \left(\frac{w_f}{L_x} + \frac{w_f}{L_y} + \frac{w_f}{L_z} \right) \tag{6.2.2}$$

where L_x, L_y, and L_z are spacing between fractures in the x, y, and z directions in ft, and w_f is fracture width in mm. In general, the effective permeability, k_e, of fractured reservoirs is approximated by

$$k_e = k_f \phi_f + k_m \tag{6.2.3}$$

where k_m is the permeability of the porous rock without fractures.

Darcy and Forchheimer equations are used to calculate viscous flow in the fractures. Relative permeabilities for each phase in the fracture are generally assumed to be linear functions of the phase volumes in the fracture, but these functions can take the shape of relative permeability in non-fractured porous rock, depending on the fracture roughness, tortuosity, and the material filling the fracture.

6.2.3 Dual-Porosity

Dual-porosity reservoirs are systems where two distinct types of porosity coexist in a representative rock volume. Naturally fractured reservoirs typify such systems. In general, fracture storage capacity is low and fracture transmissivity is high. In the adjacent porous rock (matrix) storage capacity is high and transmissivity is relatively low. Naturally fractured reservoirs are often visualized as a set of stacked sugar cubes (Warren and Root, 1963), parallel layers separated by horizontal fractures (Kazemi, 1969), or a set of vertical match stick columns (Reiss, 1980) (Fig. 6.2.1). In reality, the fracture spacing and distribution vary throughout the reservoir.

6.2.4 Triple-Porosity

The triple-porosity idealization (Abdassah and Ershaghi, 1986) assumes that the reservoir consists of fractures plus two distinct sets of matrix rock types with highly contrasting permeabilities. In this situation, the matrix with the higher permeability feeds into the fracture more effectively than does the matrix with the lower permeability. Flow between the two matrix rock types is considered negligible.

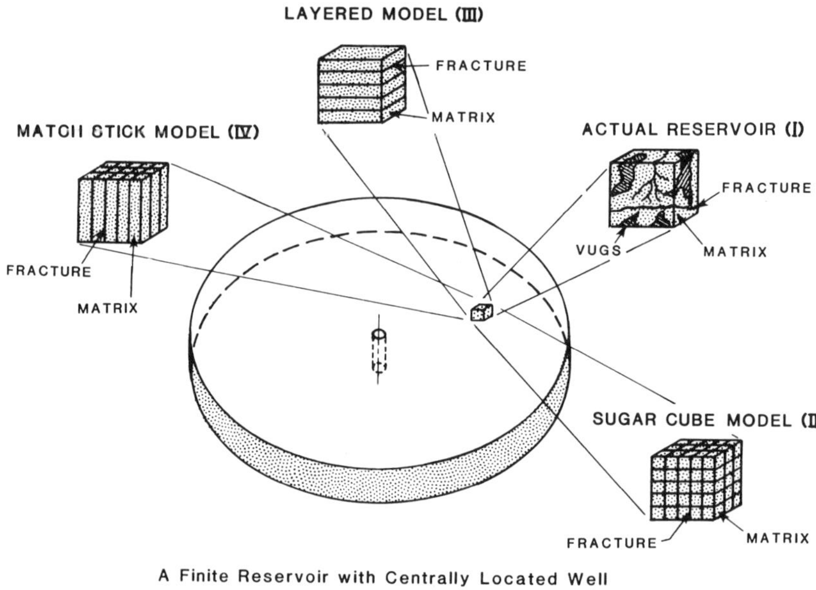

Figure 6.2.1. Idealizations of a naturally fractured reservoir (I actual reservoir; II sugar cube model; III layered model; IV match stick model) (from Kazemi, 1969; Warren and Root, 1963, courtesy SPE)

6.2.5 Dual-Permeability

In contrast to dual-porosity, the dual-permeability idealization (Barenblatt et al., 1960; Hill and Thomas, 1985) represents a reservoir system in which both the matrix and fracture network have continuity and matrix transmissivity is not insignificant compared to that of the fracture.

6.3 MATHEMATICAL FORMULATIONS

In this section we describe the mathematical equations in a finite-difference setting. We chose this approach as a matter of convenience because nearly all commercial and private large-scale simulators are formulated this way.

6.3.1 Single-Porosity Black Oil

The following equations give the finite-difference formulations for flow in a single-porosity petroleum reservoir. When water, oil, and gas are treated as three separate phases, the formulation shown below is called "black oil":

$$\underline{\text{Water}}: \quad \Delta\left[T_w(\Delta p_w - \gamma_w \Delta D)\right] + q_w = \frac{V_R}{5.6146\Delta t}\Delta_t\left(\frac{\phi S_w}{B_w}\right) \tag{6.3.1}$$

Oil: $$\Delta\left[T_o\left(\Delta p_o - \gamma_o \Delta D\right)\right] + q_o = \frac{V_R}{5.6146\Delta t}\Delta_t\left(\frac{\phi S_o}{B_o}\right) \qquad (6.3.2)$$

Gas: $$\Delta\left[T_g\left(\Delta p_g - \gamma_g \Delta D\right)\right] + \Delta\left[R_s T_o\left(\Delta p_o - \gamma_o \Delta D\right)\right]$$

$$+q_g + R_s q_o = \frac{V_R}{5.6146\Delta t}\Delta_t\left[\frac{S_g}{B_g} + \frac{R_s S_o}{B_o}\right] \qquad (6.3.3)$$

The x-direction transmissibility for the oil, water, and gas phases, $T_{\beta x}$, for Cartesian coordinates is given by

$$T_{\beta x} = 0.001127\frac{\Delta y \Delta z}{\Delta x}k_x\left(\frac{k_r}{\mu B}\right)_\beta, \beta = o, w, g \qquad (6.3.4)$$

The y- and z-direction transmissibilities are similarly defined. Relative permeability and capillary pressures are functions of phase saturations as shown below:

$$k_{rw} = k_{rw}\left(S_w\right) \qquad (6.3.5)$$

$$k_{rg} = k_{rg}\left(S_g\right) \qquad (6.3.6)$$

$$k_{ro} = k_{ro}\left(S_g, S_w\right) \qquad (6.3.7)$$

$$p_o - p_w = p_{cwo}\left(S_w\right) \qquad (6.3.8)$$

$$p_g - p_o = p_{cgo}\left(S_g\right) \qquad (6.3.9)$$

$$S_o + S_w + S_g = 1 \qquad (6.3.10)$$

6.3.2 Single-Porosity Compositional

When, in addition to phases, material balance is extended to the 'n' components of the oil and gas, the formulation shown below is called "compositional":

Hydrocarbon: $\Delta\left[T_o\, \xi_o\, x_i\left(\Delta p_o - \gamma_o\, \Delta D\right)\right] + x_i \xi_o q_o$

$$+\Delta\left[T_g\, \xi_g\, y_i\left(\Delta p_g - \gamma_g\, \Delta D\right)\right] + y_i \xi_g q_g$$

$$= \frac{V_R}{5.6146\Delta t} \Delta_t \left[\phi\left(x_i \xi_o S_o + y_i \xi_g S_g\right)\right],\ i = 1, 2, \ldots, n \tag{6.3.11}$$

The transmissibilities for the x-direction are:

$$T_{\beta x} = 0.001127 \frac{\Delta y \Delta z}{\Delta x} k_x \left(\frac{k_r}{\mu}\right)_\beta,\ \beta = o, g \tag{6.3.12}$$

assuming that the hydrocarbon gas and liquid are in phase equilibrium at all times, the other pertinent relations are:

$$\Sigma\, x_i = \Sigma\, y_i = \Sigma\, z_i = 1 \tag{6.3.13}$$

$$K_i = y_i / x_i \tag{6.3.14}$$

$$z_i = y_i\, V + x_i\, L \tag{6.3.15}$$

$$x_i = \frac{z_i}{1 + V\left(K_i - 1\right)} \tag{6.3.16}$$

$$y_i = \frac{K_i z_i}{1 + V\left(K_i - 1\right)} \tag{6.3.17}$$

$$L + V = 1 \tag{6.3.18}$$

$$V = \frac{\xi_g S_g}{\xi_g S_g + \xi_o S_o} \tag{6.3.19}$$

Hydrocarbon phase properties such as x_m, y_m, K_m, V, ξ_g, and ξ_o are normally calculated by an equation of state at each grid block.

6.3.3 Dual-Porosity Black Oil

In a dual-porosity black-oil formulation, (6.3.20) through (6.3.22) are used for the fracture network where the source terms, τ_w, τ_o, and τ_g, subtracted from the left-hand side, represent flow between the matrix and fractures.

Water: $$\Delta\left[T_{wf}\left(\Delta p_{wf}-\gamma_{wf}\Delta D_f\right)\right]-\tau_w+q_{wf}=\frac{V_R}{5.6146\Delta t}\Delta_t\left(\frac{\phi_f S_{wf}}{B_{wf}}\right) \tag{6.3.20}$$

Oil: $$\Delta\left[T_{of}\left(\Delta p_{of}-\gamma_{of}\Delta D_f\right)\right]-\tau_o+q_{of}=\frac{V_R}{5.6146\Delta t}\Delta_t\left(\frac{\phi_f S_{of}}{B_{of}}\right) \tag{6.3.21}$$

Gas: $$\Delta\left[T_{gf}\left(\Delta p_{gf}-\gamma_{gf}\Delta D_f\right)\right]+\Delta\left[R_{sf}T_{of}\left(\Delta p_{of}-\gamma_{of}\Delta D_f\right)\right]$$
$$-\tau_g+q_{gf}+R_{sf}q_{of}=\frac{V_R}{5.6146\Delta t}\Delta_t\left[\frac{S_{gf}}{B_{gf}}+\frac{R_{sf}S_{of}}{B_{of}}\right] \tag{6.3.22}$$

The source terms, τ_w, τ_o, and τ_g, are given as:

Water: $$\tau_w=0.001127\quad \sigma V_R k_m\left(\frac{k_r}{\mu B}\right)_{wmf}\left[p_{wf}-p_{wm}-\gamma_w\left(D_f-D_m\right)\right] \tag{6.3.23}$$

Oil: $$\tau_o=0.001127\quad \sigma V_R k_m\left(\frac{k_r}{\mu B}\right)_{omf}\left[p_{of}-p_{om}-\gamma_o\left(D_f-D_m\right)\right] \tag{6.3.24}$$

Gas: $$\tau_g=0.001127\quad \sigma V_R k_m\left\{\left(\frac{k_r}{\mu B}\right)_{gmf}\left[p_{gf}-p_{gm}-\gamma_g\left(D_f-D_m\right)\right]\right.$$
$$\left.+\left(R_s\frac{k_r}{\mu B}\right)_{omf}\left[p_{of}-p_{om}-\gamma_o\left(D_f-D_m\right)\right]\right\} \tag{6.3.25}$$

In the dual-porosity black oil formulation, material balance in the matrix leads to the following equations:

Water: $$\tau_w=\frac{V_R}{5.6146\Delta t}\Delta_t\left(\frac{\phi S_W}{B_w}\right)_m \tag{6.3.26}$$

Oil: $$\tau_o=\frac{V_R}{5.6146\Delta t}\Delta_t\left(\frac{\phi S_o}{B_o}\right)_m \tag{6.3.27}$$

Gas: $$\tau_g=\frac{V_R}{5.6146\Delta t}\Delta_t\left[\phi_m\left(\frac{S_g}{B_g}+\frac{R_s S_o}{B_o}\right)_m\right] \tag{6.3.28}$$

Equations (6.3.20) through (6.3.28), as well as (6.3.5) through (6.3.10) (repeated for fracture and matrix), form the basis of the solution process in dual-porosity reservoirs (Kazemi, et al., 1976; Thomas, et al., 1983; Gilman and Kazemi, 1983). Similar equations are used for the compositional formulation, as will be shown later (Coats, 1989).

6.3.4 Dual-Porosity with Empirical Matrix-Fracture Transfer Function

An empirical approach to the transfer of fluids between fracture and matrix could be based on the fact that oil recovery from a matrix block generally has the exponential form given below (Aronofsky, et al., 1958):

$$R = R_\infty\left(1 - e^{-\lambda t}\right) \tag{6.3.29}$$

where

$$R_\infty = \left(1 - S_{wcm} - S_{orm}\right)\phi_m \tag{6.3.30}$$

For waterflooding, the imbibition of water into the matrix could then be given by the following (deSwaan, 1978):

$$\tau_w = R_\infty V_R \lambda \int_o^t e^{-\lambda(t-\tau)} \frac{\partial S_{wf}}{\partial \tau} \partial\tau \tag{6.3.31}$$

Or, in finite-difference form

$$\tau_w = R_\infty \lambda \sum_{j=0}^{n} \left\{\left[S_{wf,l}^{n+1} - S_{wf,l}^{n}\right] \prod_{k=j}^{n} e^{-\lambda \Delta t_k}\right\} \quad ,1 = 1, 2, \ldots, LMAX \tag{6.3.32}$$

For incompressible flow, the oil flow from the matrix would equal the water flow into the matrix. Similar empirical equations could be used for oil drainage from the matrix due to gas in the fractures. The material presented in this section will be the basis for further theoretical developments in Sec. 6.8.2.

6.3.5 Triple-Porosity

The triple-porosity formulation has not found much application in multiphase flow (Abdassah and Ershaghi, 1986). However, the equations are similar to the dual-porosity case, with the addition of separate source terms for each matrix rock type.

6.3.6 Dual-Permeability Black Oil

The dual-permeability approach assumes two superimposed continua: fracture and matrix. Both the fracture and matrix are continuous and, therefore, equations similar to (6.3.20) through (6.3.22) are used for both the fracture and matrix with fracture-matrix flow terms given by (6.3.23) through (6.3.25). For example, the equations for the water phase are

Fracture:

$$\Delta\left[T_{wf}\left(\Delta p_{wf}-\gamma_{wf}\Delta D_f\right)\right]-\tau_w+q_{wf}=\frac{V_R}{5.6146\Delta t}\Delta_t\left(\frac{\phi_f S_{wf}}{B_{wf}}\right) \tag{6.3.33}$$

Matrix:

$$\Delta\left[T_{wm}\left(\Delta p_{wm}-\gamma_{wm}\Delta D_m\right)\right]+\tau_w+q_{wm}=\frac{V_R}{5.6146\Delta t}\Delta_t\left(\frac{\phi_m S_{wm}}{B_{wm}}\right) \tag{6.3.34}$$

where transmissibilities for the x direction are

$$T_{wfx}=0.001127\frac{\Delta y\Delta z}{\Delta x}k_{fx}\phi_f\left(\frac{k_r}{\mu B}\right)_{wf} \tag{6.3.35}$$

$$T_{wmx}=0.001127\frac{\Delta y\Delta z}{\Delta x}k_{mx}\left(\frac{k_r}{\mu B}\right)_{wm} \tag{6.3.36}$$

Equations for oil and gas will be similar.

From here on, our development of the mathematics will be in the dual-permeability formulation. This formulation reduces to the standard dual-porosity formulation by ignoring the interblock matrix transmissibility terms.

6.3.7 Dual-Permeability Compositional

For the dual-permeability compositional approach, the water equations are given by (6.3.33) and (6.3.34). The equations for the 'n' hydrocarbon components are

Fracture:

$$\begin{aligned}
&\Delta\left[\phi_f\xi_{of}S_{of}K_{oif}\,\Delta x_{if}\right]+\Delta\left[T_{of}\xi_{of}x_{if}\left(\Delta p_{of}-\gamma_{of}\Delta D_f\right)\right]+x_{if}\xi_{of}q_{of}\\
&+\Delta\left[\phi_f\xi_{gf}S_{gf}K_{gif}\,\Delta y_{if}\right]+\Delta\left[T_{gf}\xi_{gf}y_{if}\left(\Delta p_{gf}-\gamma_{gf}\,\Delta D_f\right)\right]+y_{if}\xi_{gf}q_{gf}\\
&-\tau_o\,\xi_{omf}\,x_{imf}-\tau_g\,\xi_{gmf}\,y_{imf}-\phi_m\,\xi_{omf}S_{om}K_{oimf}\left(x_{if}-x_{im}\right)\\
&-\phi_m\,\xi_{gmf}\,S_{gmf}\,K_{gimf}\left(y_{if}-y_{im}\right)\\
&=\frac{V_R}{5.6146\Delta t}\Delta_t\left[\phi_f\left(x_{if}\xi_{of}S_{of}+y_{if}\xi_{gf}S_{gf}\right)\right],\ i=1,2,\ldots,n
\end{aligned} \tag{6.3.37}$$

Matrix:

$$\Delta\left[\phi_m \xi_{om} S_{om} K_{oim}\, \Delta\, x_{im}\right] + \Delta\left[T_{om}\xi_{om} x_{im}\left(\Delta p_{om} - \gamma_{om}\Delta D_m\right)\right] + x_{im}\xi_{om} q_{om}$$

$$+\Delta\left[\phi_m \xi_{gm} S_{gm} K_{gim}\, \Delta\, y_{im}\right] + \Delta\left[T_{gm}\xi_{gm} y_{im}\left(\Delta p_{gm} - \gamma_{gm}\,\Delta D_m\right)\right] + y_{im}\xi_{gm} q_{gm}$$

$$-\tau_o\, \xi_{omf}\, x_{imf} - \tau_g\, \xi_{gmf}\, y_{imf} - \phi_m\, \xi_{omf} S_{om} K_{oimf}\left(x_{if} - x_{im}\right)$$

$$-\phi_m\, \xi_{gmf}\, S_{gmf}\, K_{gimf}\left(y_{if} - y_{im}\right)$$

$$= \frac{V_R}{5.6146\Delta t}\, \Delta_t\left[\phi_m\left(x_{im}\xi_{om} S_{om} + y_{im}\xi_{gm} S_{gm}\right)\right], i = 1, 2, \ldots, n \qquad (6.3.38)$$

where S_{gmf} is the weighted average of gas saturation between fracture and matrix region next to the fracture. Dispersion in the fracture and matrix (first and fourth terms in (6.3.37) and (6.3.38)) is generally negligible, but is included here for completeness. However, dispersion can be a significant means of mass transfer between fracture and matrix. The fracture-matrix transfer terms for oil and gas are:

Oil:

$$\tau_o = 0.001127\,\sigma V_R k_m \left(\frac{k_r}{\mu}\right)_{omf} \left[p_{of} - p_{om} - \gamma_{omf}\left(D_f - D_m\right)\right] \qquad (6.3.39)$$

Gas:

$$\tau_g = 0.001127\,\sigma V_R k_m \left(\frac{k_r}{\mu}\right)_{gmf} \left[p_{gf} - p_{gm} - \gamma_{gmf}\left(D_f - D_m\right)\right] \qquad (6.3.40)$$

The transmissibilities for x-direction flow for Cartesian Coordinates are:

Fracture:

$$T_{\beta fx} = 0.001127 \frac{\Delta y \Delta z}{\Delta x}\left(k_{fx}\phi_f\right)\left(\frac{k_r}{\mu}\right)_\beta, \beta = o, g \qquad (6.3.41)$$

Matrix:

$$T_{\beta mx} = 0.001127 \frac{\Delta y \Delta z}{\Delta x}\left(k_{mx}\right)\left(\frac{k_r}{\mu}\right)_\beta, \beta = o, g \qquad (6.3.42)$$

The dispersion coefficients for the x-direction are:

Fracture:

$$K_{\beta ifx} = \frac{\Delta y \Delta z}{5.6146 \Delta x} \left\{ \frac{D_{\beta ifx}}{\tau_f} + \frac{1}{\phi_f S_{\beta f} \left| \vec{v}_{\beta f} \right|} \left[\alpha_{Lf} v_{\beta fx}^2 + \alpha_{Tf} \left(v_{\beta fy}^2 + v_{\beta fz}^2 \right) \right] \right\} \quad , \quad \beta = o, g \quad (6.3.43)$$

Matrix:

$$K_{\beta imx} = \frac{\Delta y \Delta z}{5.6146 \Delta x} \left\{ \frac{D_{\beta imx}}{\tau_m} + \frac{1}{\phi_m S_{\beta m} \left| \vec{v}_{\beta m} \right|} \left[\alpha_{Lm} v_{\beta mx}^2 + \alpha_{Tm} \left(v_{\beta my}^2 + v_{\beta mz}^2 \right) \right] \right\} \quad , \quad \beta = o, g (6.3.44)$$

Similar terms are defined for the y- and z-directions. The dispersion coefficients for fracture-matrix flow are:

$$K_{\beta imf} = \frac{\sigma V_R}{5.6146} \left[\frac{D_{\beta im}}{\tau_m} + \alpha_m \frac{\left| v_{\beta mf} \right|}{\phi_m S_{\beta mf}} \right], \beta = o, g \quad (6.3.45)$$

In (6.3.43) through (6.3.45), τ_f and τ_m are the fracture and matrix tortuosities.

6.3.8 Dual-Permeability Chemical Transport

In the case of transport of chemicals by the liquid phases, the equations for dual-permeability are:

Fracture:

$$\Delta\left[\phi_f \rho_{of} S_{of} K_{oif} \Delta C_{oif}\right] + \Delta\left[T_{of} \rho_{of} C_{oif} \left(\Delta p_{of} - \gamma_{of} \Delta D_f\right)\right] + C_{oif} \rho_{of} q_{of}$$

$$+ \Delta\left[\phi_f \rho_{wf} S_{wf} K_{wif} \Delta C_{wif}\right] + \Delta\left[T_{wf} \rho_{wf} C_{wif} \left(\Delta p_{wf} - \gamma_{wf} \Delta D_f\right)\right] + C_{wif} \rho_{wf} q_{wf}$$

$$-\tau_o \rho_{omf} C_{oimf} - \tau_w \rho_{wmf} C_{wimf} - \phi_m \rho_{omf} S_{om} K_{oimf} \left(C_{oif} - C_{oim}\right)$$

$$-\phi_m \rho_{wmf} S_{wm} K_{wimf} \left(C_{wif} - C_{wim}\right)$$

$$= \frac{V_R}{5.6146 \Delta t} \Delta_t \left[\phi_f \left(C_{oif} \rho_{of} S_{of} + C_{wif} \rho_{wf} S_{wf}\right) + \left(1 - \phi_t\right) \rho_s C_{sif}\right] \quad (6.3.46)$$

Matrix:

$$\Delta\left[\phi_m\rho_{om}S_{om}K_{oim}\Delta C_{oim}\right]+\Delta\left[T_{om}\rho_{om}C_{oim}\left(\Delta p_{om}-\gamma_{om}\Delta D_m\right)\right]+C_{oim}\rho_{om}q_{om}$$

$$+\Delta\left[\phi_m\rho_{wm}S_{wm}K_{wim}\Delta C_{wim}\right]+\Delta\left[T_{wm}\rho_{wm}C_{wim}\left(\Delta p_{wm}-\gamma_{wm}\Delta D_m\right)\right]+C_{wim}\rho_{wm}q_{wm}$$

$$-\tau_o\rho_{omf}C_{oimf}-\tau_w\rho_{wmf}C_{wimf}-\phi_m\rho_{omf}S_{om}K_{oimf}\left(C_{oif}-C_{oim}\right)$$

$$-\phi_m\rho_{wmf}S_{wm}K_{wimf}\left(C_{wif}-C_{wim}\right)$$

$$=\frac{V_R}{5.6146\,\Delta t}\Delta_t\left[\phi_m\left(C_{oim}\rho_{om}S_{om}+C_{wim}\rho_{wm}S_{wm}\right)+\left(1-\phi_t\right)\rho_s C_{sim}\right] \tag{6.3.47}$$

where concentrations in the oil and water phases are related by a partition coefficient given as:

$$\kappa_i = C_{oi} / C_{wi} \tag{6.3.48}$$

The matrix-fracture transfer term, τ, dispersion coefficient, K, and transmissibility, T, are given by (6.3.39) through (6.3.45). C_{sif} is the adsorbed concentration of component i on the fracture surface per unit total rock weight, while C_{sim} is the adsorbed concentration of component i on the pore surface of the matrix per unit total rock weight.

6.3.9 Dual-Permeability Thermal Processes

Conduction of heat from a fracture into a matrix block proceeds much more rapidly than mass diffusion or imbibition. Thus, steam flooding has considerable potential for enhancing recovery of oil from fractured reservoirs. Warming of the matrix blocks swells the trapped oil, expelling it to the surrounding fractures. The flow of oil is also accelerated by reduced viscosity at elevated temperatures. Furthermore, elevated temperature may enhance imbibition of water into the matrix because of favorable changes in rock wettability. A pilot steamflood in the intensely fractured Lacq Superior Field in southwestern France resulted in significant incremental oil (Sahuquet and Ferrier, 1982). Effective transfer of heat from the fractures to the matrix blocks prevented early arrival of steam at producing wells. Dreher, Kenyon, and Iwere (1986) described laboratory and simulator studies for steam flooding of a sparsely fractured carbonate reservoir.

Schulte and de Vries (1982) suggest that in-situ combustion also has potential for enhancing oil recovery from fractured reservoirs with fracture spacing of less than 1 m. Diffusion of oxygen into the matrix is a limiting step in the process.

For simulation of thermal processes, the dual-permeability formulation requires mass balance equations for each component (including the water which can be in the aqueous or gas phases) and an energy balance. The component balance equations are given by (6.3.37) and (6.3.38). The energy balance is written as follows:

Fracture:

$$\Delta K_{\hat{T}f}\Delta\hat{T}_f + \Delta\left[T_{of}\xi_{of}H_{of}\left(\Delta p_{of} - \gamma_{of}\Delta D_f\right)\right] + H_{of}\xi_{of}q_{of}$$

$$+\Delta\left[T_{wf}\xi_{wf}H_{wf}\left(\Delta p_{wf} - \gamma_{wf}\Delta D_f\right)\right] + H_{wf}\xi_{wf}q_{wf}$$

$$+\Delta\left[T_{gf}\xi_{gf}H_{gf}\left(\Delta p_{gf} - \gamma_{gf}\Delta D_f\right)\right] + H_{gf}\xi_{gf}q_{gf}$$

$$+\tau_o\xi_{omf}H_{omf} + \tau_w\xi_{wmf}H_{wmf} + \tau_g\xi_{gmf}H_{gmf}$$

$$-K_{\hat{T}mf}\left(\hat{T}_f - \hat{T}_m\right) + q_{rf}$$

$$= \frac{V_R}{5.6146\,\Delta t}\Delta_t\left[\phi_f\left(\xi_{of}S_{of}U_{of} + \xi_{gf}S_{gf}U_{gf} + \xi_{wf}S_{wf}U_{wf}\right)\right] \tag{6.3.49}$$

The heat of reaction, q_r, represents kinetic reactions such as heat generation resulting from in-situ combustion or any chemical reactions.

Matrix:

$$\Delta K_{\hat{T}m}\Delta\hat{T}_m + \Delta\left[T_{om}\xi_{om}H_{om}\left(\Delta p_{om} - \gamma_{om}\Delta D_m\right)\right] + H_{om}\xi_{om}q_{om}$$

$$+\Delta\left[T_{wm}\xi_{wm}H_{wm}\left(\Delta p_{wm} - \gamma_{wm}\Delta D_m\right)\right] + H_{wm}\xi_{wm}q_{wm}$$

$$+\Delta\left[T_{gm}\xi_{gm}H_{gm}\left(\Delta p_{gm} - \gamma_{gm}\Delta D_m\right)\right] + H_{gm}\xi_{gm}q_{gm}$$

$$+\tau_o\xi_{omf}H_{omf} + \tau_w\xi_{wmf}H_{wmf} + \tau_g\xi_{gmf}H_{gmf}$$

$$-K_{\hat{T}mf}\left(\hat{T}_f - \hat{T}_m\right) + q_{rm}$$

$$= \frac{V_R}{5.6146\,\Delta t}\Delta_t\left[\phi_m\left(\xi_{om}S_{om}U_{om} + \xi_{gm}S_{gm}U_{gm} + \xi_{wm}S_{wm}U_{wm}\right)\right.$$

$$\left. + (1-\phi_t)\rho_s c_s\left(\hat{T}_m - \hat{T}_b\right)\right] \tag{6.3.50}$$

The overall enthalpy, H, and internal energy, U, of the oil and gas phases is given by (6.3.51) through (6.3.54):

$$H_o = \sum_{i=1}^{n} x_i H_{oi} \tag{6.3.51}$$

$$H_g = \sum_{i=1}^{n} y_i H_{gi} \tag{6.3.52}$$

$$U_o = \sum_{i=1}^{n} x_i U_{oi} \tag{6.3.53}$$

$$U_g = \sum_{i=1}^{n} y_i U_{gi} \tag{6.3.54}$$

The thermal conductivity coefficients for the x-direction are:

Fracture:

$$K_{\hat{T}fx} = \frac{\Delta y \Delta z}{5.6146\,\Delta x}\left[\phi_f\left(S_{wf}k_{\hat{T}wf} + S_{of}k_{\hat{T}of} + S_{gf}k_{\hat{T}gf}\right)\right] \tag{6.3.55}$$

Matrix:

$$K_{\hat{T}mx} = \frac{\Delta y \Delta z}{5.6146\,\Delta x}\left[\phi_m\left(S_{wm}k_{\hat{T}wm} + S_{om}k_{\hat{T}om} + S_{gm}k_{\hat{T}gm}\right) + \left(1-\phi_t\right)k_{\hat{T}s}\right] \tag{6.3.56}$$

Fracture/Matrix:

$$K_{\hat{T}mf} = \frac{\sigma V_R}{5.6146}\left[\phi_m\left(S_{wm}k_{\hat{T}wm} + S_{om}k_{\hat{T}om} + S_{gm}k_{\hat{T}gm}\right) + \left(1-\phi_t\right)k_{\hat{T}s}\right] \tag{6.3.57}$$

When using (6.3.49) and (6.3.50) the reader should bear in mind that the heat conduction in the fracture, $\Delta K_{\hat{T}f}\Delta\hat{T}_f$, can be neglected in favor of the convective terms because of high fluid velocity. Also, the temperature of the rock is assumed to be at the temperature of the fluid in the matrix because of the large thermal capacity of the matrix. The fracture also conducts heat to the matrix through a matrix-fracture heat transfer term, $K_{\hat{T}mf}\left(\hat{T}_f - \hat{T}_m\right)$.

6.3.10 Dual-Porosity Coal Bed Methane

For coal bed methane, gas is stored in the matrix (micropores) in an adsorbed state (King, et al., 1986). Gas flows from the micropores to the fractures by diffusion. Water does not flow in the matrix. The micropores have a large storage capacity, but do not contribute to the flow capacity of the reservoir. The formulation can therefore be cast in a dual-porosity setting, and the equations, as given by King, et al. (1986), are as follows:

Gas in Fracture:

$$\Delta\left[T_{gf}\Delta p_{gf} + \phi_f K_f\,\Delta\frac{S_{gf}p_{gf}}{Z}\right] - \frac{p_{sc}}{\hat{T}_{sc}}\hat{T}\left(\tau_g - q_g\right) = \frac{V_R}{\Delta t}\Delta_t\left(\frac{\phi_f S_{gf} p_{gf}}{Z}\right) \tag{6.3.58}$$

Water in Fracture:

$$\Delta\left[T_{wf}\,\Delta\,p_{wf}\right]+q_w=\frac{V_R}{\Delta t}\,\Delta_t\left(\frac{\phi_f\,S_{wf}}{B_{wf}}\right) \tag{6.3.59}$$

Gas in Matrix:

$$\tau_g=\frac{V_R}{\Delta t}\Delta_t\,C_{vm} \tag{6.3.60}$$

Other relevant equations are:

$$S_{gf}+S_{wf}=1 \tag{6.3.61}$$

$$p_{gf}-p_{wf}=p_{cwf}\left(S_{wf}\right) \tag{6.3.62}$$

Transmissibility and dispersivity coefficients in the fracture equations are given as follows for x-direction flow:

$$T_{gfx}=0.006328\frac{\Delta y\Delta z}{\Delta x}\frac{k_{fx}k_{rgf}}{\mu_{gf}}\frac{p_{gf}}{Z} \tag{6.3.63}$$

$$T_{wfx}=0.006328\frac{\Delta y\Delta z}{\Delta x}\frac{k_{fx}k_{rwf}}{\mu_{wf}B_{wf}} \tag{6.3.64}$$

$$K_{fx}=\frac{\Delta y\Delta z}{\Delta x}\frac{D_{gfx}}{\tau_f} \tag{6.3.65}$$

In (6.3.65), τ_f is the fracture tortuosity. The fracture-matrix transfer term is:

$$\tau_g=\frac{D_{gm}}{\tau_m}\sigma V_R\left(C_{ve}-C_{vm}\right) \tag{6.3.66}$$

where τ_m is the matrix tortuosity.

Assuming a Langmuir adsorption isotherm, the equilibrium volumetric concentration, C_{ve}, for a given fracture pressure, p_{gf}, is:

$$C_{ve}=\frac{C_{vL}p_{gf}}{p_L+p_{gf}} \tag{6.3.67}$$

C_{vL} and p_L are constants which define the Langmuir isotherm.

6.4 FINITE-DIFFERENCE GRID SYSTEM FOR VARIOUS RESERVOIR IDEALIZATIONS

Several grid systems for the mathematical formulations described in Sec. 6.3 are shown in Figs. 6.4.1 through 6.4.5. These grid systems represent different idealizations of the reservoir. Figure 6.4.1 shows a model with both vertical and horizontal fractures. Separate nodes are used for fracture and matrix. Figure 6.4.2 also shows a model that allows vertical and horizontal fractures. However, this model requires far fewer nodes because areas in which the matrix blocks behave similarly are grouped in a single node. Each gridblock may contain many matrix blocks. The matrix blocks act as sources that feed into the fractures in a gridblock. The fractures can be thought of as a system of connected pipes.

The boundary conditions used can make a dramatic difference in simulation results. Generally, we assume that only the fracture produces into the wellbore and is the path of fluid transport from one gridblock to the next.

For singe-phase flow of a gas or liquid, fluid compression and viscous forces cause fluid movement, and gravity and capillary forces are not pertinent. For multiphase flow, viscous, gravity, capillary, and diffusive forces must be properly accounted for. For this reason, we might require that the matrix blocks be further divided into a fine computational grid to obtain a better definition of temperature, phase saturation and component distribution (Fig. 6.4.3).

To account for directional tendencies other than in the x or y direction Gilman and Kazemi (1983) provided a scheme for flow to take place in the diagonal directions in the xy plane. To achieve this, the conventional five-point finite difference in the xy plane (Fig. 6.4.4) was replaced by a nine-point formulation (Fig. 6.4.5). In this "nine-point connection," the fracture permeabilities in the diagonal direction are independent of those in the x and y directions; therefore, fracture permeability is not a true tensor.

6.5 SOLUTION OF MATHEMATICAL EQUATIONS

The finite difference flow equations described in the previous sections are nonlinear. These equations are often solved by the Newton-Raphson method. This method leads to a large system of linear equations. Considerations such as optimal node ordering, matrix reduction, and innovative mesh refinement schemes are required to solve these equations efficiently. This section deals with these topics.

6.5.1 Coarse Grid

In this section, we focus on the solution of water-oil flow using a vertical dual-permeability formulation to illustrate (1) how to generate a set of linear equations from the discretization of the flow equations, (2) how to order the equations to produce the smallest bandwidth, (3) how to reduce the set of linear equations to a manageable size, and (4) how to solve the set of linear equations. Vertical dual-permeability means that the matrix blocks are allowed to be continuous in the vertical direction only. This approach will permit tracking of saturation fronts in the vertical direction both in the fracture and matrix. Furthermore, x-y plane mesh refinement in the matrix can be done to allow areal tracking of saturation and component fronts in the matrix. We

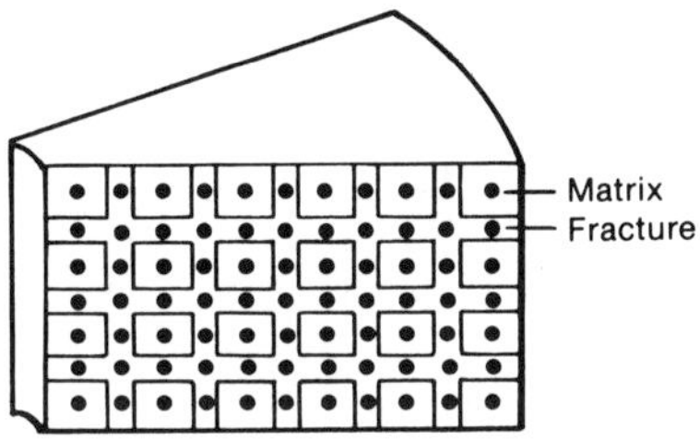

Figure 6.4.1. Fracture-matrix block model (from Gilman and Kazemi, 1983, courtesy SPE).

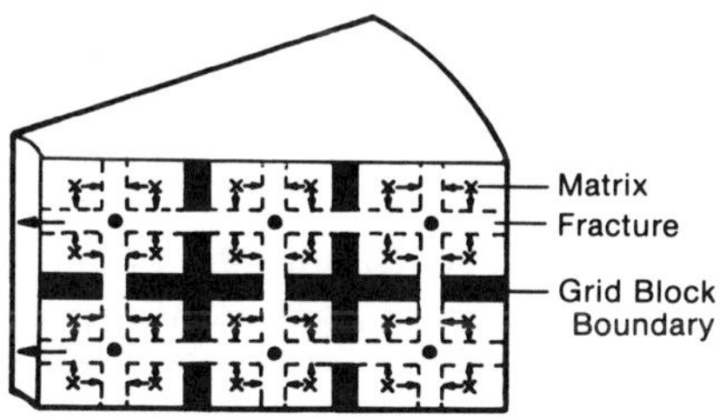

Figure 6.4.2. Dual-porosity block model (from Gilman and Kazemi, 1983, courtesy SPE).

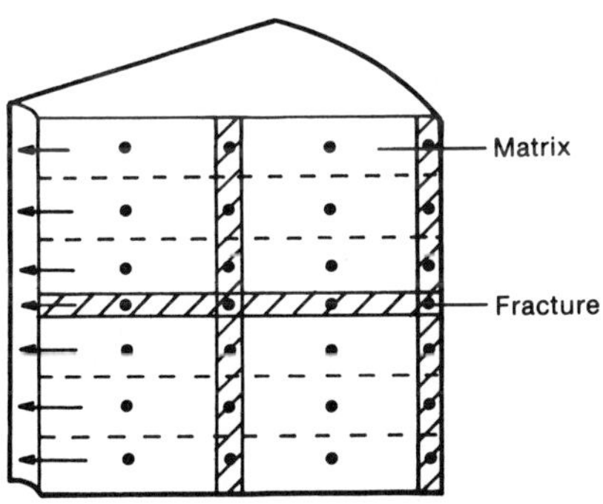

Figure 6.4.3. Multiphase block model (from Gilman and Kazemi, 1983, courtesy SPE).

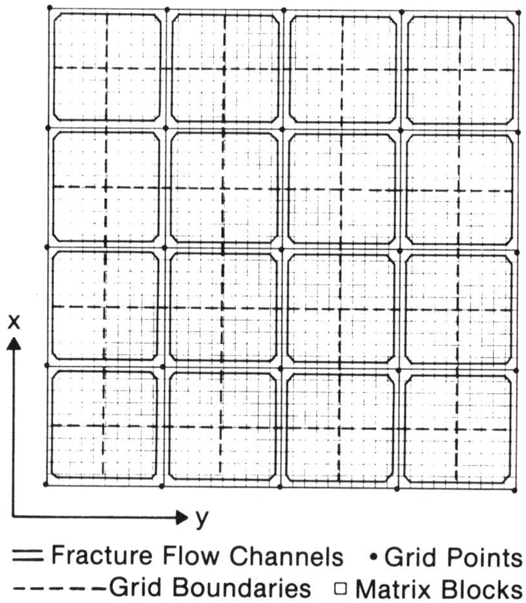

Figure 6.4.4. Five-point finite-difference grid (from Gilman and Kazemi, 1983, courtesy SPE).

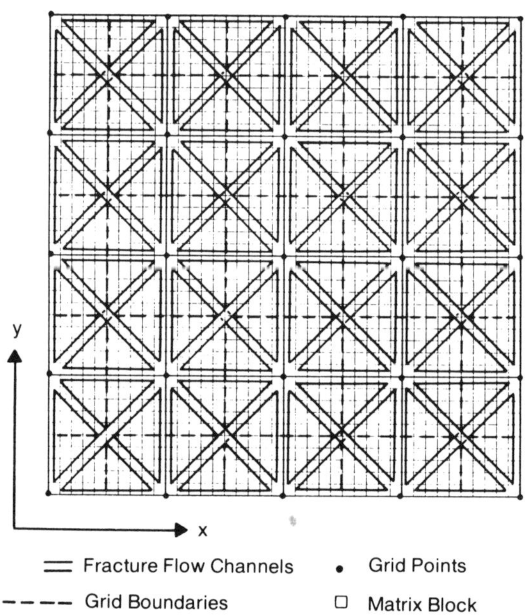

Figure 6.4.5. Nine-point connection grid (from Gilman and Kazemi, 1983, courtesy SPE).

believe that vertical dual-permeability formulation with matrix mesh refinement produces the most efficient and most rigorous solution to fracture-matrix flow problems. We discuss mesh refinement in the next section.

We illustrate the method of solution using an *x-z* vertical slice of a reservoir as shown by Fig. 6.5.1. The flow equation for a given phase generally has the following form:

$$\Delta[T\Delta p]-\tau+q=\frac{V_R}{\Delta t}\Delta_t\left(\frac{\phi S}{B}\right) \tag{6.5.1}$$

For the problem in hand, this is further expanded into

$$\Delta_x[T_x\Delta_x p]+\Delta_z[T_z\Delta_z p]-\tau+q=\frac{V_R}{\Delta t}\Delta_t\left(\frac{\phi S}{B}\right) \tag{6.5.2}$$

or

$$\begin{aligned}T^{n+1}_{x,i+1/2,k}&\left(p^{n+1}_{i+1,k}-p^{n+1}_{i,k}\right)-T^{n+1}_{x,i-1/2,k}\left(p^{n+1}_{i,k}-p^{n+1}_{i-1,k}\right)\\&+T^{n+1}_{z,i,k+1/2}\left(p^{n+1}_{i,k+1}-p^{n+1}_{i,k}\right)-T^{n+1}_{z,i,k-1/2}\left(p^{n+1}_{i,k}-p^{n+1}_{i,k-1}\right)\\&-\tau^{n+1}_{i,k}+q^{n+1}_{i,k}-\frac{V_{R,i,k}}{\Delta t}\left[\left(\frac{\phi S}{B}\right)^{n+1}_{i,k}-\left(\frac{\phi S}{B}\right)^{n}_{i,k}\right]=R^{n+1}=0\end{aligned} \tag{6.5.3}$$

For a two-phase (oil-water) system, the above equation is written for both the oil and water phases in the fracture and matrix.

The set of equations having the above structure is linearized by the Newton-Raphson method, given as:

$$-\left(\frac{\partial\vec{R}}{\partial\vec{x}}\right)^{l}\vec{\delta x}=\vec{R}^{l}(\vec{x}) \tag{6.5.4}$$

where

$$\vec{\delta x}=\vec{x}^{l+1}-\vec{x}^{l} \tag{6.5.5}$$

The partial derivative, $\partial\vec{R}/\partial\vec{x}$, is the Jacobian matrix and represents the partial derivative of all residuals with respect to the primary variables. In the five-point finite-difference formulation (6.5.3), the residuals for a given node are a function of the primary variables of that node and the four neighbor nodes. Therefore, the resulting coefficient matrix is banded. For nodes connected to a well, the residuals will also depend on wellbore pressure or flow rates.

Linearization of (6.5.3) yields (6.5.6) and (6.5.7) where *i* represents the node number in the *x*-direction and *k* represents the node number in the *z*-direction.

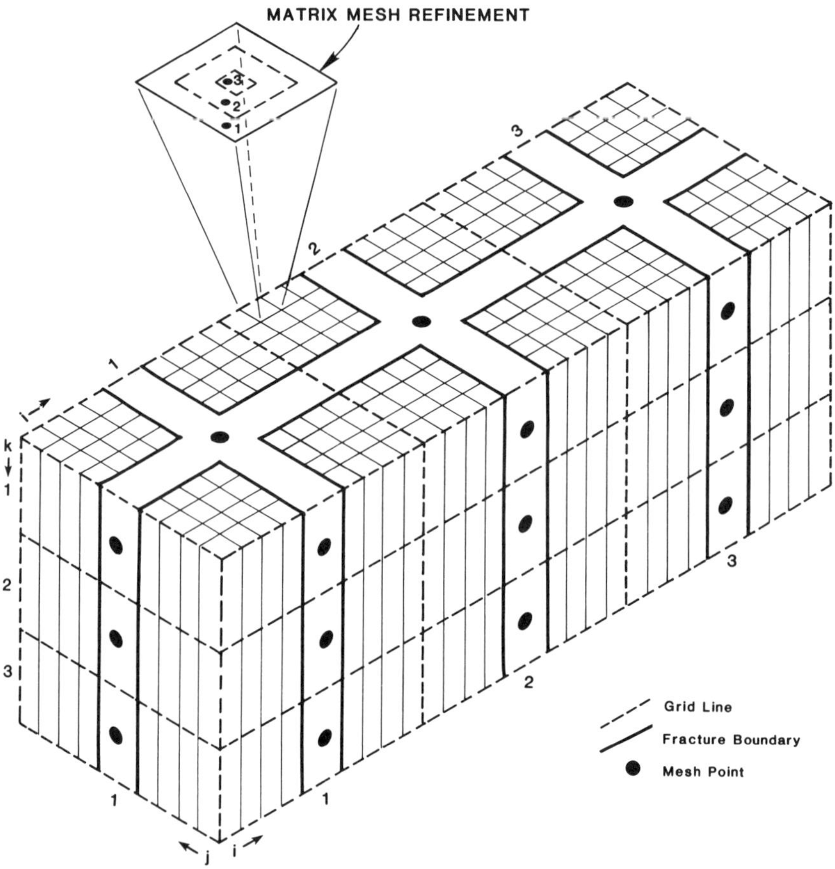

Figure 6.5.1. Vertical dual-permeability 3 × 1 × 3 grid.

Fracture:

$$\vec{R}_{f,i,k} = B_{f,i,k}\vec{X}_{f,i-1,k} + C_{f,i,k}\vec{X}_{f,i,k-1} + D_{f,i,k}\vec{X}_{f,i,k}$$

$$+E_{f,i,k}\vec{X}_{f,i,k+1} + F_{f,i,k}\vec{X}_{f,i+1,k} + G_{f,i,k}\vec{X}_{m,i,k} \qquad (6.5.6)$$

Matrix:

$$\vec{R}_{m,i,k} = C_{m,i,k}\vec{X}_{m,i,k-1} + D_{m,i,k}\vec{X}_{m,i,k}$$

$$E_{m,i,k}\vec{X}_{m,i,k+1} + G_{m,i,k}\vec{X}_{f,i,k} \qquad (6.5.7)$$

Remember that the rock matrix is allowed to be continuous only in the vertical direction (vertical dual-permeability).

The various terms in (6.5.6) and (6.5.7) are defined as

$$\vec{X}_{f,i,k} = \begin{bmatrix} \delta p_{wf} \\ \delta S_{wf} \end{bmatrix}_{i,k} \tag{6.5.8}$$

$$\vec{R}_{f,i,k} = \begin{bmatrix} R_{wf} \\ R_{of} \end{bmatrix}_{i,k} \tag{6.5.9}$$

$$C_{f,i,k} = \begin{bmatrix} \dfrac{\partial R_{wf,i,k}}{\partial p_{wf,i,k-1}} & \dfrac{\partial R_{wf,i,k}}{\partial S_{wf,i,k}-1} \\ \dfrac{\partial R_{of,i,k}}{\partial p_{wf,i,k-1}} & \dfrac{\partial R_{of,i,k}}{\partial S_{wf,i,k-1}} \end{bmatrix} \tag{6.5.10}$$

$$G_{f,i,k} = \begin{bmatrix} \dfrac{\partial R_{wf,i,k}}{\partial p_{wm,i,k}} & \dfrac{\partial R_{wf,i,k}}{\partial S_{wm,i,k}} \\ \dfrac{\partial R_{of,i,k}}{\partial p_{wm,i,k}} & \dfrac{\partial R_{of,i,k}}{\partial S_{wm,i,k}} \end{bmatrix} \tag{6.5.11}$$

$$G_{m,i,k} = \begin{bmatrix} \dfrac{\partial R_{wm,i,k}}{\partial p_{wf,i,k}} & \dfrac{\partial R_{wm,i,k}}{\partial S_{wf,i,k}} \\ \dfrac{\partial R_{om,i,k}}{\partial p_{wf,i,k}} & \dfrac{\partial R_{om,i,k}}{\partial S_{wf,i,k}} \end{bmatrix} \tag{6.5.12}$$

The B, D, E, and F matrices for both the fracture and rock matrix are similar to $C_{f,i,k}$. For nodes connected to a wellbore, additional equations are added.

The set of linear equations for the 3×1×3 system utilizing natural ordering in the z-direction is shown in Fig. 6.5.2. This can be reduced to the form shown in Fig. 6.5.3 where the primary variables for the rock matrix have been eliminated from the fracture equations. The reduced fracture equations are solved first for fracture primary variables; then, the rock matrix variables are solved for by back substitution. The solution of the reduced equations is accomplished by Gaussian elimination for small grid systems and by iterative methods (such as SOR, conjugate gradient, etc.) for large grid systems.

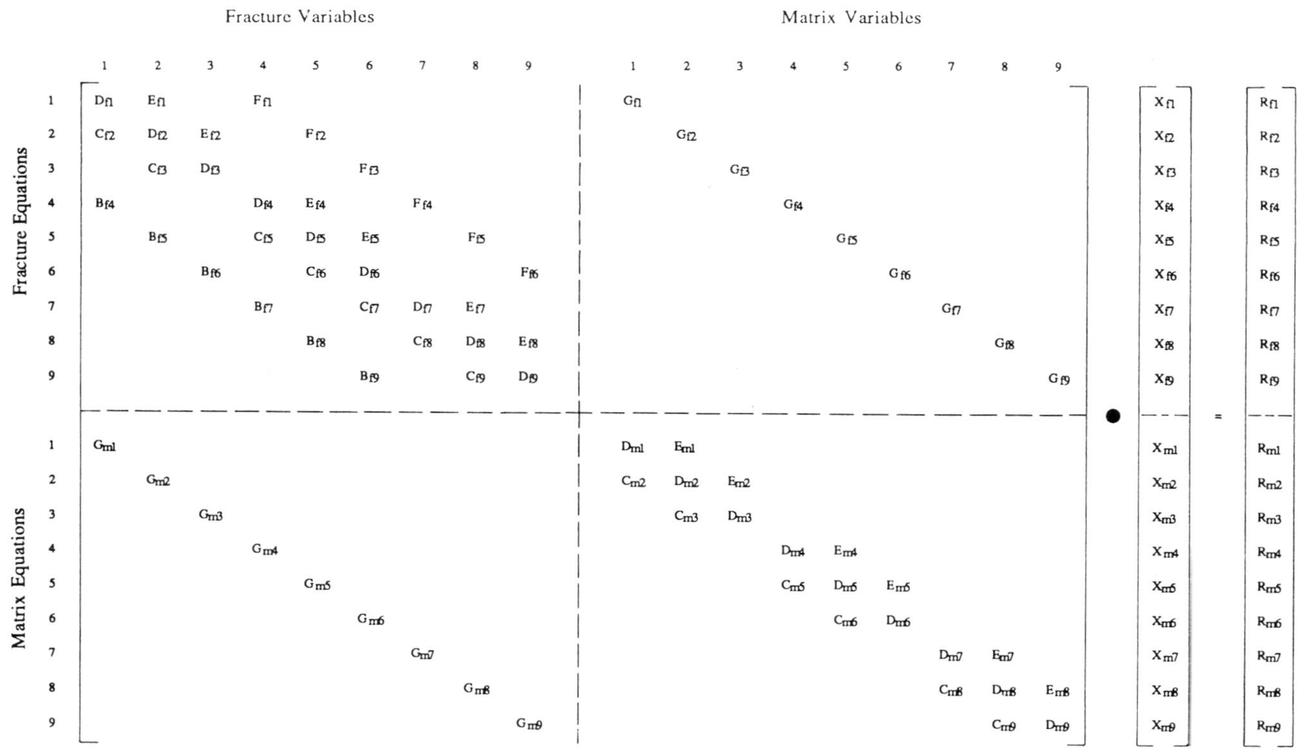

Figure 6.5.2. Equation structure for a vertical dual-permeability 3×1×3 grid.

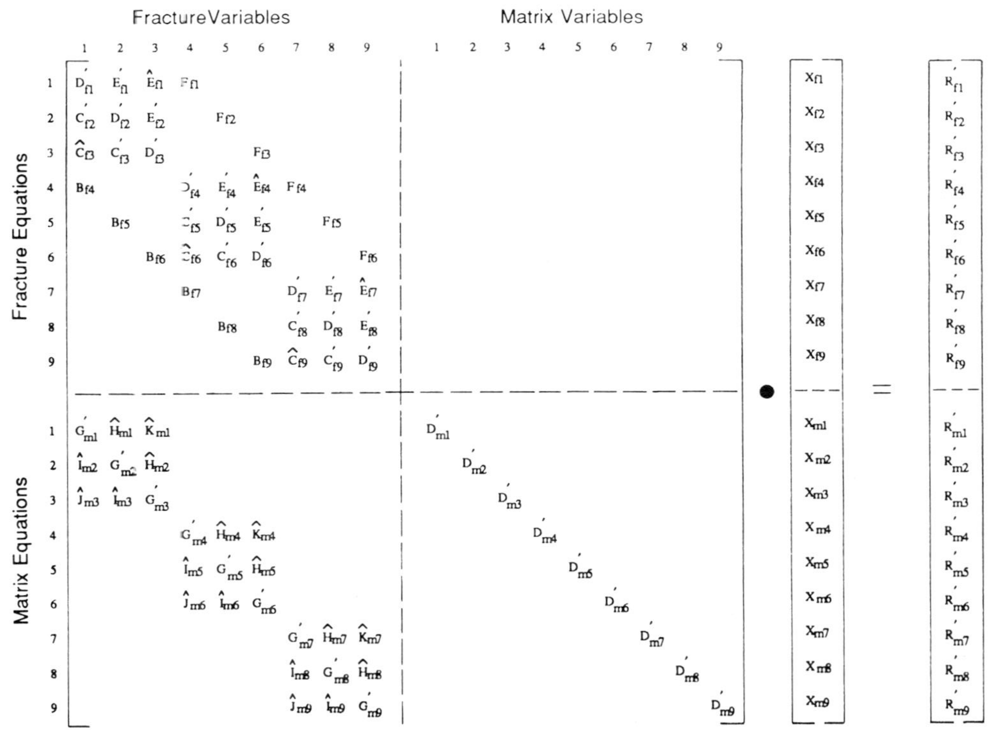

Figure 6.5.3. Reduced equation structure for a vertical dual-permeability 3×1×3 grid.

6.5.2 Matrix Mesh Refinement

In many problems of heat conduction and diffusion, we need to represent each matrix block with a fine grid system so we can track the distribution of heat or diffusion of any component with time. Specifically, this approach will find application in steam, chemical, and CO_2 flooding. In these cases, we choose to refine the matrix blocks in the areal (xy) plane. Figure 6.5.4 shows the xy view of a sugar cube or match stick grid without mesh refinement. Figure 6.5.5 is the same grid, but each matrix block is subdivided into three subdomains. All matrix blocks are assumed to behave identically within a fracture grid node. In Fig. 6.5.5 we show three fracture grids (dashed lines) with each fracture grid containing nine matrix blocks and each matrix block subdivided into three grid subdomains.

For the cases where fracture flow gradient across matrix blocks contributes to displacement, we may further subdivide the matrix blocks as shown in Fig. 6.5.6. For an example of how these grids are used in computation, see Fig. 6.5.7, where three fracture nodes and three adjacent matrix blocks are shown for illustration. Each matrix block is divided into three grid subdomains. Subscripts refer to the node number and superscripts refer to the matrix grid subdomain number. This grid arrangement was first introduced by Pruess and Narasimhan (1985) as "multiple-interacting continua (MINC)".

The linear equation arrangement resulting from the discretization of the flow equation is shown in Fig. 6.5.8. Figure 6.5.8 has the same functional form as Fig. 6.5.2. The reduction of this matrix follows the procedure discussed for Fig. 6.5.3.

Terms in Fig. 6.5.8 are defined as follows

$$C_{f,i} = \begin{bmatrix} \dfrac{\partial R_{wf,i}}{\partial p_{wf,i-1}} & \dfrac{\partial R_{wf,i}}{\partial S_{wf,i-1}} \\ \dfrac{\partial R_{of,i}}{\partial p_{wf,i-1}} & \dfrac{\partial R_{of,i}}{\partial S_{wf,i-1}} \end{bmatrix} \tag{6.5.13}$$

$$D_{f,i} = \begin{bmatrix} \dfrac{\partial R_{wf,i}}{\partial p_{wf,i}} & \dfrac{\partial R_{wf,i}}{\partial S_{wf,i}} \\ \dfrac{\partial R_{of,i}}{\partial p_{wf,i}} & \dfrac{\partial R_{of,i}}{\partial S_{wf,i}} \end{bmatrix} \tag{6.5.14}$$

$$E_{f,i} = \begin{bmatrix} \dfrac{\partial R_{wf,i}}{\partial p_{wf,i+1}} & \dfrac{\partial R_{wf,i}}{\partial S_{wf,i+1}} \\ \dfrac{\partial R_{of,i}}{\partial p_{wf,i+1}} & \dfrac{\partial R_{of,i}}{\partial S_{wf,i+1}} \end{bmatrix} \tag{6.5.15}$$

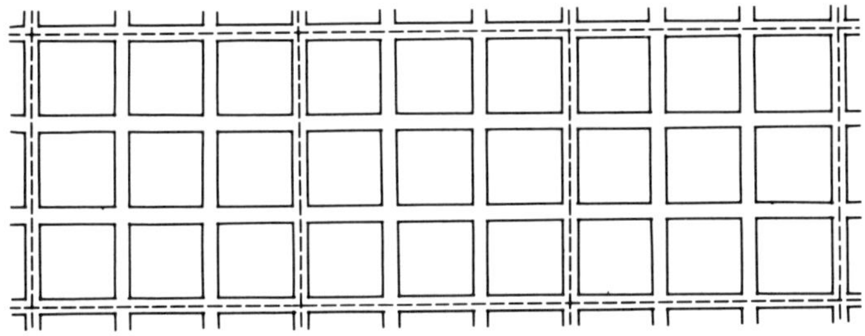

Figure 6.5.4. X-Y view of sugar cube or match stick grid without matrix refinement.

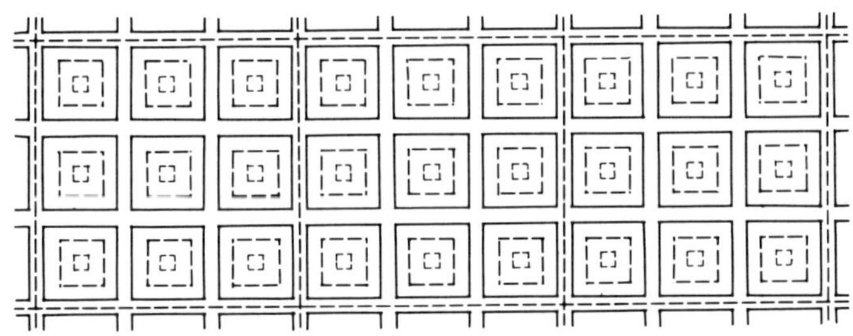

Figure 6.5.5. X-Y view of sugar cube or match stick grid with matrix mesh refinement (MINC) for use with diffusion and heat conduction calculations.

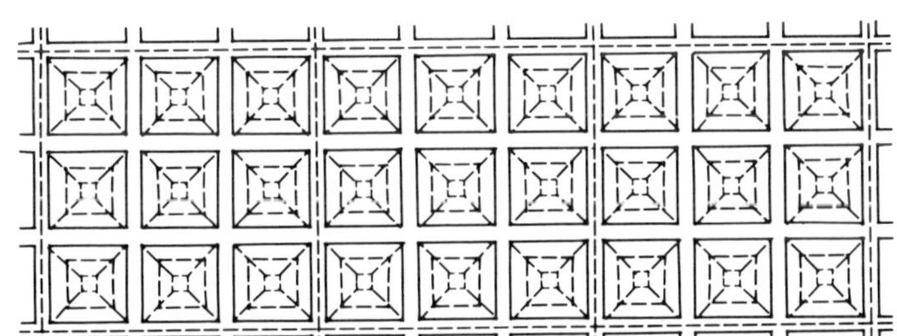

Figure 6.5.6. X-Y view of sugar cube or match stick grid with matrix mesh refinement for use with fracture flow gradient calculations.

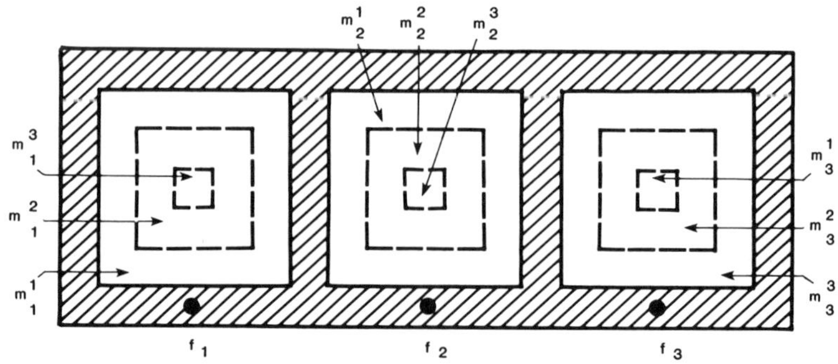

Figure 6.5.7. The multiple-interacting continua concept (MINC).

		Fracture Variables 1	2	3	Matrix Variables 1	2	3
Fracture Eqs.	1	D_{f1}	C_{f1}		G_{f1}		
	2	C_{f2}	D_{f2}	C_{f3}		G_{f2}	
	3		C_{f2}	D_{f3}			G_{f3}
Matrix Eqs.	1	G_{m1}			D_{m1}		
	2		G_{m2}			D_{m2}	
	3			G_{m3}			D_{m3}

Figure 6.5.8. Equation structure for a three-node one-dimensional grid with matrix subdomains.

$$G_{f,i} = \begin{bmatrix} \dfrac{\partial R_{wf,i}}{\partial p^1_{wm,i}} & \dfrac{\partial R_{wf,i}}{\partial S^1_{wm,i}} & \dfrac{\partial R_{wf,i}}{\partial p^2_{wm,i}} & \dfrac{\partial R_{wf,i}}{\partial S^2_{wm,i}} & \dfrac{\partial R_{wf,i}}{\partial p^3_{wm,i}} & \dfrac{\partial R_{wf,i}}{\partial S^3_{wm,i}} \\ \dfrac{\partial R_{of,i}}{\partial p^1_{wm,i}} & \dfrac{\partial R_{of,i}}{\partial S^1_{wm,i}} & \dfrac{\partial R_{of,i}}{\partial p^2_{wm,i}} & \dfrac{\partial R_{of,i}}{\partial S^2_{wm,i}} & \dfrac{\partial R_{of,i}}{\partial p^3_{wm,i}} & \dfrac{\partial R_{of,i}}{\partial S^3_{wm,i}} \end{bmatrix} \tag{6.5.16}$$

$$G_{m,i} = \begin{bmatrix} \dfrac{\partial R^1_{wm,i}}{\partial p_{wf,i}} & \dfrac{\partial R^1_{wm,i}}{\partial p_{wf,i}} \\ \dfrac{\partial R^1_{om,i}}{\partial p_{wf,i}} & \dfrac{\partial R^1_{om,i}}{\partial p_{wf,i}} \\ \dfrac{\partial R^2_{wm,i}}{\partial p_{wf,i}} & \dfrac{\partial R^2_{wm,i}}{\partial p_{wf,i}} \\ \dfrac{\partial R^2_{om,i}}{\partial p_{wf,i}} & \dfrac{\partial R^2_{om,i}}{\partial p_{wf,i}} \\ \dfrac{\partial R^3_{wm,i}}{\partial p_{wf,i}} & \dfrac{\partial R^3_{wm,i}}{\partial p_{wf,i}} \\ \dfrac{\partial R^3_{om,i}}{\partial p_{wf,i}} & \dfrac{\partial R^3_{wm,i}}{\partial S_{wf,i}} \end{bmatrix} \tag{6.5.17}$$

$$D_{m,i} = \begin{bmatrix} \dfrac{\partial R^1_{wm,i}}{\partial p^1_{wm,i}} & \dfrac{\partial R^1_{wm,i}}{\partial S^1_{wm,i}} & & & & \\ \dfrac{\partial R^1_{om,i}}{\partial p^1_{wm,i}} & \dfrac{\partial R^1_{om,i}}{\partial S^1_{wm,i}} & & & & \\ & & \dfrac{\partial R^2_{wm,i}}{\partial p^2_{wm,i}} & \dfrac{\partial R^2_{wm,i}}{\partial S^2_{wm,i}} & & \\ & & \dfrac{\partial R^2_{om,i}}{\partial p^2_{wm,i}} & \dfrac{\partial R^2_{om,i}}{\partial S^2_{wm,i}} & & \\ & & & & \dfrac{\partial R^3_{wm,i}}{\partial p^3_{wm,i}} & \dfrac{\partial R^3_{wm,i}}{\partial S^3_{wm,i}} \\ & & & & \dfrac{\partial R^3_{om,i}}{\partial p^3_{wm,i}} & \dfrac{\partial R^3_{om,i}}{\partial S^3_{wm,i}} \end{bmatrix} \tag{6.5.18}$$

6.6 EFFECT OF FRACTURE FLOW GRADIENTS ON FLUID DISPLACEMENT IN THE ROCK MATRIX

Viscous displacement in the matrix blocks caused by pressure (potential) gradient in the fracture network is generally not accounted for. Gilman and Kazemi (1988) provide a method to account for fluid displacement in the matrix resulting from flow gradients in the fracture. The transfer function for a cubic matrix block, having a uniform dimension L_m and permeability k_m, is given by (6.6.1). This equation replaces (6.3.23) through (6.3.25). The more general case of variable matrix-block dimension and anisotropic matrix permeability is given by Gilman and Kazemi (1988) as follows:

$$\tau_{\beta f} = 0.001127 \frac{4k_m V_R}{L_m^2} \Big\{ \Big[\left(\overline{\omega}_{\beta x} + \overline{\omega}_{\beta y} + \overline{\omega}_{\beta z} \right) \lambda_{\beta f}$$

$$+ \left(3 - \overline{\omega}_{\beta x} - \overline{\omega}_{\beta y} - \overline{\omega}_{\beta z} \right) \lambda_{\beta m} \Big] \Big[p_{\beta f} - p_{\beta m} - \gamma_\beta \left(D_f - D_m \right) \Big]$$

$$+ \Big[\overline{\omega}_{\beta x} \left(1 - \overline{\omega}_{\beta x} \right) g_{\beta x} + \overline{\omega}_{\beta y} \left(1 - \overline{\omega}_{\beta y} \right) g_{\beta y}$$

$$+ \overline{\omega}_{\beta z} \left(1 - \overline{\omega}_{\beta z} \right) g_{\beta z} \Big] \left(\lambda_{\beta f} - \lambda_{\beta m} \right) L_m \Big\}, \quad \beta = o, w, g \tag{6.6.1}$$

where

$$\overline{\omega}_{\beta x} = 1 \text{ if } p_{\beta f} - \gamma_\beta D_f - g_{\beta x} \left(L_m / 2 \right) > p_{\beta m} - \gamma_\beta D_m , \tag{6.6.2}$$

$$\overline{\omega}_{\beta x} = 0 \text{ if } p_{\beta f} - \gamma_\beta D_f + g_{\beta x} \left(L_m / 2 \right) > p_{\beta m} - \gamma_\beta D_m , \tag{6.6.3}$$

otherwise,

$$\overline{\omega}_{\beta x} = 0.5 . \tag{6.6.4}$$

The x-direction potential gradient in the fracture is estimated by

$$g_{\beta x,i,j,k} = \left| \frac{p_{\beta f,i+1} - p_{\beta f,i} - \gamma_\beta \left(D_{f,i+1} - D_{f,i} \right)}{\Delta x_{i+1} + \Delta x_i} \right|_{j,k}$$

$$+ \left| \frac{p_{\beta f,i} - p_{\beta f,i-1} - \gamma_\beta \left(D_{f,i} - D_{f,i-1} \right)}{\Delta x_i + \Delta x_{i-1}} \right|_{j,k} , \tag{6.6.5}$$

For $\overline{\omega}_{\beta x} = 1$, the fracture is "upstream" for both x-direction matrix block faces; for $\overline{\omega}_{\beta x} = 0$, the matrix is upstream for both x-direction matrix block faces; and for $\overline{\omega}_{\beta x} = 0.5$, the fracture is upstream on one x-direction matrix block face and downstream on the other. Similar definitions are given for $\overline{\omega}_{\beta y}$, $\overline{\omega}_{\beta z}$, $g_{\beta y}$, and $g_{\beta z}$.

Because $g_{\beta x}$, $g_{\beta y}$, and $g_{\beta z}$ are functions of fracture node pressures, the coefficient matrix resulting from (6.6.1) will yield a more dense structure. Therefore, partial elimination of fracture-matrix coefficients before solution of the entire flow equations could not be done. However, $g_{\beta x}$, $g_{\beta y}$, and $g_{\beta z}$ are not changing rapidly, and good results can be obtained by calculating them at the old time level and solving the system as in the conventional dual-porosity simulators. When flow rates are changing or new wells are added, it is likely that smaller timesteps and/or additional iterations will be required for stability because of explicit treatment of the gradient terms.

Absolute values of $g_{\beta x}$, $g_{\beta y}$, and $g_{\beta z}$ are used in the equations because saturations in the fracture surrounding a matrix block are the same at all faces. The value and sign of $(\lambda_{\beta f}-\lambda_{\beta m})$ will determine whether the gradient causes more of a given phase to move into or out of a matrix block. All matrix blocks in a given computational grid cell are assumed to behave identically; therefore, the average gradient acting on all blocks is given by (6.6.5).

For the special cases where

$$\overline{\omega}_{\beta x} = \overline{\omega}_{\beta y} = \overline{\omega}_{\beta z} = 1 \text{ or } \overline{\omega}_{\beta x} = \overline{\omega}_{\beta y} = \overline{\omega}_{\beta z} = 0 , \tag{6.6.6}$$

Equation (6.6.1) reduces to the standard dual-porosity formulation. In this case, the matrix shape factor for the cubic matrix blocks surrounded by fractures on all sides is

$$\sigma = \frac{12}{L_m^2} , \tag{6.6.7}$$

and all the gradient terms ($g_{\beta x}$, $g_{\beta y}$, and $g_{\beta z}$) will cancel. Also, for single-phase flow, $\lambda_{\beta f} \cong \lambda_{\beta m}$. Therefore, the gradient terms will cancel, and (6.6.1) will yield the form usually used in pressure transient-testing.

Figures 6.6.1 and 6.6.2 illustrate the difference between the conventional and viscous gradient formulations. The top diagrams in the figures depict three gridblocks, each containing four matrix blocks. In both formulations, the matrix blocks in a given gridblock are assumed to behave identically. Therefore, for a given gridblock, there is a single pressure for each phase in each pore type. Assuming that capillary pressure is positive in the matrix, oil pressure is greater than water pressure. Capillary pressure is near zero in the fracture, and therefore fracture oil and water pressure are equal.

In Fig. 6.6.1, the conventional approach, matrix blocks are discontinuous and therefore the pressure distribution in a gridblock can be idealized as a flat profile. Water moves into the matrix and oil moves out primarily because of capillary forces. The fracture has a potential flow gradient associated with it because the fractures are connected from one gridblock to the next.

Figure 6.6.2 depicts the improved formulation. The fracture flow gradient has been imposed across each matrix block. All matrix blocks in a given gridblock are still assumed to behave identically. When the gradient across a matrix block becomes large compared with the capillary pressure, the gradient can cause fracture fluids (oil and water) to move into the matrix block from the upstream face, while matrix oil and water flow into the fracture from the downstream face.

The additional work in generating (6.6.1) compared with the work required to generate the standard formulation given by (6.3.23) through (6.3.25) is in evaluating the gradients in the fractures, g_β, and the weighting factors, ω_β. This extra computation time is rather negligible.

If the matrix is refined into multiple subdomains (Fig. 6.5.6), the improved viscous flow calculations would require additional accounting but not much additional solution time. The

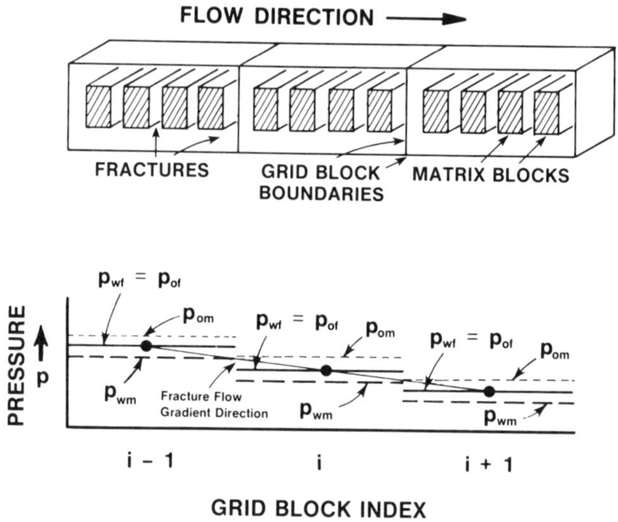

Figure 6.6.1. Conventional viscous flow idealization (from Gilman and Kazemi, 1988, courtesy SPE).

fracture potential for any phase at a given subdomain face is calculated by interpolation. If the fracture potential is greater than the matrix potential at the subdomain face, fluid would flow from fracture to matrix at this location. By doing similar calculations for every face of every subdomain and summing the flow rates through all subdomain faces, an equation similar to (6.6.1) would be obtained.

6.7 GRAVITY/CAPILLARY DISPLACEMENT CALCULATIONS IN THE ROCK MATRIX

In conventional dual-porosity simulators, fracture and matrix saturations are dispersed over the entire gridblock. Therefore, gravity displacement calculations are not rigorous in such simulators. The state of the art for gravity displacement calculations in dual-porosity simulators fall into three categories. The first method involves the use of pseudocapillary pressure functions for such calculations (Thomas, et al., 1983; Dean and Lo, 1988; Rossen and Shen, 1987; Coats, 1989). In Sec. 6.8.3, we will present a new method for generating such pseudo-functions directly from laboratory oil recovery experiments.

The second method uses fine grid in the vertical direction to account for both gravity segregation and displacement rigorously. This can be efficiently accomplished using vertical dual permeability as discussed in Sec. 6.5.1 or matrix mesh refinement as discussed in Sec. 6.5.2.

The third method relies on the assumption that all phases segregate instantaneously in both the fracture and matrix (Litvak, 1985; Sonier, et al., 1988). Because this approach has been widely used, we will present the underlying concepts in this section.

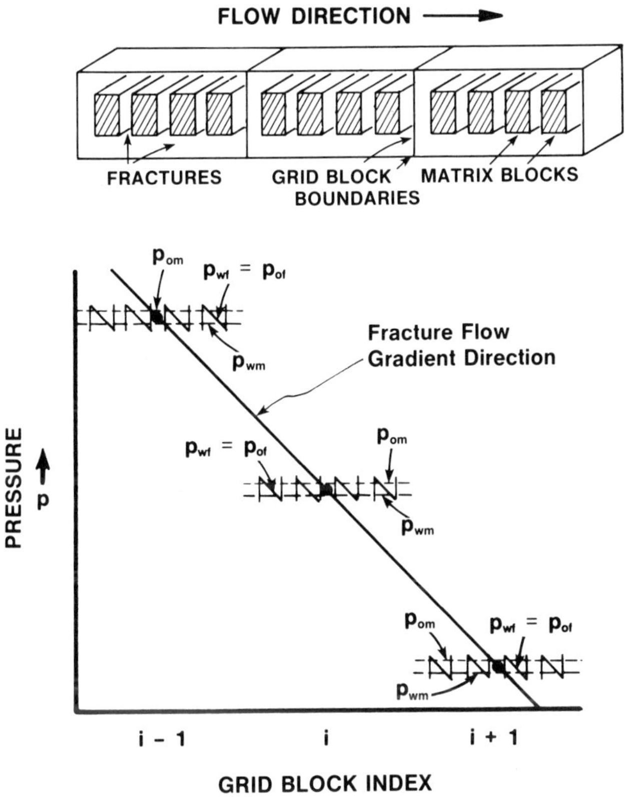

Figure 6.6.2. Idealization of viscous flow improvements (from Gilman and Kazemi, 1988, courtesy SPE).

Consider a dual-porosity grid block of height Δz containing oil and water (Fig. 6.7.1a). The water height in the fracture can be related to water saturation by

$$h_{wf} = \left(\frac{S_{wf} - S_{wcf}}{1 - S_{orf} - S_{wcf}} \right) \Delta z \quad . \tag{6.7.1}$$

A similar equation could be written for the matrix water height, h_{wm}. The difference in heights leads to a gravity potential, $(\gamma_w - \gamma_o)(h_{wf} - h_{wm})$, which is a basis for the following material.

Sonier et al. (1988) uses the matrix block height, L_z, rather than Δz in (6.7.1) to limit the gravity potential. This is idealized in Fig. 6.7.1b. Their method is to replace the matrix-fracture transfer functions, (6.3.23) through (6.3.25), by the following equation (we have added D_f and D_m for completeness):

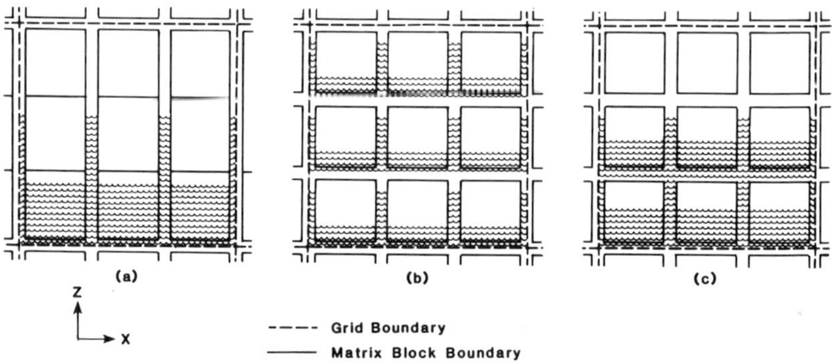

Figure 6.7.1. Three idealizations of water-oil segregation in x-z plane.

$$\tau_{\beta f} = 0.001127\, k_m \sigma V_R \left(\frac{k_r}{\mu B}\right)_{\beta mf} \left\{ p_{\beta f} + \gamma_{\beta f}\left(h_{\beta f} - D_f\right) - \left[p_{\beta m} + \gamma_{\beta m}\left(h_{\beta m} - D_m\right)\right]\right\}, \quad \beta = w, o, g \tag{6.7.2}$$

where

$$h_{\beta f} = \left(\frac{S_{\beta f} - S_{\beta irf}}{1 - S_{rf} - S_{\beta irf}}\right) L_z\,, \tag{6.7.3}$$

$$h_{\beta m} = \left(\frac{S_{\beta m} - S_{\beta irm}}{1 - S_{rm} - S_{\beta irm}}\right) L_z\,. \tag{6.7.4}$$

The residual saturation S_r is:

$$S_r = \begin{Bmatrix} S_{owr} & \text{for} & \beta = w \\ S_{lr} & \text{for} & \beta = g \\ S_{wir} & \text{for} & \beta = o \end{Bmatrix}, \tag{6.7.5}$$

Litvak (1985) calculates gravity effects using maximum gravity potential for a single matrix block and limits oil displacement in the matrix to the blocks surrounded by water (Fig. 6.7.1c). If all the matrix mobile water, S_{wm}–S_{wcm}, in a given cell is assumed to accumulate below the segregated water-oil level in the fracture, then the average matrix mobile water saturation in this region is $(S_{wm}-S_{wcm})/S_{wf}$ assuming zero irreducible saturations in the fracture. From material balance, the segregated matrix water height is then given as

$$h_{wm} = \left(\frac{S_{wm} - S_{wcm}}{1 - S_{orm} - S_{wcm}} \right) \frac{L_z}{S_{wf}} . \tag{6.7.6}$$

Below the fracture water level the matrix blocks are completely surrounded by water; therefore, the matrix-fracture water height difference for a single block is $L_z - h_{wm}$ where h_{wm} is given by (6.7.6). When this water height difference is multiplied by the fraction of matrix blocks covered in a given cell (i.e. S_{wf}), the effective height reverts to the form given by (6.7.3) and (6.7.4) if residual saturations in the fracture are accounted for. Therefore, Litvak's method for gravity calculation turns out to be the same as Sonier's.

Gilman and Kazemi (1988) applied the assumption of complete segregation to a dual-porosity model, and showed by comparison with a fine-grid model that the method can lead to erroneous results.

6.8 SPECIAL TOPICS

In this section we will discuss some relevant subjects, including shape factor, empirical matrix-fracture transfer functions, Buckley-Leverett solution, pseudo transfer functions, diagnostic flow tests, fracture compressibility, petrophysical methods for fracture delineation and fracture relative permeabilities.

6.8.1 Shape Factor

The following equations represent the transfer of material between fracture and the surrounding porous rock for a single-phase flow problem:

$$\left(\frac{\sigma k}{\mu} \right)_m \left(p_f - \bar{p}_m \right) = (\phi c)_m \frac{\partial \bar{p}_m}{\partial t} \tag{6.8.1}$$

The shape factor, σ, has a dimension of reciprocal length squared (L^{-2}). Our goal is to determine the value of σ. Carslaw and Jaeger's (1959) solution for temperature in a rectangular parallelepiped of dimension L_x, L_y and L_z with unit initial temperature and zero surface temperature is:

$$\hat{T}(x, y, z, t) = \frac{64}{\pi^3} \sum_{l=0}^{\infty} \sum_{m=0}^{\infty} \sum_{n=0}^{\infty} \frac{(-1)^{l+m+n}}{(2l+1)(2m+1)(2n+1)}$$

$$\cos\left(\frac{(2l+1)\pi x}{L_x} \right) \cos\left(\frac{(2m+1)\pi y}{L_y} \right) \cos\left(\frac{(2n+1)\pi z}{L_z} \right) e^{-\alpha_{l,m,n} t} \tag{6.8.2}$$

where

$$\alpha_{l,m,n} = \kappa \pi^2 \left[\frac{(2l+1)^2}{L_x^2} + \frac{(2m+1)^2}{L_y^2} + \frac{(2n+1)^2}{L_z^2} \right] \tag{6.8.3}$$

For large values of time, the series converges very rapidly and the temperature is very nearly given by its first term (Carslaw and Jaeger, 1959).

$$\hat{T}(x,y,z,t) - \frac{64}{\pi^3}\cos\left(\frac{\pi x}{L_x}\right)\cos\left(\frac{\pi y}{L_y}\right)\cos\left(\frac{\pi z}{L_z}\right)e^{\alpha t} \tag{6.8.4}$$

where

$$\alpha = \kappa\pi^2\left(\frac{1}{L_x^2}+\frac{1}{L_y^2}+\frac{1}{L_z^2}\right) \tag{6.8.5}$$

Applying this solution to a single matrix block of dimension L_x, L_y, and L_z surrounded by a fracture at constant pressure, p_f, (Fig. 6.8.1) we obtain

$$p_m(x,y,z,t) = p_f = \left(p_m^o - p_f\right)\frac{64}{\pi^3}\cos\left(\frac{\pi x}{L_x}\right)\cos\left(\frac{\pi y}{L_y}\right)\cos\left(\frac{\pi z}{L_z}\right)e^{-\alpha t} \tag{6.8.6}$$

where

$$\alpha = \left(\frac{k}{\phi\mu c}\right)_m \pi^2\left(\frac{1}{L_x^2}+\frac{1}{L_y^2}+\frac{1}{L_z^2}\right) \tag{6.8.7}$$

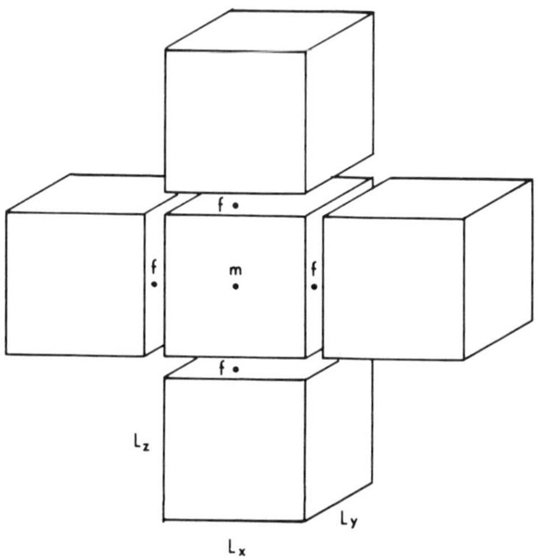

Figure 6.8.1. Idealization of a matrix block surrounded by a constant pressure fracture.

Integrating over the entire volume of the matrix

$$\frac{1}{L_x L_y L_z}\int_{-L_x/2}^{+L_x/2}\int_{-L_y/2}^{+L_y/2}\int_{-L_z/2}^{+L_z/2}\left[p_m(x,y,z,t)-p_f\right]dzdydx=$$

$$\frac{\frac{64}{\pi^3}\left(p_m^o-p_f\right)e^{-\alpha t}}{L_x L_y L_z}\int_{-L_x/2}^{+L_x/2}\int_{-L_y/2}^{+L_y/2}\int_{-L_z/2}^{+L_z/2}\left[\cos\left(\frac{\pi x}{L_x}\right)\cos\left(\frac{\pi y}{L_y}\right)\cos\left(\frac{\pi z}{L_z}\right)\right]dzdydx \quad (6.8.8)$$

$$\bar{p}_m - p_f = \frac{512}{\pi^6}\left(p_m^o-p_f\right)e^{-\alpha t} \quad (6.8.9)$$

$$\ln\left(\bar{p}_m-p_f\right)=\ln\left[\frac{512}{\pi^6}\left(p_m^o-p_f\right)\right]-\alpha t \quad (6.8.10)$$

From (6.8.1), we get

$$\frac{d\bar{p}_m}{\bar{p}_m-p_f}=-\sigma\left(\frac{k}{\phi\mu c}\right)_m dt \quad (6.8.11)$$

or

$$\ln\left(\bar{p}_m-p_f\right)=-\sigma\left(\frac{k}{\phi\mu c}\right)_m t+c \quad (6.8.12)$$

Comparing (6.8.10) and (6.8.12) leads to

$$\alpha=\sigma\left(\frac{k}{\phi\mu c}\right)_m \quad (6.8.13)$$

Comparing (6.8.7) and (6.8.13) we obtain:

$$\sigma=\pi^2\left(\frac{1}{L_x^2}+\frac{1}{L_y^2}+\frac{1}{L_z^2}\right) \quad (6.8.14)$$

The derivation of the shape factor, σ, from a standard seven-point finite difference formulation of the single-phase flow problem is shown below. Equation (6.8.15) is the pressure equation.

$$\frac{\partial}{\partial x}\left[\left(\frac{k}{\mu}\right)_m\frac{\partial p}{\partial x}\right]+\frac{\partial}{\partial y}\left[\left(\frac{k}{\mu}\right)_m\frac{\partial p}{\partial y}\right]+\frac{\partial}{\partial z}\left[\left(\frac{k}{\mu}\right)_m\frac{\partial p}{\partial z}\right]=\left(\phi c\right)_m\frac{\partial \bar{p}_m}{\partial t} \quad (6.8.15)$$

In finite difference form we discretize (6.8.15) as:

$$\frac{1}{L_x}\left(\frac{k}{\mu}\right)_m\left[\frac{p_f-\bar{p}_m}{L_x/2}-\frac{\bar{p}_m-p_f}{L_x/2}\right]+$$

$$\frac{1}{L_y}\left(\frac{k}{\mu}\right)_m\left[\frac{p_f-\bar{p}_m}{L_y/2}-\frac{\bar{p}_m-p_f}{L_y/2}\right]+$$

$$\frac{1}{L_z}\left(\frac{k}{\mu}\right)_m\left[\frac{p_f-\bar{p}_m}{L_z/2}-\frac{\bar{p}_m-p_f}{L_z/2}\right]=(\phi c)_m\frac{\partial\bar{p}_m}{\partial t} \tag{6.8.16}$$

or

$$4\left(\frac{1}{L_x^2}+\frac{1}{L_y^2}+\frac{1}{L_z^2}\right)\left(\frac{k}{\mu}\right)_m\left(p_f-\bar{p}_m\right)=(\phi c)_m\frac{\partial\bar{p}_m}{\partial t} \tag{6.8.17}$$

Comparing Eqs. 6.8.17 and 6.8.1, we get

$$\sigma=4\left(\frac{1}{L_x^2}+\frac{1}{L_y^2}+\frac{1}{L_z^2}\right) \tag{6.8.18}$$

6.8.2 Empirical Matrix-Fracture Transfer Functions

For the case in which imbibition capillary pressure is the dominant force for displacing oil from the matrix (i.e., in strongly water-wet rocks with a relatively small matrix block height), then the cumulative oil recovery by imbibition has an exponential form as shown

$$R=R_\infty\left(1-e^{-\lambda t}\right) \tag{6.8.19}$$

where

$$R_\infty=\left(1-S_{orm}-S_{wcm}\right)\phi_m \tag{6.8.20}$$

$$R=\left(S_{wm}-S_{wcm}\right)\phi_m \tag{6.8.21}$$

Here S_{or} is the minimum residual oil obtainable by imbibition alone. Equation (6.8.19) is an empirical fit of the imbibition results of Mattax and Kyte (1962). Their correlation used a dimensionless time, $t_{D,pc}$ given in consistent units by:

$$t_{D,pc}=\left[\sqrt{\frac{k}{\phi}}\left(\frac{\gamma}{\mu_w L^2}\right)\right]t \tag{6.8.22}$$

When using (6.8.22), the cumulative recovery curve will be of the form

$$R = R_\infty \left(1 - e^{-\lambda_{D,pc} t_{D,pc}}\right) \tag{6.8.23}$$

where

$$\lambda_{D,pc} = \left[\sqrt{\frac{\phi}{k}}\left(\frac{\mu_w L^2}{\gamma}\right)\right]\lambda \tag{6.8.24}$$

λ is a constant equal to the reciprocal of the time required to recover 63% of the recoverable oil.

Equations (6.8.23) and (6.8.24) can be defined in terms of the shape factor, σ, as

$$t_{D,pc} = \left[\sqrt{\frac{k}{\phi}}\left(\frac{\gamma\sigma}{\mu_w}\right)\right]t \tag{6.8.25}$$

$$\lambda_{D,pc} = \left[\sqrt{\frac{\phi}{k}}\left(\frac{\mu_w}{\gamma\sigma}\right)\right]\lambda \tag{6.8.26}$$

We used (6.8.23), (6.8.25), and (6.8.26) to correlate Mattax and Kytes' experimental data reported for alundum and sandstone cores imbibing from one end or all sides as shown in Fig. 6.8.2.

6.8.3 Buckley-Leverett Solution

Using the exponential transfer function, (6.8.19), the Buckley-Leverett displacement in a fracture surrounded by matrix rock undergoing imbibition is given by deSwaan (1978):

$$-u\frac{\partial S_{wf}}{\partial x} = \frac{\partial S_{wf}}{\partial t} + \frac{\lambda R_\infty}{\phi_f}\int_o^t e^{-\lambda(t-\tau)}\frac{\partial S_{wf}}{\partial \tau}\,d\tau \tag{6.8.27}$$

An analytic solution to this problem, using the Laplace transform, can be shown to be

$$S_{wf}(x,t) = 0; \quad t < \alpha \tag{6.8.28}$$

$$S_{wf}(x,t) = e^{-\beta}e^{-\lambda(t-\alpha)}\, I_o\left(2\sqrt{\beta\lambda(t-\alpha)}\right) +$$

$$e^{-\beta\lambda}\int_\alpha^t e^{-\lambda(\tau-\alpha)}\, I_o\left(2\sqrt{\beta\lambda(\tau-\alpha)}\right)d\tau; \quad t \ge \alpha \tag{6.8.29}$$

where

$$\beta = \frac{R_\infty \lambda}{\phi_f}\alpha \tag{6.8.30}$$

$$\alpha = \frac{x}{u} \tag{6.8.31}$$

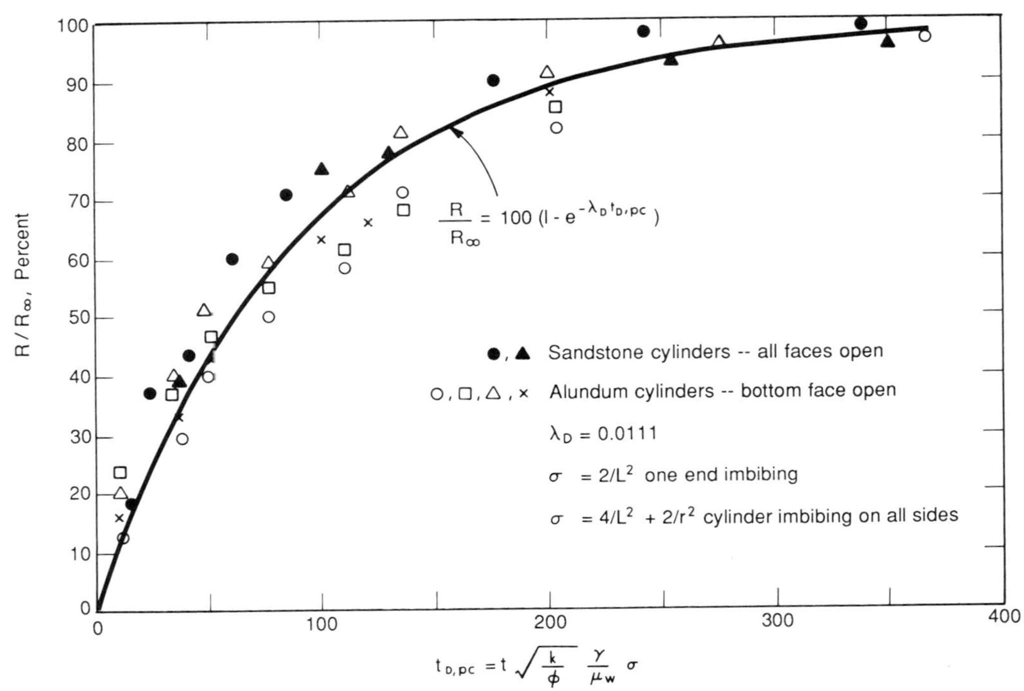

Figure 6.8.2. Mattax and Kyte imbibition oil recovery as a function of dimensionless time (from Kazemi, et al., 1992, courtesy SPE).

The finite-difference form of the integrodifferential (6.8.27) is shown below:

$$-u\left(f_{wf,i}^{n}-f_{wf,i-1}^{n}\right)/\,\Delta x_i=\left[S_{wf,i}\left(t_{n+1}\right)-S_{wf,i}\left(t_n\right)\right]/\,\Delta t+$$

$$\frac{R_\infty\lambda}{\phi_f}\sum_{j=0}^{n}\left\{\left[S_{wf,i}\left(t_{j+1}\right)-S_{wf,i}\left(t_j\right)\right]\prod_{k=j}^{n}e^{-\lambda\Delta t_k}\right\},\quad i=1,2,\ldots,IMAX \qquad (6.8.32)$$

where

$$f_{wf,i}^{n}=f_{wf}\left[S_{wf,i}\left(t_n\right)\right] \qquad (6.8.33)$$

$$\Delta t_k=t_{k+1}-t_k \qquad (6.8.34)$$

6.8.4 Pseudo Transfer Functions

Now we show how the exponential transfer function (6.8.19) can be incorporated in the standard large-scale dual-porosity formulation, (6.3.20) through (6.3.28). We follow the approach of Rossen and Shen (1987) and Coats (1989), and assume that the reservoir volumetric rate of oil recovery from the matrix is equal to the reservoir volumetric rate of water imbibition, i.e., $B_w\tau_w = -B_o\tau_o$, (6.3.23) and (6.3.24). Solving for oil pressure difference (noting that $p_w = p_o - p_c$, and assuming that gravity force is negligible compared to capillary force) we obtain

$$p_{om}-p_{of}=\frac{\lambda_{wf}}{\lambda_{wf}+\lambda_{om}}\left(p_{cwom}-p_{cwof}\right) \qquad (6.8.35)$$

Substituting (6.8.35) in (6.3.24) gives

$$\tau_o=0.001127\,\sigma V_R\,k_m\frac{\lambda_{om}\,\lambda_{wf}}{\lambda_{wf}+\lambda_{om}}\left(p_{cwom}-p_{cwof}\right) \qquad (6.8.36)$$

Combining (6.8.19) through (6.8.21), and assuming that recovery is further proportional to water saturation in the fracture, S_{wf}, we get

$$\left(S_{wm}-S_{wcm}\right)=S_{wf}\left(1-S_{orm}-S_{wcm}\right)\left(1-e^{-\lambda t}\right)\,. \qquad (6.8.37)$$

Solving for $e^{-\lambda t}$ gives

$$e^{-\lambda t}=1-\frac{S_{wm}-S_{wcm}}{S_{wf}\left(1-S_{orm}-S_{wcm}\right)} \qquad (6.8.38)$$

Then, taking the time derivative of (6.8.37) and substituting (6.8.38) we get:

$$\frac{\partial S_{wm}}{\partial t}=\lambda\left[S_{wf}\left(1-S_{orm}-S_{wcm}\right)-\left(S_{wm}-S_{wc}\right)\right] \qquad (6.8.39)$$

where it is assumed that $\partial S_{wf} / \partial t$ is negligible.

From the differential equivalent of (6.3.27)

$$\frac{\partial S_{om}}{\partial t} \approx \frac{5.6146\, \tau_o\, B_o}{\phi_m\, V_R} \tag{6.8.40}$$

or

$$\frac{\partial S_{wm}}{\partial t} = \frac{-5.6146\, \tau_o\, B_o}{\phi_m\, V_R} \tag{6.8.41}$$

Substituting for τ_o from (6.8.36) and for $\partial S_{wm} / \partial t$ from (6.8.39) in (6.8.41) we obtain:

$$p_{cwom} - p_{cwof} = \frac{\phi_m \lambda \left[S_{wf} \left(1 - S_{orm} - S_{wcm}\right) - \left(S_{wm} - S_{wcm}\right)\right]}{0.006328\, \sigma\, k_m} \left(\frac{\lambda_{wf} + \lambda_{om}}{\lambda_{wf}\, \lambda_{om}} \right) \tag{6.8.42}$$

If we substitute (6.8.42) in (6.8.36), we get

$$\tau_o = \frac{V_R\, \phi_m\, \lambda}{5.6146} \left[S_{wf} \left(1 - S_{orm} - S_{wcm}\right) - \left(S_{wm} - S_{wcm}\right)\right] \tag{6.8.43}$$

Thus for any given grid block, the transfer rate is a function of both water saturation in the fracture, S_{wf}, and the water saturation in the matrix, S_{wm}. λ is a reflection of the interaction between all forces that lead to the exponential form of the oil recovery curve. Equation (6.8.43) should be coded if the exponential transfer function is the preferred method. However, for implementing these concepts into the existing codes of dual-porosity simulators we recommend defining a set of "pseudo-capillary pressure" functions, as shown:

$$p_{cwom} = -\frac{\phi_m \lambda}{0.006328\, \sigma k_m} \left(\frac{\lambda_{wf} + \lambda_{om}}{\lambda_{wf} \lambda_{om}} \right) \left(S_{wm} - S_{wcm}\right) \tag{6.8.44}$$

$$p_{cwof} = -\frac{\phi_m \lambda}{0.006328\, \sigma k_m} \left(\frac{\lambda_{wf} + \lambda_{om}}{\lambda_{wf} \lambda_{om}} \right) \left(S_{wf}\right) \left(1 - S_{orm} - S_{wcm}\right) \tag{6.8.45}$$

Equations (6.8.44) and (6.8.45) are further simplified by setting relative permeabilities to 1:

$$p_{cwom} = -\frac{\phi_m \lambda}{0.006328\, \sigma\, k_m} \left(\mu_w + \mu_o\right) \left(S_{wm} - S_{wcm}\right) \tag{6.8.46}$$

$$p_{cwof} = -\frac{\phi_m \lambda}{0.006328\, \sigma\, k_m} \left(\mu_w + \mu_o\right) S_{wf} \left(1 - S_{orm} - S_{wcm}\right) \tag{6.8.47}$$

We have set relative permeabilities to 1 as a matter of convenience because they do not appear in (6.8.43). No physical meaning should be attributed to this substitution. When using (6.8.46) and (6.8.47), the relative permeabilities in the matrix-fracture flow terms must also be set to 1.

Again, we caution the reader that empirical relations given by (6.8.46) and (6.8.47) must be used with care because they are purely empirical and the computational results do not account for the interaction of viscous, capillary, and gravity forces.

Expressions similar to (6.8.46) and (6.8.47) can be obtained for the case of gas displacing oil by gravity.

6.8.5 Diagnostic Flow Tests

Diagnostic flow tests include pressure drawdown, pressure buildup, pressure interference, special isochronal tests, and tracer surveys. Drawdown testing consists of producing a well at a constant rate while measuring bottom hole pressure as a function of time. A semilog plot of pressure versus log of time leads to a straight line with the absolute value of the slope given by (Earlougher, 1977):

$$m = \frac{162.6\, q\, B\, \mu}{k_e\, h} \quad . \tag{6.8.48}$$

Effective permeability, k_e, can be calculated by rearranging (6.8.48). For a fracture system, k_e is often much greater than the permeability measured from a core sample.

For pressure buildup, a well producing at a constant rate is shut-in and the bottom hole pressure increase is measured with shut-in time. A semilog plot of shut-in pressure versus log[$(t_p + \Delta t)/\Delta t$] leads to a straight line with the absolute value of the slope given by (6.8.48).

Under special reservoir situations (e.g., tight matrix in an otherwise high permeability fractured system), both the drawdown and the buildup semilog plots can produce two parallel straight lines with absolute values of the slope given by (6.8.48). The vertical separation of these two straight lines, δp, is a measure of the relative fracture storage capacity, ω:

$$\omega = \exp\left[-(2.303\, \delta p / m)\right] \tag{6.8.49}$$

where:

$$\omega = \frac{(\phi c_t)_f}{(\phi c_t)_m + (\phi c_t)_f} \tag{6.8.50}$$

The time, Δt^*, corresponding to the intersection of the straight line in the transition region and second straight line is related to matrix properties and shape factor according to (Satman, 1988):

$$\Delta t^* = 1.68 \frac{(\phi c_t)_m \mu}{0.0002637\, \sigma k_m} \tag{6.8.51}$$

using $\sigma = 12/L_x^2$ in (6.8.51) one can estimate the matrix block size, L_x:

$$L_x = 0.0434 \left(\frac{k_m\, \Delta t^*}{(\phi c_t)_m \mu} \right)^{1/2} \tag{6.8.52}$$

The absolute value of the slope in the transition region is half the value given by (6.8.48) (Streltsova, 1983; Serra, et al., 1983; Satman, 1988).

Pressure interference testing using an active well (injector or producer) and several observation wells yields directional permeabilities of the reservoir resulting from fracture orientations (Earlougher, 1977).

Special isochronal tests for determining permeability, skin factor and non-Darcy flow parameters consist of flowing a well at a series of constant flow rates while recording pressure versus time (Poettmann and Kazemi, 1988). Typically, a well is flowed at a low rate first for about eight hours and then is shut in to reach static pressure. The rate is then changed to a higher choke setting for another eight hours, etc. Figure 6.8.3 shows a schematic of pressure versus flowing time for four different choke settings. The pressure drop is related to flowrate through Forchheimer's equation

$$\Delta p = a(t)q + b\,q^2 \tag{6.8.53}$$

where

$$\Delta p = \bar{p} - p_{wf}(t) \tag{6.8.54}$$

$$a(t) = \frac{162.6\,\mu B}{k_e\,h}\left[\log\left(\frac{k_e t}{\phi\,\mu\,c_t\,r_w^2}\right) - 3.23 + 2s\right] \tag{6.8.55}$$

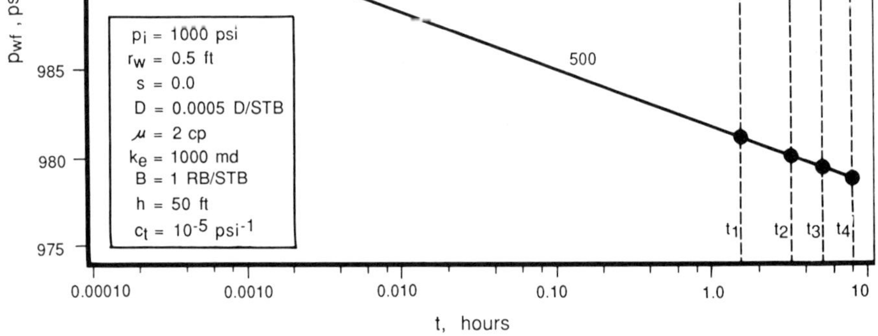

Figure 6.8.3. Pressure versus logarithm of time for a four-rate isochronal test.

$$b = \frac{141.2\,\mu B}{k_e\,h} D \tag{6.8.56}$$

and D is the non-Darcy flow coefficient. A plot of $\Delta p/q$ versus q can be used to calculate $a(t)$ and b as shown in Fig. 6.8.4. The $a(t)$ values obtained from Fig. 6.8.4 are plotted versus log of time, as shown in Fig. 6.8.5 to yield a line of slope m' given as:

$$m' = \frac{162.6\,B\,\mu}{k_e\,h} \tag{6.8.57}$$

Generally, for a fractured system, we anticipate positive values for b because of turbulent flow in the fracture.

In fractured systems, another diagnostic test involves injecting a tracer in one well and monitoring the tracer production in offset wells. Such tracer tests can be used with convection mass transport equations to determine the fracture flow characteristics, such as directional trends (Gilman and Kazemi, 1983).

6.8.6 Fracture Compressibility

Fracture pore compressibility is an important parameter in the analysis of pressure transient tests. In addition, fracture compressibility can be used to determine the dependence of fracture porosity and permeability on fluid pressure.

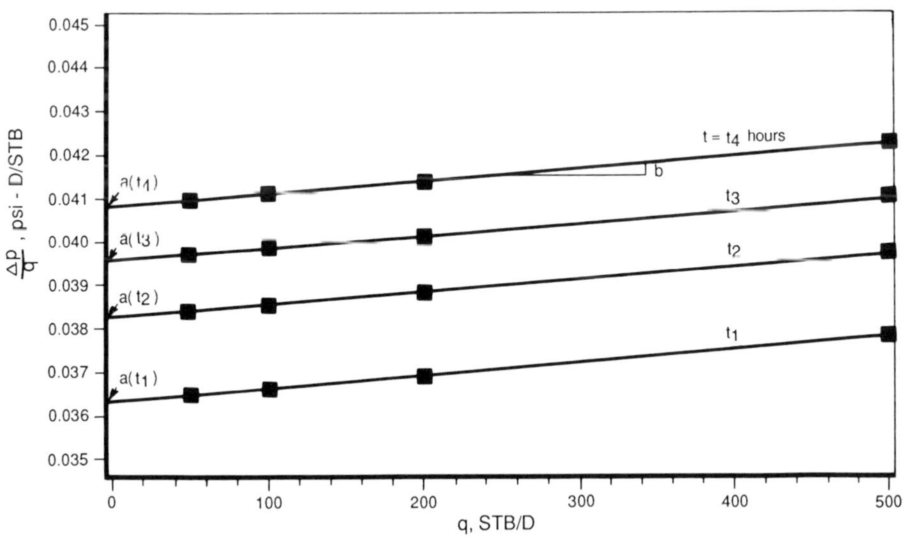

Figure 6.8.4. $\Delta p/q$ versus q for a four-rate isochronal test.

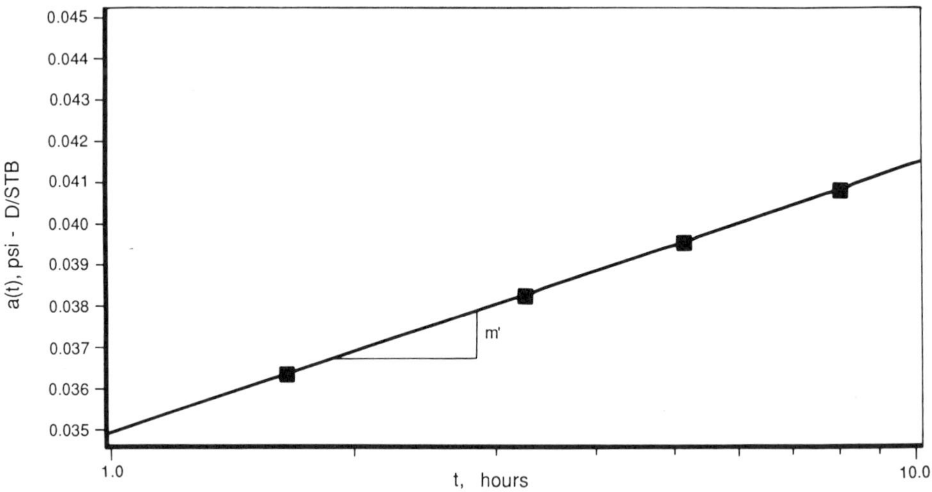

Figure 6.8.5. $a(t)$ versus logarithm of time for a four-rate isochronal test.

Assuming that the fracture network is dominated by equally spaced, parallel fractures of equal width oriented in one direction (such as the x-direction), and that k_m is negligible compared to k_e, then by combining (6.2.1), (6.2.2), and (6.2.3) we obtain (6.8.58):

$$\phi_f = 5.04 \times 10^{-5} \left[(C/96)\left(k_e / L_x^2\right) \right]^{1/3} \tag{6.8.58}$$

where

ϕ_f = fracture porosity, fraction

k_e = effective fracture permeability, md

L_x = fracture spacing, ft

C = fracture roughness measure; $96 \le C \le 400$

Alternatively (6.8.58) can be written in terms of fracture width by replacing ϕ_f by $3.28 \times 10^{-3}\, w_f/L_x$.

$$w_f = 1.54 \times 10^{-2} \left[(C/96)\, k_e\, L_x \right]^{1/3} \tag{6.8.59}$$

where

w_f = fracture width, mm

These equations can provide ballpark estimates of fracture porosity and width. Rearranging (6.8.58) gives:

$$k_e = 7.81 \times 10^{12}\ \phi_f^3\ L_x^2\ (96 / C) \tag{6.8.60}$$

For equally spaced, equally wide, parallel fractures in multiple dimensions, one should replace ϕ_f by ϕ_f/n in (6.8.58) and (6.8.60) where n = 1, 2, or 3 depending on whether the fractures are oriented in one, two, or three-directions, respectively.

Fracture porosity depends on fluid pressure according to (6.8.61)

$$\phi_f(p) = \phi_{fi}\left[1 - c_{\phi f}\left(p_i - p\right)\right] \tag{6.8.61}$$

where

ϕ_{fi} = initial porosity at initial pressure p_i

$c_{\phi f}$ = fracture compressibility, psi^{-1}

p = fluid pressure, psi

Equation (6.8.61) is valid if $c_{\phi f}(p_i - p) \ll 1$. If fracture compressibility is known, (6.8.61) can be used to calculate porosity for a given pressure. This porosity can be substituted in (6.8.60) to calculate permeability as a function of pressure. Combining (6.8.60) and (6.8.61) leads to (6.8.62):

$$k_e / k_{ei} = \left[1 - c_{\phi f}\left(p_i - p\right)\right]^3 \tag{6.8.62}$$

Jones (1975) used a triaxial loading experiment on fractured cores to develop the following equations relating porosity, permeability and compressibility to net overburden pressure and consequently to the fluid pressure. His equations are:

$$\frac{\phi_f}{\phi_{fi}} = \left(\frac{k_e}{k_{ei}}\right)^{1/3} = \frac{\log\left(p_k / p_h\right)}{\log\left(p_{ki} / p_h\right)} \tag{6.8.63}$$

$$c_{\phi f} = \frac{-0.4343}{p_k \log\left(p_k / p_h\right)} \tag{6.8.64}$$

where

p_k = net confining or overburden pressure, $p_{ob} - p$

p_{ob} = overburden pressure, psi

p = fluid pressure, psi

p_h = apparent healing pressure, psi (the pressure intercept of ϕ_f/ϕ_{fi} versus log p_k)

Using (6.8.64), Jones showed that fracture compressibility is at least an order of magnitude greater than matrix pore volume compressibility. This can also be shown to be true using linear elasticity theory.

Consider a cubic matrix block of dimension L_x similar to the schematic of Fig. 6.8.1. Assume the overburden load remains constant. When the pore pressure reduces by $\Delta p = p_i - p$, the net overburden pressure increases essentially by Δp. From the linear elastic theory, this increase in net overburden stress causes a lateral expansion of the matrix block which can be estimated from the Poisson ratio, ν, and Young's modulus as shown below:

$$\Delta L_x = \nu \frac{\Delta h}{h} = L_x = \nu \frac{\Delta p}{E} L_x \tag{6.8.65}$$

The lateral expansion of the matrix block, ΔL_x, is equal to the reduction of the width of the fracture, Δw_f. Now, using the definition of the fracture pore compressibility,

$$c_{\phi f} = \frac{1}{\phi_f} \frac{\partial \phi_f}{\partial p} \quad , \tag{6.8.66}$$

we calculate $c_{\phi f}$:

$$c_{\phi f} = \frac{1}{w_f} \frac{\Delta w_f}{\Delta p} = \frac{\nu}{E \phi_f} \tag{6.8.67}$$

Using $\nu = 0.2$, $E = 2 \times 10^6$ psi, and $\phi_f = 0.001$, we find the value for $c_{\phi f}$ to be 10^{-4} psi^{-1}, which is essentially the fracture pore compressibility reported by Jones (1975). If fracture porosity is 0.01, then $c_{\phi f} = 10^{-5}$ psi^{-1}. Furthermore, (6.8.67) indicates that fracture porosity-compressibility product, $\phi_f c_{\phi f}$, is ν/E and is constant.

Equation (6.8.67) is applicable to open fractures. In practice, however, fractures contain filler material that reduces fracture compressibility. Furthermore, when pore pressure is increased (rather than decreased) hysteresis effects may be significant.

6.8.7 Fracture Relative Permeabilities and Capillary Pressures

Relative permeabilities of water, oil, and gas in the fracture are not well understood. In fact, literature on measurement of fracture relative permeabilities is very sparse. As pointed out by Pruess (1989), the theoretical viewpoint on this subject has also received very little attention. Using theoretical flow considerations, we provide two models for fracture relative permeabilities — one for *segregated* fracture flow and one for *dispersed* fracture flow. The segregated model has appeared in the literature for many years and it assumes that relative permeability for a given phase is equal to its phase saturation. The dispersed model, however, is based on our observation that oil droplets emanating from water-wet cores as a result of water imbibition move as if they had the viscosity of the water phase. The results are summarized below.

Segregated Fracture Flow

$$k_{r\beta f} = S_{\beta f} \,, \qquad \beta = w, o, g \tag{6.8.68}$$

Dispersed Fracture Flow

$$k_{r\beta f} = S_{\beta f}\,\mu_{\beta f} \,/\, \sum_{\beta} S_{\beta f}\,\mu_{\beta f}\,, \quad \beta = w, o, g \tag{6.8.69}$$

Equation (6.8.69) satisfies the following two conditions:

i) Fractional flow for a given phase is linear with its phase saturation.

ii) Mobility for any given phase is $1/\mu_\beta$ for $S_\beta = 1.0$, $\beta = w, o, g$.

The first condition is valid for flow conditions where oil and/or gas is entrained in the water phase. The second condition is necessary to satisfy the single-phase Darcy equation.

Depending on the fracture roughness and tortuosity, and the material filling the fractures, relative permeability in the fractures can take the shape of relative permeability in nonfractured porous rock as suggested by Pruess.

Fracture Capillary Pressure

Fracture capillary pressure can be estimated from (6.8.70).

$$p_c = \frac{2\gamma \cos\theta}{w_f} \tag{6.8.70}$$

6.8.8 Petrophysical Methods for Fracture Delineation

Identification of fractures in a reservoir starts with drilling information such as core cuttings, lost circulation, caliper log skewness, and bit drop in the wellbore. Standard wireline logs may or may not reveal presence of fractures, but specialized fracture detection logs such as a formation microscanner (FMS) and borehole televiewer (BHTV) can provide direct measurement of fracture width (most often 1 mm or larger) (Schlumberger, 1987; Hurley, 1989). FMS measures electrical conductivity from arrays of buttons on two pads 90 degrees apart. It is intended to measure fracture widths as small as a fraction of 1 mm and as close together as 1 cm. FMS borehole coverage is 20% in an 8.5-inch borehole. BHTV is an acoustic wellbore scanner and can detect fracture widths larger than 1 mm and fracture spacing as close as 1 cm. BHTV provides 100% borehole coverage.

6.8.9 Emerging Technology for Fractured Reservoirs

Many old and new ideas are being applied or considered for producing the unrecovered mobile oil from fractured reservoirs. These include the use of horizontal wells, infill wells, pattern realignment, conformance improving gels and foams, surface active agents for wettability alteration, dilute surfactant, steam, and CO_2. The most promising approach is the use of horizontal wells, but so far they have been used mainly in conjunction with primary production where fluid expansion is the driving force for oil production. The application of horizontal wells in waterflooding and enhanced oil recovery (except for oil from tar sands) has not been adequately explored.

To assess and design for the practical/economic approaches that could lead to recovery of substantial amounts of oil from fractured reservoirs, one needs access to an appropriate dual-porosity simulator (Kazemi and Gilman, 1989). Another important consideration is the high priority needed in evaluating the degree of reservoir heterogeneity and directional flow tendencies as a prerequisite to design and implementation of improved recovery measures. This is usually accomplished by multi-well pressure interference testing and tracer response analysis. Both of these require appropriate simulators, too.

The scientific fundamentals for developing a viable dual-porosity simulator are well understood by the simulation experts working in the field. As computer hardware progresses (in terms of memory capacity and speed of computation), future reservoir models will be designed to utilize sufficiently fine computational grid systems to account for the major forces in the reservoir. These simulators will then be used to identify by-passed mobile oil and gas to be recovered by appropriate improved hydrocarbon recovery measures. Therefore, **fine-grid simulation** (Fig. 6.5.1) and **rigorous physics of flow** (e.g., Cuiec, et al., 1990; Keijzer and de Vries, 1990; Reis, 1990) are key words in the future of the dual-porosity simulation.

ACKNOWLEDGMENTS

The authors thank Marathon Oil Company for permission to publish and for preparation of this paper. In addition, the authors are grateful to Mrs. L.L. Fitzpatrick for her creativity in formatting the equations and for typing the manuscript.

NOTATION

a	coefficient in Forchheimer's equation, psi–D/STB
A	area, ft^2
b	coefficient in Forchheimer's equation, psi–D^2/STB^2
B	formation volume factor, RB/STB
B	coefficient matrix for i–1 unknowns
c	heat capacity, Btu/(lbm•°F)
c	compressibility, psi^{-1}
C	chemical concentration, lb/lb
C	coefficient matrix for k–1 unknowns
C	fracture roughness measure
C_v	volumetric concentration SCF/ft^3
D	non Darcy flow coefficient, D/STB
D	depth, positive down, ft
D	molecular diffusion coefficient, ft^2/D
D	main diagonal coefficient matrix
E	coefficient matrix for k+1 unknowns
E	Young's modulus of elasticity, psi
F	coefficient matrix for i+1 unknowns
G	coefficient matrix for fracture-matrix unknowns
g	potential gradient in fractures, psi/ft
H	enthalpy, Btu/(lbm-mole)
h	height, ft
k	permeability, md
$k_{\hat{T}}$	thermal conductivity, Btu/(ft·°F·D)
K	component K value, y/x
K	dispersion coefficient, ft^3/D
k_r	relative permeability, dimensionless
$K_{\hat{T}}$	thermal conductivity coefficient, Btu/(°F·D)
L	total mole fraction in liquid phase
L	matrix block size, ft
$LMAX$	maximum number of nodes
p	pressure, psi
$\overline{p}$	average pressure, psi
q	flow rate, STB/D
q_r	heat of reaction rate, Btu/D
R	recovery, fraction
$\vec{R}$	residual equation vector for Newton-Raphson
R_s	solution GOR, scf/STB
s	skin factor, dimensionless
S	saturation, fraction
t	time, days
t_p	producing time, days

$\hat{T}$	temperature, °F
T	transmissibility, STB/psi-D
U	internal energy, Btu/(lbm-mole)
v	Darcy velocity, ft/D
v_T	total Darcy velocity, ft/D
V	node volume ($\Delta x \Delta y \Delta z$), ft^3
V	total mole fraction in vapor phase
w_f	fracture width, mm
$\underset{\sim}{X}$	primary variable vector
x	component mole fraction in oil phase
y	component mole fraction in vapor phase
Z	gas super compressibility
z	overall component mole fraction
α	dispersivity, ft
α_L	longitudinal dispersivity, ft
α_T	transverse dispersivity, ft
γ	static pressure gradient, psi/ft
γ	interfacial tension, dyne/cm
Δp	pressure change, psi
Δt	shut-in time, days
Δt	time-step size, days
Δt^*	intersection time in pressure buildup, hr
Δx	gridblock length, ft
Δy	gridblock width, ft
Δz	gridblock thickness, ft
Δ	finite difference operator
Δ_t	value at time n + 1 minus value at time n
κ	partition coefficient, C_o/C_w
κ	diffusivity, ft^2/D
λ	mobility ($k_r/\mu B$), cp^{-1}
λ	exponential recovery constant, $days^{-1}$
μ	viscosity, cp
ν	Poisson's ratio
ξ	molar density, lb-mol/ft^3
ρ	density, lbm/ft^3
σ	shape factor, ft^{-2}
τ	fracture/matrix transfer term, bbl/D
τ	integration variable in the convolution integral, day
τ	tortuosity
θ	contact angle
ϕ	porosity, fraction
ω	weighting factor dimensionless
ω	relative fracture storage capacity, dimensionless

Subscripts

c	capillary

c	connate
D	dimensionless
e	effective
e	equilibrium
f	fracture
f	flowing
g	gas
i	hydrocarbon component number
i	initial
ir	irreducible
l	node index
lr	liquid residual
L	Langmuir constant
m	matrix
mf	matrix/fracture flow
o	oil
r	relative
r	reaction
r	residual
s	x direction
sc	standard conditions
t	total
w	well
w	water
x	x direction
y	y direction
z	z direction
β	phase (water, oil, gas)
ϕ	rock
∞	infinite time

Superscripts

n	time level
0	zero time

REFERENCES

Abdassah, D. and Ershaghi, I. Triple-Porosity Systems for Representing Naturally Fractured Reservoirs, *SPE Formation Evaluation*, 113–127, April 1986.

Aguilera, R. *Naturally Fractured Reservoirs*, Petroleum Publishing Company, 1980.

Ancell, K.L. and Lambert, S. Analysis of the Coalbed Degasification Process at a Seventeen Well Pattern in the Warrior Basin of Alabama, paper SPE 8971 presented at the SPE/DOE Unconventional Gas Recovery Symposium, Pittsburgh, PA, May 18–21, 1980.

Aronofsky, J.S., Massé, L. and Natanson, S.G. A Model for the Mechanism of Oil Recovery from the Porous Matrix Due to Water Invasion in Fractured Reservoirs, *Trans.* AIME, v. 213, p 17, 1958.

Barenblatt, G.I. On the Motion of a Gas-Liquid Mixture in Porous Fissured Media, *Izv. Akad. Nauk SSSR, Mekh, Machionst.*, No. 3, 47–50, 1964.

Barenblatt, G.I., Zheltov, Iu.P., and Kochina, I.N. Basic Concepts in the Theory of Seepage of Homogeneous Liquids in Fissured Rocks, *PMM*, Vol. 24, No. 5, 1960.

Beckner, B.L., Firoozabadi, A., and Aziz, K. Modeling Transverse Imbibition in Double-Porosity Simulators, paper SPE 17414 presented at the SPE California Regional Meeting, Long Beach, March 23–25, (1988).

Birks, J. A Theoretical Investigation into the Recovery of Oil from Fissured Limestone Formations by Water-Drive and Gas Cap Drive, *Proc.*, Fourth World Pet. Cong., Rome, Sec. II/F, 425–440, (1955).

Braester, C. A Shock Wave in Immiscible Displacement in a Fissured Porous Medium, *Israel J. Tech.*, Vol. 9, No. 5, 433–438, (1971).

Braester, C. Simultaneous Flow of Immiscible Liquids Through Porous Media, *Soc. Pet. Eng. J.*, 297–305, (Aug 1972).

Carslaw, H.S. and Jaeger, J.C. *Conduction of Heat in Solids*, Oxford, 2nd Ed., p. 184, Eq. 6, 1959.

Cervik, J. The Behavior of Coal-Gas Reservoirs, paper SPE 1973 presented at the SPE Eastern Regional Meeting, Pittsburgh, PA, Nov. 7–9, 1967.

Chen, W.H., Wasserman, M.L., and Fitzmorris, R.E. A Thermal Simulator for Naturally Fractured Reservoirs, SPE 16008 presented at the Ninth SPE Symposium on Reservoir Simulation, San Antonio, TX (Feb. 1–4, 1987).

Coats, K.H. Implicit Compositional Simulation of Single-Porosity and Dual-Porosity Reservoirs, paper SPE 18427 presented at the Tenth SPE Symposium on Reservoir Simulation, Houston, TX, Feb. 6–8, 1989.

Cuiec, L.E., Bourbiaux, B. and Kalaydjian, F. Imbibition in Low-Permeability Porous Media: Understanding and Improvement of Oil Recovery, paper SPE 20259 presented at the SPE/DOE Seventh Symposium on Enhanced Oil Recovery, Tulsa, OK, April 22–25, 1990.

Dean, R.H. and Lo, L.L. Simulations of Naturally Fractured Reservoirs, *SPE Reservoir Engineering*, 638–648, (May 1988).

deSwaan, A. Theory of Waterflooding in Fractured Reservoirs, *Soc. Pet. Eng. J.*,117–22, (April 1978).

Dreher, K.D., Kenyon, D.E., and Iwere, F.O. Heat Flow During Steam Injection into a Fractured Carbonate Reservoir, presented at the SPE/DOE Fifth Symposium on Enhanced Oil Recovery held in Tulsa, OK., (April 20–23, 1986).

Duguid, J.O., and Lee, P.C.Y. Flow in Fractured Porous Media, *Water Resour. Res.*, 558–566, (1977).

Earlougher, R.C., Jr. *Advances in Well Test Analysis*, Monograph Series, 5, Society of Petroleum Engineers of AIME, Dallas (1977).

Gilman, J.R. An Efficient Finite Difference Method for Simulating Phase Segregation in the Matrix Blocks in Double-Porosity Reservoirs, *SPE Reservoir Engineering*, 403–13, (July 1986).

Gilman, J.R. and Kazemi, H. Improvements in Simulation of Naturally Fractured Reservoirs, *Soc. Pet. Eng. J.*, 695–707, (Aug. 1983).

Gilman, J.R. and Kazemi, H. Improved Calculations for Viscous and Gravity Displacement in Matrix Blocks in Dual-Porosity Simulators, *J. Pet. Tech.* 60–70, (Jan. 1988).

Hamon, G. and Vidal, J. Scaling-Up the Capillary Imbibition Process From Laboratory Experiments on Homogeneous and Heterogeneous Samples, paper SPE 15852 presented at the European Petroleum Conference, London, Oct. 20–22, 1986.

Hill, A.C. and Thomas, G.W. A New Approach for Simulating Complex Fractured Reservoirs, paper SPE 13537 presented at the SPE Reservoir Simulation Symposium, Dallas, Feb. 10–13, 1985.

Hurley, N.F. Characterization of Natural Fractures in Big Horn Basin Anticlines, University of Wyoming Enhanced Oil Recovery Institute Meeting, Casper, Wyoming, May 10–11, 1989.

Iffly, R., Rousselet, D.C., and Vermeulen, J.R. Fundamental Study of Imbibition in Fissured Oil Fields, paper SPE 4102 presented at the SPE-AIME 47th Annual Fall Meeting, San Antonio, Tex., Oct. 8–11, 1972.

Jones, F.O., Jr. A Laboratory Study of the Effects of Confining Pressure on Fracture Flow and Storage Capacity in Carbonate Rocks, *J. Pet. Tech.,* 21–27, (Jan. 1975).

Jones, T.A., Wooten, S.O., and Kaluza, T.J. Single-Phase Flow Through Natural Fractures, paper SPE 18175 presented at the SPE 63rd Annual Technical Conference, Houston, TX, Oct. 2–5, 1988.

Kazemi, H. Pressure Transient Analysis of Naturally Fractured Reservoirs With Uniform Fracture Distribution, *Soc Pet. Eng. J.*, 451–462; *Trans.*, AIME, Vol. 246, (Dec. 1969).

Kazemi, H., Gilman, J.R., and El-Sharkawy, A.M. Analytical and Numerical Solution of Oil Recovery from Fractured Reservoirs Using Empirical Transfer Functions, *Soc. Pet. Eng. Res. Engr.*, 219–227 (May 1992).

Kazemi, H., Seth, M.S., and Thomas, G.W. The Interpretation of Interference Tests in Naturally Fractured Reservoirs with Uniform Fracture Distribution, *Soc. Pet. Eng. J.,* 463–472; *Trans.*, AIME, Vol. 246, (Dec. 1969).

Kazemi, H., Merrill, L.S., Porterfield, K.L., and Zeman, P.R. Numerical Simulation of Water-Oil Flow in Naturally Fractured Reservoirs, *Soc. Pet. Eng. J.* 317–26. (Dec. 1976).

Kazemi, H. and Merrill, L.S. Numerical Simulation of Water Imbibition in Fractured Cores, *Soc Pet. Eng. J.*, 175–82 (June 1979).

Keijzer, P.P.M. and de Vries, A.S. Imbibition of Surfactant Solutions, paper SPE 20222 presented at the SPE/DOE Seventh Symposium on Enhanced Oil Recovery, Tulsa, OK, April 22–25, 1990.

King, G.R., Ertekin, T., and Schwerer, F.C. Numerical Simulation of the Transient Behavior of Coal-Seam Degasification Wells, *SPE Formation Evaluation,* 165–83 (April 1986).

Kleppe, J. and Morse, R.A. Oil Production from Fractured Reservoirs by Water Displacement, paper SPE 5084 presented at the SPE-AIME 49th Annual Fall Meeting, Houston, Oct. 6–9, 1974.

Lee, B.Y.Q. and Tan, T.B.S. Application of a Multiple Porosity/Permeability Simulator In Fractured Reservoir Simulation, SPE 16009 presented at the Ninth SPE Symposium on Reservoir Simulation, San Antonio, TX (Feb. 1–4, 1987).

Litvak, B.L. Simulation and Characterization of Naturally Fractured Reservoirs, Proc., Reservoir Characterization Technical Conference, Dallas, Academic Press, New York City (1985).

Mannon, R.W., and Chilingar, G.V. Experiments on Effect of Water Injection Rate on Imbibition Rate in Fractured Reservoirs, paper SPE 4101 presented at the SPE-AIME 47th Annual Fall Meeting, San Antonio, Tex., Oct. 8–11, 1972.

Mattax, C.C. and Kyte, J.R. Imbibition Oil Recovery from Fractured, Water-Drive Reservoirs, *Soc. Pet. Eng. J.,* 177–184; *Trans.*, AIME, Vol. 225 (June 1962).

Najurieta, H.L. Interference and Pulse Testing in Uniformly Fractured Reservoirs, paper SPE 8283 presented at the SPE Annual Technical Conference and Exhibition, Las Vegas, Sept. 23–26, 1979.

Nelson, R.A. *Geologic Analysis of Naturally Fractured Reservoirs*, Gulf Publishing Co., 1985.

Odeh, A.S. Unsteady-State Behavior of Naturally Fractured Reservoirs, *Soc. Pet. Eng. J.*, 60–66; *Trans.*, AIME, Vol. 234 (March 1965).

Parsons, R.W. and Chaney, P.R. Imbibition Model Studies on Water-Wet Carbonate Rocks, *Soc. Pet. Eng. J.*, 26–34; *Trans.*, AIME, Vol 237 (March 1966).

Poettmann, F.H. and Kazemi, H. Discussion of Estimating Parameters from the Gas Pressure Equation, *Soc. Pet. Eng. Res. Eng.,* 1328–1330 (Nov. 1988).

Pruess, K. and Narasimhan, T.N. A Practical Method for Modeling Fluid and Heat Flow in Fractured Porous Media, *Soc. Pet. Eng. J.,* 15–26 (Feb. 1985).

Pruess, K. and Tsang, Y.W. Oil Relative Permeability of Rough-Walled Fractures, 14th Workshop on Geothermal Reservoir Engineering, Stanford University, January 24–26, 1989.

Quandalle, P. and Sabatheir, J.C. Typical Features of a New Multipurpose Reservoir Simulator, *Soc. Pet. Eng. Res. Engr.*, 475–480 (Nov. 1989).

Reis, J.C. Oil Recovery Mechanisms in Fractured Reservoirs During Steam Injection, paper SPE 20204 presented at the SPE/DOE Seventh Symposium on Enhanced Oil Recovery, Tulsa, OK, April 22–25, 1990.

Reiss, L.H. *The Reservoir Engineering Aspects of Fractured Formations*, Gulf Publishing Co., 1980.

Rossen, R.H. ‘Simulation of Naturally Fractured Reservoirs with Semi-Implicit Source Terms, *Soc. Pet. Eng. J.,* 201–10 (June 1977).

Rossen, R.H., and Shen, E.I. Simulation of Gas/Oil Drainage and Water/Oil Imbibition in Naturally Fractured Reservoirs, *Soc. Pet. Eng. Res. Engr.*, 464–470 (Nov. 1989).

Sahuquet, B.C. and Ferrier, J. Steam-Drive Pilot in a Fractured Carbonated Reservoir: Lacq Superieur Field, *J. Pet. Tech.,* 873–880 (April 1982).

Saidi, A. Simulation of Naturally Fractured Reservoirs, paper SPE 12270 presented at the SPE Symposium on Reservoir Simulation, San Francisco, Nov. 16–18, 1983.

Saidi, A.M. *Reservoir Engineering of Fractured Reservoirs*, Total Edition Press, 1987.

Satman, A. Solutions to Heat- and Fluid-Flow Problems in Naturally Fractured Reservoirs: Part 2 - Fluid-Flow Problems, *Soc. Pet. Eng. Prod. Engr.*, 467–473 (Nov. 1988).

Schlumberger: Fracture Detection with Logs, *The Technical Review,* 22–34, January 1987.

Schulte, W.B., and de Vries, A.S. In-Situ Combustion in Naturally Fractured Heavy Oil Reservoirs, paper SPE/DOE 10723 presented at the SPE/DOE Third Symposium on Enhanced Oil Recovery, Tulsa, OK, April 4–7, 1982.

Serra, K., Reynolds, A.C. and Raghavan, R. New Pressure Transient Analysis Methods for Naturally Fractured Reservoirs, *Jour. Pet. Tech.,* 2271–83, (Dec. 1983).

Shapiro, A.M. Transport Equations for Fractured Porous Media. *NATO Advanced Study Institute on Fundamentals of Transport in Porous Media*, J. Bear and M.Y. Corapcioglu, eds., University of Delaware, Newark, DE, 1985.

Sinnock, S., Lin, Y.T., and Brannen, J.P. Preliminary Bounds on the Postclosure Performance of the Yucca Mountain Repository Site, Southern Nevada. *J. Geophys. Res.*, 92, 7820–7842, 1987.

Sonier, F., Souillard, P., and Blaskovich, F.T. Numerical Simulation of Naturally Fractured Reservoirs, *SPE Reservoir Engineering*, 1114–1122, (November 1988).

Streltsova, T.D. Well Pressure Behavior of a Naturally Fractured Reservoir, *Soc. Pet. Eng. J.*, 769–80, (Oct. 1983).

Thomas, L.K., Dixon, T.N., and Pierson, R.G. Fractured Reservoir Simulation, *Soc. Pet. Eng. J.,* 42–54 (Feb. 1983).

van Golf-Racht, T.D. *Fundamentals of Fractured Reservoir Engineering*, Elsevier, 1982.

van Wunnik, J.N.M. and Wit, K. Improvement of Gravity Drainage by Steam Injection Into a Fissured Reservoir: An Analytical Evaluation, *Soc. Pet. Eng. Res. Engr.*, 59–66 (Feb 1992).

Warren, J.E. and Root, P.J. The Behavior of Naturally Fractured Reservoirs, *Soc. Pet. Eng. J.*, 245–255; *Trans.*, AIME, Vol. 228, (Sept. 1963).

7

UNSATURATED FLOW IN FRACTURED POROUS MEDIA

J. S. Y. Wang and T. N. Narasimhan
Earth Sciences Division
Lawrence Berkeley Laboratory
University of California
Berkeley, California

7.1 INTRODUCTION

That the thick unsaturated zone in the arid southwestern United States is potentially suitable for the disposal of high-level radioactive waste (HLW) was originally suggested by Winograd (1981) of the United States Geological Survey (USGS). The Yucca Mountain site, on and adjacent to the Nevada Test Site in southern Nevada (Fig. 7.1.1), was initially considered as a potential site for HLW disposal on the basis of host rock conditions below the water table. Later, as knowledge of the mountain grew, it was recognized that not only will the unsaturated zone itself provide an excellent host for the wastes but that the repository will derive the benefits of both the unsaturated zone and the saturated zone. Accordingly, several scientists of the USGS jointly suggested in February 1982 to the U.S. Department of Energy (DOE) that the unsaturated zone at Yucca Mountain is a potentially desirable site for HLW disposal. By 1982, the Yucca Mountain concept passed the first phase of a broad preliminary assessment and entered the next phase of testing some ideas with more detailed analysis. The authors also became actively involved at this time with the Sandia National Laboratories' (SNL) hydrological performance assessment activities in DOE's Yucca Mountain Site Characterization Project (YMP). The thoughts regarding the Yucca Mountain, as they emerged from the first phase of analysis, were summarized in Roseboom (1983) and Montazer and Wilson (1984). In December 1987, the U.S. Congress chose Yucca Mountain for site characterization, as part of the program to provide the first HLW repository. A Site Characterization Plan for Yucca Mountain was published by DOE (1989).

Thus far the focus of our unsaturated flow study with SNL has been on understanding the mechanisms and on modeling the hydrological processes in partially saturated, fractured,

FLOW AND CONTAMINANT TRANSPORT IN FRACTURED ROCK

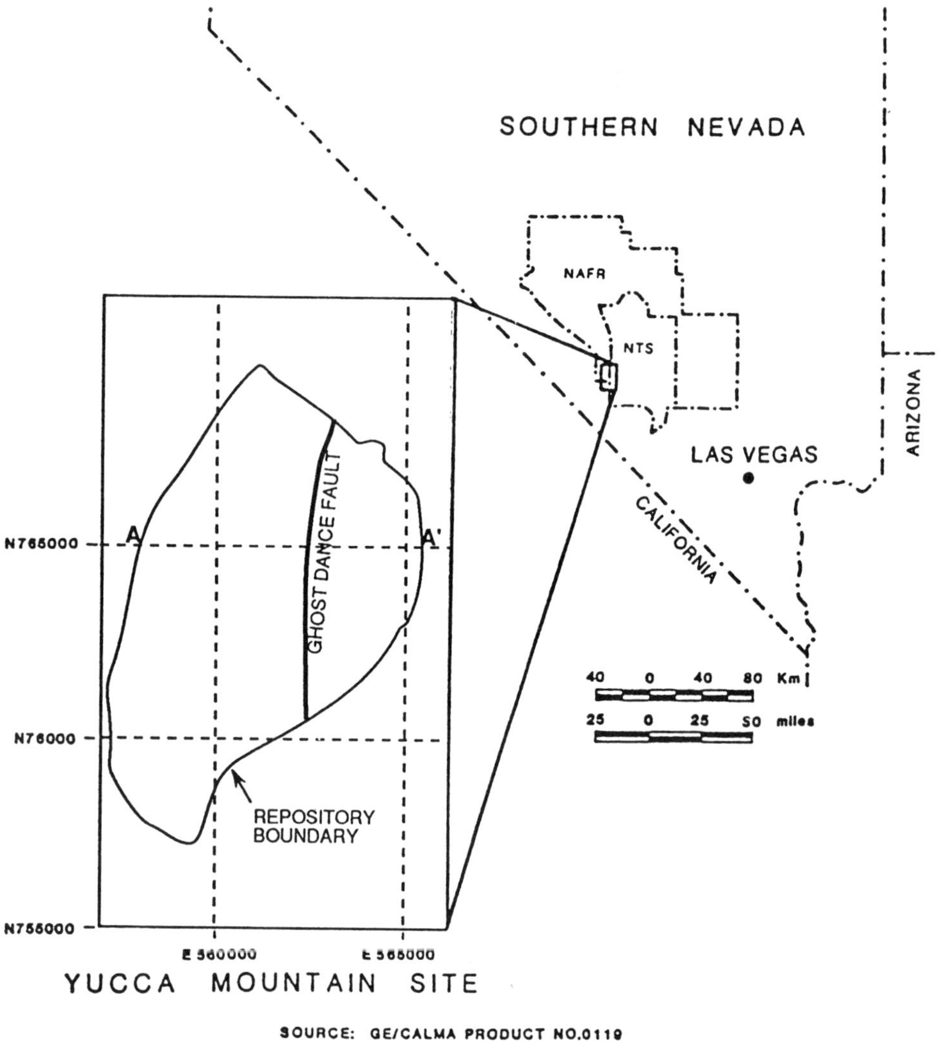

Figure 7.1.1. Location map of Yucca Mountain, Nevada, potential site for a high-level nuclear waste repository.

porous tuff at Yucca Mountain. We started with a spatial resolution in the submillimeter range for the discrete fractures and extended consideration to kilometer scale for an east-west cross section of Yucca Mountain. The time scale of interest ranged from minutes for the desaturation of fractures to thousands of years for infiltration through the partially saturated zones. The objective of assessing the hydrological processes at Yucca Mountain is to predict the three-dimensional responses of the whole system, from the repository to the accessible environment, over thousands of years. Starting with the small scale, this study has gradually evolved to the large scale, which is of ultimate interest for performance assessment. The broad pattern of this evolution can be summarized as shown in Fig. 7.1.2. After the sections on unsaturated flow processes in different scales, we discuss the parameter variations and correlations of tuff properties and the potential use of fractal geometry for future modeling studies. The chapter concludes with a brief discussion of general hydrological issues that extend beyond our conceptual analysis of unsaturated flow in fractured, porous tuff (Narasimhan and Wang, 1992).

7.2 CONCEPTUAL MODEL FOR MOISTURE MOVEMENT IN FRACTURED, POROUS TUFF: FRACTURE FLOW VS. MATRIX FLOW

During 1983, the first year of the study, we focused on the development of a conceptual model to understand the hydrological mechanisms governing fluid flow in a partially saturated, fractured porous medium. The Topopah Spring welded tuff formation, the candidate host rock for the repository at Yucca Mountain, is highly fractured, with tuff matrix blocks of sub-meter sizes surrounded by discrete fractures. The main question we addressed was whether, under partially saturated conditions, water moves in the fractures or in the matrix. This is one of the most important hydrological questions in determining the flow pattern and travel time through fractured, porous units. This single question will remain as one of the key issues to be resolved and quantified throughout the course of YMP investigations.

7.2.1 Capillary Mechanism: Preferential Saturation of Fine Pores

We recognized from the onset of this study that, while fractures with large apertures will drain relatively easily, water is likely to remain in the matrix. This intuition stems from the knowledge that large pores cannot sustain large capillary suction forces. For instance, in a heterogeneous system where sand and clay coexist, the sand with large pores can be easily drained and the clay will remain wetter than the sand under partially saturated conditions. This can be readily explained by the nature of the capillary mechanism. When we insert a capillary tube into a beaker of water, the height of water rise along the capillary tube is inversely proportional to the radius of the tube. A dry towel with fine texture can soak up more water than a rag with coarse pores. Capillary theory predicts that the larger pores will desaturate more easily than smaller pores. If this simple capillary model is valid, we can infer, firstly, that fractures in a fractured porous medium will tend to remain dry under partially saturated conditions and, secondly, that liquid water is held by capillarity in the finer pores of the matrix.

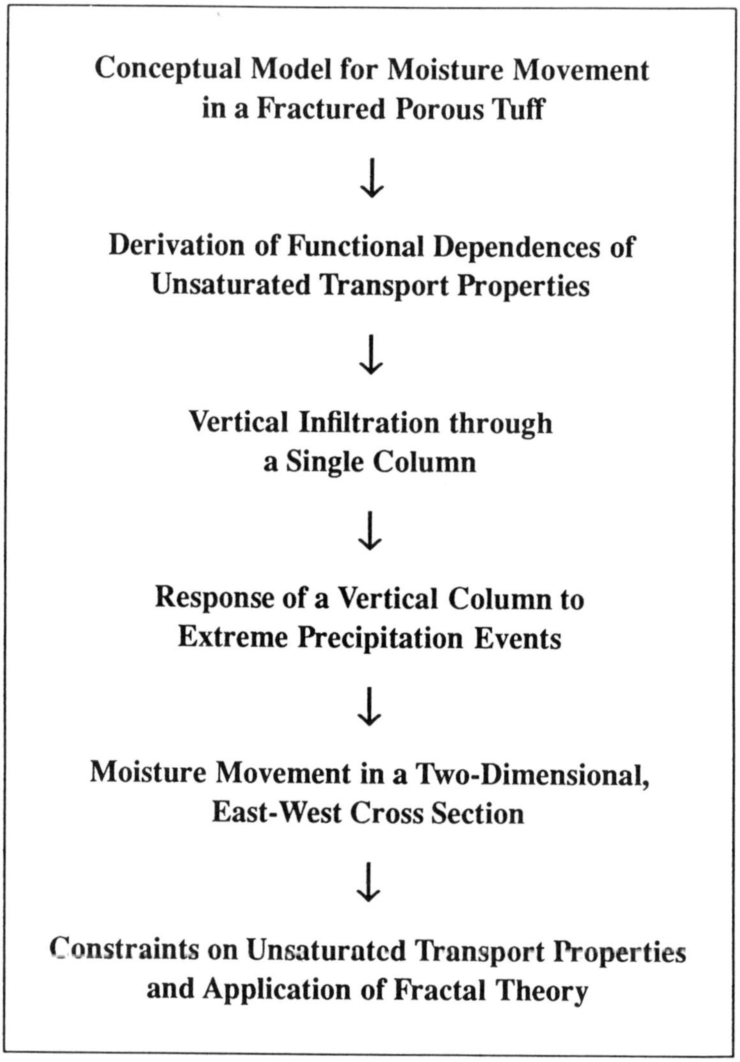

Figure 7.1.2. Evolution of performance assessment task.

The well-known capillary equation for a tube with height of water rise $|h|$ and radius r,

$$|h| = \frac{2\gamma \cos(\theta)}{r\rho g} \quad , \tag{7.2.1}$$

is also applicable to a parallel-plate fracture with aperture b replacing the radius r. The surface tension γ is the surface energy per unit area, or equivalently, surface force per unit length. For a pair of parallel plates with width L much greater than the aperture b, the upward force component along two liquid-air-solid interfaces with contact angle θ is $\gamma \cos(\theta) \times (2L)$, and the downward weight of the liquid rise is $\rho g \times (Lb|h|)$. From force balance, we obtain the capillary equation relating the capillary rise $|h|$ with fracture aperture b.

Before we use capillary theory to describe the response of fractured tuff, we must point out that the capillary mechanism has a limited range of validity. If the fracture aperture or capillary radius is large, the radius of curvature of the meniscus will be infinite and the capillary effect will be negligible. For a wetting fluid with contact angle $\theta = 0$, the hemispherical surface at top of the rise can no longer be defined when b or r is greater than the height $|h|$. Therefore the maximum capillary size (with $b = |h|$ or $r = |h|$ in Eq. 7.2.1) is

$$b_{\max} = r_{\max} = \left(\frac{2\gamma}{\rho g}\right)^{1/2} \tag{7.2.2}$$

For $\gamma = 0.072$ kg/sec^2, density $\rho = 1000$ kg/m^3, and gravitational acceleration $g = 9.8$ m/sec^2, the maximum capillary size is 3.83 mm. For fractures tens to hundreds of meters below the ground surface, we expect that a majority of fractures will be closed under overburden stress and have apertures smaller than this maximum capillary size. For fracture aperture and macropore radius greater than this size, a capillary-based conceptual model is not valid.

In capillary theory, we also ignore the small fraction of water molecules adsorbed by the solid wall above the meniscus. However, the slow movement of adsorbed water outside the liquid region may be important in extremely dry conditions. The capillary limit and adsorptive mechanism are two hydrological issues which may be important for studying the validity of capillarity-based conceptual model.

7.2.2 Natural Fracture Model: Importance of Wall Roughness

In quantifying the capillary mechanism in fractured tuff, the parallel-plate model frequently used in saturated fracture flow studies is too crude and simplistic to represent natural fractures under partially saturated conditions. In a highly fractured formation such as the Topopah Spring welded tuff unit, the fractures divide the tuff matrix into blocks surrounded and bounded on all sides by the fractures. If the fractures were represented by parallel plates, then the fractures would either be completely filled with liquid water or completely drained. Liquid water within the porous blocks, isolated by the drained fractures, would not be able to flow from one matrix block to another. With parallel-plate fractures, liquid water in the matrix blocks could not move under partially saturated conditions, and travel time will effectively be infinite. This is simply too idealistic a model for performance assessment of flow and transport through a partially saturated, fractured porous formation.

Natural fractures are characterized by rough surfaces. In fractures at depth under compressional stress, the walls are pressed together, with a finite fraction of fracture surfaces in contact, such contacts being known as asperities. The aperture of a fracture is likely to be smaller near the asperities and larger in the open channels within the fracture plane. As a fracture desaturates, the open region with larger apertures will be drained easily and liquid water will remain as pendular islands around the contact areas. The wetted regions centered around the contact areas provide the bridges for liquid water to flow across the fracture plane from one matrix block to its neighboring matrix blocks. Figure 7.2.1 and Fig. 7.2.2 illustrate the matrix-to-matrix flow pattern and fluid distribution in a fractured porous medium. Figure 7.2.1 is in a cross section cut normal to the fracture plane showing the pores in matrix blocks and rough fractures with spatially varying apertures. Figure 7.2.2 looks down on a fracture surface, and reveals the distribution of contact areas and fluid configurations. With portions of fractures dry, liquid tends to flow across fractures from one matrix block to another instead of flowing along the fractures. One may expect the flow lines to circumvent dry portions of the fractures as illustrated in Fig. 7.2.1. As a fracture desaturates, the liquid phase configuration changes from continuous phase at high saturation to discontinuous phase at low saturation, with liquid forming rings around contact areas. With the idealized array of contact areas in Fig. 7.2.2, a sharp cutoff transition will separate the continuous liquid phase configuration from the discontinuous configuration.

Capillary theory, together with the rough fracture model, forms the basis of our conceptual model for a fractured porous medium under partially saturated conditions.

7.2.3 Aperture Distribution: An Approach to Quantifying a Rough Fracture

Mathematically, a rough fracture may be characterized by an aperture distribution function. As the fracture desaturates under a negative pressure head h, the portions of the fractures with aperture greater than the saturation cutoff aperture b_S, where

$$b_S = -\frac{2\gamma cos(\theta)}{\rho g h} \quad , \qquad (7.2.3)$$

will be desaturated and unable to transmit liquid flow. We have derived relationships for fracture saturation $S(h)$, fracture relative permeability $k_r(h)$, and effective flow area for the fraction of wet fracture surface $\sigma(h)$ as functions of fluid pressure head h, in terms of integrals over the aperture distribution function $f(b)$:

$$S(h) = \frac{\langle b \rangle_S}{\langle b \rangle_1} \quad ; \qquad (7.2.4)$$

$$k_r(h) = \tau(h)\frac{\langle b^3 \rangle_S}{\langle b^3 \rangle_1} \quad ; \qquad (7.2.5)$$

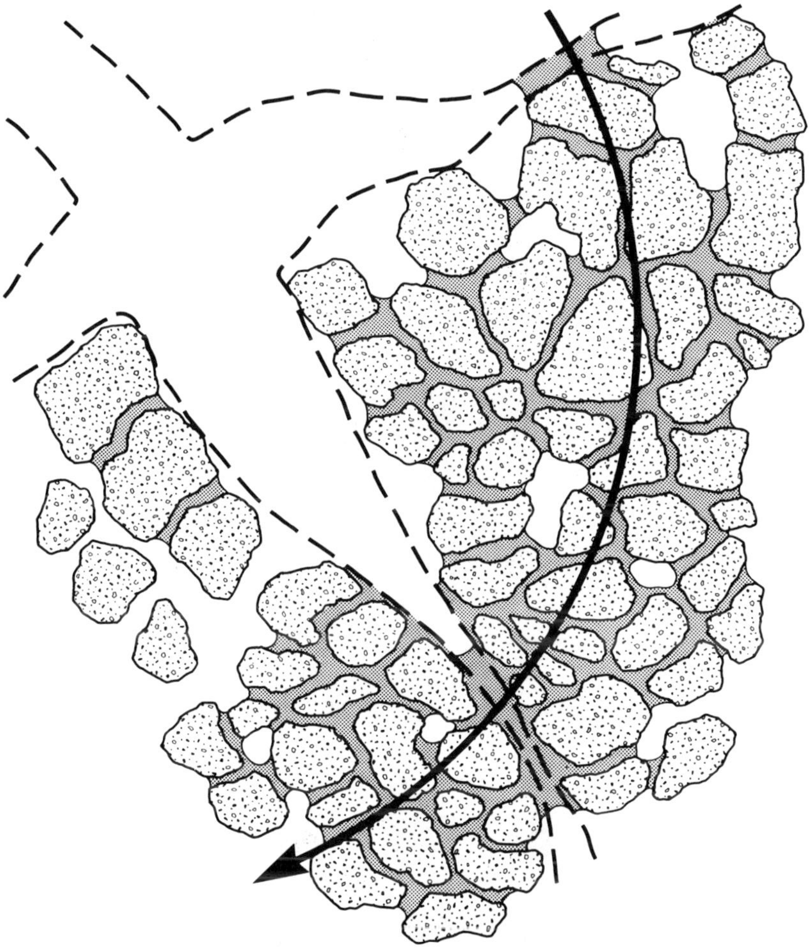

Figure 7.2.1. Schematic representation of water flow in a fractured porous medium.

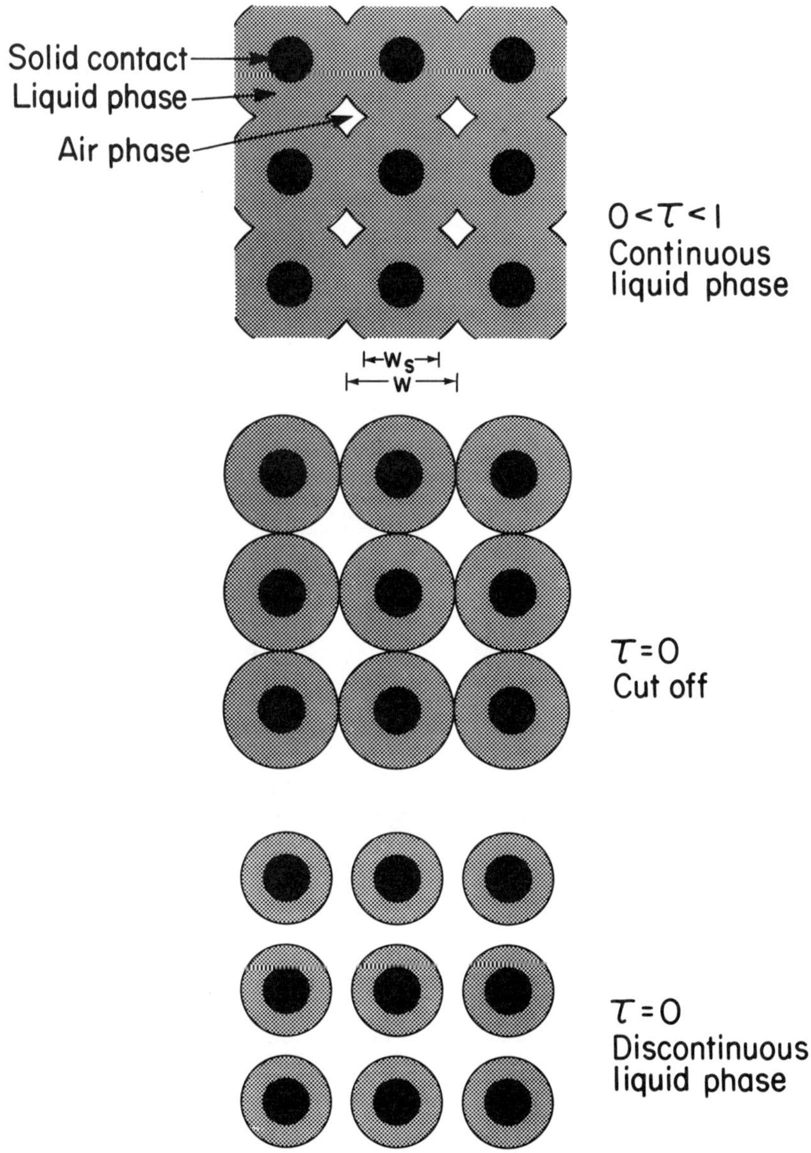

Figure 7.2.2. Schematic plan view of a fracture plane showing asperities (solid circles), liquid water (gray) and air (blank).

where

$$\langle b^n \rangle_S = \int_0^{b_S} b^n f(b)\, db \quad , \quad n = 1, 3 \quad ; \tag{7.2.6}$$

and

$$\sigma(h) = \omega + \int_0^{b_S} f(b)\, db \quad , \tag{7.2.7}$$

where ω is the fraction of contact area.

The phase-constriction factor $\tau(h)$ in Eq. 7.2.5 accounts for the distortion of flow by the air pockets and sets the fracture permeability and flow to zero when the liquid phase is discontinuous. For the simple geometry shown in Fig. 7.2.2,

$$\tau(h) = \frac{W_S}{W} \tag{7.2.8}$$

can be related to the effective fraction of area $\sigma(h)$ occupied by the liquid water and contact area (Eq. 7.2.7). When the liquid flow path width W_S decreases to zero and the effective flow area is smaller than the area occupied by touching pendular rings, we have $\tau(h) = 0$ and $k_r(h) = 0$. If we neglect the phase-constriction effect by setting $\tau = 1$ for any h values, the relative permeability is determined by the average of aperture cubes in Eq. 7.2.5. This is equivalent to a capillary bundle model with all flow channels parallel to the direction of flow.

The general relationships given by Eqs. 7.2.4, 7.2.5, and 7.2.7 can be used with any aperture measurement and distribution function. The detailed derivations and results to quantify the conceptual model are given in our first report (Wang and Narasimhan, 1984) and an archival paper (Wang and Narasimhan, 1985). For one simple aperture distribution function, the gamma distribution, we integrated explicitly to obtain algebraic relationships for the fracture characteristic curves.

The gamma distribution function

$$f_0(b_0) = \beta^2 b_0 e^{-\beta b_0} \quad , \tag{7.2.9}$$

has only one parameter, β. $f_0(b_0)$ describes the variation of aperture b_0 of a fracture under zero stress. Under zero stress, the fracture surfaces are in point contact and the contact area is zero. In practice, $f_0(b_0)$ can be measured by scanning the open fracture surfaces and tracing the roughness profiles. The mismatch between the roughness profiles of two surfaces yields the aperture distribution function.

For in situ fractures where the surfaces are in contact under nonzero stress, the fraction of contact area ω of the total area of the fracture can be expressed as

$$\omega = \int_0^{b_c} f_0(b_0)\, db_0 \quad . \tag{7.2.10}$$

All the portions of fracture with initial aperture less than the contact cutoff aperture b_c will be in contact. The aperture under stress will be $b = b_0 - b_c$ in the open sections of the fracture. The aperture distribution function of the fracture under stress $f(b)$ in Eqs. 7.2.6 and 7.2.7 is

$$f(b) = f_0(b + b_c) = \beta^2 (b + b_c) e^{-\beta(b+b_c)} \quad , \tag{7.2.11}$$

Table 7.2.1 summarizes the algebraic relationships for unsaturated fracture characteristic curves, from integrations over the gamma distribution function with finite contact area ω and nonzero contact cutoff aperture b_c. Evans and Huang (1983) have constructed a fracture network model of parallel-plate fractures with the well-known lognormal aperture distribution. With lognormal distribution, they numerically evaluated the integrals in Eq. 7.2.6. Although the classical Evans and Huang model is for fracture network, the results can be reinterpreted for a single fracture with lognormal aperture distribution and zero contact area.

Very little data have been available to quantify the aperture distribution function for fractures in Yucca Mountain tuff. The simplicity of the gamma distribution function, and especially the fact that there is only one parameter to quantify, were on our minds when we chose it as the basis for detailed study. Another reason was that the limited analysis available at that time from a French study of natural granite fractures by Gentier (1986) indicated that $f_0(b_0)$ takes on a skewed shape, with long tails toward large apertures (Tsang, 1984). The gamma distribution can fit these aperture measurements well.

We should emphasize that we did not use granite data to represent tuff fractures. We used the gamma distribution function only to represent the functional form of the aperture distribution. Without actual aperture measurements on fractures in tuff, we used available fracture spacings and bulk permeability of tuff to inversely determine the β parameters. The assumptions and procedures used to determine β parameters from available tuff data will be discussed

Table 7.2.1. Fracture Characteristic Curves

Property	Algebraic Relationship*
contact area	$\omega = I_1(\beta b_c)$
effective flow area	$\sigma(h) = I_1(\beta b_c + \beta b_s)$
saturation	$S(h) = \dfrac{I_2(\beta b_S) + \beta b_c I_1(\beta b_S)}{I_2(\infty) + \beta b_c I_1(\infty)}$
relative permeability	$k_r(h) = \tau(h) \dfrac{I_4(\beta b_S) + \beta b_c I_3(\beta b_S)}{I_4(\infty) + \beta b_c I_3(\infty)}$

*$I_1(x) = 1 - (1 + x)\, e^{-x}$

$I_2(x) = 2 - (2 + 2x + x^2)\, e^{-x}$

$I_3(x) = 6 - (6 + 6x + 3x^2 + x^3)\, e^{-x}$

$I_4(x) = 24 - (24 + 24x + 12x^2 + 4x^3 + x^4)\, e^{-x}$

in Sec. 7.3.1. The fracture distribution functions derived for the vertical fracture and for the horizontal fracture are plotted in Fig. 7.2.3, with distribution parameters derived from Topopah Spring welded tuff data. The distribution functions are normalized, with the areas under the curves set equal to unity. Both vertical and horizontal fractures have hydraulic apertures in the sub-millimeter range according to these indirectly derived aperture distribution functions.

7.2.4 Flow Across Fractures: Importance of Effective Flow Area

For fractures underground, we expect that a fraction of the apertures will be closed under stress. We do not have data for the fraction of *in situ* contact area. However, borehole USW-G4 core analyses (Spengler and Chornack, 1984) indicate that about 12% of the fracture surfaces have coatings of zeolite, clay, and calcite, which are minerals likely to be precipitated from water. We assumed that the fracture coating area of 12% corresponds to the *in situ* contact area ω on the premise that contact areas remain wetted for a long time and allow hydrochemical alterations to occur in contact areas so that coatings form. This assumed relation of fracture coatings to fracture contacts remains to be checked for consistency against other geochemical and geomechanical data. We believe that the evolution of fracture coatings can shed light on the long-term processes relating mechanical deformation and stability, chemical dissolution and precipitation, and hydrological flow and transport in fractured, porous tuff under partially saturated conditions.

If both the matrix blocks and the fractures are fully saturated with liquid water, the whole fracture-matrix interface will be available for water to flow from the matrix into the fracture and vice versa. As the fractures drain, flow from one matrix block to another block across a fracture can occur only through a fraction of the area in the fracture surfaces. Figure 7.2.4 shows that the effective flow area at fracture-matrix interfaces decreases, quickly approaching the limiting contact area fraction of 12% as the pressure head decreases. With liquid flow between matrix blocks restricted, crossing the interfaces through reduced areas, the flow lines will converge toward and then diverge away from the contact areas. Therefore, flow paths will be more tortuous in a partially saturated, fractured porous medium. The pressure dependence of effective fracture-matrix area is not taken into account in conventional numerical models. We incorporate this effective area dependence into the TRUST code (Narasimhan et al., 1978) to properly simulate the flow in fractured tuff systems.

7.2.5 Transport Properties of Fractured Tuff: Contrasts between Fracture and Matrix Characteristics

The relation between water saturation and pressure head, and the relation between permeability and pressure head are two physical functions that are required for modeling water movement in the fractured tuff. Figure 7.2.5 compares the fracture and tuff matrix characteristic curves of saturation versus head for the Topopah Spring welded tuff unit. The tuff matrix curve was provided by SNL from laboratory measurement of a representative core sample (Peters et al., 1984). The thermocouple psychrometer matrix data were fitted with the equation suggested by van Genuchten (1980). Figure 7.2.5 shows that a small suction of –0.1 m can

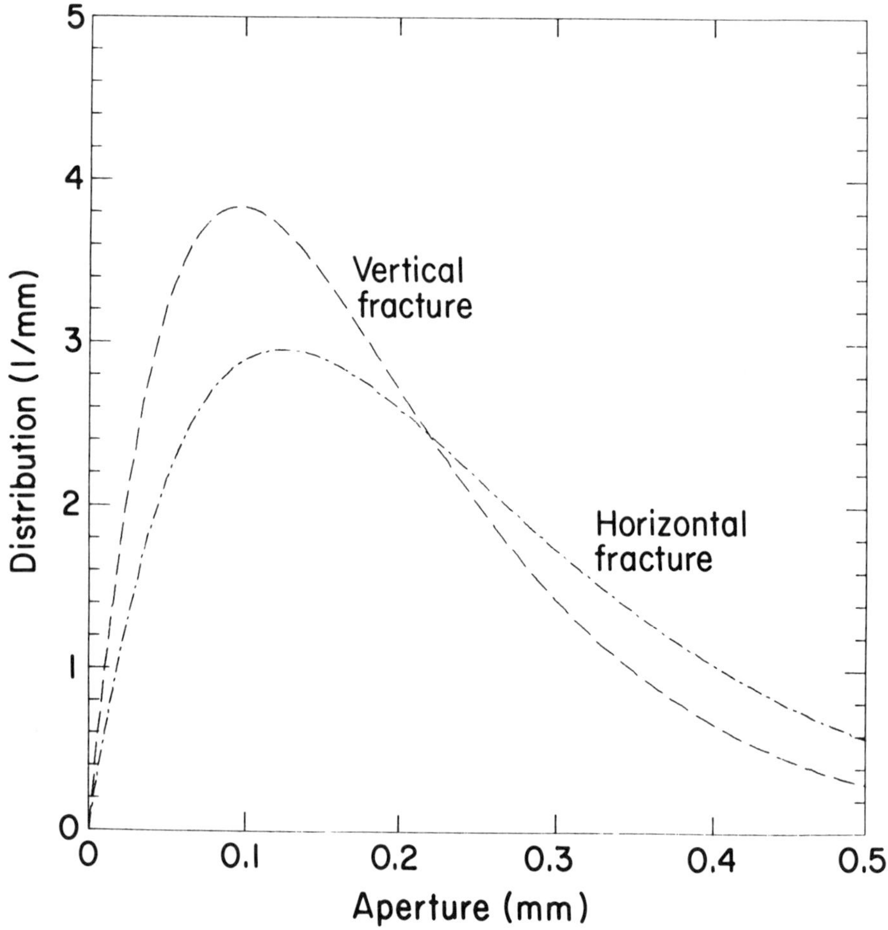

Figure 7.2.3. Aperture distributions for the Topopah Spring welded tuff unit (distribution parameter: vertical fracture - $\beta_v = 10.4\ mm^{-1}$; horizontal fracture - $\beta_h = 8.04\ mm^{-1}$).

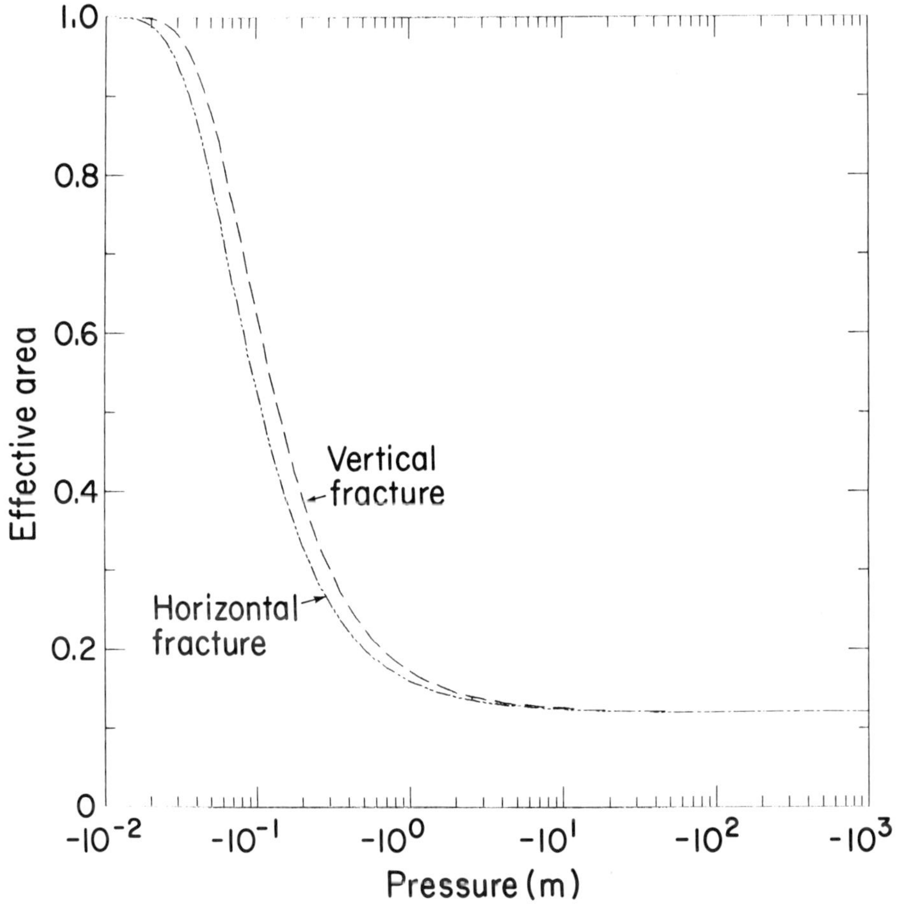

Figure 7.2.4. Effective flow area versus pressure head for flow across the fracture plane.

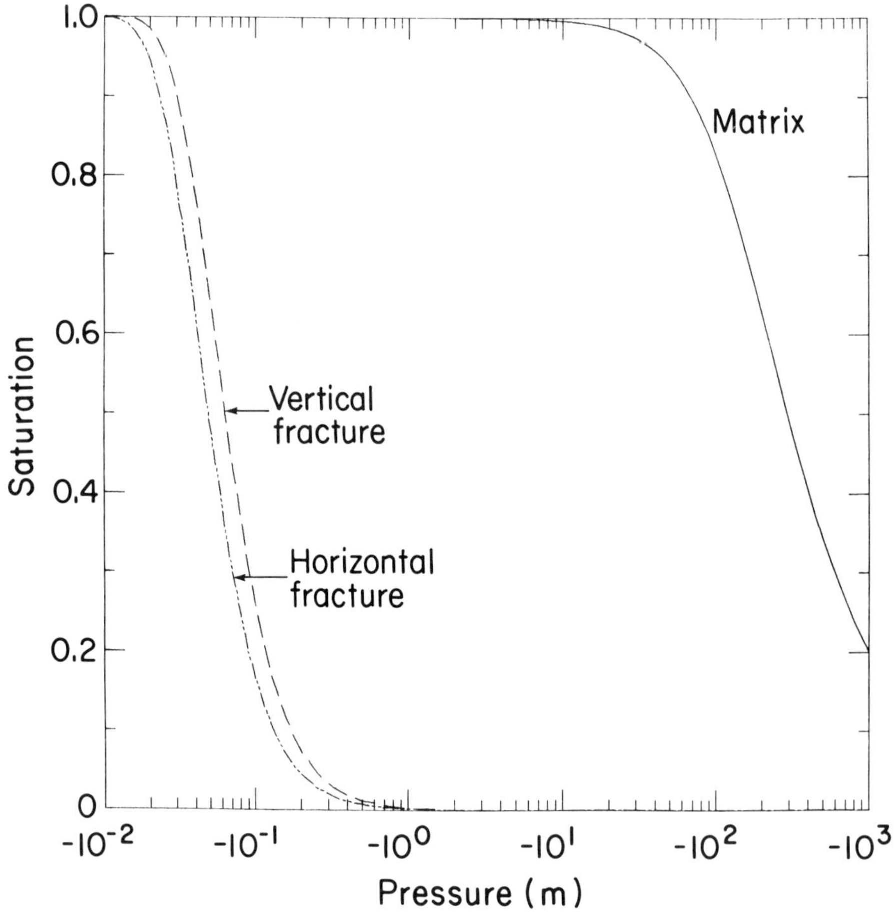

Figure 7.2.5. Saturation characteristic curves of fractures and tuff matrix.

easily desaturate the fractures, while the saturation in the matrix remains high. If the in situ matrix saturation in Topopah Spring welded tuff unit is 0.8, then the negative pressure head in this unit is over –100 m. With such a large suction, the fracture saturations will be essentially zero.

The relations between effective permeability and pressure head are given in Fig. 7.2.6, and demonstrate that the fracture permeabilities are very sensitive to suction. The fracture permeabilities are also sensitive to whether the phase-constriction factor $\tau(h)$ is included in the calculations (Eq. 7.2.5) to take into account the effects of blockage of flow by air pockets, the flow-path distortion around the air pockets, and the transition from continuous to discontinuous liquid phase associated with desaturation. If fully saturated, each discrete fracture's permeability is eight orders of magnitude greater than the matrix permeability. With small suctions in the range of –0.1 to –10 m, the discrete fracture permeabilities decrease drastically, especially for the curves with the phase-constriction factor. The matrix permeability decreases much more gently than fracture permeability as the pressure head becomes more negative.

Near fully saturated conditions, the fractures control the fluid flow. As desaturation proceeds and the fracture permeability of each discrete fracture becomes less than the matrix permeability, the matrix will control the flow. With *in situ* suction of over –100 m, fracture flow diminishes and matrix flow dominates.

The drastic change in the role played by the fractures, from the active main conduits for flow and transport in saturated conditions to passive dry pores in unsaturated conditions, is the most important result arising from our conceptual model. If the role reversal between fracture flow and matrix flow does not occur under ambient partially saturated conditions, we should find that the highly fractured tuff unit will transmit liquid water very quickly through the connected fracture networks, and that travel time is very short. Substantiating our findings of a role reversal between fracture flow and matrix flow is crucial for the performance assessment of the partially saturated flow field at Yucca Mountain.

7.3 DRAINAGE RESPONSES OF A FRACTURED TUFF COLUMN: DYNAMIC INTERACTION BETWEEN DISCRETE FRACTURES AND POROUS MATRIX

With the development of a conceptual model based on capillary theory and aperture distribution of rough-walled fractures, we used the fracture characteristic curves derived in 1983 to study the detailed responses of a fractured, porous tuff column in 1984, explicitly taking into account the vertical and horizontal fractures and tuff matrix blocks. Our finding that fracture pressure and matrix pressure are approximately equal after the column is drained provided a rational basis for the development of a composite medium model at SNL by Peters and Klavetter (1988). In addition, the dominance of liquid flow in the matrix, together with the findings by Pruess et al. (1985) that gas flow occurs mainly through the fractures, has provided incentive for the concurrent development of a multiphase composite medium model. This model is known as an effective medium model. Detailed studies with discrete fractures are valuable for better understanding of system behaviors and for developing useful approximations for large-scale analyses.

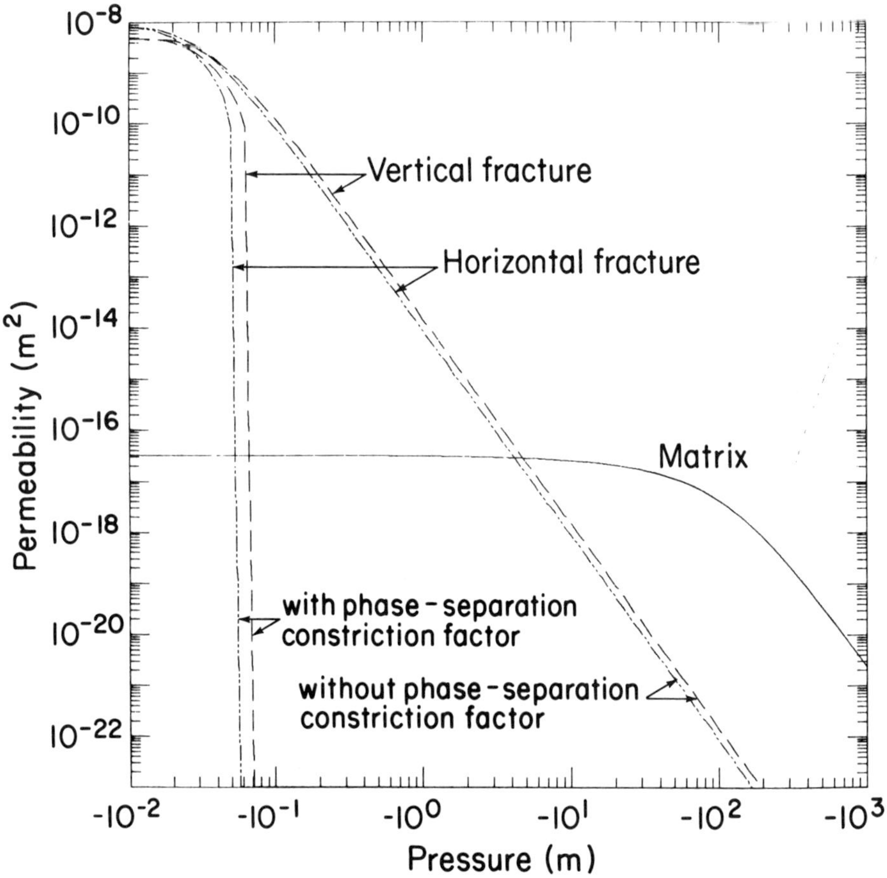

Figure 7.2.6. Permeabilities of partially saturated fractures and tuff matrix.

7.3.1 Fracture Description: Adaptation of Field Data

To construct the detailed model representing the Topopah Spring welded tuff unit, we first analyzed available data on fracture orientation and spacing at Yucca Mountain. There are two nearly vertical fracture sets, one striking roughly north-northwest and the other, north-northeast (Scott et al., 1983). Fracture inclination data from borehole USW-G4 (Spengler and Chornack, 1984) also indicate the presence of a minor, nearly horizontal fracture set. By grouping the fractures as either nearly vertical or nearly horizontal according to their dip inclinations, and using the measured fracture frequency of 4.36/m, we determined the average fracture spacings of the sets as shown in Fig. 7.3.1. We assumed that the two nearly vertical sets had equal spacing and that they contributed equally to the equivalent fracture continuum permeability (cube of aperture divided by spacing).

Because, in 1984, data for the equivalent fracture continuum permeability was not directly available, we deduced the equivalent permeability value $k_f = 1.02 \times 10^{-11}$ m^2 from well testing in well J-13 (Thordarson, 1983). Well J-13 is located in a low-standing structural block east of Yucca Mountain in which part of the Topopah Spring welded tuff unit is below the water table. We assumed that the water pumped out during testing had moved through a connected network of fractures that provide effective pathways for fluid movement under saturated conditions. We also assumed that equivalent fracture continuum permeability was isotropic.

Along a given direction in three-dimensional space, two fracture sets contribute to the equivalent fracture continuum permeability k_f. We assume that

$$\frac{k_f}{2} = \frac{\langle b^3 \rangle_1}{12d} \tag{7.3.1}$$

for each set with spacing d. Applying gamma distribution Eq. 7.2.11 in Eq. 7.2.6, we have

$$\langle b^3 \rangle_1 = \frac{24 + 6\beta b_c}{\beta^3} e^{-\beta b_c} \quad . \tag{7.3.2}$$

Because βb_c can be determined by the parameter ω (Table 7.2.1 and Sec. 7.2.4), we can use Eq. 7.3.1 to determine the distribution parameter β for each fracture set, as shown in Fig. 7.2.3. All other fracture characteristic curves in Table 7.2.1 can then be evaluated.

The steps outlined above show that the parameters used in our simulations were derived from available tuff data and no other parameter values were assumed or adjusted in the tuff fracture desaturation curves. The simplifying assumptions about the fracture network can easily be relaxed, and the procedures of fracture network analyses can be generalized to determine anisotropic fracture spacings and non-parallelepiped blocks if additional data is provided on anisotropic equivalent continuum permeability tensors. In the course of this study, we have continuously updated our models as more realistic data were provided by SNL and have incorporated them into later simulations.

7.3.2 Discrete-Fracture-Porous-Matrix Model: Nested Matrix Elements

With the spacings of the fracture sets and the sizes of the tuff matrix blocks determined, we constructed a small discrete-fracture-porous-matrix model to simulate vertical drainage of a

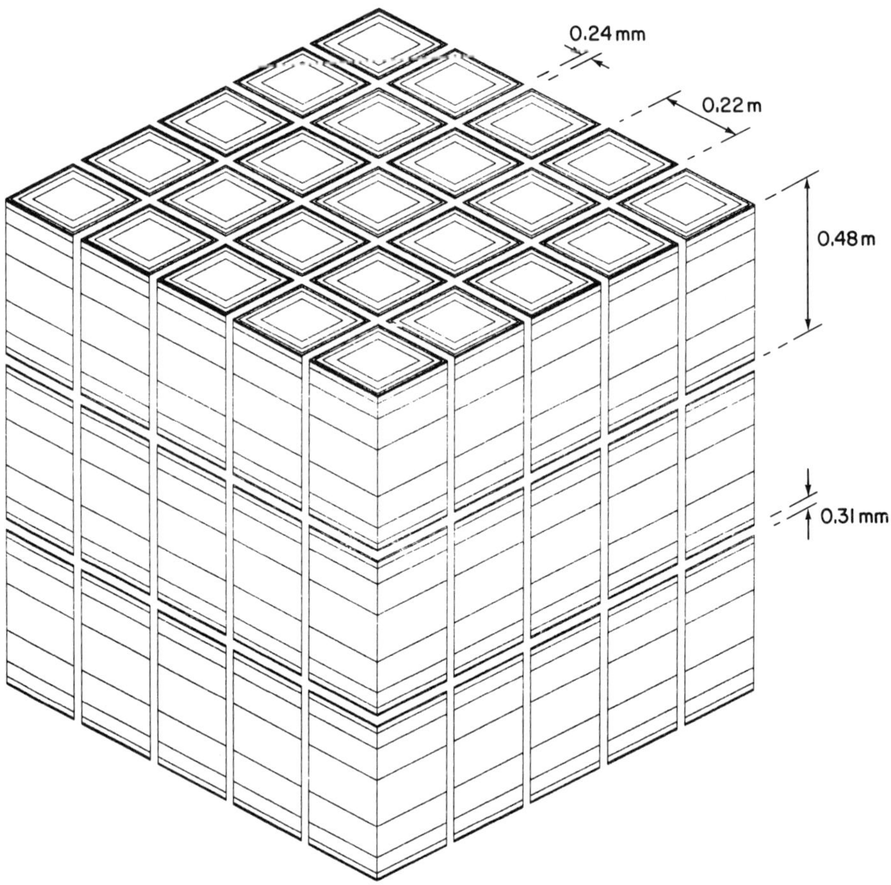

Figure 7.3.1. Schematic representation of three fracture sets for the Topopah Spring welded tuff unit.

fractured tuff column (Fig. 7.3.1). Two horizontal fractures, normal to the direction of general flow, are explicitly simulated. On the local scale, lateral flow is allowed between the vertical fracture and the matrix blocks. The horizontal cross section of a matrix block is further partitioned into six nested elements and the vertical cross section of each block is sliced into nine sections. The grid elements are small near the fractures and large toward the middle of the matrix block. The partition of horizontal nested elements is similar to that used in the multiple interacting continuum idealization (Pruess and Narasimhan, 1985). Such a nested grid allows simulation of convergent flow from the fractures into the matrix block and divergent flow from the matrix block into the fractures with relatively few elements. This is sufficient if we are primarily interested in the fracture-matrix interflows and not the detailed flow field near the diagonals within the matrix block. The nested elements also account for the geometrical configuration, with each horizontal cross section of a matrix block surrounded by fractures on all four sides. The usual semi-analytic treatments of fracture-matrix flow use one-dimensional approximations, and thus do not properly account for effects on multiple sides. In the simulations, we did not group together the matrix elements from different blocks situated at different elevations. The two horizontal fractures and the three matrix blocks within each vertical column are explicitly simulated. In this respect, we did not carry out the simulation in the multiple interacting continuum grouping approximation; we account for the effects of discrete fractures in detail.

7.3.3 Tool of Analysis: The TRUST Code

In 1984 we modified the TRUST code so that it accounts for the effective flow area, with the available area for flow across the fracture-matrix interfaces being determined by the fracture saturations. At the limit at which the fractures are completely desaturated, the available area for matrix flow from one block to the next is the contact area between adjacent blocks. In the simulations, the outermost matrix node for one block is connected to the fracture and is not directly connected to the matrix node in the next block. It is important to note that fracture permeabilities are intrinsically defined only along the fracture planes. Normal to the fracture-matrix interfaces, the fractures will not introduce any resistance under either saturated conditions or partially saturated conditions, with flow proceeding through finite wetted areas around contact points. Using the integrated finite difference method (Narasimhan and Witherspoon, 1976) in the TRUST code, this absence of resistance can be easily taken into account by setting the distances from the fracture nodes to the fracture-matrix interfaces to zero, so that normal flow from matrix blocks to fractures is controlled by matrix permeability. Conventional finite difference and finite element codes do not usually take this intrinsic anisotropy of fracture permeability into account. If the proper treatment of fracture permeability is not taken into account, the zero permeability of the fracture elements under completely desaturated conditions will block the flow from one matrix block to the next. Physically, we expect that flow will cross one matrix block to another matrix block through the finite contact areas.

In 1984, we used the usual harmonic mean approximation to determine the conductances at the element interfaces from permeability values of neighboring nodes weighted by the distances from the node center to the interfaces. If we use the geometry-imbedded Darcy's law (Narasimhan, 1985), we can probably reduce further the number of nested elements from six to one or two for each horizontal matrix block cross section and vertical grid. The development of geometry-imbedded path-integration modeling techniques can increase the efficiency and accuracy of modeling discrete-fracture-porous-matrix systems.

7.3.4 Major Findings: Flow Dominance by Matrix

We have simulated the transient changes in the fractured tuff column from an initial fully saturated condition to a final partially saturated condition. Initially the column is filled with water with pressure determined by hydrostatic equilibrium. The hydraulic head is zero throughout and fluid is stationary. At time $t = 0^+$ a negative suction head is applied to the lower boundary to induce drainage and transient changes in the fluid flow field throughout the fractured, porous tuff column. The boundary suction head is over –100 m, which corresponds to a final ambient matrix saturation of $S = 0.8$.

The transient pressure drops at different locations in the column are shown in Fig. 7.3.2 for three cases. We studied five cases for flow in the column: (1) a fractured, porous tuff matrix column with the phase-constriction factor taken into account for the discrete fractures ($1 \leq \tau(h) \leq 1$); (2) a fractured, porous column without the phase-constriction factor ($\tau = 1$); (3) a matrix column without fractures; (4) a discrete fracture column without a matrix and with the τ factor; and (5) a discrete fracture column without a matrix and without the τ factor. The locations of interest are the middle of the matrix block (point A), the middle of the vertical fracture (point B), the middle of the upper fracture (point C), and the middle of the lower horizontal fracture (point D), as shown in the inserts of the plots.

The pressure drops for the first three cases at different locations in Fig. 7.3.2 are essentially the same after an initial desaturation period of 10^{-2} yr. Within this short period, the onset of fracture desaturation depends on whether the phase-constriction factor is taken into account as shown in the small differences between the top two plots in Fig. 7.3.2. As pressures become negative following initial desaturations, the pressure changes are controlled mainly by the matrix. At the same elevation, the matrix pressure at point A and the vertical fracture pressure at point B are approximately the same; the fracture-matrix pressure equilibrium has been reached. The lower fracture, point D, is closer to the suction boundary and has more negative pressure than the points at higher elevation. The results of the simulation for the matrix only, in the third plot in Fig. 7.3.2, are nearly identical to the results when discrete fractures are simulated. It is clear, therefore, that pressure changes can be simulated using only matrix properties, if the transition from saturated conditions to unsaturated conditions is not of interest. We also simulated the changes in the fractured column without the matrix (case 4 and case 5) and obtained very unrealistic results. This demonstrates the importance of porous matrix blocks as flow channels for fluid subjected to a change in pressure (Wang and Narasimhan, 1985).

Associated with the pressure drop, the fracture saturations drop abruptly. The matrix changes more slowly from the initial fully saturated condition of $S = 1$ to the final ambient saturation of $S = 0.8$. These changes can be observed in the simulated results. They can also be deduced from the characteristic curves in Fig. 7.2.5. The permeability changes confirm the role reversal between fracture flow at saturated condition and matrix flow after the desaturations. The effective fracture-matrix flow area changes abruptly from 100% at full saturation to 12%, the contact area fraction. The simulated results suggest that the ambient flow field is through the matrix blocks, the fractures are dry, and the flow lines are distorted in crossing through fracture contact areas from one matrix block to the next matrix block.

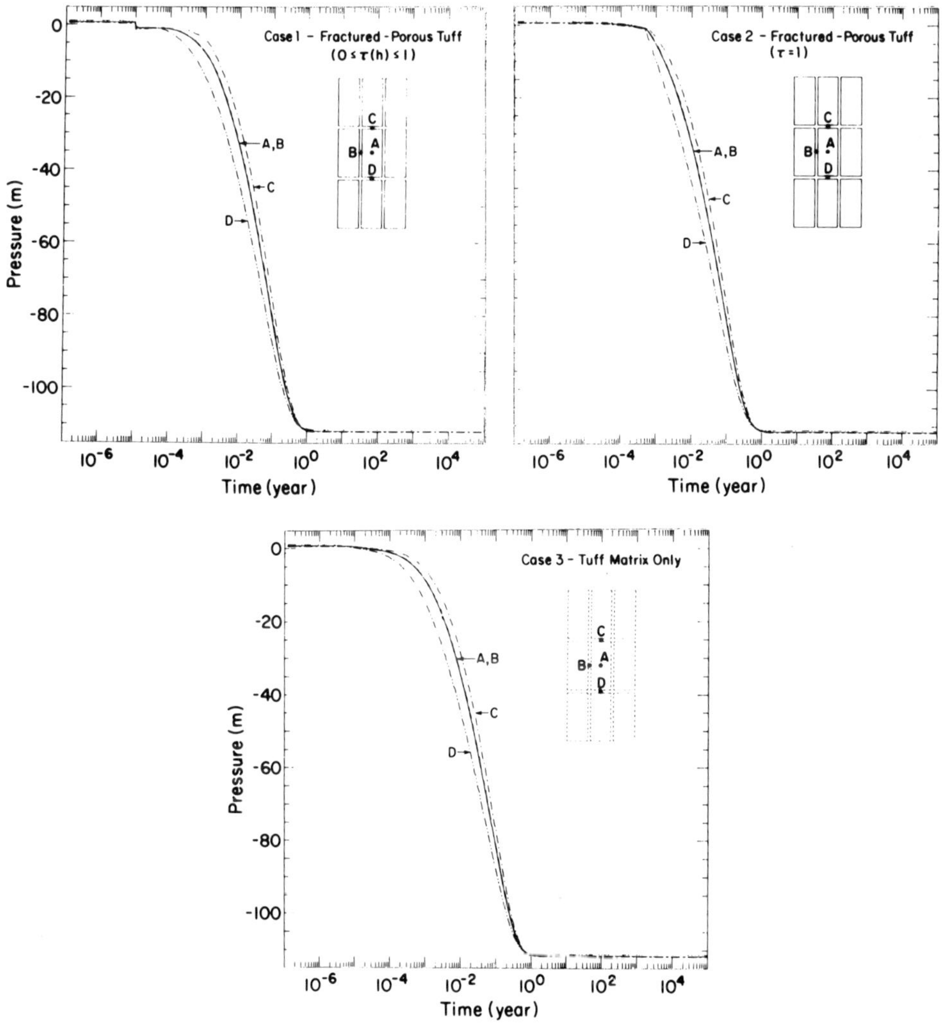

Figure 7.3.2. Unsaturated drainage in a tuff column: comparison of pressure changes for three different conceptualizations.

In addition to pressure, saturation, permeability, and effective fracture-matrix flow area changes, we have also calculated the Darcy velocity versus time at different locations. Because they depend on both permeability and pressure gradient, the Darcy velocity changes are very sensitive to the details of fracture-flow characteristics. Before the fractures completely desaturate, we have the flows along the fractures change from an initially static condition to, firstly, large flows associated with fast fracture drainage, and then to abrupt flow stoppage as the fractures lose the water within the small fracture volume. The fracture-matrix interflows are also sensitive to fracture characteristics. Figure 7.3.3 illustrates the changes in Darcy velocities across the fracture-matrix interfaces at the midpoints of fractures. Before the fractures desaturate, the fluid in the matrix flows toward the fractures, supplying fluid to the fractures which is then drained rapidly by the suction. At the saturated-desaturated transition, these fracture-matrix flows change drastically. After the transition, the fluid mainly moves vertically from upper matrix blocks, across horizontal fractures, into lower matrix blocks. The horizontal flow also reverses direction from an initial matrix-to-fracture flow to a small but significant fracture-to-matrix flow. After the transition, the fractures no longer transmit fluid, but become a passive and small fluid storage source to feed the matrix blocks as they drain.

After the fractures desaturate, the Darcy velocities from one matrix block across the fracture into the next matrix block are essentially the same as those calculated by a matrix-only model, as illustrated in Fig. 7.3.3, case 3, for a tuff matrix column without fractures. The agreement among the velocities in cases 1, 2, and 3 strongly indicates that the fluid flow field of a partially saturated, porous formation can be simulated without taking fractures into account, if the transition region from saturated to unsaturated condition is not of concern.

In quasi-steady-state flow, the dry fractures limit the available areas through which flow can cross from one matrix block to another matrix block, but the total flux through the fractured column is insensitive to local distortion near the fracture-matrix interfaces. In the 1984 study, we did not simulate the flow field at the scale of individual fracture contacts. The effects of the small-scale heterogeneity are taken into account by the statistical average procedures used in deriving the relative permeabilities and effective fracture-matrix flow areas. Although the total quasi-steady-state flow is not sensitive to the flow path distortion through the fracture-matrix interfaces, the particle velocity field and solute transport are expected to be affected by the reduction of fracture-matrix flow areas. We need to study the fundamental process at small scale around individual contact areas at fracture-matrix interfaces to achieve a better understanding of the particle movement through fractured, porous tuff systems.

7.3.5 A Useful Corollary: The Composite Model

One of our important findings in 1984 was that the fracture pressure and matrix pressure were approximately the same. This suggests that one may simplify the simulation by considering only one pressure field. The fracture-matrix pressure equilibrium, or, more generally, thermodynamic equilibrium, including temperature as well as pressure, is one of the essential assumptions of the composite medium-effective continuum models (Peters and Klavetter, 1988; Pruess et al., 1985, 1990). With the same pressure defined for both the fractures and the porous matrix, one governing equation is sufficient for a fractured, porous matrix system. In the composite medium-effective continuum governing equation, the equivalent saturation is expressed as a sum

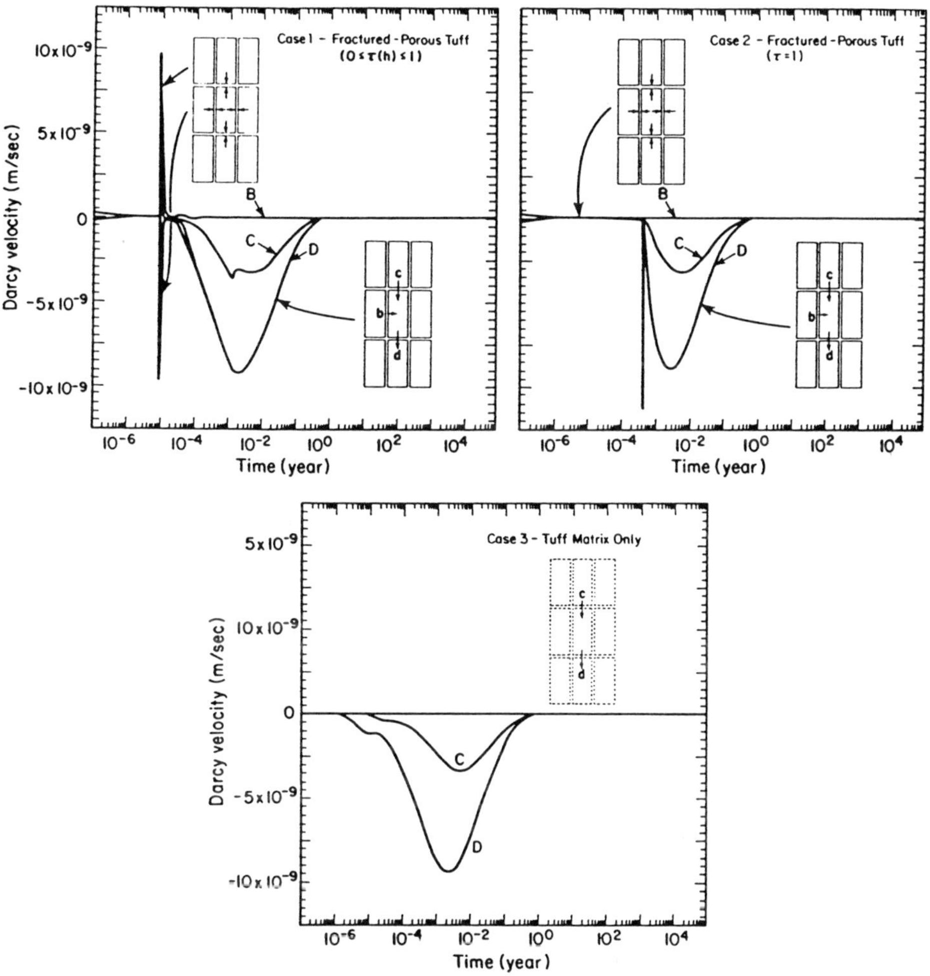

Figure 7.3.3. Unsaturated drainage in a tuff column: patterns of Darcy velocities (sign convention: horizontal fracture — flow into fracture is positive; vertical fracture — upward is positive).

of fracture saturation S_f and matrix saturation S_m (Eq. 7.3.3), weighted by continuum porosities. The equivalent permeability is also expressed similarly by a sum of fracture permeability $k_f k_{rf}$ and matrix permeability $k_m k_{rm}$ (Eq. 7.3.4).

$$S_{fm}(h) = \frac{S_f(h)\phi_f V_f + S_m(h)\phi_m V_m}{\phi_f V_f + \phi_m V_m} = \frac{S_f(h)\overline{\phi}_f + S_m(h)\overline{\phi}_m}{\overline{\phi}_f + \overline{\phi}_m} \quad . \tag{7.3.3}$$

$$k_{fm}(h) = k_f k_{rf}(h) + k_m k_{rm}(h) \quad . \tag{7.3.4}$$

The subscripts f and m refer to the fracture and matrix domains, respectively. ϕ is the porosity (void fraction) of intact domain and $\overline{\phi}$ is the equivalent continuum porosity of the voids in one domain as a fraction of total volume $V = V_m + V_f$.

There is no conceptual difficulty in accepting the assumption that the equivalent saturation is a porosity-weighted sum of fracture saturation and matrix saturation. Saturation is a scalar quantity that is defined as the ratio of saturated volume to pore volume. The permeability, on the other hand, is not scalar. By summing over fracture permeability and matrix permeability, the fracture flow channel and the matrix flow channel are intrinsically assumed to be parallel in the composite medium-effective continuum models. In view of their popularity and usage in benchmark studies, we have programmed these composite saturation and permeability characteristic functions into the TRUST code. We also used these models later in this study for some of the large-scale simulations. However, there is a need to check the results of simpler models from time to time with the results generated from detailed discrete fracture-porous-matrix models. More important, the assumptions and limits of simpler models must be kept in mind when we use and interpret the results from these models. For example, the composite model is adequate for vertical infiltration studies if we only consider vertical fracture sets so that both fracture flow and matrix flow are vertical. For multidimensional simulations with potential lateral flows crossing the fractures, it remains to be determined if composite models are adequate. The development of large-scale models to account for both fractures and porous matrix blocks remains an important task for the studies of a partially saturated flow field of a fractured, porous system.

7.4 VERTICAL INFILTRATIONS THROUGH WELDED-NONWELDED UNITS: MAINTENANCE OF UNSATURATED STATE

In 1985, we extended our models from a short column with a few fractures to a tall column over 500 m in depth to represent the whole unsaturated zone from ground surface to the water table. Yucca Mountain consists of alternating units of welded and nonwelded tuff. The infiltration of water through these alternating layers was studied under steady and transient conditions. We wanted to determine if spatial heterogeneities associated with welded-nonwelded interfaces and temporal variations associated with extreme flooding on the ground surface can induce significant fracture flow in the Topopah Spring welded tuff unit. For performance assessment, the flow velocity field from the repository in the Topopah Spring welded tuff unit to the water table determines the groundwater travel time. The onset of significant fracture flow would shorten the travel time drastically.

7.4.1 Fracture and Matrix Data and Model: Observational Basis

To construct models representing the welded and nonwelded units, we analyzed the fracture orientation and spacing data from borehole USW-G4 (Spengler and Chornack, 1984) for all the unsaturated units. For the welded units, Tiva Canyon (TCw) and Topopah Spring (TSw), fracture densities are high and fracture spacings are in the sub-meter range. For the nonwelded units, Paintbrush (PTn), vitric Calico Hill (CHnv), and zeolitic Calico Hill (CHnz), fracture densities are low and fracture spacings are in the meter and tens-of-meters ranges. It is generally assumed that flow of water in the nonwelded units is dominated by matrix flow, and fractures can be ignored. Within fracture planes, we assumed that the contact areas were identical to the areas where mineral coatings occurred in the fractured cores. The fraction of fracture coatings, which are predominately clay, calcite, and zeolite, increases with depth. This is consistent with fractures closing with depth as a result of overburden stress. The equivalent fracture continuum permeabilities for the welded and nonwelded units were based on the estimates of Thordarson (1983) and Sinnock et al. (1987). For the tuff matrix, we used the characteristic curves of the representative samples chosen by Peters et al. (1984).

Our simulations used either a discrete fracture-porous medium model oı a composite medium model. Before we discuss the detailed models, we will briefly review the results from one-dimensional composite medium simulations. One-dimensional infiltration problems have been extensively studied, especially the steady-state cases. In a one-dimensional column, we can represent the different units with nonuniform grids, as illustrated in Fig. 7.4.1. The thicknesses of the units correspond to the values determined from borehole USW-G4 (Ortiz et al., 1984).

7.4.2 Estimate of Ambient Infiltration Rate: Observational Constraints

Although the infiltration rate is a given input parameter and controls the simulation results, it has not been determined by direct measurements, but rather determined inversely by models, ambient saturation, and other characteristic parameters. Most importantly, the effective matrix conductivity under partially saturated conditions determines the infiltration rate. The saturated conductivity for the TSw matrix is about 1 mm/yr for the representative sample used in our 1985 analyses. Our calculations showed that for the 0.1-mm/yr case (Fig. 7.4.2) and for the 0.5-mm/yr case (Fig. 7.4.3) the saturation profiles were uniform and the corresponding fluid pressure was constant in the upper part of the TSw unit.

In the range of the constant pressure profile, the pressure gradient is, of course, zero, and the hydraulic potential gradient is unity from the gravity component alone. In the range of the unit hydraulic gradient, the partially saturated hydraulic conductivity is equal to the Darcy velocity, which is equal to the steady infiltration rate. The solution in the thick TSw unit is essentially determined by demanding that the saturation attains a uniform value at which the infiltration rate exactly matches the partially saturated, effective hydraulic conductivity.

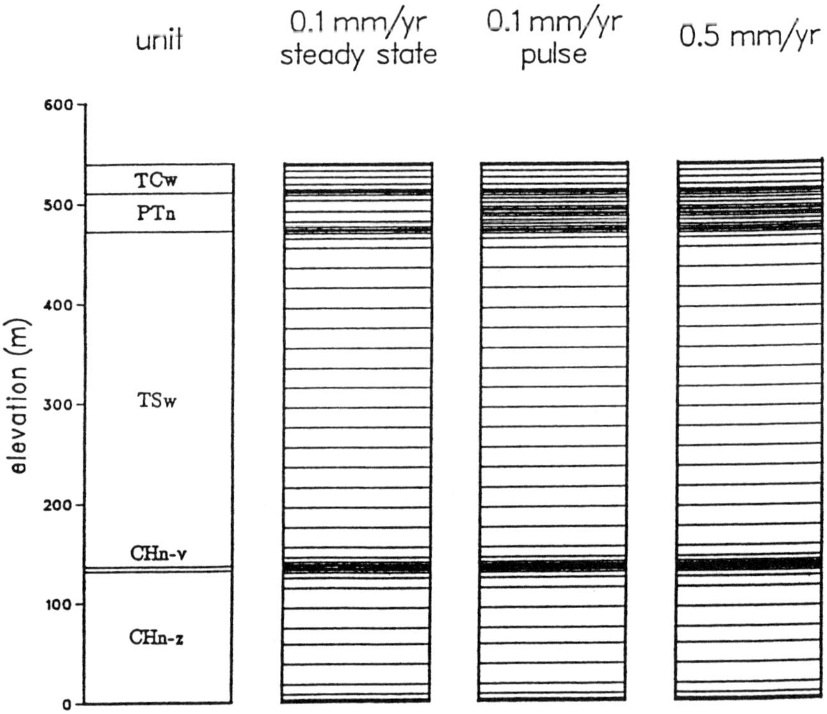

Figure 7.4.1. Nonuniform grids used for one-dimensional simulations.

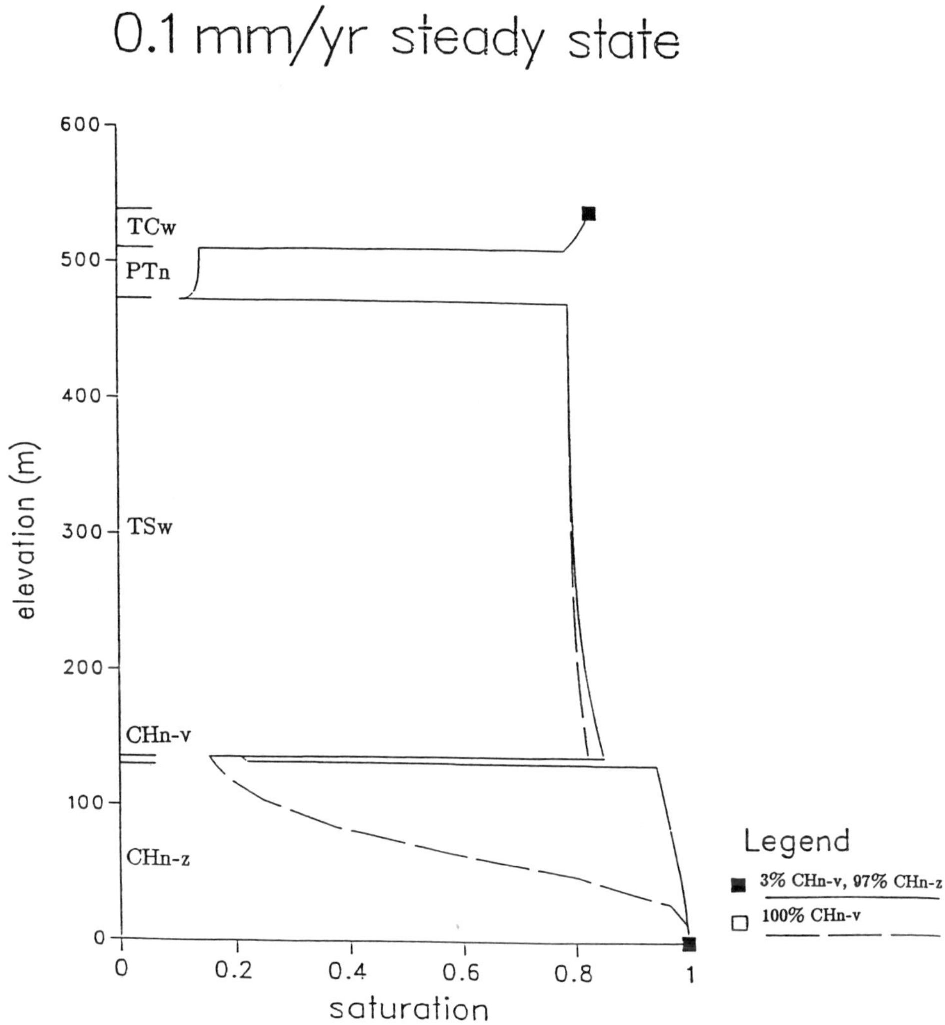

Figure 7.4.2. Saturation profiles for 0.1-mm/yr steady-state infiltration.

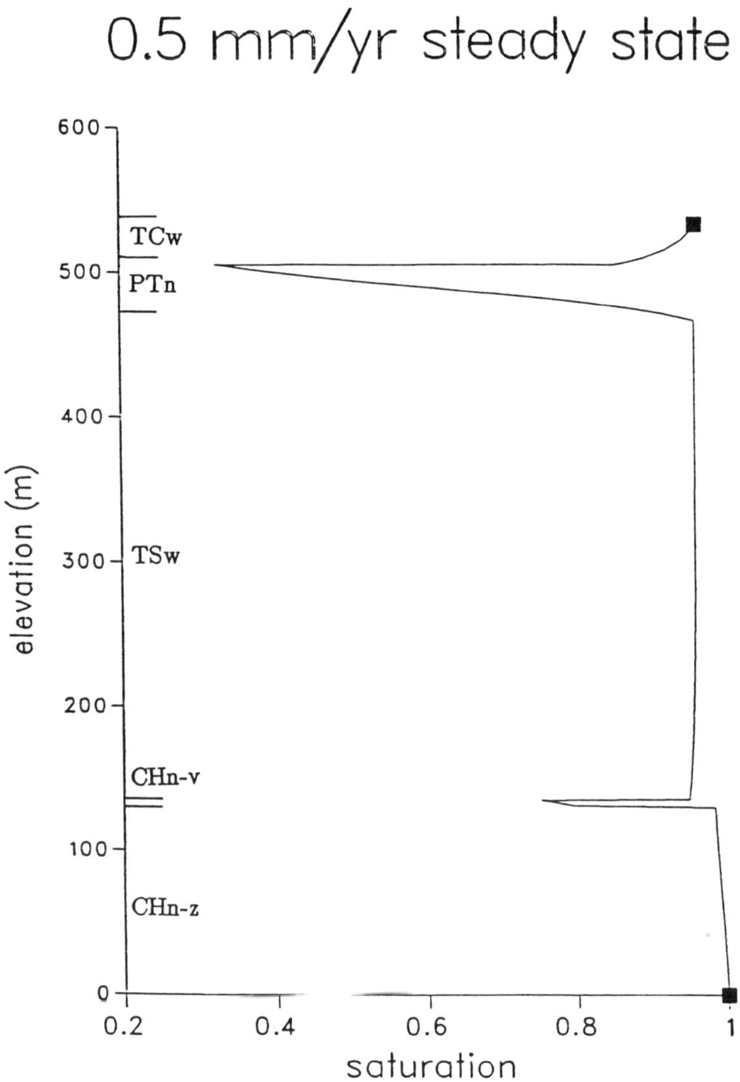

Figure 7.4.3. Saturation profile for 0.5-mm/yr steady-state infiltration.

The ambient saturation for TSw was initially estimated by SNL to be 0.8 in 1983, a figure later reduced to 0.69 ± 0.15 (Sinnock et al., 1987; Reference Information Base (RIB), SNL, 1986). With an ambient saturation of 0.8 or lower, the modeling results indicate that one can only allow the system to have 0.1 mm/yr or less water flowing through the system. If one increases the infiltration rate over 0.1 mm/yr, the TSw saturation will be higher than the ambient saturation. The ambient saturation can be measured easily in the field by weighing analyses of core samples. Net infiltration rates, on the other hand, cannot be easily and directly measured. We are not aware of any practical experimental technique to directly measure low infiltration rate under partially saturated conditions. To reproduce the ambient condition for the saturation, we inversely determine the infiltration rate of 0.1 mm/yr or less as a reasonable estimate.

Because infiltration simulation is usually cast as a boundary value problem with infiltration rate as an input parameter, reviewers tend to ask what the effects of infiltration rate changes are. We can indeed create alarmingly fast fracture flows with infiltration of over 1 mm/yr which seems to be still a very small infiltration rate. However, the onset of fracture flow is also associated with nearly full saturation in the tuff matrix. At Yucca Mountain, with a water table over 500 m deep, tuff units are in partially saturated states and are not on the verge of becoming fully saturated. If we accept the findings that the tuff units are partially saturated, with an ambient matrix saturation of less than one, and that the matrix characteristic curves can be measured in the laboratories with reasonable accuracy, then we can set a limit on the infiltration rate. This is an important but subtle issue which is frequently raised in analyzing the results.

7.4.3 Welded-Nonwelded Flow Interaction: Effect of a Thick Unit on Thin Units

The 1985 results indicate very low saturation values for the nonwelded units (Fig. 7.4.2). The thick TSw unit at Yucca Mountain is bounded above by a PTn unit and below by a CHn unit. The CHn unit is further divided into an upper vitric subunit and a lower zeolitic subunit. The relative thicknesses of CHnv and CHnz subunits vary throughout Yucca Mountain, and the CHnv is very thin at well USW-G4 (4.6 m or 3% of the total CHn thickness, Ortiz et al., 1985). The calculated saturation values for PTn and for CHnv do not agree with the ambient saturation of 0.56 ± 0.17 for PTn (Sinnock et al., 1987), 0.9 for CHnv, and 0.91 ± 0.06 for CHnz (SNL, 1986). These discrepancies were reevaluated in 1986 by sensitivity studies for the nonwelded matrix parameters (see Sec. 7.5.4). The PTn and the CHnv units exhibit low saturations because of one particular feature in the matrix characteristic curves of the representative samples chosen for these two units (Peters et al., 1984): for the representative PTn and CHnv matrix curves, most of the saturation changes occur over a relatively narrow pressure range around –100 m. This indicates that the pores in these nonwelded PTn and CHnv representative samples are nearly uniform in size. For the 0.1-mm/yr case the thick TSw unit stabilizes the pressure profile to –150 m. Since –150 m is more negative than –100 m, most of the pores in these two nonwelded units are drained and the calculated saturations are nearly equal to their residual saturation values.

The thick TSw unit effectively decouples the upper units from the lower units. The saturation, pressure and potential profiles in the upper units are insensitive to profiles in the lower units with or without CHnz below CHnv. The thick TSw unit damps out boundary effects from both the upper and the lower boundaries. This suggests that we may use an approximation to decouple the studies of the upper units from those of the lower parts. We used this approximation in some of the pulse infiltration studies.

7.4.4 Approximate Model for Steady-State Profiles: Elementary Path Integration

For the transient simulations, the initial conditions used are appropriate steady-state conditions. Computationally, the steady-state conditions are obtained by letting the transient process evolve until changes in conditions are negligibly small. If we can estimate the pressure distribution well, we can easily obtain the steady-state. Before we discuss the transient pulse simulations, we will briefly describe a simple and efficient procedure we use to estimate one-dimensional steady-state pressure distribution without using sophisticated numerical codes. For a vertical one-dimensional column with steady-state infiltration flux q everywhere constant, one can explicitly integrate Darcy's equation,

$$q = -K(h)\left(\frac{dh}{dz} + 1\right) , \tag{7.4.1}$$

from a boundary with given elevation z and pressure head h. At the lower water table boundary $z = z_0 = 0$, we know the pressure head $h = h_0 = 0$. With pressure known at that elevation, we can calculate the corresponding conductivity value $K(h_0)$. The pressure at the next node Δz above the boundary can be estimated from

$$\Delta h = -\left(1 + \frac{q}{K(h_0)}\right)\Delta z . \tag{7.4.2}$$

We can calculate the conductivity $K(h_1)$ at $h_1 = h_0 + \Delta h$, and then calculate the pressure head at the next higher node. If Δz and Δh are small enough so that the conductivity value does not change significantly between two neighboring nodes, we can explicitly calculate the pressure profile node-by-node from the bottom boundary upward. We have incorporated this simple path-integration procedure in a mesh generator program as a preprocessor for the TRUST code. We used this procedure in the one-dimensional column simulations and, in 1986, in two-dimensional multicolumn simulations. Simple procedures such as the one outlined above and the code LLUVIA, developed at SNL, can be useful and practical tools for sensitivity and statistical analyses.

7.4.5 Response to Extreme Infiltration Pulses: Damping by Near-Surface Units

The purpose of the transient simulations is to study the responses of the fractured, porous units to extreme infiltration pulses of high rates and short durations. We first focus on the upper layers in Yucca Mountain. We investigate possible scenarios with short-transient infiltration pulses having periodic rates much higher than the saturated matrix conductivity. Figure 7.4.4 is a discrete-fracture-porous-matrix model with the PTn unit sandwiched between two 10-m layers of fractured welded tuff. The discrete model for the welded units contains both vertical fracture grids and horizontal connections from the vertical fractures into the matrix. The PTn unit is treated as a composite medium.

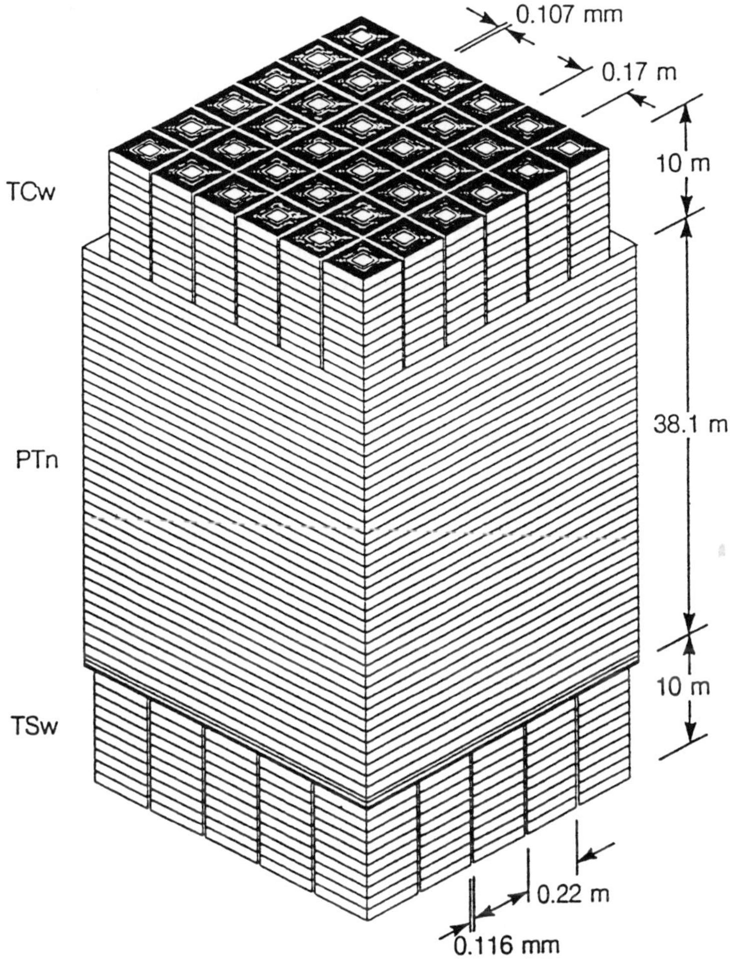

Figure 7.4.4. Three-dimensional mesh used for studying column response to extreme infiltration pulses.

For simulations involving infiltration pulses, the initial conditions correspond to the steady-state solution, with a constant infiltration source of 0.1 mm/yr on the top TCw node and a sink of 0.1 mm/yr on the bottom TSw node. The infiltration pulses are applied to the fracture continuum at the top. The thickness of the TCw unit is greater than 10 m and varies with the topography at Yucca Mountain. At well USW-G4, TCw is approximately 30 m thick. By modeling only the lower 10-m section and applying infiltration pulses to the fractures, we assume that the fractures near the surface in the upper part of TCw have apertures larger than the capillary limit and that water can penetrate 20 m of TCw without retention and delay by the tuff matrix. By bypassing the upper portion of TCw, we also neglect the complex near-surface evaporation-transpiration and runoff processes. These processes determine the percentage of precipitation (1) returning upward to the atmosphere, (2) moving laterally away from the mountain, and (3) infiltrating downward into the ground. We are primarily interested in the effects of net infiltration into the TSw and lower units, and the travel time from repository to water table. We assume that the water reaching the top of the lower 10-m section of TCw is the net infiltration.

To systematically study this pulse infiltration problem, we first fix the duration of the pulse to 0.2 yr, corresponding to an intense rainfall or snowmelt period. The pulses are applied periodically. The periods are 1 yr, 5 yr, 50 yr, 500 yr, and 5,000 yr. For the 1-yr pulse, the intensity of the infiltration is 0.5 mm/yr, which corresponds to 1 yr × 0.1 mm/yr, divided by 0.2 yr. For the 5,000-yr pulse, the intensity of infiltration is 2,500 mm/yr, which corresponds to 5,000 yr × 0.1 mm/yr, divided by 0.2 yr. For the 5,000-yr pulse, all the infiltration in 5,000 yr is applied to the system within 0.2 yr. After 0.2 yr, the infiltration rate is zero until the start of the next infiltration pulse, 4999.8 yr later. The average infiltration rate over any 5,000-yr period is the same as the constant infiltration rate of the initial steady-state condition.

For the mild and frequent pulses (1 yr, 5 yr, and 50 yr), the matrix was found to always remain partially saturated and the fractures essentially dry. The effects of the pulses are completely damped out in TCw. These results are consistent with the findings of Travis et al. (1984) that a slug of water moving through fractures with apertures as small as 0.1 mm will be unable to penetrate more than a few meters, at most, through fractured welded tuff with matrix saturations as high as 0.9. Repeated pulsing does not change the capacity of the partially saturated matrix of TCw to absorb all the pulsed water by capillarity. Even for a 50-yr pulse with more than 10 cycles of repeat pulsing, the transient changes are limited within TCw, and the matrix remains partially saturated.

For the 500-yr pulse, the intensity of the infiltration pulse (250 mm/yr for 0.2 yr) is high enough to saturate the top 2 m of TCw. Figure 7.4.5 illustrates the horizontal saturation profiles within the topmost matrix block of TCw. The pulse is applied to vertical fractures surrounding the four sides of the matrix block. At time t = 0, the first (topmost) discrete fracture continuum node starts to receive all the water at a rate corresponding to 250 mm/yr infiltration per unit surface area. Although the water enters the fracture node first, it does not accumulate in the fractures, which remain partially saturated. The water is quickly drawn into the matrix by capillarity and distributed nearly evenly throughout the matrix block, which is simulated with 10 nested nodes. The matrix saturations near the matrix-fracture interfaces, 0.085 m (half the vertical fracture spacing of TCw) away from the matrix block center, are only slightly higher than the matrix saturations in the center of the matrix block.

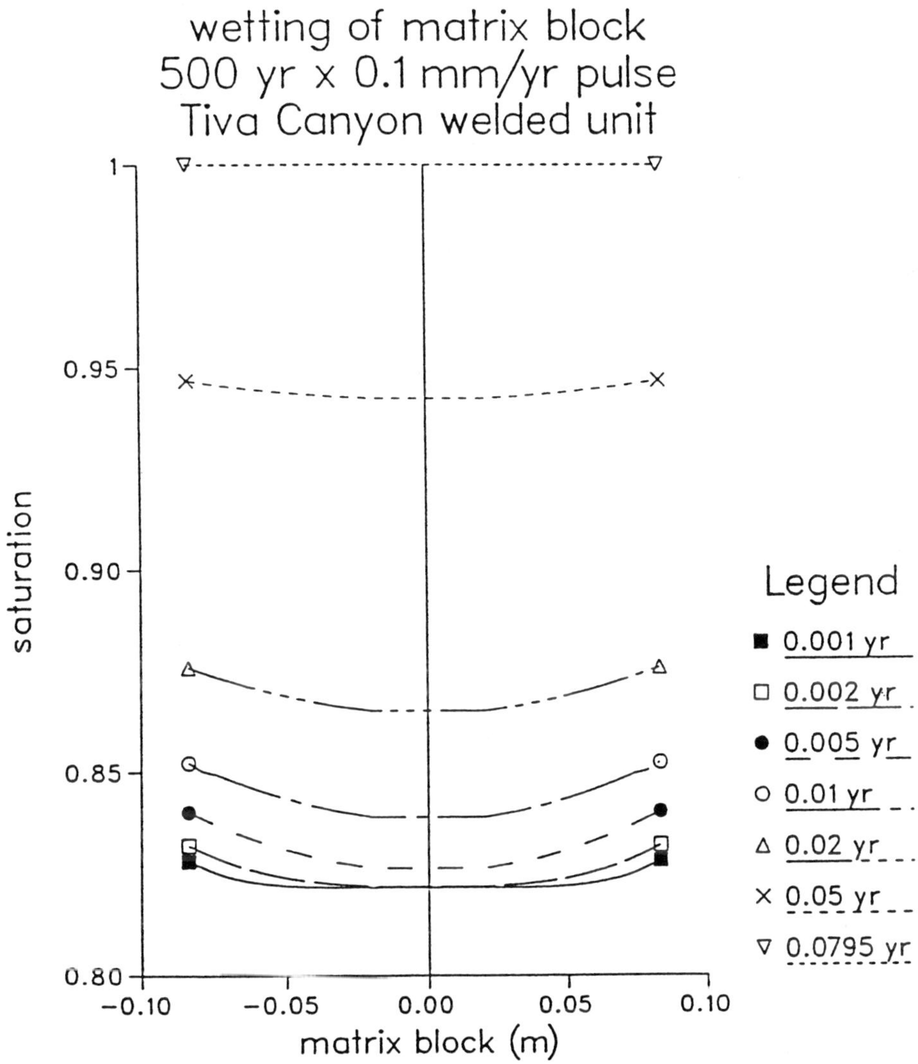

Figure 7.4.5. Horizontal saturation profiles within matrix block of Tiva Canyon welded tuff unit.

At $t = 0.08$ yr, the matrix block becomes fully saturated ($S = 1$). The total amount of water infiltrated from $t = 0$ to $t = 0.08$ yr is approximately equal to the initial available pore spaces, $\phi(1 - S)$, in the topmost matrix block. For TCw with porosity of $\phi = 0.1$ and an initial saturation of $S = 0.8$, the available pore space in the 1-m thick block will be filled in $0.1 \times (1 - 0.8) \times 1$ m divided by 250 mm/yr, or 0.08 yr. As the matrix approaches full saturation and the negative pressure head approaches zero, the saturation in the fractures increases and water moves down through fractures to the next fracture node. The process of horizontal wetting of the second matrix block follows essentially the same fast-suction, even-wetting behavior of the first matrix block shown in Fig. 7.4.5. Because we can calculate the times required for each block to become fully saturated by the available pore volume and the intensity of pulse without detailed calculations of the horizontal wetting and vertical sharp wetting front movements, the pulse calculations can be greatly simplified as described by Wang and Narasimhan (1986).

The results in Fig. 7.4.6 for the first cycle for a 500-yr pulse showed that changes in saturation occur mainly in the 10-m section in TCw. Full saturation occurs only in the upper part of TCw during the infiltration period. The pulse effects penetrate slightly into the upper part of PTn but do not reach TSw. Effects of a second cycle are similar to those of the first cycle. For a 5,000-yr pulse, the 10-m section of TCw is completely saturated during the infiltration period. After infiltration of 2,500 mm/yr over 0.2 yr, the infiltrating water moves through PTn and reaches TSw in about 1,000 yr. Even for this large pulse, PTn and TSw remain partially saturated. Because the 5,000-yr pulse effects do reach TSw, one may question if the lower boundary used in the simulations was perhaps too close to the PTn-TSw interface. We did another set of simulations for the 5,000-yr pulse cases using an extended grid.

In the extended grid (see Fig. 7.4.1), we use the composite-medium model to represent all welded and nonwelded units from the surface to the water table. For the 5,000-yr × 0.1-mm/yr pulse case, an infiltration rate of 2,500 mm/yr is applied for 0.2 yr to TCw below the fully saturated interval, which has all its available pore space occupied by infiltrated water. The total amount of water applied to the system in 0.2 yr corresponds to all the infiltration in 5,000 yr. The results of transient profiles for the first 5,000 yr showed that the zone of full saturation does not penetrate the PTn. Similar results are obtained for additional cycles of pulse infiltration. The damping of the pulse and the penetration of the transient effects for this composite-medium, all-unit simulation agrees with the earlier discrete-fracture, upper-unit results with finer meshes. The effects of the pulsing reach the top of TSw in about 1,000 yr and penetrate only the upper part of TSw.

Most of the damping of the pulse occurs in PTn. A large 5,000-yr pulse can saturate the matrix, initiate fracture flow, and generate a complex flow field in TCw. When the transient effects move into PTn, its large pore volume and highly nonlinear unsaturated characteristics can effectively accommodate and absorb the highly transient and intense water pulses. Only mild perturbations reach TSw. Figure 7.4.7 illustrates the propagation and damping of the pulse in PTn. The large mass flux passing through the top TCw-PTn interface at 100 yr is damped out as it moves down PTn. The mass flux crossing the lower PTn-TSw interface into TSw is much smaller in magnitude and later in time.

In this extended-grid, composite-medium column simulation, we also use a fixed-gradient approximation to estimate the potential lateral flow, which may divert a fraction of infiltration from the vertical direction along the column. The results in Fig. 7.4.7 include the effects of

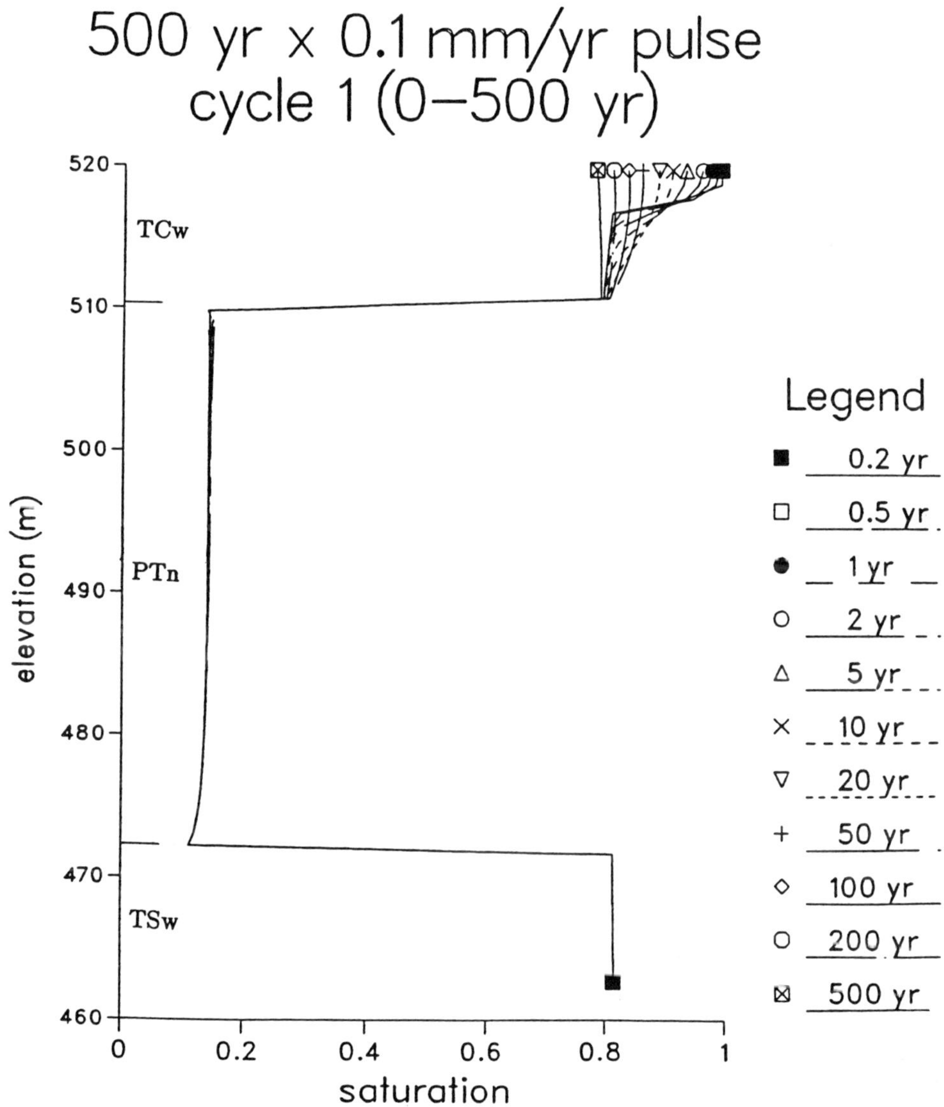

Figure 7.4.6. Saturation profiles for the first cycle of 500-yr × 0.1-mm/yr transient infiltration.

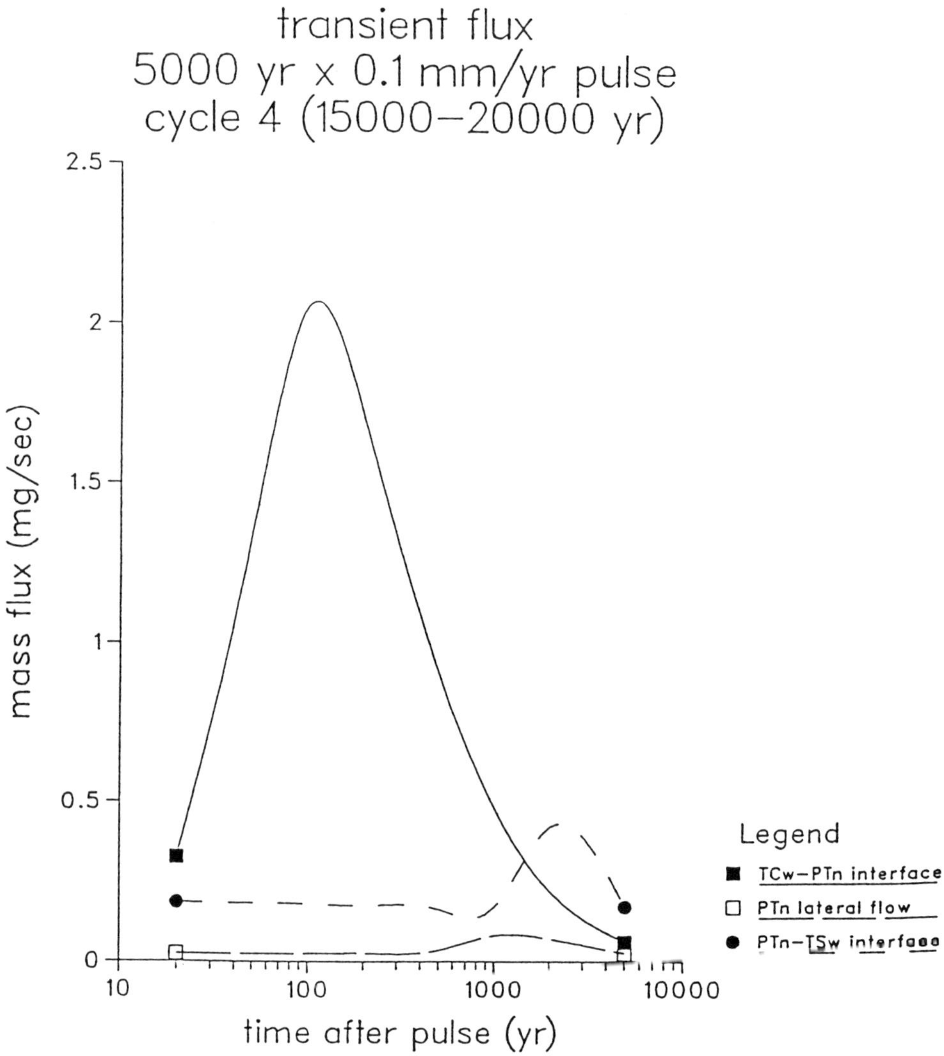

Figure 7.4.7. Response to 5,000-yr infiltration pulse: damping of fluxes within the Paintbrush nonwelded tuff unit.

lateral leakage from anisotropic PTn. The usefulness and limitation of the fixed gradient approximation are discussed by Wang and Narasimhan (1986). Because we went beyond the single vertical column to more accurately estimate the effect of lateral flow in 1986, we will omit further discussion of the fixed-gradient approximation.

7.4.6 Ambient Infiltration: An Estimate

Our 1985 results support a conclusion that water flow occurs mainly through the tuff matrix if the average net infiltration rate is less than the saturated conductivity of the matrix. Spatial heterogeneities associated with welded-nonwelded interfaces and temporal variations associated with large infiltration pulses do not change the flow field at TSw significantly. Based on laboratory measurements, the saturated conductivity of the welded tuff matrix is very small. With small saturated conductivity in the range of 1 mm/yr, the net infiltration rate is estimated to be 0.1 mm/yr or less, so that the simulated saturations agree with the ambient values of 0.6 to 0.8 for the welded units. If the onset of fracture flows could be demonstrated to be unlikely under ambient conditions and under extreme flooding events, and if the dominant flow direction is vertical, performance assessment could be easily and convincingly carried out for the partially saturated, fractured tuff at Yucca Mountain.

7.5 DISTRIBUTION OF FLUID NEAR A FAULT ZONE: POSSIBILITY OF LATERAL FLOW

In 1986, we extended our models from a single vertical column to a two-dimensional east-west cross section bounded by a vertical fault. Yucca Mountain is approximately 2,000 m wide. A minor fault, the Ghost Dance Fault (GDF), inside the potential repository area, will be a major hydrological concern if the fault flow is an important transport mechanism. Without experimental data on the unsaturated characteristics of the fault zone material, we focused our attention in 1986 on the distribution and movement of water in the unsaturated formation near GDF and assessed the potential for water to overcome capillary forces and exit laterally through the formation-fault interface. If water can be diverted away from dispersed vertical flow through the unsaturated formation to concentrated flow through the localized fault zone, the fault zone would probably constitute the fastest flow path from the repository to the water table. On the other hand, the absence of significant fault flow and the dominance of matrix flow will result in long travel time through the unsaturated tuff units.

7.5.1 Unsaturated Fault: Different Hydrological Possibilities

The distribution of fluid flow in the unsaturated zone is governed by the balance between the capillary forces and gravity. The external gravitational force moves the water downward, and the internal capillary forces hold the water in place within the pores. The magnitude of capillary force is inversely proportional to pore size, and the heterogeneity of rock pore structure controls distribution and movement of water through partially saturated formations. Within the meter scale, the heterogeneity consists of tuff matrix blocks separated by discrete fractures. The large fracture openings cannot maintain the strong capillary forces required to hold water within discrete fractures under ambient partially saturated conditions. In the much larger scale, covering the mountain width, the faults are the heterogeneous feature.

In modeling the hydrology of Yucca Mountain, we must assess the impacts of faults that traverse the tuff formation. If they are barriers to flows, faults will partition the system into separate hydrologic blocks, and we can focus on the study of flows through the interior of blocks. If faults are the main conduits for fluid, then we should focus on the study of fault flows. In saturated systems, there is a clear distinction between these two possibilities. A closed fault is treated as a no-flow boundary, and an open fault as a constant potential boundary if it is connected to a constant-head reservoir. In the latter case, faults may help transport water from a far-removed source or may help remove water to a deeper sink. In partially saturated systems, the distinction between an open and a closed fault is not as clear, and the following three possibilities present themselves:

- If a fault zone is made up of broken-up mass because of brecciation, it might have very coarse openings that drain easily. If the openings are dry, there is no fluid to be transported. On the other hand, a dry opening is not a barrier to fluid flow. If water enters the opening, it will have high permeability and fluid will encounter low resistance to sliding down the walls of large openings (noncapillary flow). Under partially saturated conditions, an "open" fault becomes a passive boundary that can transport water if water enters it but cannot keep water, because large openings have a very weak capillary force to hold water.

- If a fault is sealed with gouge material and fine-grained materials, it can retain water and may even attract water with the strong capillary force associated with small openings. On the other hand, small openings have low permeability and cannot be good conduits for transporting fluid. Under partially saturated conditions, a "closed" fault becomes a relatively wet zone that passively retains water in nearly static conditions.

- There is a third, remote, possibility that the fault may have a unique combination of saturated and unsaturated characteristics. If the fault material can maintain both strong capillary attractive force and large permeability, then the fault can pull water away from the formation and transport it quickly away from the system. We cannot rule out the existence of this combination, by which fault flow would control the partially saturated hydrology.

7.5.2 How to Model the Fault: The Seepage Face Idealization

There are no data for faults that clearly establish whether faults in tuff formations have open, closed, or exotic characteristics. In the 1986 study, we did not use material properties to represent the fault. Rather, we treat the fault-formation interface as a seepage boundary. It is well known from hillside studies that water will seep out of the formations if the potential at the open face is lower than the potential inside the hill. We assume that the fault is so open that capillary force is zero and the face of the fault behaves like a hillside open surface exposed to atmospheric pressure. We focus on the interior of the formation to determine whether water can overcome capillary forces holding water inside the tuff formation. If water does not exit through the fault-formation boundary, the seepage boundary is essentially a closed boundary. If water remains inside the formations, distribution of fluid flow will be independent of whether the fault is open or closed. The only possibility we do not model is the third possibility, in which the fault acts as a strong capillary suction sink as well as a high-permeability conduit under partially saturated conditions. Within the tuff formation, the lateral component of the gravity gradient caused by the tilting of the units has been proposed as a mechanism to divert water laterally (Montazer and Wilson, 1984; Rulon et al., 1986). The stratigraphic units at Yucca Mountain generally tilt eastward at 5° to 7° (Scott and Bonk, 1984). For an average 6° tilting, the tangent of 6° is 0.1 and the external driving force along a lateral direction is 10% of gravity. The partially saturated flow movement is determined by the balance of external force with internal capillary forces. We expect that a weak external force associated with the lateral component of gravity will demand a mild readjustment in the lateral variation in saturation, capillary pressure head, and potential so that the capillary force balances out the external force.

7.5.3 Stratigraphy, Fracture, Matrix Data: An Update

Before we present our results from the two-dimensional cross section models, we briefly discuss several preliminary steps and the systematic approaches we adapt in this study. In 1986, a Reference Information Base (RIB) was developed at SNL. Y.T. Lin of SNL provided us with digitized data for section A-A' from RIB. Section A-A', illustrated in Fig. 7.5.1, is one of the sections used by Sinnock et al. (1987) in a multicolumn unit-gradient vertical-flow stochastic simulation to study the groundwater travel time distributions. To focus on the impact of the GDF on the two-dimensional flow behavior in section A-A', we first repeat the single column simulations with thicknesses of the units corresponding to digitized values at GDF instead of borehole USW-G4 as in our earlier studies. We also update the matrix (Table 7.5.1) and fracture properties (Table 7.5.2) according to RIB and other new SNL reports. For examples, the average fracture orientation and frequency data from five boreholes (Bauer, 1986) was used instead of the early analyses with USW-G4 data alone. The main difference between the updated results and early analyses is that there are relatively fewer horizontal fractures than vertical fractures, especially for the welded units. For the matrix, the RIB saturated conductivity values for TSw and CHn matrix are lower than corresponding values in earlier references. With lower saturated conductivity, the calculated saturation in TSw is approximately 5% higher than the 1985 result for the same infiltration of 0.1 mm/yr.

In the RIB (SNL, 1986), the characteristic parameters are not tabulated. We still used the data of Peters et al. (1984) for these parameters. Until 1986, we had been focusing attention on understanding fracture characteristics and relied on SNL to provide matrix characteristics. In 1986, we also started to analyze available matrix data. The comprehensive report by Peters et al. (1984) contains much data and analyses of laboratory measurements, both of small intact and fractured tuff core samples. In our 1986 simulations, the matrix characteristic curve parameters for TSw1, TSw2, CHnz, and PPw units correspond to the representative samples chosen by Peters et al. (1984). For PTn and CHnv, we used two other samples with wider pore size distributions which were also from Peters et al. (1984). These two nonwelded units may have multimodal pore structure as shown in later detailed PTn study by Peters et al. (1987). Recent study by P.G. Kaplan of SNL indicates that the CHnv unit is very heterogeneous, with many subunits of different characteristics.

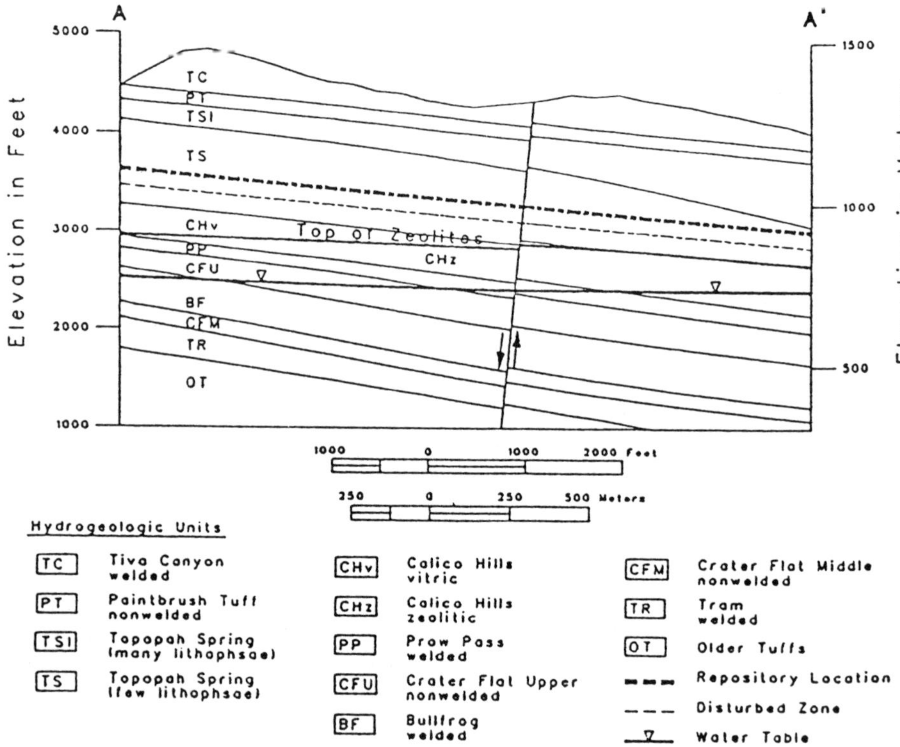

Figure 7.5.1. Idealized cross section of Yucca Mountain.

Table 7.5.1. Matrix properties.

Unit	Saturated Permeability k_s (Conductivity[1]) m^2 (mm/yr)	Porosity[2] ϕ	van Genuchten Parameters $\alpha^{(3)}$ (1/m)	$n^{(3)}$	$S_r^{(4)}$	Compressibility[5] (1/m)
TCw	2.55×10^{-18} (0.789)	0.114	8.21×10^{-3}	1.558	0.002	6.2×10^{-7}
PTn	2.45×10^{-15} (757.385)	0.448	3.05×10^{-2}	1.22	0.040	8.2×10^{-6}
TSw1	2.33×10^{-18} (0.722)	0.143	5.67×10^{-3}	1.798	0.091	1.2×10^{-6}
TSw2	2.33×10^{-18} (0.722)	0.1168	5.67×10^{-3}	1.798	0.091	5.8×10^{-7}
CHnv	3.46×10^{-16} (107.168)	0.3541	4.40×10^{-2}	1.496	0.0852	3.9×10^{-6}
CHnz	1.73×10^{-18} (0.535)	0.3064	3.08×10^{-3}	1.602	0.1211	2.6×10^{-6}
PPw	2.84×10^{-16} (87.742)	0.2557	1.41×10^{-2}	2.639	0.0686	1.5×10^{-6}
CFUn	6.99×10^{-17} (21.637)	0.3239	3.82×10^{-3}	1.9455	0.2282	2.4×10^{-6}
BFw	3.83×10^{-16} (118.439)	0.2391	2.025×10^{-2}	3.2025	0.0584	1.7×10^{-6}

(1) TCw, PTn values from Sinnock et al. (1987), other values from RIB (SNL, 1986) Sec. 1.1.4.2

(2) TCw, PTn, TSw1 values from RIB 1.3.1.2.1-4, other values calculated from data in RIB 1.1.8.1 and 1.1.3.1-2. TCw, PTn, TSw1 values from RIB 1.3.1.2.1-4, other values calculated from data in RIB 1.1.8.1 and 1.1.3.1-2.

(3) From Peters et al. (1984) with sample code: TCw:G4-1, PTn:G4-2, TSw1:G4-6, TSw2:G4-6, CHnv:GU3-15, CHnz:G4-11, PPw:G4-18, CFUn:G4-19/20, BFw:G4-21/22.

(4) TCw, PTn values from Peters et al. (1984), other values from RIB 1.1.3.1-2.

(5) Nimick et al. (1984).

Table 7.5.2. Fracture properties.

Unit	Bulk Saturated Permeability (Conductivity[1]) m^2 (mm/yr)	Fracture[2] Spacing d (m) Vertical	Horizontal	Effective Aperture $\bar{b}$ (mm) Vertical	Horizontal	Discrete Fracture Permeability k_s (m^2) Vertical	Horizontal
TCw	1.18×10^{-12} (3.65×10^5)	0.18	0.42	0.109	0.144	9.87×10^{10}	1.72×10^{-9}
PTn	2.42×10^{-13} (7.5×10^4)	1.02	1.12	0.114	0.118	1.09×10^{-9}	1.15×10^{-9}
TSw1	1.18×10^{-12} (3.65×10^5)	0.21	1.26	0.114	0.208	1.09×10^{-9}	3.59×10^{-9}
TSw2	1.18×10^{-12} (3.65×10^5)	0.21	1.26	0.114	0.208	1.09×10^{-9}	3.59×10^{-9}
CHnv	2.42×10^{-13} (7.5×10^4)	2.12	4.49	0.146	0.187	1.77×10^{-9}	2.91×10^{-9}
CHnz	2.42×10^{-13} (7.5×10^4)	2.12	4.49	0.146	0.187	1.77×10^{-9}	2.91×10^{-9}
PPw	1.18×10^{-12} (3.65×10^5)	1.49	2.73	0.219	0.268	4.01×10^{-9}	6.00×10^{-9}
CFUn	2.42×10^{-13} (7.5×10^5)	5.07	4.49	0.195	0.187	3.16×10^{-9}	2.91×10^{-9}
BFw	1.18×10^{-12} (3.65×10^5)	0.393	3.39	0.141	0.289	1.65×10^{-9}	6.94×10^{-9}

Unit	Contact[4] Cutoff Aperture b_c (mm) Vertical	Horizontal	Aperture[4] Distribution Parameter β (mm^{-1}) Vertical	Horizontal	Fracture[5] Surface Contact Area ω	Discrete[6] Fracture Compressibility (1/m)
TCw	0.010	0.014	24.8	18.8	2.8%	1.32×10^{-6}
PTn	0.027	0.028	21.7	21.1	11.8%	1.90×10^{-7}
TSw1	0.027	0.050	21.7	11.9	12%	5.60×10^{-8}
TSw2	0.027	0.050	21.7	11.9	12%	1.20×10^{-7}
CHnv	0.041	0.053	16.6	13.0	15% (no zeolite)	2.80×10^{-8}
CHnz	0.132	0.169	12.7	9.91	50% (w. zeolite)	2.80×10^{-8}
PPw	0.130	0.159	9.53	7.79	35%	1.20×10^{-7}
CFUn	0.115	0.110	10.7	11.2	35%	2.80×10^{-8}
BFw	0.083	0.170	14.9	7.24	35%	1.20×10^{-7}

(1) Representative values from well J-13, Thordarson (1983); Sinnock et al. (1987).
(2) Derived from average fracture spacings and orientation data in well USW G-1, UE-25a#1, USW GU-3/G-3, USW G-4, see Bauer (1986).
(3) Derived from bulk saturated permeability and fracture spacings, assuming two vertical sets and one horizontal fracture set.
(4) Derived with gamma distribution from permeability and surface contact area for rough-walled fractures, Wang and Narasimhan (1985).
(5) Assumed to be equal to the fraction of fracture surface coated with clay, calcite, and/or zeolite in well USW G-4, Spengler and Chornack (1984).
(6) Peters et al. (1984).

7.5.4 The Ghost Dance Fault: A Column Profile

In our 1986 study, the use of characteristic curves representing wider pore size distributions yielded simulated saturation values closer to measured ambient saturations for the nonwelded units, as shown in the GDF column results in Fig. 7.5.2. The evolution of the 1986 results from the 1985 results represents our efforts to update our data and improve our model predictions. All our model parameters are determined from available data. In 1985 and earlier, we had the wrong impression that the nonwelded units are porous and homogeneous. The predictions of low saturations in nonwelded units obtained using matrix characteristic curves with narrow pore size distribution bothered us. In 1986, new studies and observations suggested that the nonwelded units are heterogeneous, and the updated predictions reduced the discrepancies between model predictions and experimental measurements. We expect that the continuing process of crosschecking between models and measurements will validate the models for use in performance assessment.

Before implementing two-dimensional grids, we conducted tests with several one-dimensional cases to study the effects of mesh refinement and spatial averaging of permeability on the solution. Our calculations showed that our choice of the mesh refinement and our method of spatially averaging permeability did not introduce any serious errors.

7.5.5 Multicolumn Models: Fluid Redistribution within the Tuff Formation

To focus on the effects from eastern tilting of the units and from GDF on the two-dimensional flow behavior, we modeled the region west of GDF, treating GDF as a seepage boundary. The location of the western boundary was treated as variable, ranging from 190 m to 648 m. Five meshes were used in the two-dimensional studies to evaluate the effects of mesh refinement and the effects of system boundaries. Figure 7.5.3 is a 9-column, slanted mesh used in the 648-meter-wide model. Our results showed that numerical uncertainties associated with lateral mesh refinements were small. With mild variations in saturation, pressure, and potential in the lateral direction, coarse gridding can be used to generate acceptable results (Wang and Narasimhan, 1988).

The two-dimensional simulations showed that the eastern tilting of the units results in slightly higher matrix saturations in the eastern side of the cross section. Consequently, pressure heads along eastern columns closer to the seepage boundary are slightly higher than pressure heads along western columns. Pressure head values remain negative and deviations of two-dimensional results from one-dimensional one-column results were found to be small. The equal potential lines in most units are nearly horizontal, as illustrated in Fig. 7.5.4. Judging from matrix saturation, pressure head, and potential results of the two-dimensional multicolumn model, tilting of the units changed vertical distributions very little.

The matrix saturation and pressure head values of the nodes in the middle of TSw2 unit were found to be higher in the east than in the west. From saturation differences between the easternmost and westernmost columns, we calculated that the rate of linear saturation change varied from 10 to 17%/km, depending on the number of columns in the model; the smaller the number of columns, the larger the change. From results of the 9-column, 648-m section

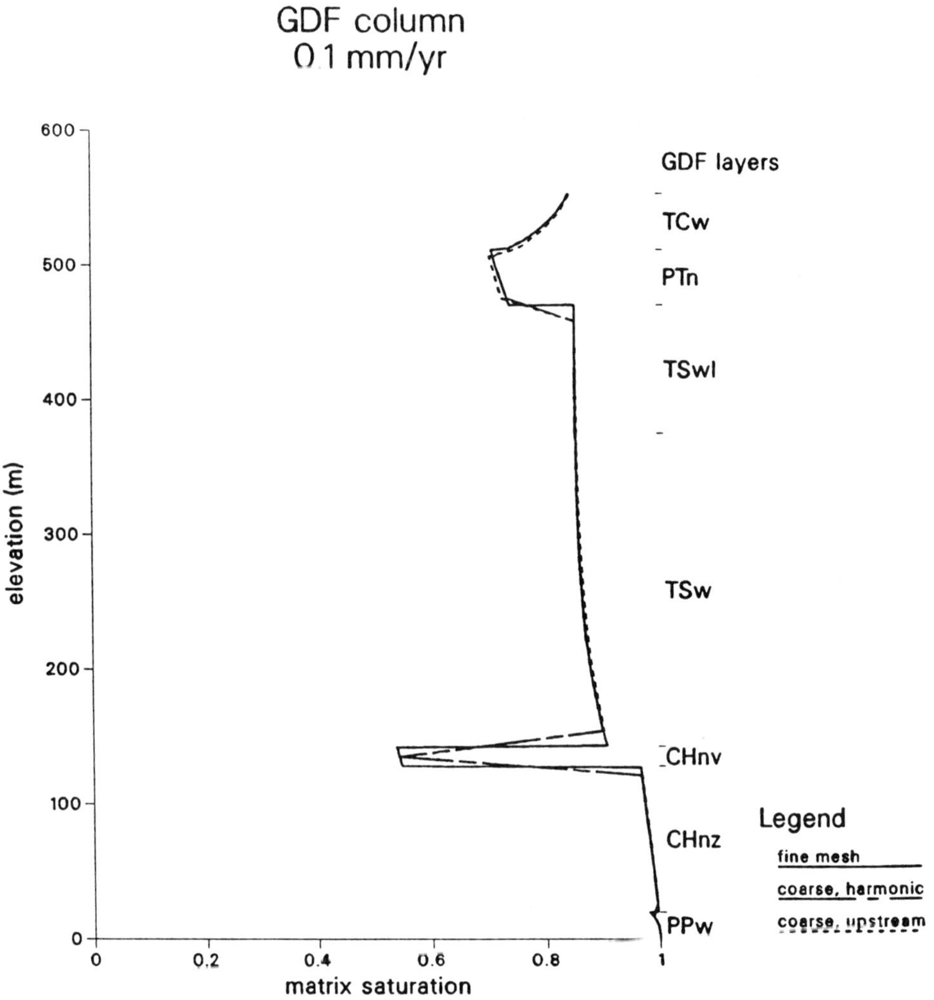

Figure 7.5.2. Saturation profiles for 0.1-mm/yr steady-state infiltration for fine and coarse meshes at Ghost Dance Fault.

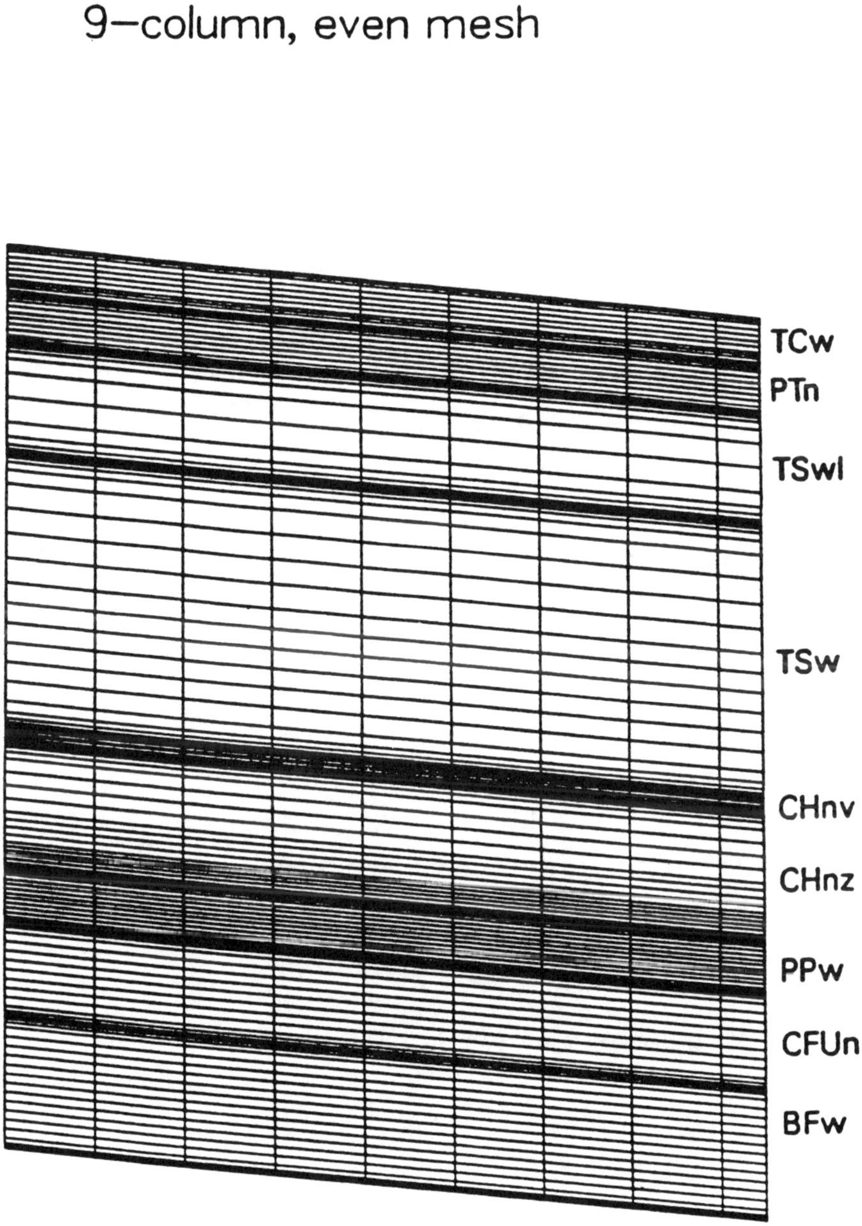

Figure 7.5.3. Two-dimensional grid for vertical and lateral flow model.

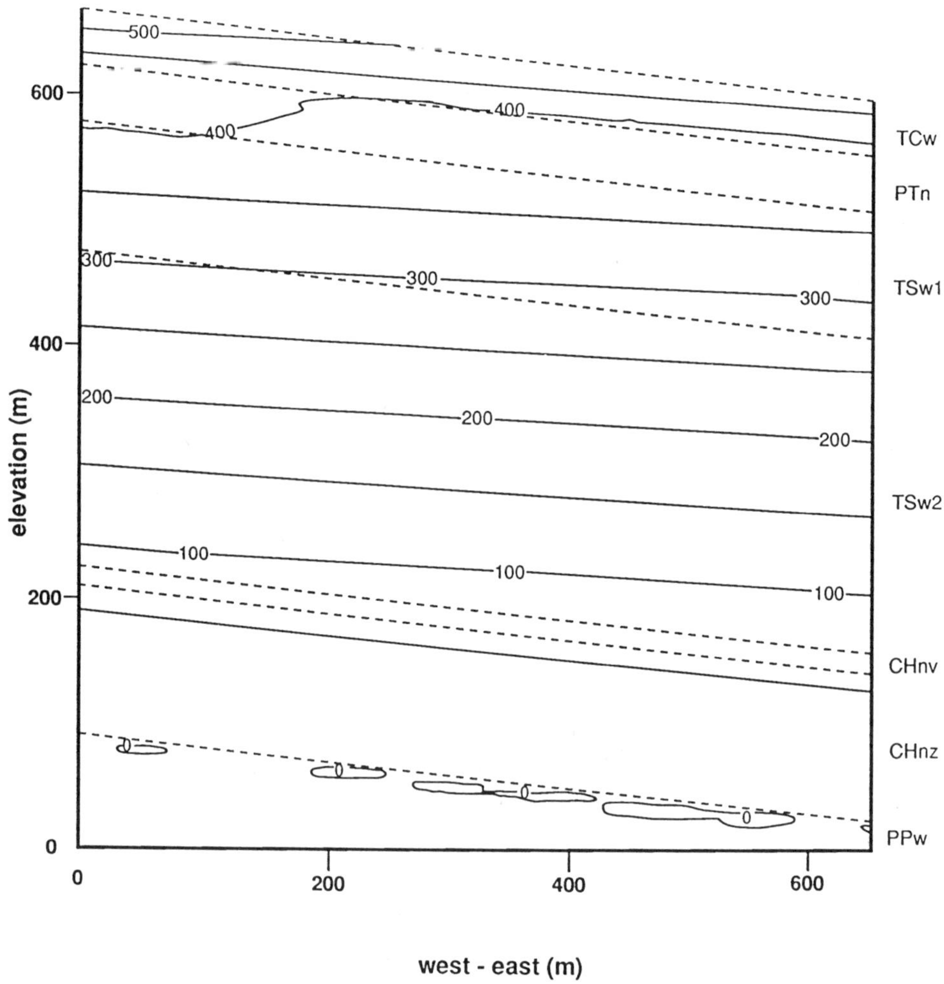

Figure 7.5.4. Potential distribution in a 645-m section west of Ghost Dance Fault.

model, we can also estimate wider section upper bounds for the change rate of other variables: pressure head; 60 m/km; potential, 44 m/km; vertical Darcy velocity, 0.085 mm/yr/km; and vertical matrix pore velocity, 0.79 mm/yr/km.

7.5.6 Vertical and Lateral Flow: Nonwelded Unit as Lateral Channel for Fluid Redistribution

Calculations using the 9-column, 645-m model showed that vertical Darcy velocity along columns closest to the GDF boundary is approximately 25% higher than the 0.1-mm/yr infiltration rate. Flow in column 5, in the middle, is approximately 0.1 mm/yr. Flow in column 9, away from the GDF boundary, is approximately 25% lower than 0.1 mm/yr (Fig. 7.5.5). From the vertical matrix velocity, we can calculate the travel time from the repository to the water table. For the 9-column, 648-m model, the travel time along the easternmost column is approximately 50% shorter than the travel time along the westernmost column. If the width of the section is doubled, we expect the groundwater travel time to vary in the range of a factor of 2 from different points in the repository. The fluid flow redistribution effects induced by the tilting units are unlikely to introduce orders-of-magnitude changes in travel time for the 0.1-mm/yr infiltration cases.

In slanted meshes used in this study, lateral fluxes between two neighboring columns are evaluated from both pressure and elevation differences of neighboring nodes. Because elevation differences are included in flux calculations, a lateral velocity is not a horizontal velocity. Lateral velocities measure flux magnitudes along layers parallel to tilted interfaces. A careful evaluation of the lateral fluxes between the columns showed that lateral Darcy velocities are smaller than vertical Darcy velocities.

When it does occur, lateral flow occurs mainly within the PTn unit. Lateral flow is the mechanism that redistributes vertical infiltration by channeling water from the west to the east, so that vertical Darcy flux in the east is higher than in the west and saturation is higher in the east than in the west. Magnitudes of lateral flow were found to be higher in the middle of the cross section than near the side boundaries (Fig. 7.5.5). In this simulation, water does not exit through the fault-formation boundary. The lateral gravity gradient associated with tilting of interfaces cannot overcome capillary forces that hold water in partially saturated formations.

7.5.7 Onset of Seepage Flow: A Hypothetical Nearly-Saturated System

For the 0.1-mm/yr cases, representing the ambient saturation conditions, no seepage flow occurs through the formation-fault boundary. The onset of seepage flows is simulated in the 0.5-mm/yr infiltration cases. With 0.5-mm/yr infiltration, the tuff units are nearly saturated, and seepage flow was found to occur only at the CHnv-CHnz interface for a 5-column, 343-m section model with 0.5-mm/yr infiltration. In terms of mass fluxes, we have approximately 8% lateral diversion; i.e., 8% of infiltration on the ground surface exits at the CHnv-Chnz interface out of the two-dimensional section, and the remaining 92% moves down to the water table. For this case, the amount of water flux driven by the lateral gravity gradient parallel to tilted interfaces is larger than the matrix saturated conductivity, and fracture flows are initiated at the seepage exit point. With the material properties used in the 1986 study, we observed seepage flow only in the CHnv-CHnz interface for the 0.5-mm/yr cases. If we use different characteristic curves for the units, we

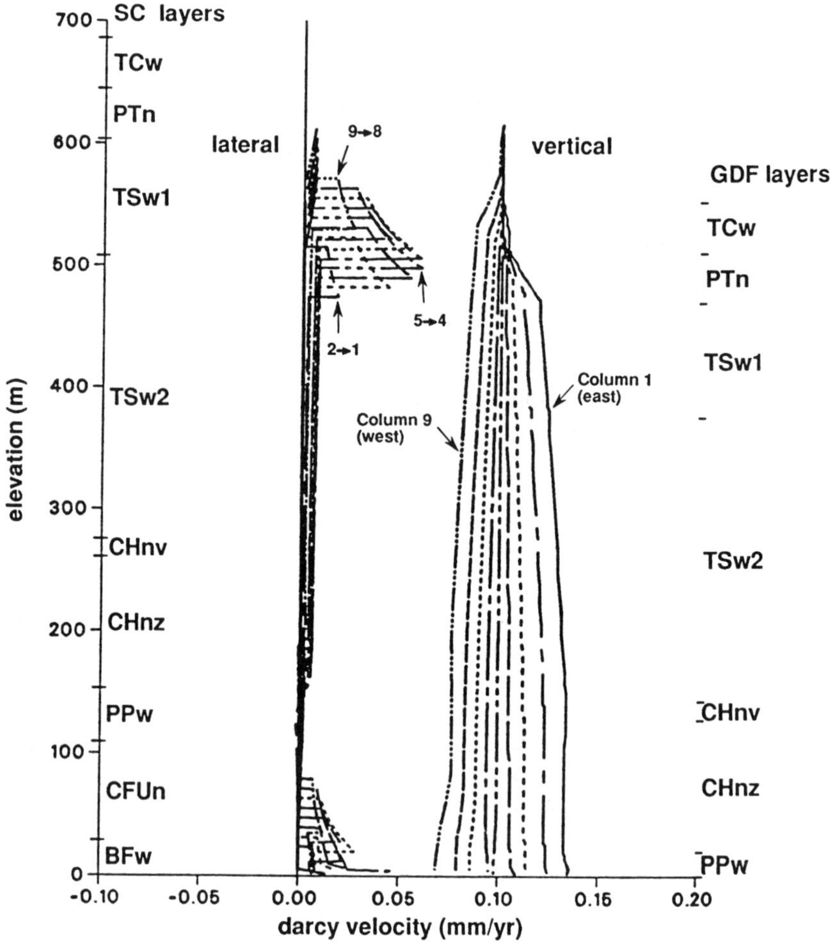

Figure 7.5.5. Darcy velocity profiles for vertical and lateral flows with 0.1-mm/yr steady-state infiltration.

may also generate results with seepage flow in other units if the units have pressure values of nearly zero and if redistribution effects are strong enough to increase pressure values to nonnegative ranges. We should reemphasize that saturation and pressure distributions with 0.5-mm/yr infiltration do not represent ambient partially saturated conditions at Yucca Mountain. Therefore, predictions of the existence of seepage flows in the 0.5-mm/yr cases were mainly modeling exercises.

7.5.8 Ambient Flow Field: Effect of Stratigraphic Tilting

Based on results with two-dimensional cross section models for alternating layers tilted toward the seepage boundary on the east, we may conclude the following for 0.1-mm/yr infiltration:

- Eastern tilting of units contributes to redistribution of flow, resulting in higher saturations and higher vertical velocities in the eastern side of a block bounded by faults. This redistribution effect is approximately proportional to cross-section width.

- Nonwelded units are major channels in redistributing vertical infiltration. Magnitudes of lateral flow are higher in the middle of the cross section than near side boundaries.

- Tilting itself is not a strong enough mechanism to induce the drastic changes in saturation, pressure, and potential distributions needed to drive flows into fault zones.

The mild lateral variations in saturation, pressure, and potential indicate that one-dimensional results are fairly good estimates of ambient conditions. The lateral tilting mechanism is a second-order perturbation that does not globally change the downward, gravity-driven flow patterns. So long as fluid flow remains inside partially saturated units, redistribution effects caused by tilting will not introduce order-of-magnitude differences in fluid transport predictions. If the system is on the verge of making a phase transition from partially saturated to fully saturated condition, as in the case of 0.5-mm/yr infiltration, the lateral tilting mechanism can perturb the system and change a fraction of fluid flow from dispersed slow flows percolating through the formations to concentrated fast flows rushing down the faults. If fault flow is important, the early-arrival portion of the groundwater travel time distributions may be controlled by the fault characteristics. For ambient conditions with large negative pressure values in all partially saturated units, it is unlikely that the lateral tilting mechanism can perturb the pressure and saturation field significantly enough to overcome negative capillary forces holding water in formations and thereby change groundwater travel time distributions drastically.

7.6 STATISTICAL ANALYSES OF TUFF AND SOIL DATA: VARIABILITY OF HYDROLOGICAL PARAMETERS IN DIFFERENT MEDIA

In 1987, we systematically analyzed the hydrological parameters of tuff matrix and fractures. Because many hydrological parameters are determined from microscopic pore structures, together with fluid properties and rock textures, correlations among different parameters may manifest themselves in the measured values. To elucidate the parameter correlations, we compile and compare saturated and unsaturated parameters of tuff matrix with watershed soil data. There are two reasons for us to analyze the tuff matrix and soil data in detail. Firstly, we are concerned with the proper range of the fault and fracture parameters, which are difficult to measure but may control the system response in the fractured tuff formations. Secondly, the determination of groundwater travel time distribution requires statistical quantification of model parameters. We conclude this section with a discussion of the application of fractal theory to the analysis of rough-walled fractures and heterogeneous media, and of the relationships between fractal theory and geostatistics.

7.6.1 Parameter Distributions: Difference between Tight Tuff Matrix and Permeable Soil

Because many unsaturated flow models were developed in soil studies, it is important to analyze the similarity and difference between soil and tuff rock characteristics. Some of the soil data have also been used to represent fractures and faults in tuff formations. We compared parameter distributions of tuff matrix data with watershed soil data. Specifically, the tuff matrix data are from the SNL studies by Peters et al. (1984) and Klavetter and Peters (1987), fitted with the van Genuchten (1980) model. The watershed data are from the United States Department of Agriculture study by Holtan et al. (1968), fitted by Panian (1987) of Desert Research Institute, also with the van Genuchten model.

In the van Genuchten model, the dependence of effective saturation S_e on suction pressure head h is described by the equation,

$$S_e(h) = \left[1 + \left(|\alpha h|\right)^n\right]^{-1+\frac{1}{n}} \quad . \tag{7.6.1}$$

Figure 7.6.1 compares the van Genuchten curves with the popular Brooks and Corey (1966) model,

$$S_e(h) = \begin{cases} 1 & if|\alpha h| < 1 \\ |\alpha h|^{-\lambda} & if|\alpha h| \geq 1 \end{cases} \quad . \tag{7.6.2}$$

In the range of large suction, the van Genuchten equation is asymptotically equivalent to the Brooks and Corey power function with the pore-size distribution index $\lambda = n - 1$. The inverse of the parameter α, $|h_{ae}| = \alpha^{-1}$, is frequently referred to in the soil literature as the air entry value or bubbling pressure. This is one interpretation for the parameter α or h_{ae}. When suction pressure head h overcomes the air entry pressure h_{ae} ($|h| \geq |h_{ae}|$), air enters the pores and water flows out of the medium to initiate desaturation. This interpretation is based on the cutoff separating the S_e = 1 saturated region from the $S_e = |\alpha h|^{-\lambda}$ desaturated region in the Brooks and Corey model. Another interpretation for α is to treat it as a scaling parameter. The saturation depends on αh together and not on α and h separately. Characteristic curves of soils with different α, but same λ, can be scaled into one curve. Therefore, we refer to α as the air entry scaling factor. If flow channels are represented by capillary tubes, we can define air entry radius by the capillary equation

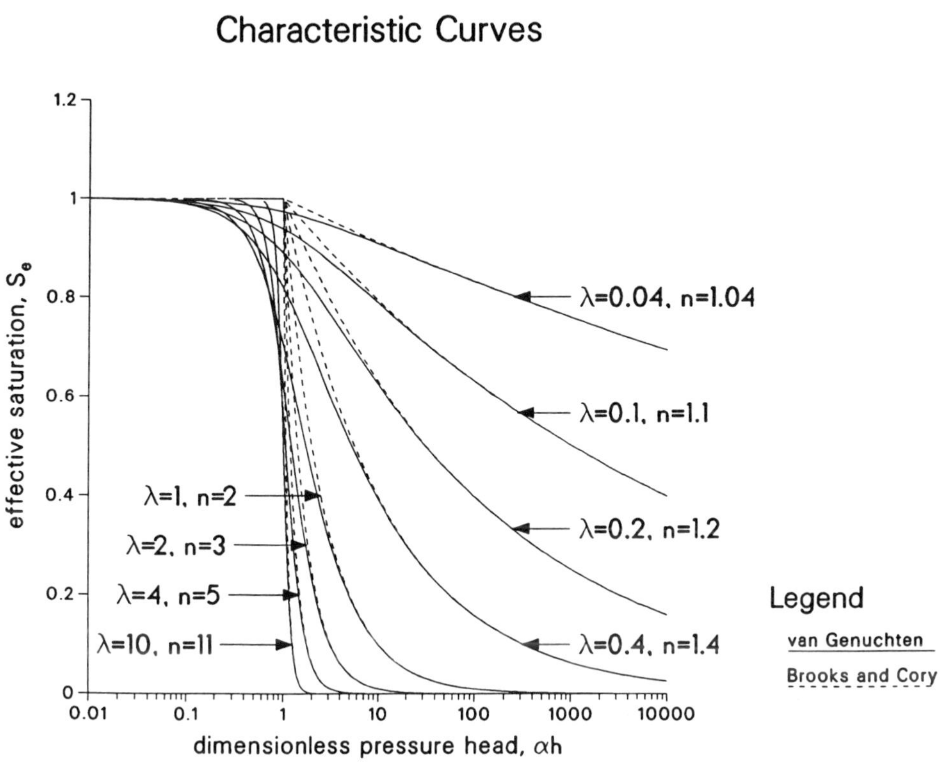

Figure 7.6.1. Comparison of characteristic curve of the van Genuchten model and the Brooks-Corey model.

$$r_{ae} = -\frac{2\gamma}{\rho g h_{ae}} = \frac{2\gamma}{\rho g}\alpha \quad . \tag{7.6.3}$$

With surface tension γ, water density ρ, and gravitational acceleration g as constants, r_{ae} is proportional to α. We can treat the air entry radius r_{ae} as a parameter equivalent to the air entry scaling factor α.

We have analyzed the pore-size distribution index and the air entry scaling factor, together with saturated permeability, porosity, and residual saturation. Detailed analyses of the distributions and the correlations of soil and tuff parameters are given in the report by Wang and Narasimhan (1989). In this review, we focus on two parameters: the saturated permeability and the unsaturated air entry scaling factor. Figure 7.6.2 compares the distributions of log air entry scaling factor of watershed soils and of tuff matrix. The important observation is that the average air entry scaling factor for watershed soils is two to three orders of magnitude larger than both the welded tuff and nonwelded tuff. Because air entry radius is proportional to air entry scaling factor, the order of magnitude differences simply indicate that the soils have much larger pores that can be easily desaturated, and tuff matrix has smaller pores that are more difficult to desaturate. Figure 7.6.3 shows that the average saturated permeability for watershed soils is five to seven orders of magnitude larger than those for welded/zeolitized tuffs.

7.6.2 Parameter Correlations: Relationship between Hydrological Parameters and Flow Geometry

The correlation between saturated permeability and air entry radius, based on mean values of these parameters for different soil texture categories and different tuff stratigraphic units, are plotted in Fig. 7.6.4. If a porous medium is represented by uniform circular tubes with radius r_{ae}, the maximum permeability of this idealized air entry tube model is

$$k_{\max} = \frac{r_{ae}^2}{8} \quad . \tag{7.6.4}$$

Figure 7.6.4 shows that the correlation between saturated permeability and air entry radius follows approximately the square functional dependence when soil data are compared with welded/zeolitized tuff matrix data. Wang and Narasimhan (1989) discuss other data sets in the literature on this and other parameter correlations.

In the literature, the correlation between permeability and porosity has been extensively studied. If the saturated permeability does correlate with other hydrological parameters, it probably correlates better with air entry radius than with porosity. Physically, this correlation with air entry radius makes more sense than a correlation with porosity. The saturated permeability measures the flow through connected channels in a medium. The air entry radius measures the desaturation of fluid from the interior of the medium through connected channels to the outside. Neither parameter depends on dead-end pores with stagnant fluid isolated from the connected channels. The porosity measures the volume of both connected channels and dead-end pores. In the literature, there is considerable interest in the permeability-porosity relationship. The correlation between permeability and air entry radius, observed in soil and tuff matrix data, also deserves attention in natural geological medium studies. This correlation of saturated

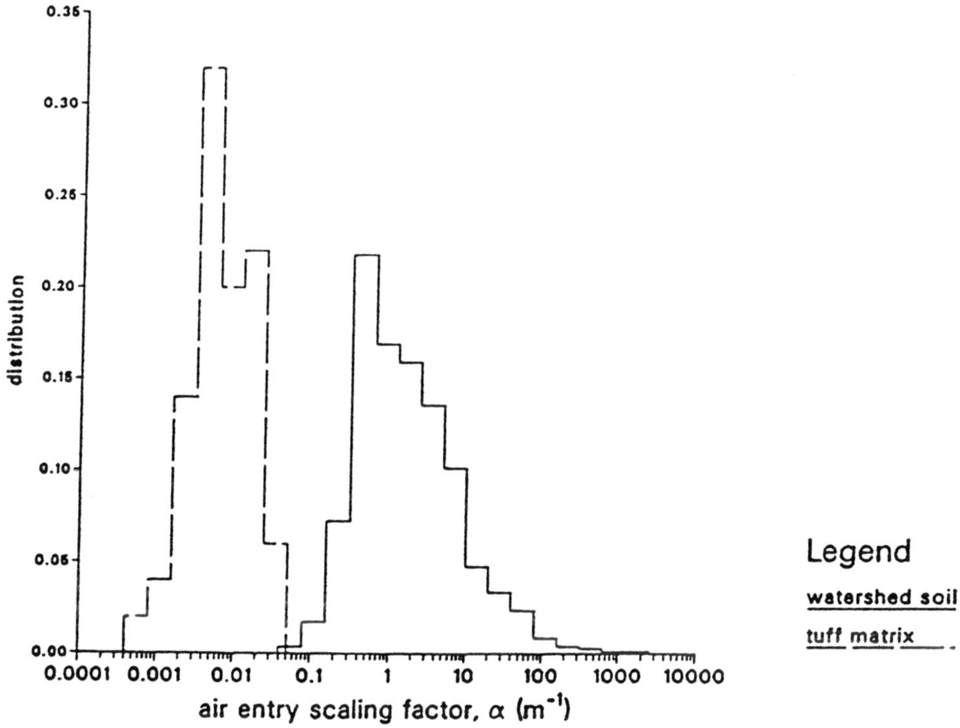

Figure 7.6.2. Distribution of air-entry scaling factor for natural soils and tuff matrix.

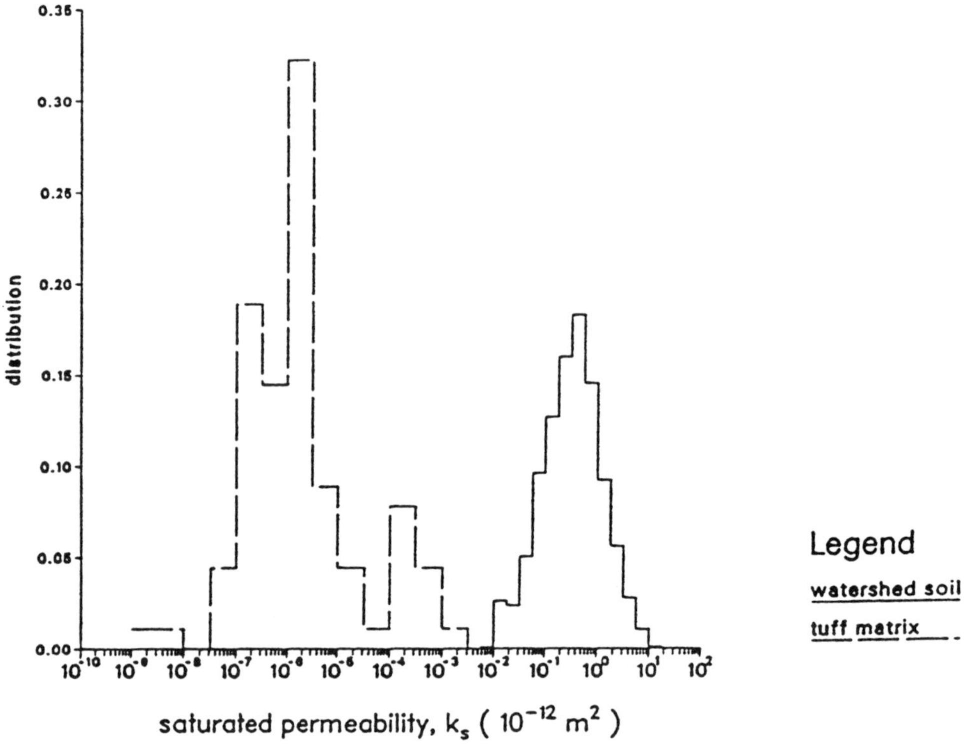

Figure 7.6.3. Distribution of saturated hydraulic conductivity for natural soils and tuff matrix.

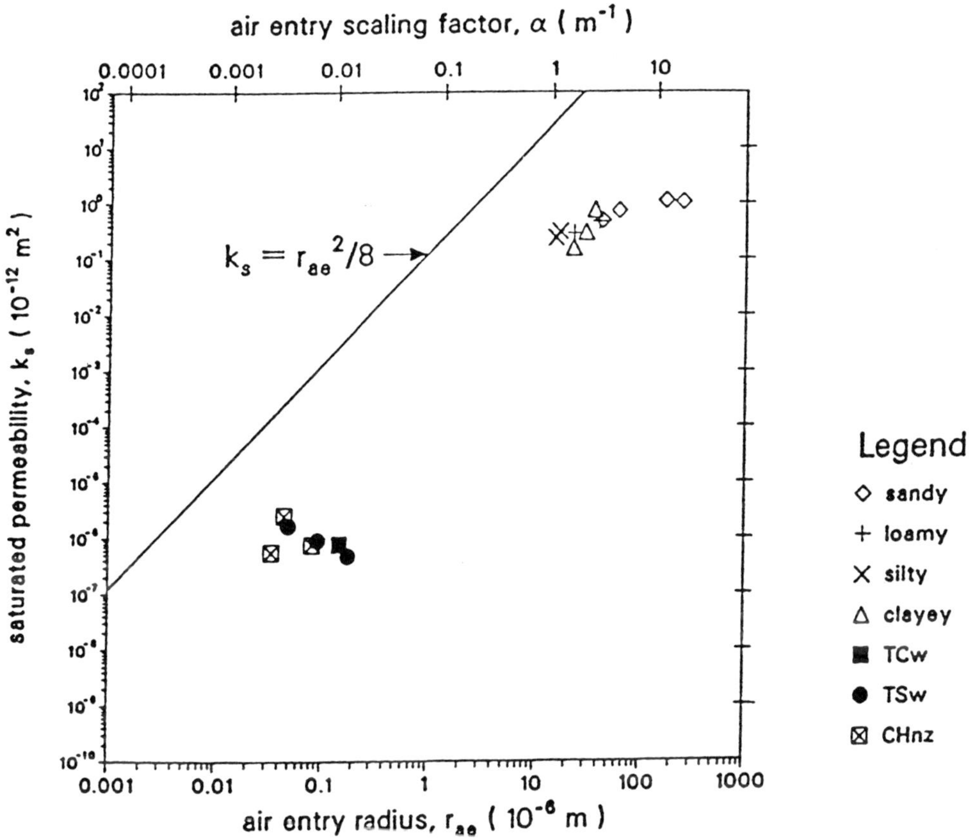

Figure 7.6.4. Correlation between saturated hydraulic conductivity and air-entry scaling factor for natural soils and tuff matrix.

permeability can be more easily demonstrated with air entry radius than with porosity. The air entry radius, or, equivalently, the air entry scaling factor, varies over several orders of magnitude among different geological media. The porosity varies over a much narrower range among different geological media. With wider variations, the air entry radius may correlate better with saturated permeability, which has the widest range of variation among all hydrological parameters.

7.6.3 Fracture and Fault Parameters: Physical Constraint

We need hydrologic data for the fractures and faults. There are only a few measurements on saturated permeability of fractured cores (Peters et al., 1984) and none on unsaturated characteristic curves of fractures. Only theoretical models have been derived or assumed to represent the fracture and fault characteristic curves. Wang and Narasimhan (1985) derived the fracture curves in terms of measurable bulk conductivity, fracture orientations and spacings, and fractions of fracture coatings. Peters et al. (1984) assumed that the fractures behave like permeable sand. Rulon et al. (1986) used a parallel channel model (Harrold et al., 1985) to derive anisotropic fracture and fault characteristic curves. Figure 7.6.5 compares the model parameters in the correlation crossplot of saturated permeability and air entry radius. Clearly the model parameters are very scattered, and some of the models have saturated permeability greater than the maximum capillary tube value $r_{ae}^2 / 8$.

Even though theoretical models can be derived for fractures and faults, we need to have experimental measurements to validate the theoretical models. The contrast of hydrological parameters between fractures and tuff matrix in meter scale determines if the fluid flow is fast through fractures or slow through matrix under ambient partially saturated conditions. The contrast of hydrological parameters between stratigraphic units and bounding faults in kilometer scale determines if the net infiltration is dispersed through the unsaturated tuff formations or concentrated through a few major faults from ground surface to water table. The lack of data in fractures and faults is one of the reasons we survey the soil data and study the parameter variations and correlations among different natural geological media. If constraints and correlations can be shown to govern and scale among different geological media, we can evaluate whether the theoretical model parameters are reasonable and physical, and establish better confidence in our calculation results.

If correlation between saturated permeability and air entry radius is important, we need to take this correlation into account. For stochastic modeling of alternating units of welded and nonwelded tuff or of formation blocks bounded by fault zones, we suggest imposing correlations and constraints on average values of both saturated permeability and air entry scaling factor between highly different units to avoid unlikely combinations of parameters and predictions. Within a given stratigraphic unit, this physically meaningful correlation is likely to be masked by data variability, and uncorrelated parameters with wide distributions may be acceptable.

7.6.4 Fractal Theory: Spatial Correlation and Scaling of Rough-Walled Fractures

With parameter distributions and correlations determined from laboratory measurements, we need additional information on spatial correlation and scale dependence to properly model

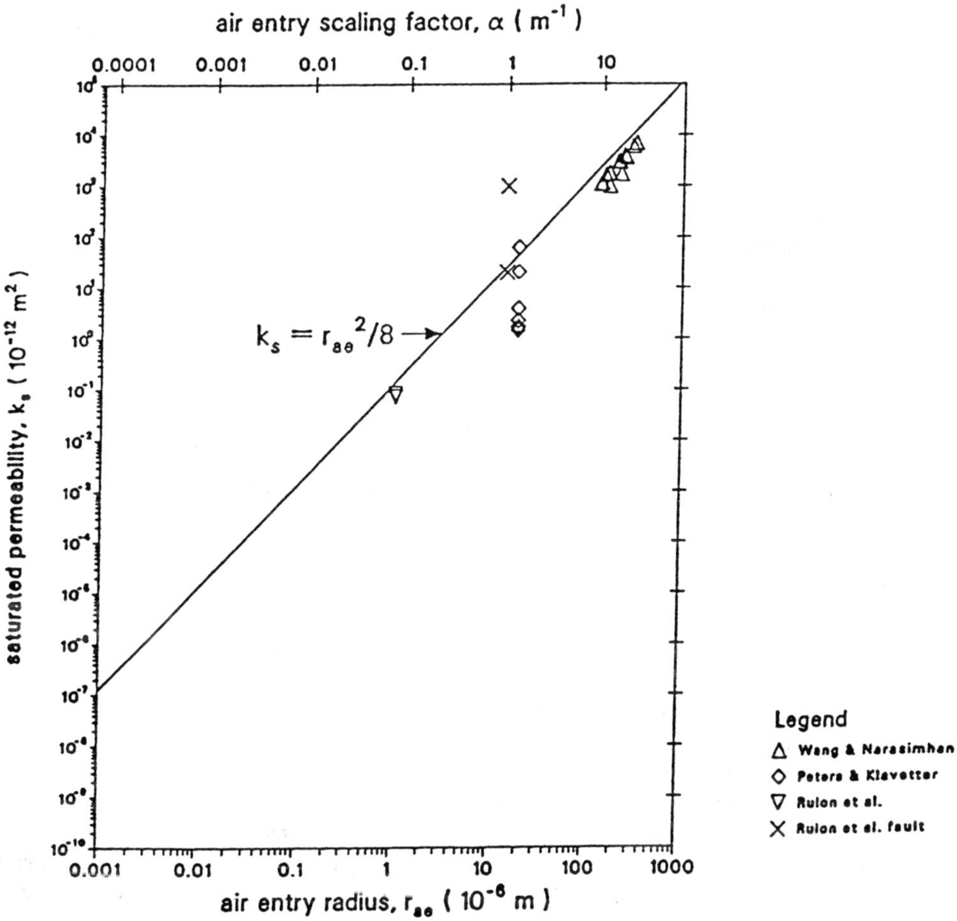

Figure 7.6.5. Correlation between saturated hydraulic conductivity and air-entry scaling factor for tuff fracture and fault models.

field-scale hydrological processes. The spatial correlation structure determines the grid block sizes to represent a heterogeneous formation and the scaling law relates the laboratory measurements to field experiments. We have constructed a fractal fracture model (Wang et al., 1988) to represent a rough-walled fracture. From that particular model study, we also recognize that the fractal scaling approach can be used for other parameters. Even more importantly, fractal scaling is more general than traditional geostatistical analyses that use empirical variograms to represent spatial correlation.

A very useful and interesting aspect of the fractal is its scaling behavior. A rough profile measured with a low-resolution measuring device over a large distance may be geometrically similar to the profile measured with a high-resolution device over a much smaller scale on the same rock sample. If this scaling property can indeed be demonstrated, we can deduce structure information at small scales by measuring at large scales, and vice versa. The fractal dimension D can be regarded as a scaling parameter. Figure 7.6.6 illustrates the steps in a recursive subdivision algorithm (Fournier et al., 1982; Brown, 1987) to generate a fractal surface. The procedure starts with a 3×3 mesh having z values of zero on the mesh points to define a reference flat surface. At the second level, a 5×5 mesh, the z value of a new point is computed by a Gaussian random variable function, with the expected value determined by its four known neighbors (two on the boundaries) from the 3×3 mesh and the standard deviation scaled by the fractal dimension D and distance to the neighbors. We can continue this generation procedure level-by-level to obtain a finer mesh. A low-D surface will have the standard deviation scaled down faster than a high-D surface as one proceeds from one level to another level to fill the mesh (Fig. 7.6.7).

We can use a similar procedure to generate representations of heterogeneous media with spatially varying and correlated parameters. We simply replace the elevations by the property values and the Gaussian distribution by the parameter distribution. If data values are known at a given number of locations, we can start with these locations as the first-level mesh points and generate refinements in accordance with the parameter distribution and spatial correlation structure. The heterogeneous field generated by this procedure will take into account the spatial correlation, because the mean value at a mesh point is determined by the values of its neighbors. With the generation of each subdomain following the same procedure used for the larger domain, domains of different sizes are scaled. We can use a low-level or low-resolution mesh to create an approximate model of the field, and continue upgrading the results with higher level, more refined meshes.

The fractal dimension D can be determined by log-log type curve analyses from variogram data of the parameter distribution. The log-log slope near the origin of zero spatial separation is equal to $6 - 2D$, as shown in Fig. 7.6.8 from a derivation for the fractal fracture model (Wang et al., 1988). The spherical model and many other empirical variogram models widely used in geostatistical analyses have a log-log slope of 1. If we reanalyze many variogram data in a log-log plot instead of linear plot, we can deduce the fractal dimension and use the fractal dimension as a scaling parameter to generate geostatistical representations.

If a heterogeneous field is generated with parameter distribution, correlation, and scaling properly taken into account, we can then proceed to solve flow equations to obtain the distributions in pressure, flux, and velocity field. We need to quantify uncertainties to obtain groundwater travel time distributions. We are still perfecting our approaches and models so that the groundwater travel time calculations can be efficiently and accurately calculated.

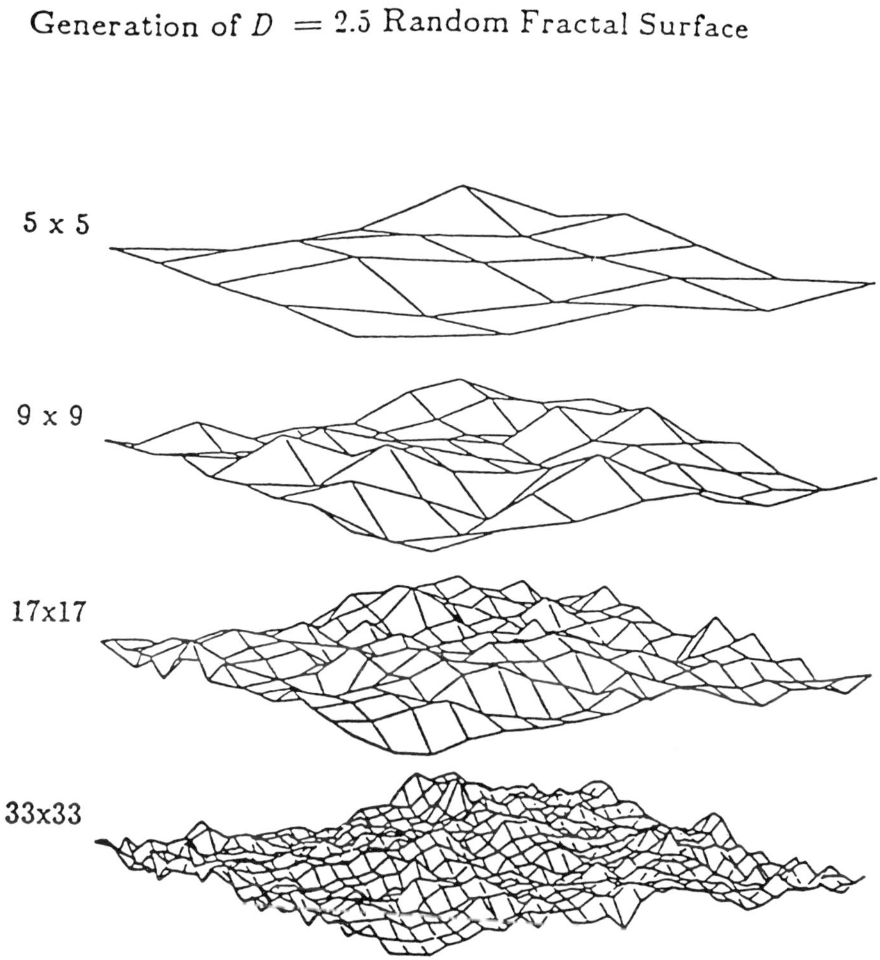

Figure 7.6.6. Steps involved in generating a rough fracture with fractal dimensions.

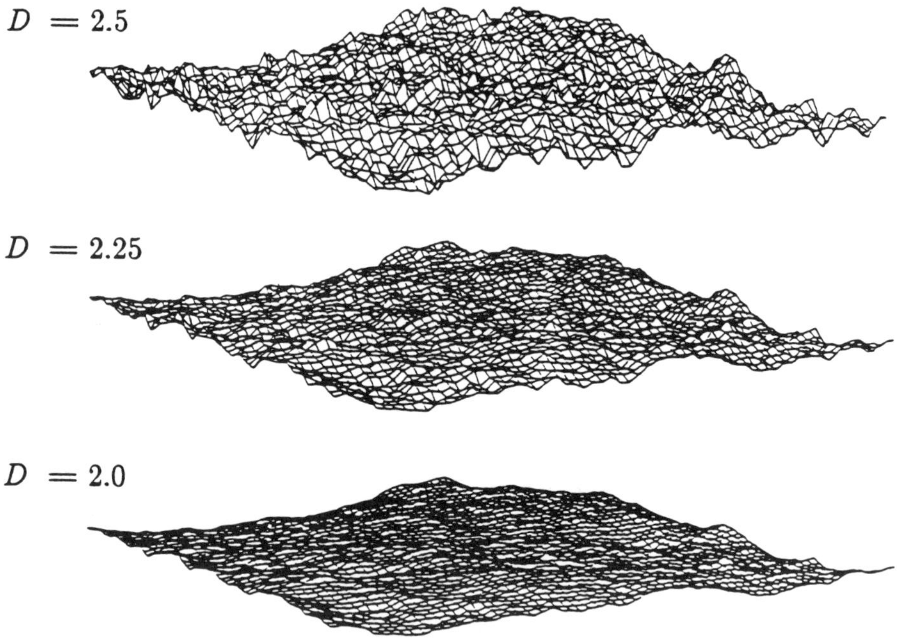

Figure 7.6.7. Representation of rough fractures with fractal surfaces.

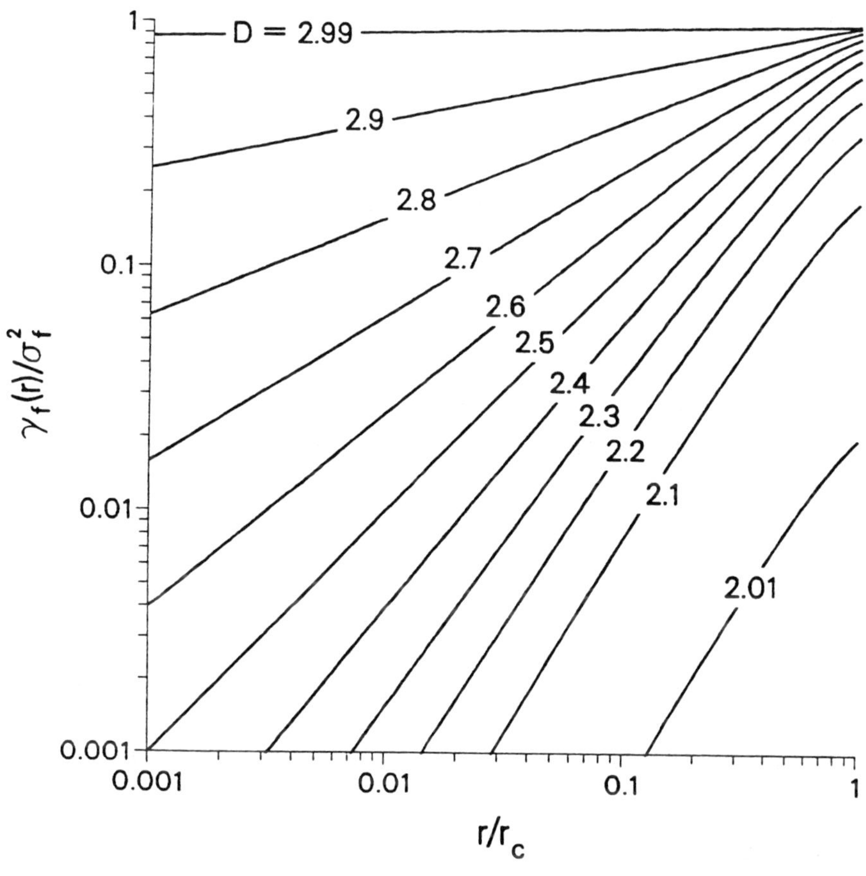

Figure 7.6.8. Relation between variograms and fractal dimensions

7.7 SUMMARY

From analysis of the capillarity-based conceptual model (Sec. 7.2 and 7.3, Wang and Narasimhan, 1984, 1985), we conclude that fractures can be desaturated more easily than the matrix if fracture apertures are larger than matrix pore sizes. With fractures drained, liquid water could flow from one matrix block to another across contact areas. Reda (1987) and Eaton and Bixler (1987) measured and analyzed the imbibition of liquid water into a tuff core that contained several microfractures transversely oriented to the direction of the wetting-front propagation. The local minima in saturation were shown to occur near the fractures. A fracture could also impede liquid front movement, whereas vapor-phase transport might occur across a fracture. The dominance of matrix flow was illustrated by the following two tests. Haldeman et al. (1991) showed that, when liquid water under suction was applied to the top of a fractured tuff block from Apache Leap, Arizona, the wetting front moved mainly through the unsaturated matrix, and not much through the vertical fracture. Gvirtzman et al. (1988) examined the partitioning of downward percolation between slow bulk flow through the matrix and fast flow through the fractures in calcareous sandstone, using environmental tritium as a natural tracer.

Pruess and Tsang (1990) studied the fracture flow for both the liquid phase and the gas phase and showed that the interference between fluid phases flowing in a fracture was strong, with the nonwetting phase flow easily choked off when the wetting phase occupied small aperture regions. Fracture flow can be important when the fracture is in nearly-saturated conditions. If water ponding occurs at the ground surface at the entrance of a vertical fracture, thus guaranteeing a continuous slug of water along the fracture, the wetting front along the fracture may be ahead of that in the matrix (Nitao and Buscheck, 1989). Capillary hysteresis in the matrix may also enhance vertical fracture flow penetration (Niemi and Bodvarsson, 1988). If the supply of water is finite, as in the case of cyclic infiltration pulses (Sec. 7.4, Wang and Narasimhan, 1986, 1990), the water entering the fractures may easily imbibe into the matrix. When a fracture is filled with water, capillary force will draw water into the unsaturated matrix blocks. Zimmerman et al. (1990) showed that the rate of absorption into the matrix is proportional to the ratio of surface area to volume for irregularly shaped blocks. The presence of a low-permeability skin at the fracture-matrix interface might delay, but not entirely eliminate, the absorption of water into the matrix blocks (Pruess and Wang, 1987).

Composite-porosity models, effective medium models and multiple interacting continua methods were used in recent field-scale simulation studies to assess the impact of siting a high-level nuclear waste repository at Yucca Mountain. Sinnock et al. (1987) estimated the release rate of radionuclides from the potential repository with a multiple-column unit-gradient stochastic model. Kaplan et al. (1989) illustrated the sensitivity of groundwater travel time calculations to particle tracking and conservative tracer concentration criteria. Bloomsburg et al. (1989) demonstrated the need to include multidimensional analyses for heterogeneous unsaturated systems. The concern for lateral flows and fault flows (Rulon et al., 1986; Montazer and Wilson, 1984) motivated the two-dimensional modeling studies by Prindle (1989) and Wang and Narasimhan (1988, 1990) (Sec. 7.5). Multidimensional modeling results are sensitive to the boundary conditions and to the contrast in hydrological parameters of fault and tuff units. In addition to liquid flow, gas flow can be significant in unsaturated, fractured media. Weeks (1987) and Kipp (1987) measured and modeled the seasonal variation of high-rate air flows in boreholes at the crest of the mountain. Tsang and Pruess (1987) estimated that the buoyancy-driven gas phase convection velocity is orders of magnitude larger than the liquid velocity.

Statistical, geostatistical, and fractal analyses (Sec. 7.6, Wang et al., 1988; Wang and Narasimhan, 1989) will be needed to characterize the fractured, tuff formations and to scale the hydrological parameters. As site characterization activities at Yucca Mountain (DOE, 1989) proceed, the uncertainties about the ambient conditions will be better quantified and models for performance assessment will be better calibrated and validated. In addition to ambient conditions, we need to understand the long term evolution of the system to make long term predictions more credible. This chapter mainly addresses the conceptual analysis of the nature of unsaturated flow regime in a fractured, porous tuff system. For the assessment of the unsaturated repository concept, we need to account for the contact of flow with waste packages and the migration and retardation of radionuclides (Fig. 7.7.1). The amenability to long term monitoring and the retrievability of the wastes for a repository in the unsaturated zone are also important components in finding a rational solution to this challenging problem. These general issues are addressed by Narasimhan and Wang (1992) with a hierarchy approach to identify the hydrological issues that need resolution to support the performance assessment of Yucca Mountain.

ACKNOWLEDGMENTS

We would like to thank Scott Sinnock, Tom Lin, and Paul Kaplan of Sandia National Laboratories for continued discussions and review throughout the course of this study. Interactions with our colleagues Karsten Pruess, Yvonne Tsang, and Chin-Fu Tsang helped greatly in developing our models. The work was supported by the Sandia National Laboratories for the Yucca Mountain Site Characterization Project of the U.S. Department of Energy under contract DE-AC03-76SF00098.

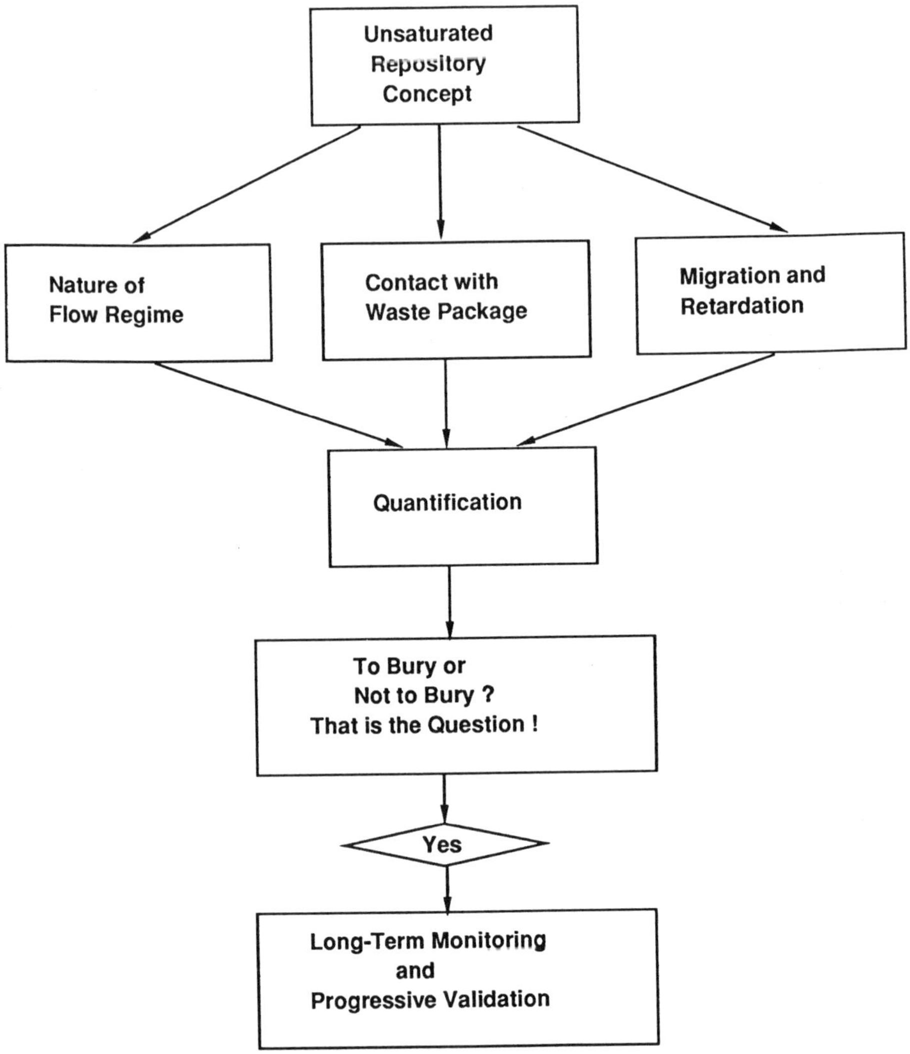

Figure 7.7.1. Framework of fundamental questions in performance assessment.

ABBREVIATIONS

BFw	Bullfrog welded tuff unit
CFUn	Crater Flat upper nonwelded tuff unit
CHnv	Calico Hill nonwelded, vitric tuff unit
CHnz	Calico Hill nonwelded, zeolitic tuff unit
DOE	United States Department of Energy
GDF	Ghost Dance Fault
HLW	High Level Radioactive Wastes
PPw	Pah Pass welded tuff unit
PTn	Paintbrush nonwelded tuff unit
RIB	Reference Information Base
SC	Solitario Canyon
SCP	Site Characterization Plan
SNL	Sandia National Laboratories
TCw	Tiva Canyon welded tuff unit
TSw	Topopah Spring welded tuff unit
USGS	United States Geological Survey
YMP	Yucca Mountain Site Characterization Project

NOTATION

b	aperture	$[L]$
d	fracture spacing	$[L]$
D	fractal dimension	
$f(b)$	aperture distribution function	$[L^{-1}]$
g	gravitational constant	$[Lt^{-2}]$
h	pressure head	$[L]$
k	permeability	$[L^2]$
K	hydraulic conductivity	$[Lt^{-1}]$
L	fracture width	$[L^2]$
n	van Genuchten's parameter	
q	flow rate	$[Lt^{-1}]$
r	radius	$[L]$
S	saturation	
V	volume	$[L^3]$
W	width of flow channel	$[L]$
z	elevation	$[L]$
α	air entry scaling factor	$[L^{-1}]$
β	gamma distribution parameter	$[L^{-1}]$
γ	surface tension	$[Mt^{-2}]$
λ	pore-size distribution index	
ω	contact area fraction	
ϕ	porosity	
ρ	density	$[ML^{-3}]$

σ	effective flow area
τ	phase-constriction factor
θ	contact angle

Subscript:

ae	air entry value
c	contact
e	effective value
f	fracture
h	horizontal
m	matrix
max	maximum value
r	relative or residual value
S	condition with saturation S
v	vertical
0	zero stress or zero elevation
1	condition with saturation $S = 1$

REFERENCES

Bauer, S.J. Preliminary Report on Yucca Mountain Fracture Characteristics, Memorandum to J.R. Tillerson dated March 27, 1986, Appendix D, Attachment 1 of the Site Characterization Plan Conceptual Design Report, SAND84-2641, compiled by H.R. MacDougall, L.W. Scully and J.R. Tillerson, Sandia National Laboratories, Albuquerque, NM, 1986.

Bloomsburg, G., Williams, R.E., and Osiensky, J.L. Distribution of Downward Flux in Unsaturated Heterogeneous Hydrogeology Environments, *Geol. Soc. Am. Bull.*, 101, 1623–1630, 1989.

Brooks, R.H., and Corey, A.T. Properties of Porous Media Affecting Fluid Flow, *J. Irrigation and Drainage Div., Proc. Am. Soc. Civ. Eng.*, 92, IR2, 61–88, 1966.

Brown, S.R. Fluid Flow Through Rock Joints: the Effect of Surface Roughness, *J. Geophys. Res.*, 92, 1337–1347, 1987.

DOE, Site Characterization Plan Public Handbook, DOE/RW-0206, Office of Civilian Radioactive Waste Management Program, U.S. Department of Energy, Washington, DC, 1989.

Eaton, R.R., and Bixler, N.E. Analysis of a Multiphase, Porous-Flow Imbibition Experiment in Fractured Volcanic Tuff, in *Flow and Transport through Unsaturated Fractured Rock*, Geophysical Monograph 42, American Geophysical Union, edited by D.D. Evans and T.J. Nicholson, 91–97, 1987.

Evans, D.D., and Huang, C.H. Role of Desaturation on Transport through Fractured Rock, Role of the Unsaturated Zone in Radioactive and Hazardous Waste Disposal, edited by J. Mercer, P.S. Rao and I.W. Marine, Ann Arbor Science Publisher, Ann Arbor, MI, pp. 165–178, 1983.

Fournier, A., Fussell, D., and Carpenter, L. Computer Rendering of Stochastic Models, *Comm. Asso. Computing Machinery*, 25, 371–384, 1982.

Gentier, S. Morphologie et Comportement Hydromecanique d'une Fracture Naturalle dans un Granite Sours Contrainte Normale, Doctoral Thesis, Univ. d'Orl'eans, France, 1986.

Gvirtzman, H., Magaritz, M., Kanfi, Y., and Carmi, I. Matrix and Fissure Water Movement through Unsaturated Calcareous Sandstone, *Transport in Porous Media*, 3, 343–356, 1988.

Haldeman, W.R., Chuang, Y., Rasmussen, T.C., and Evans, D.D. Laboratory Analysis of Fluid Flow and Solute Transport through a Fracture Imbedded in Porous Tuff, *Water Resour. Res.*, 27(1), 53–66, 1991.

Harrold, P., Barton, C.C., and Montazer, P. Statistical Analysis of Fracture Roughness Profiles in Volcanic Tuffs, Yucca Mountain, Nevada, EOS, *Trans. Amer. Geophys. Union*, 66 (46), p. 883, 1985.

Holtan, H.N., England, C.B., Lawless, G.P., and Schumaker, G.A. Moisture-Tension Data for Selected Soils on Experimental Watersheds, ARS 41-144, Agricultural Research Service, U.S. Department of Agriculture, Beltsville, MD, 609 p, 1968.

Kaplan, P., Klavetter, E., and Peters, R. Approaches to Groundwater Travel Time, Proc. Waste Manag. '89, Tucson, AZ, edited by R.G. Post, v. I, pp. 493–495, 1989.

Kipp, K.L., Jr. Effect of Topography on Gas Flow in Unsaturated Fractured Rock: Numerical Simulation, in *Flow and Transport through Unsaturated Fractured Rock*, Geophysical Monograph 42, American Geophysical Union, edited by D.D. Evans and T.J. Nicholson, 171–176, 1987.

Klavetter, E.A., and Peters, R.R. An Evaluation of the Use of Mercury Porosimetry in Calculating Hydrologic Properties of Tuffs from Yucca Mountain, Nevada, SAND86-0286, Sandia National Laboratories, Albuquerque, NM, 1987.

Montazer, P., and Wilson, W.E. Conceptual Hydrologic Model of Flow in the Unsaturated Zone, Yucca Mountain, Nevada, USGS-WRI-84-4345, U. S. Geological Survey, Lakewood, CO, 55 p., 1984.

Narasimhan, T.N. Geometry-Imbedded Darcy's Law and Transient Subsurface Flow, *Water Resour. Res.*, 21, 1285–1292, 1985.

Narasimhan, T.N., and Wang, J.S.Y. Conceptual, Experimental and Computational Approaches to Support Performance Assessment of Hydrology and Chemical Transport at Yucca Mountain, SAND89-7018, Sandia National Laboratories, Albuquerque, NM, LBL-28613, Lawrence Berkeley Laboratory, Berkeley, CA, 1992.

Narasimhan, T.N., and Witherspoon, P.A. An Integrated Finite Difference Method for Analyzing Fluid Flow in Porous Media, *Water Resour. Res.*, 12, 57–64, 1976.

Narasimhan, T.N., Witherspoon, P.A. and Edwards, A.L. Numerical Model for Saturated-Unsaturated Flow in Deformable Media, 2. The Algorithm, *Water Resour. Res.*, 14, 255–261, 1978.

Niemi, A., and Bodvarsson, G.S. Preliminary Capillary Hysteresis Simulations in Fractured Rocks, Yucca Mountain, Nevada, *J. Contaminant Hydro.*, 3, 277–291, 1988.

Nimick, F.B., Bauer, S.J., and Tillerson, J.R. Recommended Matrix and Rock-Matrix Bulk, Mechanical and Thermal Properties for Thermomechanical Stratigraphy of Yucca Mountain, Appendix O of the Site Characterization Plan - Conceptual Design Report, SAND84-2641, compiled by H.R. MacDougall, L.W. Scully, and J.R. Tillerson, Sandia National Laboratories, Albuquerque, NM, 1984.

Nitao, J.J., and Buscheck, T.A. On the Infiltration of a Liquid Front in an Unsaturated, fractured porous medium, Proc. Nuc. Waste Iso. Unsat. Zone, Focus '89, *Am. Nuc. Soc.*, pp. 381–397, 1989.

Ortiz, T.S., Williams, R.L., Nimick, F.B., Whittet, B.C., and South, D.L. A Three-Dimensional Model of Reference Thermal/Mechanical and Hydrological Stratigraphy at Yucca Mountain, Southern Nevada, SAND84-1076, Sandia National Laboratories, Albuquerque, NM, 1984.

Panian, T.F. Unsaturated Flow Properties Data Catalog, Volume II, DOE/NV/10384-20, Desert Research Institute, University of Nevada, 493 p., 1987.

Peters, R. R. and Klavetter, E.A. A Continuum Model for Water Movement in an Unsaturated Fractured Rock Mass, *Water Resour. Res.*, 24, 416–430, 1988.

Peters, R.R., Klavetter, E.A., Hall, I.J., Blair, S.C., Heller, P.R., and Gee, G.W. Fracture and Matrix Hydrologic Characteristics of Tuffaceous Materials from Yucca Mountain, Nye County, Nevada, SAND84-1471, Sandia National Laboratories, Albuquerque, NM, 188 p., 1984.

Peters, R.R., Klavetter, E.A., George, J.T., and Gauthier, J.H. Measuring and Modeling Water Imbibition into Tuff, *Proc. Symp. Flow and Transport through Unsaturated Fractured Rock*, Geophysical Monograph 42, Am. Geophys. Union, edited by D.D. Evans and T.J. Nicholson, pp. 99–106, 1987.

Prindle, R.W. On a Sensitivity Analysis of Flow through Layered, Fractured Tuff: Implications for Performance Allocation and Performance-Assessment Modeling, Proc. Nuc. Waste Iso. Unsat. Zone, Focus '89, *Am. Nuc. Soc.*, pp. 294-300, 1989.

Pruess, K., and Narasimhan, T.N. A Practical Method for Modeling Fluid and Heat Flow in Fractured Porous Media, *Soc. Pet. Eng. J.*, 25, 14–26, 1985.

Pruess, K., and Tsang, Y.W. On Two-Phase Relative Permeability and Capillary Pressure of Rough-Walled Rock Fractures, *Water Resour. Res.*, 26(9), 1915–1926, 1990.

Pruess, K., and Wang, J.S.Y. Numerical Modeling of Isothermal and Non-Isothermal Flow in Unsaturated Fractured Rock—a Review, in *Flow and Transport through Unsaturated Fractured Rock*, Geophysical Monograph 42, American Geophysical Union, edited by D.D. Evans and T.J. Nicholson, 11–21, 1987.

Pruess, K., Wang, J.S.Y., and Tsang, Y.W. On Thermohydrological Conditions Near High-Level Nuclear Wastes Emplaced in Partially Saturated Fractured Tuff, 2. Effective Continuum Approximation, *Water Resour. Res.*, 26(6), 1249–1261, 1990.

Pruess, K., Tsang, Y.W., and Wang, J.S.Y. Modeling of Strongly Heat-Driven Flow in Partially Saturated Fractured Porous Media, *Proc. Hydrogeology of Rocks of Low Permeability, Memoires 17*, Int. Asso. Hydrogeologists, pp. 486–497, 1985.

Reda, D.C. Influence of Transverse Microfractures on the Imbibition of Water into Initially Dry Tuffaceous Rock, in *Flow and Transport through Unsaturated Fractured Rock*, Geophysical Monograph 42, American Geophysical Union, edited by D.D. Evans and T.J. Nicholson, 83–90, 1987.

Roseboom Jr., E.H. Disposal of High-Level Nuclear Waste Above the Water Table in Arid Regions, Circular Number 903, U.S. Geological Survey, 1983.

Rulon, J., Bodvarsson, G.S., and Montazer, P. Preliminary Numerical Simulations of Groundwater Flow in the Unsaturated Zone, Yucca Mountain, Nevada, LBL-20553, Lawrence Berkeley Laboratory, Berkeley, CA, 91 p., 1986.

Scott, R.B., and Bonk, J. Preliminary Geologic Map of Yucca Mountain, Nye County, Nevada, with Geologic Sections, USGS-OFR-84-494, U. S. Geological Survey, Denver, CO., 1984.

Scott, R.B., Spengler, R.W., Diehl, S., Lappin, A.R., and Chornack, M.P. Geologic Character of Tuffs in the Unsaturated Zone at Yucca Mountain, Southern Nevada, in Role of the Unsaturated Zone in Radioactive and Hazardous Waste Disposal, edited by J. Mercer, P.S. Rao, and I.W. Marine, Ann Arbor Science Publisher, Ann Arbor, MI, pp. 289–335, 1983.

SNL, The Nevada Nuclear Waste Storage Investigations Project Reference Information Base, Sandia National Laboratories, Albuquerque, NM, 1986.

Sinnock, S., Lin, Y.T., and Brannen, J.P. Preliminary Bounds on the Expected Postclosure Performance of the Yucca Mountain Repository Site, Southern Nevada, SAND84-1492, Sandia National Laboratories, Albuquerque, NM, *J. Geophys. Res.*, 92(B8), 7820–7842, 1987.

Spengler, R.W., and Chornack, M.P. Stratigraphic and Structural Characteristics of Volcanic Rocks in Core Hole USW-G4, Yucca Mountain, Nye County, Nevada, with a Section of Geophysical Logs by D.C. Muller and J.E. Kibler, USGS-OFR-84-789, U.S. Geological Survey, Denver, CO, 1984.

Thordarson, W. Geohydrologic Data and Test Results from Well J-13, Nevada Test Site, Nye County, Nevada, USGS-WRI-83-4171, U.S. Geological Survey, Denver, CO, 1983.

Travis, B.J., Hodson, S.W., Nuttall, H.E., Cook, T.L., and Rundberg, R.S. Numerical Simulation of Flow and Transport in Fractured Tuff, *Proc. Scientific Basis for Nuclear Waste Management VII, Mat. Res. Soc. Sym. Proc.*, 26, 1039–1046, 1984.

Tsang, Y.W. The Effect of Tortuosity on Fluid Flow Through a Single Fracture, *Water Resour. Res.*, 20, 1209–1215, 1984.

Tsang, Y.W., and Pruess, K. A Study of Thermally Induced Convection Near a High-Level Nuclear Waste Repository in Partially Saturated Fractured Tuff, *Water Resour. Res.*, 23(10), 1958–1966, 1987.

van Genuchten, M.Th. A Closed-Form Equation for Predicting Hydraulic Conductivity of Unsaturated Soils, *Soil Sci. Soc. Am. J.*, 44, 892–898, 1980.

Wang, J.S.Y., and Narasimhan, T.N. Fluid Flow in Partially Saturated, Welded-Nonwelded Tuff Units, *Geoderma*, 46(1-3), 155–168, 1990.

Wang, J.S.Y., and Narasimhan, T.N. Processes, Mechanisms, Parameters, and Modeling Approaches for Partially Saturated Flow in Soil and Rock Media, SAND88-7054, Sandia National Laboratories, Albuquerque, NM, LBL-26224, Lawrence Berkeley Laboratory, Berkeley, CA, 305 p., 1989.

Wang, J.S.Y., and Narasimhan, T.N. Hydrologic Modeling of Vertical and Lateral Movement of Partially Saturated Fluid Flow near a Fault Zone at Yucca Mountain, SAND87-7070, Sandia National Laboratories, Albuquerque, NM, LBL-23510, Lawrence Berkeley Laboratory, Berkeley, CA, 122 p., 1988.

Wang, J.S.Y., and Narasimhan, T.N. Hydrologic Mechanisms Governing Partially Saturated Fluid Flow in Fractured Welded Units and Porous Non-welded Units at Yucca Mountain, SAND85-7114, Sandia National Laboratories, Albuquerque, NM, LBL-21022, Lawrence Berkeley Laboratory, Berkeley, CA, 126 p., 1986.

Wang, J.S.Y., and Narasimhan, T.N. Hydrologic Mechanisms Governing Fluid Flow in a Partially Saturated, Fractured, Porous Tuff at Yucca Mountain, *Water Resour. Res.*, 21, 1861–1874, 1985.

Wang, J.S.Y., and Narasimhan, T.N. Hydrologic Mechanisms Governing Fluid Flow in a Partially Saturated, Fractured, Porous Tuff at Yucca Mountain, SAND84-7202, Sandia National Laboratories, Albuquerque, NM, LBL-18473, Lawrence Berkeley Laboratory, Berkeley, CA, 1984.

Wang, J.S.Y., Narasimhan, T.N., and Scholz, C.H. Aperture Correlation of a Fractal Fracture, *J. Geophys. Res.*, 93, 2216–2224, 1988.

Weeks, E.P. Effect of Topography on Gas Flow in Unsaturated Fractured Rock: Concepts and Observations, in *Flow and Transport through Unsaturated Fractured Rock*, Geophysical Monograph 42, American Geophysical Union, edited by D.D. Evans and T.J. Nicholson, 165–170, 1987.

Winograd, I.J. Radioactive Waste Disposal in Thick Unsaturated Zones, *Science*, 212 (4502), 1457–1464, 1981.

Zimmerman, R.W., Bodvarsson, G.S., and Kwicklis, E.M. Adsorption of Water into Porous Blocks of Various Shapes and Sizes, *Water Resour. Res.*, 26(11), 2797–2806, 1990.

8

SIMULATION OF FLOW AND TRANSPORT IN FRACTURED POROUS MEDIA

G. F. Pinder
College of Engineering and Mathematics
University of Vermont
Burlington, Vermont

P. S. Huyakorn
Hydrogeologic
Reston, Virginia

E. A. Sudicky
University of Waterloo
Waterloo, Canada

In this chapter we turn our attention to the strategies employed in the solution of the equations describing fluid flow and mass and energy transport in fractured porous media. Strategies, as used in this chapter, involve not only mathematical techniques, but also equation formulations that exploit the physical characteristics of the problem.

Because strategies depend upon the physical system under consideration, different strategies will evolve dependent upon the particular class of subsurface problems under investigation. Thus, to some degree, one can catalogue strategies based upon the problem environment. In the case of subsurface flow and contaminant transport the literature reveals that several related yet distinctly different simulation strategies evolved in parallel. While any subdivision of strategies by problem type is somewhat arbitrary, one approach is to group subsurface simulation problems into those associated with groundwater hydrology, soil physics, petroleum reservoir engineering and geothermal reservoir engineering. Taking this approach, the role of each of these disciplines in the development of simulation strategies will be presented.

Even within a given discipline, there are two distinctly different approaches to solving fractured reservoir problems. Although both schools of thought envision a fractured porous medium as a set of relatively impermeable blocks separated one from the other by discrete fractures, each school sees the role of the individual fractures quite differently. One group believes that each fracture must be taken into account separately. Thus it is necessary to determine from the field the geometry, orientation and fluid-mechanical properties of each fracture, and to incorporate this information into the mathematical model.

The alternative approach is to assume that the fractures are sufficiently ubiquitous and distributed in such a manner as to allow them to be meaningfully described statistically. In this approach the importance of the individual fractures is secondary to the significance accorded the average fracture properties. One can describe these average properties as macroscopic and those associated with the individual fractures as microscopic. While the advantage of viewing the world from the macroscopic perspective is the ability to use macroscopic parameters that do not require detailed knowledge of individual fractures, the price that is paid is the inability to determine reservoir behavior at the individual fracture (microscopic) level. Drawing from the nomenclature used to formally develop the macroscopic level equations, the ubiquitous fracture model is often referred to as the overlapping continuum approach since the governing equations for both the fractures and the porous bloc are assumed to exist everywhere. Because of the form of the resulting equations, which will be discussed later, this approach is also called a two-porosity or dual-porosity model.

In fact, naturally occurring reservoirs have a hierarchy of fracture sizes that suggest that neither of the above models completely represent the physics of the system. The above models are applicable only when two and only two well-defined scales of behavior are represented (that is, the pores and the fractures). When a multiplicity of scales is represented the double porosity and discrete-fracture models should be combined. This is indeed the strategy proposed in recent publications (see, for example, Huyakorn, et al., 1983b).

The evolution of the strategies for fractured reservoir simulation can also be examined from the point of view of the mathematical techniques. A review of the literature reveals that both analytical and numerical methods have played an important role in the development of simulators. The numerical procedures that have been employed to date include finite-difference methods, finite-element methods, integrated finite-difference methods, boundary-element methods, and a approach based upon particle tracking. In our discussion of the various problem types, the role of the various mathematical techniques will also be considered.

8.1 HISTORICAL PERSPECTIVE

Let us now briefly recapitulate the various types of fractured reservoir problems that have been simulated. Although this exposition will be presented in terms of the disciplines described above, unfortunately, as would be expected, there is unavoidable overlap between them.

8.1.1 Oil Reservoir Engineering

It has long been recognized that one can assume the permeability of the porous blocks in a fractured reservoir to be very low relative to the permeability of the fractures. It is Barenblatt (1960), however, that is credited with putting this concept into a mathematical-physical statement. This concept was adopted in the classic paper of Warren and Root (1963) wherein it was assumed that the matrix blocks were arranged in a "systematic array of identical, rectangular parallelepipeds." These blocks provided flow to the fractures which, in turn, transported the fluid to the well.

Odeh (1965) attempted to generalize the concept of Warren and Root to accommodate a fractured reservoir in which the pattern of fractures was not known. Interestingly, this conceptual jump is consistent with the fundamental postulates of modern continuum theory, but was ignored by many later researchers. In his paper he stated his assumptions as follows:

1. The matrix blocks act like sources which feed the fractures with fluid;

2. The net fluid movement toward the wellbore obtains only in the fractures; and

3. The fractures' flow capacity and the degree of fracturing of the reservoir are uniform. By the degree of fracturing is meant the "fracture bulk volume per unit reservoir bulk volume."

The discussion between Warren and Root and Odeh presented in the Odeh (1965) paper gives a rare and candid picture of researchers in conflict.

Warren and Root and Odeh used these conceptual models to develop analytical solutions. These solutions were subsequently extended by Kazemi et al. (1969). In their theoretical development they state the assumptions for which their conceptual model holds. They are similar to but extend those of Odeh:

1. The matrix porosity has a high storage capacity but low flow capacity. The converse is true of the fracture porosity;

2. Radial flow to the wellbore is only through the fracture network. Matrix flow is into the fractures;

3. Matrix flow to the fractures is quasi-steady state. Fracture flow to the wellbore is unsteady state;

4. The reservoir is horizontal and infinitely large;

5. The flow rate is constant and gravity effects are negligible; and

6. Vertical pressure gradients are negligible.

It is interesting that the "Rubix-cube-like" diagram used to help visualize the fracture network and generally identified with the model of Warren and Root was absent from Odeh's paper, but appeared in this later work of Kazemi et al. (1969). Also of interest is the statement by Kazemi et al. that "the analytical solutions cited ...were developed for cross-checking the numerical results." The numerical solution to which they refer is also presented in the paper.

In the same year Kazemi published a paper (Kazemi, 1969) that described, apparently for the first time, a "layer cake" model wherein the porous blocks and fractured matrix each occupied a different layer. This model is shown in Fig. 8.1.1, which also includes the "Rubix cube" conceptualization. In stating the general assumptions for this model he relaxes some of the constraints of his other model, as follows:

1. There is single phase flow;

2. The matrix with high storage and extremely low flow capacity produces into the fracture, which has low storage and high flow capacity. The fracture, in turn, produces into the wellbore;

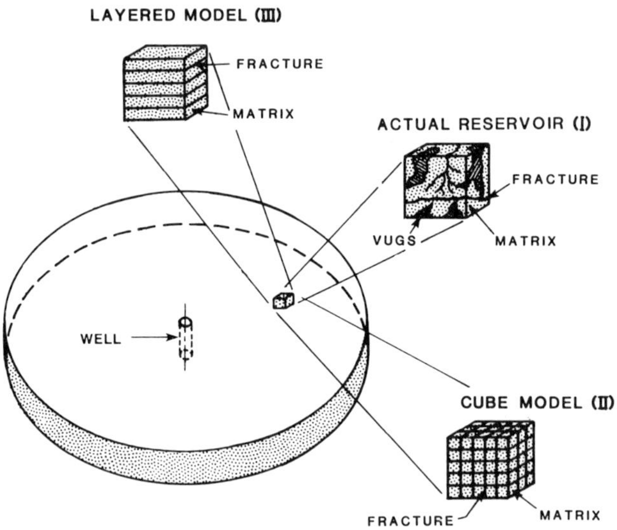

Figure 8.1.1. Idealization of a naturally fractured heterogeneous porous medium: II, Warren-Root model; III, Kazemi model (from Kazemi, 1969).

3. Flow occurs in both radial and vertical directions;
4. Flow is unsteady state;
5. The reservoir is horizontal and the matrix and the fracture are each homogeneous and isotropic; and
6. The well is centrally located in a finite circular reservoir.

Of significance in this model formulation is the recognition that porous medium flow is not only orthogonal to the fractures (although practically speaking it may nearly be) but responds as a continuum over the entire model. As a result of this conceptualization of the system two mass balance equations appear, one for the fractures and one for the porous blocks. This is consistent with the continuum mechanical equation first developed and presented by O'Neill (1977) and later extended by Shapiro (1981).

The equations describing the layer cake model of Kazemi (1969) were solved numerically. An iterative alternating-direction implicit procedure was employed on a finite-difference representation of the fractured porous medium equations.

Braester (1971,1972) was also among the first to develop analytical solutions to the fractured media problem. His conceptual model was similar to that of Barenblatt et al. (1960), but his mathematical formulation was different. The source function in his model that described the connection between the fractures and the blocks was defined in terms of the potential gradient in the fractures, the capillary pressure difference between the liquid in the fractures and the matrix blocks, and the density difference between liquid phases. Streltsova (1983) extended earlier analytical work by incorporating the assumption that matrix to fracture cross flow was the result of a diffusion mechanism of fluid transfer through the matrix.

In contrast to the formulations described above which use the dual-porosity concept in their theoretical development, Gringarten et al. (1974, 1975) developed analytical solutions for problems dominated by single discrete fractures. These solutions, as those developed earlier, were intended to provide insight into well test behavior.

At about the same time as Gringarten was developing his discrete-fracture formulations, considerable work was being done to extend the dual porosity concepts to large scale reservoir simulators. Bossie-Codreanu et al. (1985) report that the first attempt to apply the dual-porosity approach to simulate multiphase flow in fractured reservoirs was presented by Reiss et al. (1973) and Lefebvre du Prey and Bossie-Codreanu (1975). In their approach a conventional three-dimensional three-phase model was used to simulate the fracture behavior. Fluid movement from the blocks to the fractures was accomplished by the introduction of time-dependent transfer functions.

Another early paper describing a two-phase water-oil flow model was published by Kazemi et al. (1976). This model employed a three-dimensional finite-difference formulation to solve the resulting equations. When the finite-difference cell was large relative to the matrix blocks all the blocks in the cell were assumed to have the same pressure and saturation. Although gravity segregation within individual matrix blocks was not accommodated unless there were several finite-difference cells in the block, the overall gravity segregation process from grid cell to grid cell was represented. This model formulation generated a conservation equation for the blocks and one for the fractures. The two equations were coupled by assuming the matrix flux to the fractures to be proportional to the pressure difference between matrix and fracture. This assumption, which has been called the "pseudo steady-state" or "lumped-parameter" assumption, was inherent in the early formulation of Barenblatt et al. (1960) and used by nearly all subsequent researchers during this period.

In the years since the above work was published, a number of more sophisticated models have been developed. Some of these models considered the blocks simply as source terms (see Rossen, 1977) while others (see Gilman and Kazemi, 1983) used a separate field equation for the fracture and the blocks. One of the most sophisticated multiphase models was presented by Thomas et al. (1983). Bossie-Codreanu et al. (1985) introduced a novel "Checker Model" that is based on a checkerboard pattern for fracture and block numerical cells. The presentation by Gilman (1986) is notable inasmuch as it introduces explicitly the concept of using a numerical solution within the matrix blocks to determine the fluid transfer rate to the fractures. A similar strategy was introduced by Wu and Pruess (1988). They called their formulation the multiple interacting continua method and combined it with an integrated finite-difference model to simulate oil recovery in a fractured medium.

8.1.2 Groundwater Flow

While the history of flow and contaminant transport through fractured media simulation is shorter in groundwater hydrology than in petroleum reservoir engineering, significant progress has been made and several novel concepts have been introduced. For example, the most sophisticated mathematical-physical developments of the equations describing flow and transport through fractured media appear to be those of O'Neill (1977) and Shapiro (1981, 1987). In addition, there has been significantly more use of novel numerical schemes, largely based upon finite-element theory or the integrated finite-difference method. There have also been noteworthy contributions to the suite of analytical solutions applicable to fractured media flow and transport.

One of the earliest papers to examine water flow in fractured porous media was presented by Wilson and Witherspoon in 1974. They developed two finite-element models to describe steady-state seepage in fracture systems beneath a dam. In one formulation they represented both the fractures and the intervening blocks with arrays of triangular finite-elements. The other model utilized linear one-dimensional finite-elements to represent the fractures in an otherwise impermeable medium. As we will see, some recent models have employed line elements for fractures in conjunction with two-dimensional elements for the fractured media to form a hybrid formulation.

Gureghian (1975) presented a finite-element model for fluid flow in a three-dimensional fractured porous medium. He used tetrahedral elements for the matrix and triangular elements for the fracture planes. The triangles that represent the fractures are made to correspond to the faces of selected tetrahedrons. While formulated from a quite different perspective, this model is conceptually similar to that presented by Wilson and Witherspoon (1974).

A similar formulation was developed by Baca et al. (1984), to address flow and transport in a fractured porous medium. Having formulated two finite-element equations, one for the fractures and another for the blocks, they combined the resulting equations by summing the Galerkin weighted residuals. A diagrammatic representation of their conceptual model is presented as Fig. 8.1.2. Among several recent papers devoted to discrete-fracture modeling we note those of Khaleel (1987) and Eaton and Bixler (1987). In the latter work an effort is made to reproduce experimental results.

An aquifer model based upon the conceptual model of Warren and Root (1963) and Odeh (1965) was presented by Closmann (1975). Unique to this formulation was the recognition that different rock qualities could be encountered in the field and should be accommodated by the model. To achieve this, two rock types were assumed to exist in the blocks. Thus the model

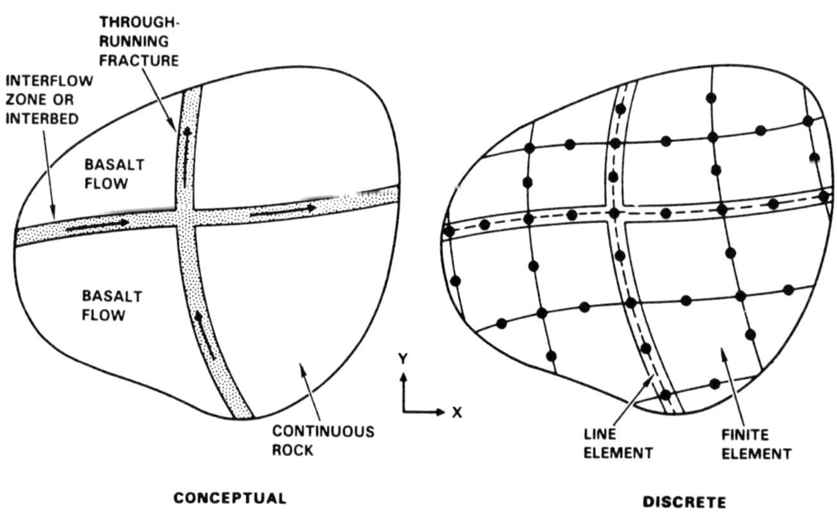

Figure 8.1.2. Finite-element representation of continuum and discrete fractures; from Baca et al. (1984).

permitted quasi-steady-state leakage from two rock types which he denoted as "good rock" and "poor rock". Good rock he designated as rock with a porosity greater than 12 percent and poor rock had a porosity between 2 and 12 percent.

A finite-element formulation based upon the Galerkin method and a dual-porosity formulation was presented by Duguid and Abel (1974). This work was later extended by Duguid and Lee (1977) wherein they used fundamental fluid- and solid-mechanical principles to derive governing equations with secure mathematical-physical underpinnings. In their formulation they extended earlier formulations by incorporating transient leakage within the matrix blocks.

Huyakorn et al. (1983) examined the various mathematical vehicles available to accommodate matrix-to-fracture leakage in a finite-element model of fractured media flow. The three analytical models available in the literature at the time of their analysis were:

1. The quasi-steady-state fracture flow model,
2. the unsteady parallel-fracture model, and
3. the unsteady block-fracture model.

In the implementation of these models the authors present a formulation that reportedly improves the computational efficiency of the model.

In addition to the formulations based upon the analytical solutions outlined above, Huyakorn et al. (1983) also described a model that combined a finite-element description of the fractured matrix with a finite-difference representation of the transient behavior in the blocks. The objective was to avoid the convolution integral inherent in the analytical solutions used for the description of the transient block behavior (this is discussed in more detail in a later section).

Three conceptual models for the geometry of the blocks were considered. In the simplest case they considered the quasi-steady-state case and found the numerical results compared favorably with the analytical solution. Next they used the finite-difference-finite-element approach to represent unsteady flow in a parallel fracture network configuration. Finally unsteady matrix flow in blocks of spherical geometry were considered. This was conveniently described mathematically using a spherical coordinate system. To assure accuracy it was found necessary to use a variable nodal spacing in the one-dimensional finite-difference representation of the spherical flow equations. Finally, to illustrate the capability of the dual-porosity model to treat discrete fractures, the authors simulated a single vertical fracture problem.

It appears that there has been relatively little work done in the groundwater field to simulate flow in partially saturated fractured porous media. Wang and Narasimhan (1985) report modeling unsaturated fracture flow using a discrete-fracture approach. Recent work of Dougherty and Pinder (1989a,b) also addresses saturated-unsaturated flow in fractured porous media, in this case employing a dual-porosity formulation.

The boundary element method was recently used to simulate single-phase fluid flow in fractured media. Elsworth (1986,1987) used this mathematical approach to describe the movement of fluids within the fractures (Elsworth, 1986) and later, in combination with a finite-element formulation for the fracture, to simulate the entire fractured medium. Recently Rasmussen (1987) has also developed a flow model based upon a boundary-element formulation. He has included in his presentation some one-dimensional results for mass transfer.

Analytical solutions for flow in porous media have been developed for problems involving groundwater flow in porous media. In Streltsova (1976) the general differential equations for flow in a fractured formation of uniform anisotropy were developed. Solutions were provided for fissure flow to a well, whether or not the fissure is filled with a finer material. Also provided is a solution for flow in a porous block.

8.1.3 Contaminant Transport

In response to keen interest in problems of contaminant and radionuclide transport, the past decade has seen the rapid development of new techniques for the simulation of solute transport in fractured porous media. Grisak and Pickens (1980) presented a finite-element model of reactive and nonreactive solute transport including advection, mechanical dispersion and diffusion in a single fracture hydrodynamically connected to porous blocks. This analysis is very similar in concept to earlier geothermal simulation work by Pinder et al. (1979) (to be described in the next section). Grisak and Pickens showed solutions for varying aperture size, water velocity in the fracture, matrix porosity, matrix distribution coefficient, and dispersivity in the fracture.

Bibby (1981) developed a dual-porosity finite-element model based on the parallel fracture concept of Kazemi (1969). In Bibby's model both flow and transport are assumed to occur in the joints or fractures while water in the blocks is assumed to be static. Thus the blocks have no influence on groundwater flow but do affect the transport of solutes because of the effect of molecular diffusion. The influence of the blocks is treated analytically along the lines of earlier work introduced in petroleum engineering.

An integrated-finite-difference model of similar conceptual design was presented by Rasmuson (1982). Using the TRUMP computer code, originally formulated for heat flow, in conjunction with analytical solutions used as transfer functions between the blocks and the fractures, the authors simulated the following problems:

1. one-dimensional transport with advection and dispersion in a uniform fracture,
2. advection and dispersion in a set of parallel fractures with diffusion in microfissured matrix (early time solution),
3. advection and dispersion in a set of parallel fractures with diffusion in microfissured matrix (solution for large times),
4. case 2 with radioactive decay, and
5. case 3 with radioactive decay.

Noorishad and Mehran (1982) reported a finite-element model similar in concept to that of Grisak and Pickens (1980) wherein discrete fractures are coupled to matrix blocks directly and the governing transport equations are solved in both media. Unusual about this model is the introduction of one-dimensional elements for the fractures in a domain otherwise subdivided into two-dimensional elements. The two-dimensional system represents convective and dispersive flow in the porous medium. The one-dimensional finite-elements are introduced to represent the discrete fractures wherever they are encountered. It is worth noting that dispersion and advection in the transverse direction within any fracture is neglected and that the porous blocks are considered as a continuum so that all blocks are perceived as interconnected in some sense.

A quite different approach to fracture-transport simulation was presented by Schwartz et al. (1983). In their model the fractures were considered discrete and the blocks passive. Particle tracking was used to simulate transport through a fracture network that consists of two orthogonal fracture sets of finite length, oriented at various angles to the direction of the mean hydraulic gradient. Further work along the same lines, but dedicated to the investigation of the importance of fracture connectivity and aperture on transport, was presented by Smith and Schwartz (1984).

A related, yet distinctly different approach, was described by Hull et al. (1987). In their model they used a discrete-fracture approach in conjunction with a rectangular finite-difference grid of unit thickness. Fractures were permitted to connect any two adjacent nodes, vertically, horizontally or diagonally, provided no more than eight fractures converged at a node. A velocity profile was allowed to develop within an individual fracture. Solute transport was simulated using marker particles. To accommodate diffusion, particles were permitted to move across the velocity profile within the fracture through a random walk process.

Building on the work of Bibby (1981), Huyakorn et al. (1983) developed a dual-porosity model that used a two-dimensional finite-element mode to approximate transport in the fractures and a one-dimensional finite-element model to provide leakage from the blocks to the fractures. The specific assumptions associated with this development are:

(i) Transport in the fractures is two-dimensional and is controlled by both hydrodynamic dispersion and advection.

(ii) Local sorption equilibrium is reached instantaneously, and the sorption equilibrium isotherm is linear.

(iii) Chemical reaction of radioactive decay effects can be described by a first-order term.

(iv) The fractured porous medium is microscopically uniform and can be approximated by the parallel fracture-prismatic block conceptual model (Fig. 8.1.3) or by the spherical-block model (Fig. 8.1.4).

(v) Transport in the matrix blocks is dominated by molecular diffusion, and this can be described by the well-known, one-dimensional diffusion equation in Cartesian or spherical coordinates.

(vi) For the spherical model the block diameter is small in comparison with overall transport distance.

Note that the concept of the blocks as spheres inherent in assumption (iv) appears to be a novel aspect of this work. The coordinate system associated with this block model is illustrated in Fig. 8.1.3. In Huyakorn et al. (1983c) the authors extend their earlier work to include nuclide decay chains. They also incorporate the one-dimensional finite-element representation of discrete fractures, mentioned earlier. This appears to be the first time the fractured porous medium represented by a dual-porosity formulation was combined with the one-dimensional finite-element model of a discrete fracture.

Wilson and Dudley (1987) presented a one-dimensional finite-difference model for radionuclide transport in an unsaturated, fractured porous medium. In the development of their

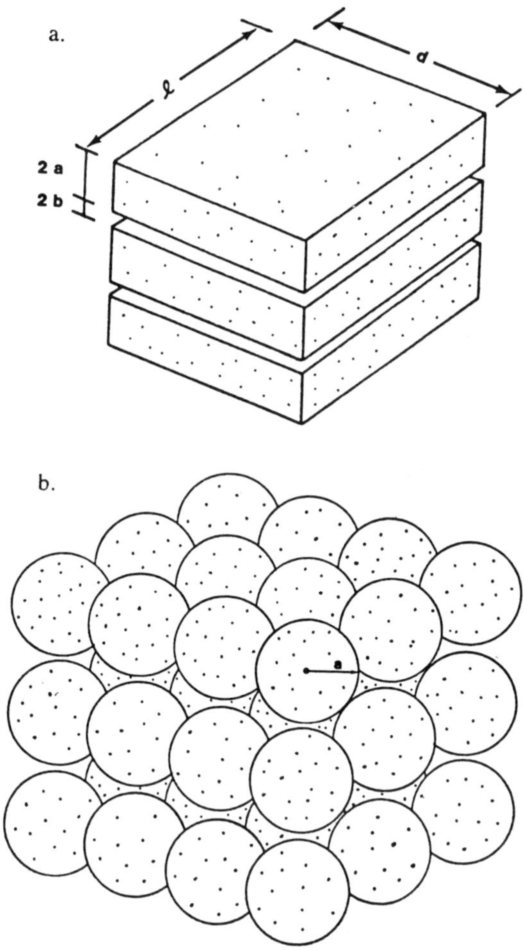

Figure 8.1.3. Dual-porosity fractured media: (a) prismatic blocks, (b) spherical blocks; from Huyakorn et al. (1983).

model they included the advective transport of nuclides in the blocks. This they achieved by using two continuum equations throughout the domain, one for the fractures and the other for the porous blocks. Thus all blocks are connected via the transport equation for the blocks and all fractures are connected via a transport equation for the fractures. The two continuum equations are coupled by a leakage term describing transport between them.

Although analytical solutions are generally only available for idealized geometries and boundary conditions, and assume uniform material properties, they are useful in the determination of solute-penetration distances along a fracture and the associated response times for worst-case or geologically simple scenarios. Example situations where analytical solutions have yielded

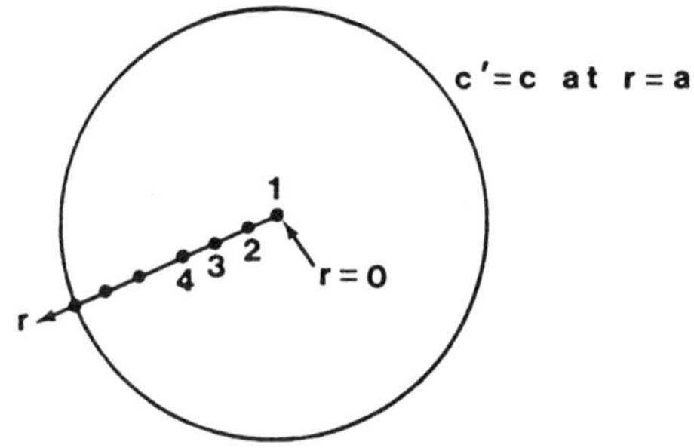

Figure 8.1.4. Discretization of spherical block; from Huyakorn et al. (1983).

valuable insight include the case of a single dominant fracture intersecting a subsurface waste repository (Neretnieks, 1980) and the interpretation of tritium profiles through vertically-fractured clays to establish groundwater ages (Day, 1977; Cherry et al., 1979; D'Astou et al., 1989). Other useful applications of analytical solutions include the interpretation of laboratory tests (Sudicky et al., 1985), interpreting organic solute transport data from field experiment (Goltz and Roberts, 1989), and evaluation of the accuracy of numerical solutions to the problem of transport through fractured porous media (e.g. Huyakorn et al., 1983).

Two conceptually simple scenarios amenable to analytical solution techniques include the case of a single fracture bounded by an infinitely thick porous matrix and the case of a set of identical fractures whose axes are parallel and equally spaced. Analytical solutions applicable to the single fracture problem have been developed by Neretnieks (1980) and Grisak and Pickens (1981). These authors ignored the effect of dispersion along the fracture axis. Tang et al. (1981) take into account longitudinal dispersion in the fracture. Solutions for one-dimensional advection-dispersion in a system of discrete parallel fractures and transverse matrix dispersion have been presented by Barker (1982), Lever et al. (1982), and Sudicky and Frind (1982). Solutions for radial advection-dispersion in a single fracture and transverse matrix diffusion have been presented by Chen (1986) and Feenstra et al. (1984).

8.1.4 Geothermal Engineering

Physically and mathematically sophisticated modeling has been done in fractured geothermal reservoir simulation. As in the case of the preceding discussions, discrete and dual-porosity formulations have been presented.

One of several papers on this topic presented as part of the series identified with the Stanford Geothermal Reservoir Engineering workshop is that of Kasameyer and Schroeder (1975). In this model the fractures in the porous formation are modeled by a series of channels between parallel plates. The heat transfer from the block to the fracture is given by the steady-state conduction equation wherein the average temperature difference between the rock and the fracture fluid is used. The reservoir is represented by several volumes in series and the fluid is assumed to move through the reservoir without mixing. The temperature decline is obtained independently for each volume. Three coupled differential equations, one for the rock, one for the interstitial pore space, and one for the fluid in the fractures are required to represent this system.

A dual-porosity single-phase fractured-reservoir model was developed by Gray et al. (1976). This model coupled mass and heat transfer into one code and assumed that the pressures in the fractures and the blocks were near equality. A linear relationship was used to couple the temperature transport in the fractures and the pores. The numerical simulator used isoparametric Hermitian finite-elements in three space dimensions. Figure 8.1.5 shows the simulated propagation of the temperature front in the fractures and porous blocks. The two sets of curves are for two different values of the coupling temperature coefficient between the two media, denoted in the figure as h.

A discrete multiphase steam-water fracture model was introduced by Coats (1977). In this finite-difference model a number of vertical and horizontal fractures were constructed around blocks that were discretized into several finite-difference cells. In simulating the fractures he used sufficiently large permeabilities so as to render viscous forces negligible relative to gravitational forces.

Pinder et al. (1979) presented a single-block model surrounded by fractures to investigate the detailed behavior of multiphase flow in a parallel plate model and a model wherein the blocks were represented by spheres. Because of the block geometries selected, this numerical experiment could be conducted using a one-dimensional finite-difference model.

A model similar to that of Coats (1977) was introduced by Pruess (1983). In this multiphase formulation the transient behavior in the blocks is once again represented by a numerical calculation, such as used by Coats (1977), but is different inasmuch as the integrated finite-difference method used to solve the resulting set of equations was extended to blocks of irregular geometry. Subsequent work utilizing this conceptual model is found in Pruess and Narasimhan (1985) and Bodvarsson (1985). A review paper by Pruess and Wang (1987) presents a summary of the state of the art in geothermal modeling of multiphase flow in fractured porous media.

8.1.5 Soil Physics

In the soil horizon it is possible to encounter problems akin to those encountered in geothermal, petroleum and groundwater reservoirs. Indeed, infiltration into both macropores and cracks in clayey soil is of considerable importance to the soil physicist.

One of the early papers in this area of simulation was written by Edwards et al. (1979). These researchers used a standard unsaturated flow model to calculate the flow from a vertical hole into surrounding soil, using a cylindrical coordinate system. At about the same time Hoogmoed and Bouma (1980) developed a model to examine the pattern of infiltration in cracked clay soil. A quasi-two-dimensional formulation was developed that employed two one-dimensional models. One model was used to simulate vertical infiltration into the blocks between the cracks and the other to calculate flow from the fractures into the blocks in the horizontal direction. Davidson (1985) presented a model for saturated/unsaturated flow into a cracked soil and applied it to a problem similar to that considered by Edwards et al. (1979). A standard two-dimensional finite-difference formulation based upon the Kirchhoff transformation was used to examine infiltration in a soil containing regularly spaced, water-filled, vertical cracks. Germann and Beven (1985) used a kinematic wave approximation to simulate infiltration into soils exhibiting macropores. They employed the kinematic wave approach to describe vertical water flow along the macropores. Water is lost to the porous blocks via a sink function. This represents a dual-porosity formulation along the conceptual lines of work described above. Finally, a recent paper

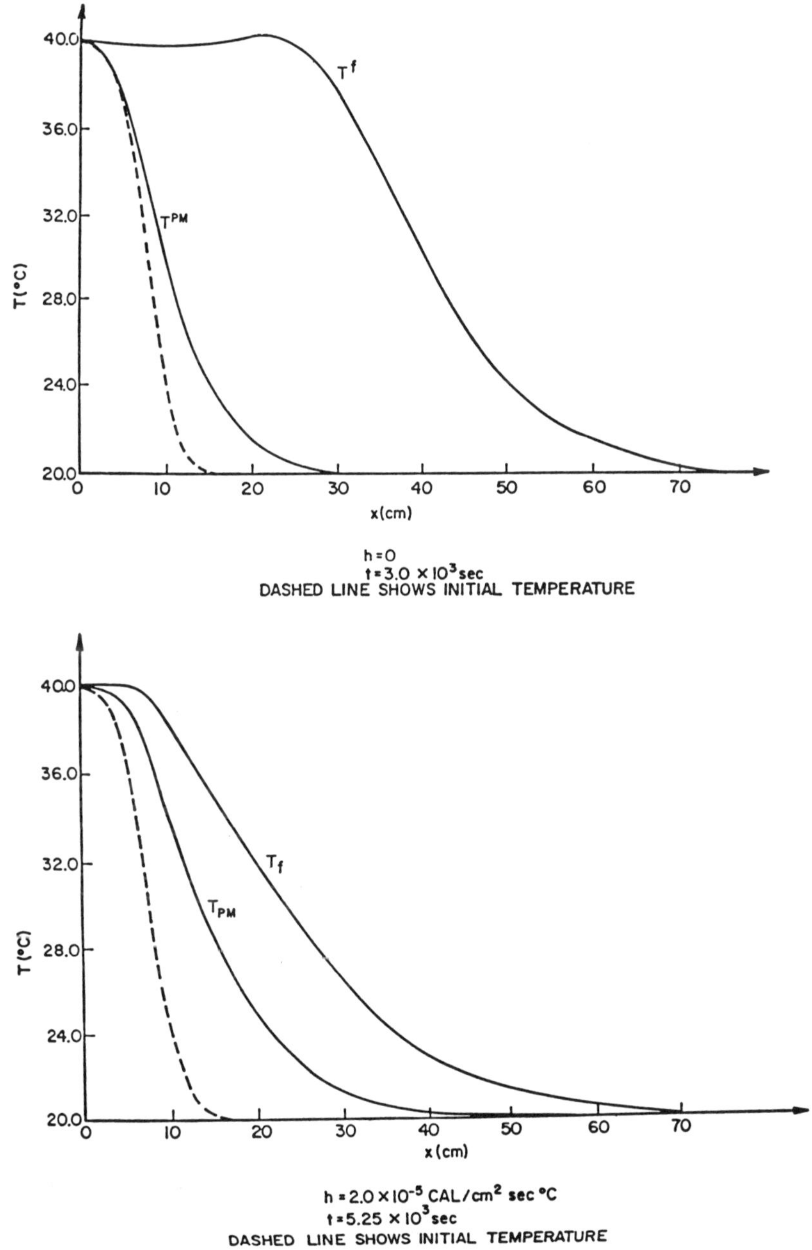

Figure 8.1.5. Relationship between temperature and distance for fractures (T_f) and blocks (T_{pm}) at a specified time: a) when there is no thermal coupling between blocks and fractures ($h = 0$), and b) when there is thermal coupling between the blocks and fractures (h = .00002 cal/cm^2 sec degree C.); from Gray et al. (1976).

by Huyakorn (1988) presents a review of modeling approaches to water flow and solute transport in unsaturated soils and macropores. He includes a discussion of discrete versus dual-porosity formulations.

8.1.6 Dual-Porosity, Discrete-Fracture, and Single-Porosity Models

In the foregoing description of the various simulation methods that have been developed, both discrete and dual-porosity formulations were encountered. Recently two attempts have been made to determine the circumstances under which each of these formulations is applicable and when either or both can be replaced by a single-porosity counterpart.

Berkowitz et al. (1988) presented an analysis wherein they used one-dimensional equations to describe flow and transport of a solute in fracture and two-dimensional equations to represent flow and transport in the porous blocks. The two sets of equations were coupled via the mass exchange term across the fracture walls. Continuity of both mass flux and potential is required at these boundaries. A finite-element formulation was used to approximate the governing equations.

Employing this mathematical formulation, solute transport through fractured media with two fracture orientations was simulated. The two-dimensional concentration distributions calculated were used to generate flux-averaged breakthrough curves that could be compared with the equivalent single-porosity model. It was found that the ability to represent the discrete-fracture system with the single-porous-medium model depended up the orientation of the fracture system relative to the flow field. More specifically it was found that when the flow was at an angle of 45 degrees to the fractures, it was possible to obtain parameters for the porous medium model that would reproduce the discrete-fracture results. However, in the case of flow parallel to the fractures, the problem was much more difficult. The authors point out that this difficulty may be due in large part to the particular geometry of the fractures and the reservoir.

Dougherty and Pinder (1989) investigated the same basic question as Berkowitz et al. (1988), using a different theoretical protocol. In their work they developed not only the dual-porosity formulation for the fractured porous medium but also the corresponding single-equation porous medium representation. They found, as did Berkowitz et al. (1988), that it was possible to represent the dual-porosity simulations using a single-equation formulation when suitable constitutive parameters were employed. It is worth noting that in both of the above-described works, convective transport between fractures and blocks was accommodated.

8.2 MATHEMATICAL FORMULATION

The preceding discussion documents two distinctly different approaches to the simulation of flow and transport in fractured porous media. In one approach the fractures are recognized as distinct entities whose geometry and fluid mechanical properties must be known along with the fluid mechanical properties of the porous blocks in order to properly define the simulation problem. In the other approach one recognizes the fractures as members of a statistically meaningful set which can be described mathematically as a continuum, with a second overlapping continuum representing the flow and transport in the blocks. Within these two generic approaches there are many specific formulations which exploit the ingenuity of the analyst and the admissible simplifications of the problem at hand. In the following discussion only the generic formulations will be presented. Details on specific applications can be obtained from the references discussed in the preceding sections and the additional references provided at the end of the chapter. The presentation will begin with mathematical statements of the problem. This will be followed by a discussion of the analytical and numerical techniques that can and have been brought to bear on the solution of problems presented in the literature.

8.2.1 Governing Equations

As is evident from the foregoing, there are very complex and sophisticated fracture flow models described in the literature. In this presentation we will consider for clarity of presentation only saturated flow and a single, non-reacting solute. The complete description of single phase fluid flow and solute transport in a fractured porous medium requires two governing equations. In the discrete-fracture model, one equation describes fluid flow and mass transport in the fracture and the other describes fluid flow and mass transport in the blocks. The dual-porosity model on the other hand employs two continuum equations, one for the fractures and the other for the blocks. These two equations are coupled together by a transfer function that describes mass transfer between the blocks and the fractures.

8.2.2 Fluid Flow

Flow in the blocks of the discrete-fracture model is described in two space dimensions by

$$\frac{\partial}{\partial x_i}\left(K_{ij}\frac{\partial h}{\partial x_j}\right) = S_s\frac{\partial h}{\partial t} - q \quad i,j = 1,2 \tag{8.2.1}$$

where K_{ij} is the hydraulic conductivity of the porous blocks,

h is the hydraulic head,

S_s is the specific storage, and

q is the external source of fluid due, for example, to injection wells.

Flow in each individual fracture can be reduced, under certain simplifying assumptions to give

$$\frac{\partial}{\partial s}\left(K^f\frac{\partial h^f}{\partial s}\right) = S_s^f\frac{\partial h^f}{\partial t} - q^f \tag{8.2.2}$$

where s is the local coordinate along the fracture line,

K^f is the hydraulic conductivity in the fracture in the direction of the fracture line,

S_s^f is the specific storage coefficient for the fracture, and

q^f is once again the flow rate of the source per unit volume of the fractured medium.

In the case of the dual-porosity model, flow in the fractures is described by

$$\frac{\partial}{\partial x_i}\left(K_{ij}^f \frac{\partial h^f}{\partial x_j}\right) = S_s^f \frac{\partial h^f}{\partial t} - \varphi^f \Lambda - \varphi^f q^f i, j = 1,2 \tag{8.2.3}$$

where Λ is the volumetric rate of fluid transfer per unit area from the porous matrix blocks to the fractures. The hydraulic conductivity and specific storage are now continuum parameters, and φ^f is fracture porosity defined as the volume of fractures per unit volume of the formation. In the above formulation it is assumed for clarity that the model does not explicitly incorporate the water table.

The equation for flow in the blocks in the case of the dual-porosity model appears as (8.2.1), with the addition of the porous medium porosity where appropriate, and with the Λ term to account for the interflow between the fractures and the blocks. In the above formulation, there is now the question of how to select Λ. As outlined earlier, there are three popular models for defining this parameter. We will consider each of these in sequence.

Quasi-Steady Fracture-Flow Model

This, the simplest of the models, assumes a linear expression for Λ in (8.2.3), that is

$$\Lambda = \Pi\sigma^2\left(\bar{h} - h^f\right) \tag{8.2.4}$$

here $\bar{h}$ is the average head in the rock matrix, σ is the specific surface of the fractures and is defined as the surface area of fractures per unit volume of the entire porous medium, and Π is a leakage parameter assumed to be constant. The porous block flow model is simplified to give

$$S_s \frac{\partial h}{\partial t} = \Pi\sigma^2\left(h^f - \bar{h}\right) = -\Lambda \tag{8.2.5}$$

or

$$\frac{\partial \bar{h}}{\partial t} + \omega\left(\bar{h} - h^f\right) = 0 \tag{8.2.6}$$

where ω is given by

$$\omega = \Pi\sigma^2/S_s \quad .$$

This equation can be solved to yield

$$\Lambda = -S_s\omega\int_0^t \frac{\partial h^f}{\partial \tau} e^{-\omega(t-\tau)} d\tau \quad .$$

Note that this expression is predicated on the assumption that leakage is dependent only on the difference between the head in the fracture and the average head in the matrix. Because there is no reference to block geometry, this expression is consistent with continuum theory that assumes no specific knowledge at the sub-continuum level.

Unsteady Parallel Fracture-Flow Model

The parallel fracture model for leakage from the fractures assumes a low-permeability prismatic block with linear flow perpendicular to the fractures. It is further assumed that the fractures are equally spaced and of constant aperture. Using the nomenclature indicated in Fig. 8.2.1, the expression obtained by solving the one-dimensional flow equation for the blocks yields

$$\Lambda = \left[\frac{-2K}{a(a+b)}\right]\left[\sum_{n=0}^{\inf} \int_0^t \frac{\partial h^f}{\partial \tau} e^{-\alpha_n (t-\tau)} d\tau\right] \tag{8.2.7}$$

where b is half the fracture aperture, a is half the block aperture, and

$$\alpha_n = \pi^2 (2n+1)^2 K / \left(4 S_s a^2\right) \ .$$

Unsteady Blocky Fracture Flow Model

In this conceptual model the blocks of the matrix are envisioned as spheres. Flow in the blocks is assumed to be governed by the flow equations defined, in the case of two dimensional flow, in radial coordinates. The solution to this equation that is relevant to our simulation problem is given by

$$\Lambda = \left[\frac{-6K}{a(a+b)}\right]\left[\sum_{n=1}^{\inf} \int_0^t \frac{\partial h^f}{\partial \tau} e^{-\alpha_n (t-\tau)} d\tau\right] \tag{8.2.8}$$

and

$$\alpha_n = \pi^2 n^2 K / \left(S_s a^2\right)$$

where r, the radius of the sphere, is defined as a.

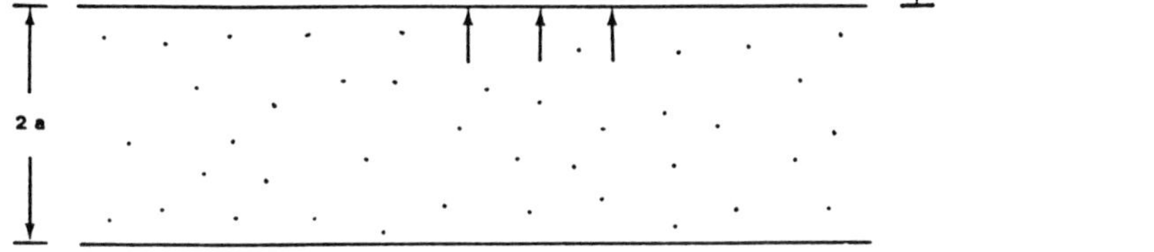

Figure 8.2.1. Dual-porosity parallel-fracture model; from Huyakorn et al., (1983a).

8.2.3 Finite-Element Formulation

The brief review of the literature presented earlier indicates that there are a number of methods that can be used to approximate the equations governing flow in fractured media. To keep this chapter to a manageable length we will consider only the finite-element numerical method. In this presentation we select the approach of Huyakorn et al. (1983a) as an example.

Two Porosity Formulation

The point of departure is the fracture flow equation with the choice of Λ defined by one of the three block models indicated above. Choosing the parallel fracture flow model we obtain

$$\frac{\partial}{\partial x_i}\left(K_{ij}^f \frac{\partial h^f}{\partial x_j}\right) - S_s^f \frac{\partial h^f}{\partial t} + \varphi^f q^f - \varphi^f \beta \sum_{n=0}^{\inf} I_n = 0 \qquad (8.2.9)$$

where

$$\beta = \frac{2K}{a(a+b)} \qquad (8.2.10)$$

$$I_n = \int_0^t \frac{\partial h}{\partial \tau} e^{-\alpha_n (t-\tau)} d\tau \qquad (8.2.11)$$

Application of Galerkin's method yields the following set of equations

$$\int_R K_{ij} \frac{\partial N}{\partial x} I \frac{\partial N}{\partial x} J\, h_j^f \, dR + \int_R S_s^f N_I N_J \frac{dh^f}{dt} J \, dR - \int_B N_I \varphi^f q^f dB - \varphi^f Q_I^f$$

$$+ \left[\int_R \varphi^f \beta N_I N_J \, dR\right]\left[\sum_{n=0}^{\inf} I_{nJ}\right] = 0 \quad I = 1,2,\ldots,M \qquad (8.2.12)$$

where R and B are the flow region and the boundary of the region respectively, N_I ($I = 1,2,\ldots,M$) are the basis functions, M is the number of nodes in the approximation of the region by finite-elements and I_{nJ} is the value of the convolution integral at node J which is given by

$$I_{nJ} = \int_0^t \frac{dh^f}{d\tau} J \, e^{-\alpha_n (t-\tau)} d\tau$$

and Q_I^f is the flow rate of a well located in the fractured medium.

Equation (8.2.12) represents a set of ordinary-differential equations in time. It is customary in finite-element work to approximate the time derivative using a standard finite-difference approximation. Employing the superscript k as the representation of time, that is $t = k\Delta t$ (8.2.12) can be written

$$\left[\theta G_{IJ} + \frac{H_{IJ}}{\Delta t}\right] h_J^{f,k+1} = \left[(\theta - 1) G_{IJ} + \frac{H_{IJ}}{\Delta t_{k+1}} h_J^{f,k} + \theta \varphi^f Q_I^{f,k+1}\right.$$

$$+ (1-\theta)\varphi^f Q_I^{f,k} - \left(H_{IJ}^*\right) \sum_{n=0}^{\inf} \left(\theta I_{nJ}^{k+1} + (1-\theta) I_{nJ}^k\right) \tag{8.2.13}$$

where θ is the time-weighting factor. The matrices represented in (8.2.13) are given by the elements

$$G_{IJ} = \int_R K_{ij}^f \frac{\partial N}{\partial x_i} I \frac{\partial N}{\partial x_j} J \, dR$$

$$H_{IJ} = \int_R S_s^f \, N_I \, N_J \, dR$$

$$H_{IJ}^* = \int_R \varphi^f \beta \, N_I \, N_J \, dR$$

$$I_{nJ}^{k+1} = \int_0^t k + 1 \frac{dh^f}{d\tau^J} e^{-a_n(t-\tau)} d\tau$$

The integrations indicated in G_{IJ}, H_{IJ} and H_{IJ}^* are performed using the standard techniques which may involve either numerical or analytical procedures. The integral indicated by I_{nJ}^{k+1} is evaluated using (see Huyakorn et al. 1983a)

$$I_{nJ}^{k+1} = e^{-\alpha_n \Delta t_{k+1}} \, I_{nJ}^k$$

$$+ (1/\alpha_n) \left(\Delta h_J / \Delta t\right)_{k+1} \left(1 - e^{-\alpha_n \Delta t_{k+1}}\right) \tag{8.2.14}$$

where

$$\left(\Delta h_J / \Delta t\right)_{k+t} = \left(h_J^{f,k+1} - h_J^{f,k}\right) / \left(t_{k+1} - t_k\right)$$

The flux type boundary condition appearing in (8.2.14) would vanish if no flux boundary is encountered and would appear as a known value in the matrix equation should there be a non-zero flux. Constant-head boundaries are incorporated by replacing the appropriate value of h_J^f.

As indicated in our review presented earlier, an alternative strategy to the use of the analytical solution indicated above to represent the transient flux from the block to the fracture is to solve the transient one-dimensional block equation numerically. Thus the equation

$$K \frac{\partial^2 h}{\partial z^2} = S_s \frac{\partial h}{\partial t}$$

is solved subject to the conditions

$$h = h_0 \quad \text{at} \quad t = 0$$

$$h = h^f \quad \text{at} \quad z = 0$$

$$\frac{\partial h}{\partial z} = 0 \quad \text{at} \quad z = a$$

where the coordinate system in use is presented in Fig. 8.2.2. The appropriate finite-difference equations are given by

$$h_1^{k+1} = h_J^{f,k+1} \tag{8.2.15a}$$

$$\alpha_i \, h_{i-1}^{k+1} + \beta_i h_i^{k+1} + \zeta_i \, h_{i+1}^{k+1} = d_i \qquad i = 2, 3, \ldots. N-1 \tag{8.2.15b}$$

$$\alpha_N h_{N-1}^{k+1} + \beta_N h_N^{k+1} = d_N \tag{8.2.15c}$$

where N is the number of finite-difference nodes in the matrix block approximation. The terms in this equation are defined as

$$\alpha_i = -K \Big/ \left(\Delta z_i \, \Delta z_{i-\frac{1}{2}} \right)$$

$$\zeta_i = -K \Big/ \left(\Delta z_i \, \Delta z_{i+\frac{1}{2}} \right)$$

$$\beta_i = -(\alpha_i + \zeta_i) + S_s / \Delta t_{k+1}$$

$$d_i = S_s \, h_i^k / \Delta t_{k+1}$$

$$\alpha_N = -K \Big/ \left(\Delta z_N \, \Delta z_{N-\frac{1}{2}} \right)$$

$$\beta_N = \alpha_N + S_s / (2 \Delta t_{k+1})$$

$$d_N = \left(S_s / (2 \Delta t_{k+1}) \right) h_N^k$$

$$\Delta z_i = z_{i+1} - z_i, \quad \Delta z_{i-\frac{1}{2}} = \frac{1}{2} (\Delta z_i + \Delta z_{i-1}) \, .$$

Given a solution to the above set of equations it is possible to determine the flux term Λ_J from the relationship

$$\Lambda_J^{k+1} = \frac{K}{\Delta z_1} \left(h_J^{k+1} - h_J^{f,k+1} \right) - \frac{\Delta z_1 S_s}{2 \Delta t_{k+1}} \left(h_J^{k+1} - h_J^k \right) \tag{8.2.16}$$

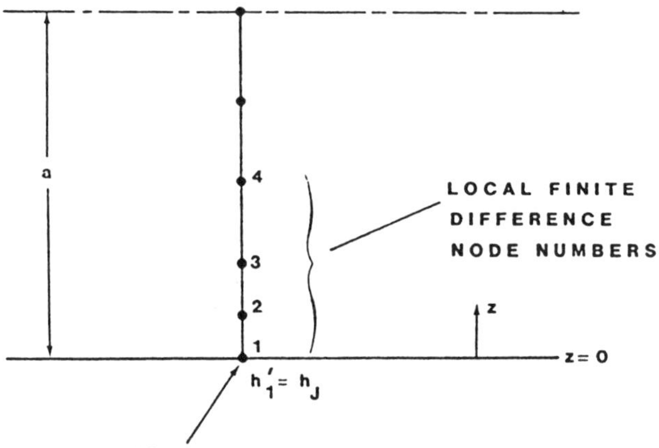

Figure 8.2.2. Finite-difference discretization of one-dimensional flow region in a prismatic matrix block (from Huyakorn, 1983a).

where

$$h_J^{f,k+1} = h_1^{k+1}$$

$$h_J^{k+1} = h_2^{k+1}$$

$$\Delta z_1 = z_2 - z_1$$

and the value $h_J^{f,k+1}$ is the value of the head in the matrix block immediately above the fracture node J. There are a number of possible strategies for the solution of (8.2.13 through 8.2.16). Huyakorn et al. (1983a) used a sequential solution strategy.

Discrete-Fracture Formulation

The finite-element method is particularly well-suited to dealing with discrete fractures, whether or not a continuum formulation is employed. The formulation employs equations (8.2.1 through 8.2.2). The block equation is either identical to or a simplification of the continuum equation discussed in the preceding section. Thus the finite-element approximation of this equation is not required and we will focus on the discrete-fracture equation (8.2.2).

While there are all levels of sophistication with respect to the equation selected to solve fracture flow, in practice it is likely that the simple form presented in (8.2.3) will be used. In fact it is common to further simplify this expression to a steady form (Huyakorn et al., 1983a), that is

$$\frac{\partial}{\partial s}\left[K^f \frac{\partial h^f}{\partial s}\right] + q^f = 0 \tag{8.2.17}$$

Application of Galerkin's method to this equation yields, after using integration by parts,

$$\int_S K^f \frac{dN^f}{ds}{}_I \frac{dN^f}{ds}{}_J h_J^f \, ds + K^f \frac{\partial \hat{h}^f}{\partial s} N_I^f \Big|_B - Q_I^f = 0 \qquad (8.2.18)$$

where N_I^f are basis or shape functions defined along the one-dimensional coordinate system s and $\hat{h}$ is the approximation to h. The second term in (8.2.18) represents the boundary conditions at the ends of the fracture. If linear basis functions are employed, the integrations indicated in this equation are easily achieved and the resulting expressions are the same as the standard finite-difference approximation for the same differential equation.

In application, (8.2.18) is assembled into the global two-dimensional system of equations in a straightforward manner. The entire system is then solved for all the heads. Note that it is no longer necessary to include a fluid-transfer term because both the block and fracture equations have common nodal degrees of freedom. The same approach can be used for two-dimensional fractures in a three-dimensional formulation.

8.2.4 Analytical Solutions to Flow Problems

Analytical solutions for flow in fractured media can be catalogued according to whether the discrete-fracture model or continuum model form the conceptual framework of the analysis. As an example of the dual-porosity approach, consider the work of Streltsova (1976). Considering flow to a well in a fractured medium the governing dual-porosity equation for the drawdown, $\$$, that can be determined through combination of the flow and fracture equations takes the form

$$T\left[\frac{\partial^2 \$^f}{\partial r^2} + \frac{1}{r}\frac{\partial \$^f}{\partial r}\right] = S_s^f \frac{\partial \$^f}{\partial t} + S_s \frac{\partial \$}{\partial t} \qquad (8.2.19)$$

where T is the aquifer transmissivity, and r is the radial distance from the center of the extraction well. For a suitable transformation of coordinates, this equation can represent flow in an anisotropic aquifer. If the quasi-steady fracture flow model is used for flow in the fracture (8.2.6), then, recognizing that Λ is equivalent to $S_s \frac{\partial \$}{\partial t}$, the solution to (8.2.6) can be used to obtain the expression

$$T\left[\frac{\partial^2 \$^f}{\partial r^2} + \frac{1}{r}\frac{\partial \$^f}{\partial r}\right] = S^f \frac{\partial \$^f}{\partial t} + \omega S \int_0^t e^{-\omega(t-\tau)} \frac{\partial \$}{\partial t} \, d\tau \qquad (8.2.20)$$

Solution of this expression subject to the boundary condition at the well of

$$-r\frac{\partial \$}{\partial r} = Q/(2\pi T) \qquad r \to 0$$

is

$$\$^f = \frac{Q}{2\pi T} W_f \left[\frac{r}{B}, \omega t, \eta\right]$$

where

$$W_f = \int_0^{\inf} J_0\left[\frac{r}{B}\frac{x}{v}\right]\left[2 - \frac{A}{1+a_1} - \frac{C}{1+a_2}\right.$$

$$\left. - \frac{\eta(1-x^2)}{\hat{q}}\left[\frac{A}{1+a_1} - \frac{C}{1+a_2}\right]\right]\frac{dx}{x}$$

and

$$A = e^{a_1 \omega t} + a_1 e^{-\omega t}, a_1 = 0.5\left[\tilde{q} - \eta(1+x^2)\right], C = e^{a_2 \omega t} + a_2 e^{-\omega t},$$

$$a_2 = -0.5\left[\tilde{q} + \eta(1+x^2)\right], \text{ and } \tilde{q} = \left[\eta^2(1+x^2) - 4\eta x^2\right]^{\frac{1}{2}}.$$

This solution is used for well test evaluation, and type curves for the case of η = 10 and various assumed values of parameters r/B and ωt are given in Fig. 8.2.3.

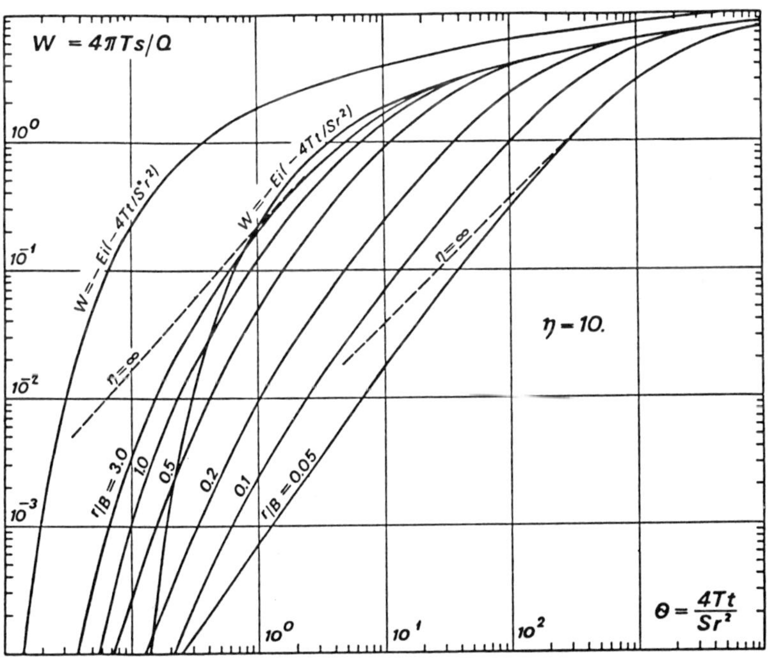

Figure 8.2.3. Type curves of the fissure flow drawdown function W_f plotted on double logarithmic graph paper for η=10 (from Streltsova (1976).

8.2.5 Contaminant Transport

Now that we have examined, albeit briefly, the determination of groundwater heads in fractured porous media, we now turn our attention to groundwater transport. As in the preceding sections, we will begin with an examination of the basic equations, followed by a discussion of their solution.

8.2.6 Governing Equations for Contaminant Transport

While substantial work has been done in the area of groundwater transport and relatively complex models exist, to achieve clarity in the following presentation we will consider the

simplified formulation provided in Huyakorn et al. (1983b). This formulation assumes the following:

1. Transport in the fissures is two-dimensional and is controlled by both hydrodynamic dispersion and advection.
2. Local sorption equilibrium is reached instantaneously, and the sorption equilibrium isotherm is linear.
3. Chemical reaction or radioactive decay effects can be described by a first-order term.
4. The fractured porous medium is macroscopically uniform and can be approximated by the parallel fracture-prismatic block conceptual model.
5. Transport in the blocks is dominated by molecular diffusion, and this can be described by the well-known one-dimensional diffusion equation in Cartesian or spherical coordinates.
6. For the spherical model the block diameter is small in comparison with overall transport distance.

Employing these assumptions, the transport equation can be written (Huyakorn et al., 1983b)

$$\frac{\partial}{\partial x_i}\left[D_{ij}^f \frac{\partial c^f}{\partial x_j}\right] - V_i^f \frac{\partial c^f}{\partial x_i} =$$

$$\frac{\partial c^f}{\partial t} + \lambda c^f - q^f\left(c^* - c^f\right) - \left[\frac{1-\varphi^f}{\varphi^f}\right]\Gamma \qquad (8.2.21)$$

where

c^f is concentration of solute expressed in terms of mass per unit volume of solution,

D_{ij}^f is the hydrodynamic dispersion tensor,

V_i^f is the average velocity in fractures along the x_i component of the Cartesian coordinate system,

λ is the first-order decay constant,

c^* is the concentration of the injected fluid, and Γ is the rate of solute transfer from matrix block to fracture per unit volume of the matrix block.

While it is generally recognized that dispersion in a fractured medium is physically different than dispersion in a porous medium, the porous medium form of the dispersion tensor is generally used for lack of a more appropriate expression. This relationship is given as

$$D_{ij}^f = \alpha_T |V| \delta_{ij} + (\alpha_L - \alpha_T) \frac{V_i V_j}{|V|} + \tau D^*$$

where α_L and α_T are the longitudinal and transverse dispersivities respectively, δ_{ij} is the Kronecker delta, D^* is molecular diffusion, and τ is the tortuosity.

As in the case of flow in a dual-porosity medium, there are several models for transport from blocks to fractures. For the case of the parallel fracture model with fracture half-thickness of b and block half-thickness of a, the fracture porosity can be described by

$$\varphi^f = b/(a+b).$$

With this definition in mind, the mass flux from the blocks to the fractures due only to molecular diffusion is given by

$$\Gamma = -\frac{1}{a}\left[D_m \frac{\partial c}{\partial z}\bigg|_{z=a} \right] \tag{8.2.22}$$

where D_m is the molecular diffusion coefficient. The coordinate z is a local coordinate with center at the origin of the block.

8.2.7 Finite-Element Formulation of the Transport Equations

As in the case of groundwater flow, it is possible to view groundwater transport from either a dual-porosity or discrete-fracture perspective. The dual-porosity approach will be considered first and then extended for the case of discrete fractures. The formulation follows the work of Huyakorn et al. (1983b).

Dual-Porosity Formulation

The development of the finite-element formulation of the transport equation follows along the same lines as that employed in approximating the flow equations. The principal difference is the utilization of upstream weighting to control numerical oscillations in the concentration field near steep fronts.

The approximating equations are obtained by application of a weighted residual method wherein the weighting function for the convective and dispersive terms is not the basis function but rather is an asymmetric quadratic polynomial with adjustable coefficients (see Huyakorn and Pinder, 1983). Assuming the concentration in the fractures is given by

$$c^f \approx \hat{c}^f = N_J c_J^f \quad J = 1,2,\ldots m \tag{8.2.23}$$

where N_J are the m spatially dependent basis functions. Note that m may be less than M since the transport region may be smaller than the flow region. Letting W_I be the asymmetric weighting function, the weighted residual approximation becomes

$$\int_R W_I \left[\frac{\partial}{\partial x_i} \left[D_{ij}^f \frac{\partial c^f}{\partial x_j} \right] - V_i^f \frac{\partial c^f}{\partial x_i} \right] dR$$

$$-\int_R N_I \left[\frac{\partial c^f}{\partial t} + \lambda \hat{c} - q^f (c * - \hat{c}) - \xi \Gamma \right] dR = 0$$

$$I = 1,2,\ldots m \tag{8.2.24}$$

where $\xi = (1 - \varphi^f)/\varphi^f$.

Application of Green's theorem to the second-order term appearing in (8.2.25), representation of the flux term q^f by a well discharge Q_I^f, as was done in the flow equation, and the introduction of a general third-type boundary condition

$$\left[V_i^f c^f - D_{ij}^f \frac{\partial c^f}{\partial x_j} \right] n_i = -q_n^f \tilde{c}^f$$

where n_i is the outward directed normal vector and $\tilde{c}^f$ is the concentration of the solute crossing the flow boundary yields a matrix equation very similar to that generated for the flow equation

$$\left[\theta S_{IJ} + M_{IJ} / \Delta t + \theta \lambda M_{IJ} \right] c_J^{f,k+1} - \xi M_{IJ} \Gamma_J^{k+\theta}$$

$$+\theta Q_I^f c_I^{f,k+1} + \theta \tilde{M}_{IJ} c_J^{f,k+1} = F_I \quad I = 1,2,\ldots,m \tag{8.2.25}$$

where $t = k\Delta t$ as in the flow development and θ once again is the time weighting. The coefficients in the matrices indicated in (8.2.25) are defined as follows

$$S_{IJ} = \int_R \left[D_{ij}^f \frac{\partial W_I}{\partial x_i} \frac{\partial N_J}{\partial x_j} + V_i^f W_I \frac{\partial N_J}{\partial x_i} \right] dR$$

$$M_{IJ} = \int_R N_I N_J \, dR$$

$$\tilde{M}_{IJ} = \int_R q_n^f W_I N_J \, dB$$

$$F_I = (\theta - 1)(S_{IJ} + \lambda M_{IJ}) c_J^{f,k} + (M_{IJ} / \Delta t) c_J^{f,k} + (\theta - 1) Q_I^f c_I^{f,k}$$

$$+ \tilde{M}_{IJ} c_J^{f,k} (\theta - 1) + Q_I^f c * + \tilde{M}_{IJ} \tilde{c}_J^{f,k}$$

$$\Gamma_J^{k+\theta} = \theta \Gamma_J^{k+1} + (1 - \theta) \Gamma_J^k$$

In the above equations it is necessary to have a value for $\Gamma^{k+\theta}$ in order to solve the equations.

Although there are several approaches to obtaining this coupling function, the method proposed by Huyakorn et al. (1983b), which relies on finite-element representation of the behavior in the blocks, will be presented. Consider the one-dimensional form of the transport equation

$$\varphi R\frac{\partial c}{\partial t}-\frac{\partial}{\partial z}\left[D\frac{\partial c}{\partial z}\right]+\lambda\,\varphi Rc=0\,. \tag{8.2.26}$$

One can apply Galerkin's method to this equation directly since there is no first-order convective term and obtain, after application of integration by parts

$$\int_0^a D\frac{\partial N_I}{\partial z}\frac{\partial N_J}{\partial z}c_J\,dz+\int_0^a \varphi RN_I\,N_J\frac{dc_J}{dt}\,dz$$

$$+\int \lambda\varphi RN_I\,N_J\,c_J\,dz+\Gamma=0 \quad I=1,2\ldots p \tag{8.2.27}$$

where

$$\Gamma=-D\frac{\partial\hat{c}}{\partial z}\bigg|_{z=a}$$

Recall that $z = a$ is the edge of the matrix block. The matrix equation generated via (8.2.27) is analogous to the preceding finite-element matrix equations and the required integrations can be performed in a straightforward way.

Two strategies are presented in Huyakorn et al. (1983b) for the solution of the set of two matrix equations represented by (8.2.25) and (8.2.27). The first and most obvious is to solve the equations iteratively. More interesting is the idea of using the Thomas algorithm to reduce the tridiagonal set of equations represented by (8.2.27) until only the leakage parameter remains. This expression can then be substituted into (8.2.25) and the resulting equations solved directly. Once the concentration values c_J^{k+1} are obtained the concentration in the blocks can be calculated by back substitution in the second step of the Thomas algorithm. Details of both solution techniques are provided in Huyakorn et al. (1983).

While the above presentation applies to the parallel fracture model, a similar development can be formulated using the spherical model.

Discrete-Fracture Formulation of the Transport Equation

The formulation of the discrete-fracture transport equations is analogous to that presented earlier for flow, except for the inclusion of a convective term in the governing equation. The one-dimensional transport equation written along the fracture is given by (8.2.26) with the recognition that the coordinate z, now defined along the fracture, is analogous to the coordinate s used in the description of the flow equation. The fracture must correspond to a two-dimensional element boundary in the overall region. The velocity required for the solution of the discrete-fracture equation is obtained through numerical differentiation of the flow solution. Asymmetric weighting functions are used to achieve upstream weighting in the weighted residual formulation.

As in the case of flow, the element level matrices obtained are incorporated directly into the global matrix and the system of equations is solved simultaneously.

8.2.8 Analytical Solutions to the Transport Equation

In our consideration of the analytical solution of the equations governing transport in fractured porous media, let us begin by revisiting the equations governing solute behavior in the fractures and the pores. The distribution of solute along a fracture is given by

$$\frac{\partial c^f}{\partial t} + \frac{V^f}{R}\frac{\partial c^f}{\partial z} - \frac{D^f}{R}\frac{\partial^2 c^f}{\partial z^2} + \lambda c^f + \frac{\Gamma}{R^f b} - = 0 \qquad (8.2.28)$$

where the coordinate system and fracture and block dimensions are as indicated in Fig. 8.2.4. The parameter $\tilde{\Gamma}$ represents the diffusive loss of mass from the fracture.

The differential equation describing the diffusive transport of solute in the rock matrix in the direction perpendicular to the fracture is

$$\frac{\partial c}{\partial t} - \frac{D}{R}\frac{\partial^2 c}{\partial x^2} + \lambda c = 0 \qquad (8.2.29)$$

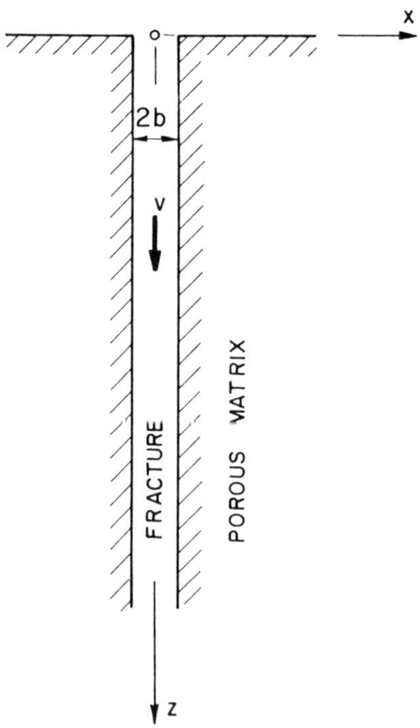

Figure 8.2.4. Fracture-matrix system.

The diffusive flux term $\tilde{\Gamma}$ in (8.2.28) is represented, as before, using a Fickian representation at the fracture-matrix boundary. Thus we have

$$\tilde{\Gamma} = -\varphi D \frac{\partial c}{\partial x}\bigg|_{x=b} \tag{8.2.30}$$

The matrix porosity φ is included to account for the fact that diffusion will occur only through the porous area of the interface.

In the work of Tang et al. (1981) the following initial and boundary conditions were considered:

$$c^f(z,0) = 0 \qquad (8.2.31a)]$$

$$c^f(0,t) = c_0^f \tag{8.2.31b}$$

$$c^f(\text{inf},t) = 0 \tag{8.2.31c}$$

The initial and boundary condition at $x = b$ for the porous matrix equation (8.2.29) in the single and parallel fracture cases are

$$c(x,z,0) = 0 \tag{8.2.32a}$$

$$c(b,z,t) = c^f(z,t). \tag{8.2.32b}$$

The remaining boundary condition for equation (8.2.29) is

$$c(\text{inf},z,t) = 0 \tag{8.2.33}$$

for the single fracture scenario whereas, in the case of multiple parallel fractures, this condition is replaced by

$$\frac{\partial c}{\partial x}(L,z,t) = 0. \tag{8.2.34}$$

Equations (8.2.32b) and (8.2.30), when substituted into (8.2.28), express the coupling between a fracture and the porous matrix. Because of the linear nature of the governing equations, solutions for a time-variable source concentration, $c_0(t)$, can be obtained by superposition of solutions for a constant source strength.

Single Fracture Solutions

Tang et al. (1981) present a general transient solution in which all processes are considered, a non-dispersive transient solution in which dispersion along the fracture axis is neglected ($D^f = 0$) and a steady-state solution that includes all of the processes. Analytical solutions for both the fracture and the porous matrix have been derived. For conciseness, only the fracture solutions will be presented here.

The general transient solution expressing the concentration in the fracture takes the form

$$\frac{c^f}{c_0^f} = \left(e^{\nu t}/\sqrt{\pi}\right)\int_{\ell}^{\inf} e^{\left(-\xi^2 - \frac{\nu^2 z^2}{4\xi^2} - \frac{R^f}{4}\frac{\lambda}{D^f}\frac{z^2}{\xi^2}\right)}$$

$$\cdot\left[e^{-\sqrt{\lambda}\,Y}\mathrm{erfc}\left(\frac{Y}{2\sqrt{T}} - \sqrt{\lambda}\,\sqrt{T}\right) + e^{\sqrt{\lambda}\,Y}\mathrm{erfc}\left(\frac{Y}{2\sqrt{T}} + \sqrt{\lambda}\,\sqrt{T}\right)\right]d\xi \qquad (8.2.35)$$

where

$$\nu = V^f/\left(2D^f\right)$$

$$Y = \frac{\phi\,\sqrt{RD}\,z^2}{4bD^f\xi^2}$$

$$T = t - \frac{R^f\,z^2}{4\,D^f\xi^2}$$

$$\ell = \frac{z}{2}\sqrt{R^f/\left(D^f t\right)}$$

Equation (8.2.35) is easily integrated using Gauss-Legendre quadrature. Prior to integration, Tang et al. (1981) first scan the integrand to find the interval over which it is numerically significant and then locate the Gauss points within this interval.

The non-dispersive solution ($D^f = 0$) is useful for demonstrating the effect of neglecting longitudinal dispersion along the fracture. In certain instances, for example when the advective flux of solute in the fracture is large, it is reasonable to neglect longitudinal dispersion along the fracture axis (Tang et al. 1981; Sudicky and Frind, 1982). For this case, equation (8.2.28) is reduced to a first-order equation whose solution is

$$\frac{c^f}{c_0^f} = \frac{1}{2}e^{-\left(R^f\lambda z/V^f\right)}\left[e^{-\sqrt{\lambda}\,\omega}\,\mathrm{erfc}\left(\omega/\left(2\sqrt{T_0}\right) - \sqrt{\lambda}\,\sqrt{T_0}\right)\right.$$

$$\left.\cdot\, e^{\sqrt{\lambda}\,\omega}\mathrm{erfc}\left(\omega/2\sqrt{T_0} + \sqrt{\lambda}\,\sqrt{T_0}\right)\right] \qquad T_0 > 0 \qquad (8.2.35a)$$

$$\frac{c^f}{c_0^f} = 0 \qquad T_0 < 0 \qquad (8.2.36b)$$

where

$$\omega = \frac{\phi\sqrt{R^f D^f z}}{bV^f}$$

$$T_0 = t - R^f z/V^f$$

It can be seen that (8.2.36) is much easier to evaluate than the general solution because it involves no integration.

Parallel Fracture Solutions

The analytical solutions describing transport in a system of parallel fractures are given by Sudicky and Frind (1982). The solutions include a general transient solution, a non-dispersive solution and the corresponding steady-state solution for each of these cases. Once again, only the solutions describing transport in the fractures will be given here.

The general transient solution is a double integral taking on the form

$$\frac{c^f}{c_0^f}=\frac{2}{(\pi)^{3/2}}e^{vz}\int_{\ell}^{\inf} e^{\left[-\xi^2-\frac{v^2z^2}{4\xi^2}-\frac{R^f\lambda z^2}{4D^f\xi^2}\right]}\int_0^{\inf}\frac{\varepsilon}{\lambda^2+\frac{\varepsilon^4}{4}}e^{\varepsilon_R}$$

$$\cdot\left[e^{-\lambda T}\left[\frac{\varepsilon^2}{2}\sin(\varepsilon_I)\big|_T-\lambda\cos(\varepsilon_I)\big|_T\right]\right. \qquad (8.2.37)$$

$$\left.+\frac{\varepsilon^2}{2}\sin(\Omega)+\cos(\Omega)\right]d\varepsilon\, d\xi$$

where

$$\Omega=\frac{Y\varepsilon}{2}\left[\frac{\sinh(\sigma\varepsilon)+\sin(\sigma\varepsilon)}{\cosh(\sigma\varepsilon)+\cos(\sigma\varepsilon)}\right]$$

$$\varepsilon_R=-\frac{Y\varepsilon}{2}\left[\frac{\sinh(\sigma\varepsilon)+\sin(\sigma\varepsilon)}{\cosh(\sigma\varepsilon)+\cos(\sigma\varepsilon)}\right]$$

$$\varepsilon_I=\frac{\varepsilon^2 t}{2}-\frac{Y\varepsilon}{2}\left[\frac{\sinh(\sigma\varepsilon)+\sin(\sigma\varepsilon)}{\cosh(\sigma\varepsilon)+\cos(\sigma\varepsilon)}\right]$$

$$\sigma=\sqrt{R/D}\,(L-b).$$

The variables v, ℓ, Y, and T have been defined earlier.

The strategy of Barker (1982) and Hodgkinson et al. (1983) for evaluating the concentration in a system of parallel fractures avoids the problematic numerical integrations associated with the solution (8.2.37). This is done by numerically inverting the Laplace transformed solution which has the following form

$$\frac{\bar{c}^f}{c_0^f}=\frac{1}{P-\lambda}e^{vz}e^{\left[-vz\left(1+\kappa^2\left[\left(\sqrt{P}/A\right)\tanh\left(\sigma\sqrt{P}\right)+P\right]^{\frac{1}{2}}\right)\right]} \qquad (8.2.38)$$

where

$$\kappa^2=4R^fD^f/\left(v^f\right)^2$$

$$A=bR^f/\left[\phi\sqrt{RD}\right]$$

and $$P = p + \lambda .$$

The relationship between the Laplace transformed concentration $\bar{c}^f$ and the real concentration c^f is defined according to

$$\bar{c}^f(z,p) = \int_0^{\text{inf}} c(z,t)e^{-pt}\,dt . \tag{8.2.39}$$

where p is a parameter in the Laplace transformation that is in general complex-valued. The numerical inversion of (8.2.38) to obtain $c^f(z,t)$ can be efficiently and accurately performed using algorithms such as those developed by Crump (1976). Generally, these methods require the evaluation of the Laplace transform concentration for a series of discrete values of the Laplace transform parameter, p_k. Using the inversion formula developed by Crump (1976), the concentration $c^f(z,t)$ is given by

$$c^f(z,t) \approx 1/T^* e^{p_0 t}\left[\frac{1}{2}\bar{c}^f(z,p_0)\right.$$
$$\left. + \sum_{k=1}^{2N+1}\left[\mathrm{Re}\left[\bar{c}^f(z,p_k)\right]\cos(k\pi t/T^*) - \mathrm{Im}\left[\bar{c}^f(z,p_k)\sin(k\pi t/T^*)\right]\right]\right] \tag{8.2.40}$$

where $2T^*$ is the period of the Fourier series approximating the inverse function on the interval $[0,2T^*]$ and Re and Im denote the real and imaginary parts, respectively, of the complex $\bar{c}^f$ function. The values of the discrete Laplace parameter are defined according to

$$p_0 = -\ell n(E)/(2T^*)$$

$$p_k = p_0 + \frac{k\pi i}{T^*}$$

where E is an absolute error tolerance and $i = \sqrt{-1}$. Crump recommends using $E = 10^{-6}$ and has found that $T^* = 0.8\, t_{max}$, where t_{max} is the maximum simulation time, gives near optimal results. The accuracy of the inversion represented by (8.2.40) increases with increasing N, although $N = 7$ has been found to produce acceptable results for a wide range of input parameters. To enhance convergence of the series, Crump recommends use of the epsilon partial sums algorithm.

Because of the simple form of the Laplace transformed concentration, evaluation of the transformed concentration in the porous matrix, $\bar{c}(x,z,p)$, is also straightforward. It is given by (Sudicky and Frind, 1982):

$$\bar{c} = \bar{c}^f \frac{\cosh\left[G\sqrt{P}(L-x)\right]}{\cosh\left[\sigma\sqrt{P}\right]} \tag{8.2.41}$$

where $G = \sqrt{R/D}$. Inversion is carried out by replacing $\bar{c}^f$ in (8.2.40) with $\bar{c}$.

Rasmuson (1984) also presents an analytical solution applicable to the parallel fracture case, but unlike Sudicky and Frind (1982) and Barker (1982), uses a dual-porosity formulation.

Special-case solutions have been developed to handle the problem involving an injection well that intersects a horizontal planar fracture into which wastes are injected (Feenstra et al. 1984). Chen (1985, 1986) extended the solution by Feenstra et al. (1984) to take into account longitudinal dispersion in the fracture and radioactive decay. The Laplace-transformed concentration is inverted numerically by Chen for intermediate and large time. Chen (1986) concludes that the effects of longitudinal dispersion diminish with increasing injection time and virtually vanish at steady-state conditions.

8.3 SUMMARY

Simulation of fluid flow and solute transport in fractured porous media is a problem of importance in a number of disciplines including groundwater hydrology, petroleum engineering, soil physics and geothermal reservoir engineering. Progress has been made in each of these disciplines although it is not evident that each discipline has been aware of the progress being made in others.

Fractured reservoir simulation equations can be based upon the concept of dual continua, also called dual-porosity, or upon the concept of discrete fractures. Some models incorporate both of these concepts in one formulation. Finite-element, finite-difference, boundary-element, and analytical methods have been employed to solve the equations resulting from these two approaches.

REFERENCES

Avdonin, N.A. Some formulas for calculating the temperature field of a formation during thermal injection, Izv. Vyssh. Ucheb. Zaved. Neft. Gaz. 7(3), 37–41, 1964.

Baca, R.G., Arnett, R.C., and Langford, C.W. Modeling of fluid flow in fractured -porous rock masses by finite-element techniques, *Int. J. Num Meth. Fluids*, 4, 337–348, 1984.

Baca, R.G., Arnett, R.C., and King, I.P. Numerical modeling of flow and transport in fractured-porous rock system, Rep. RHO-BWI-SA-113, Rockwell Hanford Operations, Richland, Wash., 1981.

Barenblatt, G.E., Zheltov, Iu. P., and Kochina, N. Basic concepts in the theory of seepage of homogeneous liquids in fissured rocks, *Prikl. Mat. Mekh.*, 24, 852–864, 1960.

Barenblatt, G.E., and Zheltov, Y.P. Basic flow equations of homogeneous fluids in fissured rocks, Dokl. Akad. Nauk SSSR, 132, 545–548, 1960.

Barenblatt, G.I., and Jeltov, U.P. On fundamental equations of homogeneous liquids filtration through fractured rocks, Dan USSR, 132(3), 545–548, 1960.

Barker, J.A. Laplace transform solutions for solute transport in fissured aquifers, *Adv. Water Resour.*, 5(2), 98–104, 1982.

Berkowitz, B., Bear, J., and Braester, C. Continuum models for contaminant transport in fractured porous formations, *Water Resour. Res.* 24(8), 1225–1236, 1988.

Beven, K.J. and Germann, P.F. Water flow in soil macropores II. A combined flow model, *J. Soil Science*, 32, p. 383–388, 1981.

Beven, K.J., and Clarke, R.T. On the variation of infiltration into a homogeneous soil matrix containing a population of macropores, *Water Resour. Res.*, 22(3), p. 383–388, 1986.

Bibby, R., Mass transport of solutes in dual-porosity media, *Water Resour. Res.*, 17, 1075–1081, 1981.

Bodvarsson, G.S., Pruess, K., and O'Sullivan, M.J., Injection and energy recovery in fractured geothermal reservoirs, *Soc. Pet. Eng. J.* 303–312, 1985.

Bossie-Codreanu, D., Bia, P.R., and Sabathier, J-C. The "checker model," improvement in modeling naturally fractured reservoirs with a tridimensional, triphasic, black-oil numerical model, *Soc. Pet. Eng. J.*, 743–756, 1985.

Braester, C. A shock wave in immiscible displacement in a fissured porous medium, *Israel J. Tech.* 9 5), 433–438, 1971.

Braester, C. Simultaneous flow of immiscible liquids through porous media, *Soc. Pet. Eng. J.*, 297–305, 1972.

Chen, C.S. Analytical and approximate solutions to radial dispersion from an injection well into a hydrogeologic unit with simultaneous diffusion into adjacent strata, *Water Resour. Res.*, 21(8), 1069–1076, 1985.

Chen, C.S. Solutions for radionuclide transport from an injection well into a single fracture in a porous formation, *Water Resour. Res.*, 22(4) 508–518, 1986.

Chen, H-Z. Numerical simulation of coning behavior of a single well in naturally fractured reservoir, *Soc. Pet. Eng. J.*, 879–884, 1983.

Chen, Z.-X. and You, J. The behavior of naturally fractured reservoirs including fluid flow in matrix blocks, *Transport in Porous Media*, 2 145–163, 1987.

Cherry, J.A, Desaulniers, D.E., Frind, E.O., Fritz, P., Gevaert, D.M., Gillham, R.W., and LeLievre, B. Hydrogeologic properties and pore-water origin and age: Clayey till and clay in south-central Canada, paper presented at the Workshop in Low-Flow Low-Permeability

Measurements in Largely Impermeable rocks, OECD Nucl. Energ. Agency and Int. At. Energ. Agency, Paris, March , 1979.

Closmann, P.J. An aquifer model for fissured reservoirs, *Soc. Pet. Eng. J.*, 15, 385–398, 1975.

Coats, K.H. Geothermal reservoir modeling, SPE 6892, presented at 52 Annual Fall Technical Conference and Exhibition of the Soc. of Petrol. Eng., Denver, Colorado, October, 1977.

Crump, K.S. Numerical inversion of Laplace transforms using a Fourier Series approximation, *J. Assoc. Comput. Mach.*, 23 (1), 89–96, 1976.

D'Astou, A.Y., Ruland, W.W., Bruce, J.R.G., Cherry, J.A., and Gillham, R.W. Fracture effects in the shallow groundwater zone in weathered Sarnia-area clay, *Canadian Geotech. J.*, 26(1), 43–56, 1989.

Davidson, M.R. A numerical study of saturated-unsaturated infiltration in a cracked soil, *Water Resour. Res.*, 21(5), 709–714, 1985.

Day, M.J. Analysis of movement and hydrochemistry of groundwater in the fractured clay and till deposits of the Winnipeg area, Manitoba, M.Sc. thesis, Univ. of Waterloo, Waterloo, Ont. 1977.

de Swaan, A. Theory of waterflooding in fractured reservoirs, *Soc. Pet. Eng. J.*, 117–122, 1978

Dougherty, D.E., and Pinder, G.F. Equivalent porous medium models for solute transport in fractured porous media, submitted to *Water Resour. Res.*, 1989a.

Dougherty, D.E., and Pinder, G.F. Equivalent porous medium models for two phase flow in fractured porous media, submitted to *Water Resour. Res.*, 1989b.

Duguid, J.O., and Abel, J. Finite element Galerkin method for flow in fractured porous media, in Finite Element Methods in Flow Problems, J.T. Oden, O.C. Zienkiewicz, R.H. Gallagher, and C. Taylor, eds., 599–615, University of Alabama Press, Huntsville, 1974.

Duguid, J.O., and Lee, P.C.Y. Flow in fractured porous media, *Water Resour. Res.*, 13(3), 558–566, 1977.

Duguid, J.O. Flow in fractured porous media, Ph.D. dissertation, Dept Civil Engineering, Princeton University, Princeton, N.J., 1973.

Eaton, R.R., and Bixler, N.E. Analysis of a multiphase, porous-flow imbibition experiment in a fractured volcanic tuff, in *Flow and Transport through Unsaturated Fractured Rock*, Geophysical Monograph 42, D.D. Evans et al., eds., 91–97, 1987.

Edwards, W.M., van der Ploeg, R.R., and Ehlers, W. A numerical study of effects of noncapillary-sized pores upon infiltration, *Soil Sci. Soc. Am. J.*, 43, 5, 851–856, 1979.

Eica, A.S., and Ershaghi, I. Modeling of gas injection in fractured reservoirs, paper SPE 10758 presented at the 1982 SPE California Region Meeting, San Francisco, March 24–26, 1982.

Elsworth, D., and Goodman, R.E. Hydromechanical modeling of fractured rock masses using coupled numerical schemes, in Coupled Processes Associated with Nuclear Waste Repositories, C. F. Tsang, ed., 637–654, Academic Press, 1987.

Elsworth, D. A boundary element-finite element procedure for porous and fractured media flow, *Water Resour. Res.*, 23(4), 551–560, 1987

Elsworth, D. A model to evaluate the transient hydraulic response of three-dimensional sparsely fractured rock masses, *Water Resour. Res.*, 22(13), 1809–1819, 1986.

Feenstra, S., Cherry, J.A., Sudicky, E.A., and Haq, Z. Matrix diffusion effects on contaminant migration from an injection well in fractured sandstone, *Ground Water*, 22(3), 307–316, 1984.

Freeze, R.A., and Cherry, J.A. Groundwater , Prentice-Hall, Englewood Cliffs, N.J., 1979.

Germann, P.F., and Beven, K.J. A distribution function approach to water flow in soil macropores based on kinematic wave theory, *J. Hydrology*, 8 173–183, 1986.

Germann, P.F., and Beven, K. Kinematic wave approximation to infiltration into soils with sorbing micropores, *Water Resour. Res.*, 21(7), 990–996, 1985.

Gilman, J.R. An efficient difference method for simulating phase segregation in the matrix blocks in double-porosity reservoirs, *Soc. Pet. Eng. J.*, July, 403–413, 1986.

Gilman, J.R., and Kazemi, H. Improvements in simulation of naturally fractured reservoir, *Soc. Pet. Eng. J.*, 695–707, 1983.

Goltz, M.N., and Roberts, P.V. Interpreting organic solute transport data from a field experiment using physical nonequilibrium models, *J. Contam. Hydrol.*, 1, 77–93, 1986.

Gray, W.G., O'Neill, K., and Pinder, G.F. Simulation of heat transport in fractured single-phase reservoirs, Summaries of Second Workshop Geothermal Reservoir Engineering, P. Kruger and H.J. Ramey Jr. Eds., SGP-TR-20, 1976.

Gringarten, A.C., Ramey, H.J. Jr., and Raghavan, R. Applied pressure analysis for fractured wells, *J. Pet. Technol.*, 887–892, 1975.

Gringarten, A.C., Ramey, H.J. Jr., and Raghavan, R. Unsteady-state pressure distributions created by a well with a single infinite conductivity vertical fracture, *Soc. Pet. Eng. J.*, 14, 347–360, 1974.

Gringarten, A.C., Witherspoon, P.A., and Onishi, Y. Theory of heat extraction from fractured hot dry rocks, *J. Geophys. Res.*, 80, 1120–112 1975.

Grisak, G.E., and Pickens, J.F. An analytical solution for solute transport through fractured media with matrix diffusion, *Journal of Hydrology*, 52 47–57, 1981.

Grisak, G.E., and Pickens, J.F. Solute transport through fractured media, 1, The effect of matrix diffusion, *Water Resour. Res.*, 16(4), 719–730, 1980.

Gureghian, A.B. A study by the finite element method of the influence of fractures in confined aquifers, *Soc. Pet. Eng. J.*, 15, 181–191, 1975.

Heber-Cinco, L., Sameniego, V.F., and Dominquez, A.N. Transient pressure behavior for a well with a finite conductivity vertical fracture, Conf. Pap. SPE 6019, p12, Soc. Pet. Eng. Dallas, Tex. 1976.

Herbert, A.W., Hodgkinson, D.P., Lever, D.A., Rae, J., and Robinson, P.C. Mathematical modeling of radionuclide migration in groundwater, Paper presented at the meeting of Exotic Uses of Aquifers, Geol. Soc., London, Oct. 9, 1984.

Hodgkinson, D.P., and Maul, P.R. One-dimensional modeling of radionuclide migration through permeable and fractured rock for arbitrary length decay chains using numerical inversion of Laplace transforms, UK Atomic Energy Authority Report, AERE-R. 11880, 1985.

Hodgkinson, D.P., Lever, D.A., and England, T.H. Mathematical modeling of radionuclide transport through fractured rock using numerical inversion Laplace transforms: Application to INTRACOIN Level 3., UK Atomic Energy Report, AERE-R. 10986, 1983.

Hoogmoed, W.B., and Bouma, J. A simulation model for predicting infiltration into cracked clay soil, *Soil Sci. Soc. AM. J.*, 44(3), 458–461, 1980.

Hull, L.C., Miller, J.P., and Clemo, T.M. Laboratory and simulation studies of solute transport in fracture networks, *Water Resour. Res.*, 23(8), 1505–1513, 1987.

Huyakorn, P.S. FTRANS: A two dimensional code for simulating fluid flow and transport of radioactive nuclides in fractured rock for repository performance assessment. Report ONWI-426, Office of Nuclear Waste Isolation, Battelle Memorial Institute, Columbus, Ohio, p. 199, 1983.

Huyakorn, P.S. Review of modeling approaches to water flow and solute transport in unsaturated soils and macropores , Prepared for U.S. Environmental Protection Agency, Environmental Research Laboratory, Athens, Georgia, 30613, 1988.

Huyakorn, P.S., and Pinder, G.F. Computational Methods in Subsurface Flow, Academic Press, p. 473, 1983.

Huyakorn, P.S., Lester, B.H., and Faust, C.R. Finite element techniques for modeling groundwater flow in fractured aquifers, *Water Resour. Res.* 19(4), 1019–1035, 1983a.

Huyakorn, P.S., Lester, B.H., and Mercer, J.W. An efficient finite element technique for modeling transport in fractured porous media, 1, Single-species transport, *Water Resour. Res.*, 19(3), 841–854, 1983b.

Huyakorn, P.S., Lester, B.H., and Mercer, J.W. An efficient finite element technique for modeling transport in fractured porous media 2. Nuclide decay chain transport, *Water Resour. Res.* v. 19, no. 5, p. 1286–1296, 1983c.

Huyakorn, P.S., White, H.O. Jr., and Wadsworth, T.D. TRAFRAP-WT A two-dimensional finite element code for simulating fluid flow and transport of radionuclides in fractured porous media with water table boundary conditions, Rpt. by Hydrogeologic, Inc., p. 92, 1987.

Kasameyer, P.W., and Schroeder, R.C. Thermal depletion of liquid dominated geothermal reservoirs with fracture and pore permeability, Workshop Rep (SGP-TR-12) for Workshop of Geothermal Reservoir Eng., at Stanford U., Stanford, Calif., Dec. 15–17, 1975.

Kazemi, H., Merril, L.S. Jr., Porterfield, K.L., and Zeman, P.R. Numerical simulation of water-oil flow in naturally fractured reservoirs, *Soc. Pet. Eng. J.*, 16, 317 –326, 1976.

Kazemi, H., Seth, M.S., and Thomas, G.W. The interpretation of interference tests in naturally fractured reservoirs with uniform fracture distribution, *Soc. Pet. Eng. J.*, December, 1969, p.463–472, 1969.

Kazemi, H. Pressure transient analysis of naturally fractured reservoirs with uniform fracture distribution, *Soc. Pet. Eng. J.*, 9, 451–462, 1969

Khaleel, R. Porous continuum models for fractured basalts, in *Flow and Transport through Unsaturated Fractured Rock*, Geophysical Monograph 42, D.D. Evans et al., eds., 141–148, 1987.

Klavetter, E., and Peters, R.R. Flow in a fractured rock mass , Sandia National Laboratories, Report SAND85-0855, Albuquerque, NM, March 1986.

Kleppe, I., and Marse, R.A. Oil production from fractured reservoirs by water displacement , paper SPE 5084 presented at the 1964 SPE Annual meeting, Houston, Oct. 6 9, 1964.

Lefebvre du Prey, E.J., and Bossie-Codreanu, D. Simulation numerique de l'exploitation des reservoirs fissures, Proc., Ninth World Pet. Cong., Tokyo, PD 13(5), 1975.

Lever, D.A., Bradbury, M.H., and Hemingway, S.J. Modeling the effect of diffusion into the rock matrix on radionuclide migration, UK Atomic Energy Report, AERE R. 10614, 1982.

Lowell, F.P. Contaminant transport in a single fracture: Periodic boundary and flow conditions, *Water Resour. Res.*, 25(5), 774–780, 1989.

Mattax, C.C., and Kyte, J.R. Imbibition oil recovery from fractured water drive reservoirs, *Soc. Pet. Eng. J.* 225, 177–184 1962.

Menouar, J., and Knapp, R.M. Numerical simulation of the imbibition process in fractured reservoirs, paper SPE 9370 presented at the 1980 SPE Annual Technical Conference and Exhibition, Dallas, Sept. 21–24, 1980.

Mishra, S., Bodvarsson, G.S., and Attanayake, M.P. Injection and falloff test analysis to estimate properties of unsaturated fractures, in Flow Transport through Unsaturated Fractured Rock, Geophysical Monograph 42, D.D. Evans et al., eds., 149–156, 1987.

Moench, A.F. Double-porosity models for a fissured groundwater reservoir with fracture skin, *Water Resour. Res.*, 20(7), 831–846, 1984.

Moench, A.F. Well test analysis in naturally fissured, geothermal reservoirs with fracture skin, Proceedings, Ninth Workshop on Geothermal Reservoir Engineering, Stanford University, Stanford, CA., December 1983

Muralidjar, K. and Long, J.C.S. A new approach to characterizing flow in single fractures, in *Flow and Transport through Unsaturated Fractured Rock*, Geophysical Monograph 42, D.D. Evans et al., eds., 115–120, 1987.

Neretnieks, I. Diffusion in the rock matrix: An important factor in radionuclide retardation, *Journal of Geophysical Research*, 85(B8), 4379–4397, 1980.

Neretnieks, I., Abelin, H., Birgersson, L., Moreno, L., Rasmuson, A., and Skagius, K. Chemical transport in fractured rock. in Fundamentals of Transport in Porous Media, J. Bear and M.Y. Corapcioglu, eds., 473–550, 1987.

Neretnieks, I., and Rasmuson, A. An approach to modeling radio-nuclide migration in a medium with strongly varying velocity and block sizes along the flow path, *Water Resources Research*, 20(12), 1823–1836, 1984.

Noorishad, J., and Mehran, M. An upstream finite element method for solution of transient transport equations in fractured porous media, Water Resour. Res., 18(3), 588–596, 1982.

O'Neill, K., and Pinder, G.F. A derivation of the equations for transport of liquid and heat in three dimensions in a fractured porous medium, *Adv. Water Resourc.*, 4, 1981.

O'Neill, K. The transient three-dimensional transport of liquid and heat in fractured porous media, Ph.D. dissertation, Dept. of Civil Engineering, Princeton University, 1977.

Odeh, A.S. Unsteady-state behavior of naturally fractured reservoirs, *Soc. Pet. Eng. J.*, 5, 60–66, 1965.

Pankow, J.F., Johnson, R.L., Hewetson, J.P., and Cherry, J.A. An evaluation of contaminant migration patterns at two waste disposal sites on fractured porous media in terms of the equivalent porous medium (EPM) model, *J. Contam. Hydrol.*, 1, 65–76, 1986.

Papay, J. Numerical modeling of mixed porosity reservoirs-part 1-description of the model, Koolaj Foldgaz (Dec. 1982), 14(114), 372–37 1982.

Pinder, G.F., Ramey, H.J. Jr., Shapiro, A., and Abriola, L. Block response to reinjection in a fractured geothermal reservoir, Proceedings Fifth Workshop Geothermal Reservoir Engineering, H.J. Ramey and P. Kruger, Eds SGP-TR-40, 1979.

Pollock, D.W. Simulation of fluid flow and energy transport processes associated with high-level radioactive waste disposal in unsaturated alluvium, *Water Resour. Res.*, 22(5), 765–775, 1986.

Powell, R.P. Contaminant transport in a single fracture: Periodic boundary conditions, *Water Resour. Res.*, 25(5), 774–780, 1989.

Pruess, K., and Narasimhan, T.N. A practical method for modeling fluid and heat flow in fractured porous media, *Soc. Pet. Eng. J.* 14–26, 1985.

Pruess, K., and Wang, J.S.Y. Numerical modeling of isothermal and nonisothermal flow in unsaturated fractured rock—a review, in *Flow and Transport through Unsaturated Fractured Rock*, D.D. Evans et al., Geophysical Monograph 42, 11–21, 1987.

Pruess, K. Heat transfer in fractured geothermal reservoirs with boiling, *Water Resour. Res.*, 19(1), 201–208, 1983.

Pruess, K., Wang, J.S.Y., and Tsang, Y.W. Effective continuum approximation for modeling fluid flow in fractured porous tuff, Sandia National Laboratories, in preparation, Report SAND86-7000, Albuquerque, NM, 1986.

Pruess, K., Tsang, Y.W., and Wang, J.S.Y. Modeling of strongly heat driven flow in partially saturated fractured porous media, Inter. Assoc. of Hydrogeologists, (ed.), Memoires, Vol. XVII, 486–497, 1985

Raghavan, R. Pressure behavior of wells intercepting fractures, paper presented at the Invitational Well-Testing Symposium, Lawrence Berkeley Laboratory, Berkeley, California, Oct. 19–21, 1977.

Rasmuson, A., and Neretnieks, I. Migration of radionuclides in fissured rock: The influence of micropore diffusion and logitudinal dispersion, *Journal of Geophysical Res.*, 86(B5), 3749–3758, 1981.

Rasmuson, A., and Neretnieks, I. Exact solution of a model for diffusion in particles and logitudinal dispersion in packed beds, AIChE J., 26, 6 1980.

Rasmuson, A. Diffusion and sorption in particles and two-dimensional dispersion in a porous medium, *Water Resour. Res.*, 17(2), 321–328, 1981.

Rasmuson, A. Migration of radionuclides in fissured rock: Analytical solutions for the case of constant source strength, *Water Resour. Res.*, 20(10), 1435–1442, 1984.

Rasmuson, A., Narasimhan, T.N., and Neretnieks, I. Chemical transport in a fissured rock: Verification of a numerical model, *Water Resour. Res.*, 18(5), 1479–1492, 1982.

Rasmussen, T.C. Computer simulation model of steady fluid flow and solid transport through three-dimensional networks of variably saturated, discrete fractures, in *Flow and Transport through Unsaturated Fractured Rock*, Geophysical Monograph 42, D.D. Evans et al., eds., 107–113, 1987.

Rasmussen, T.C., and Evans, D.D. Unsaturated flow and transport through fractured rock related to high-level waste repositories, NUREG/CR-4655, p. 474, U.S. Nuclear Reg. Comm., Washington, D.C., 1987.

Reiss, L.H., Bossie-Codreanu, D., and Lefebvre du Prey, E.J. Flow in fissured reservoirs, paper SPE 4343 presented at the 1973 SPE European Petroleum Conference, London, April 2–3, 1973.

Roegiers, J.-C., Curran, J.H., and Bawden, W.F. Numerical modeling of flow in fractured rock masses: Discontinuous versus continuum approaches, Publ. 79-02, Dept of Civil Eng., Univ. of Toronto, Ont. , Canada, 1979.

Rossen, R.J. Simulation of naturally fractured reservoirs with semi-implicit source terms, *Soc. Pet. Eng. J.*, 17, 201–210, 1977.

Rouleau, A. Statistical characterization and numerical simulation of a fracture system—application to groundwater flow in the Stripa granite, Ph.D. dissertation, Univ. of Waterloo, Ont., Canada. 1984.

Schwartz, F.W., Smith, L., and Crowe, A.S. A stochastic analysis of macroscopic dispersion in fractured media, *Water Resour. Res.*, 19(5), 1253–1265, 1983.

Shapiro, A.M. Transport equations for fractured porous media, in *Fundamentals of Transport in Porous Media*, J. Bear and M.Y. Corapcioglu, eds. 405–472, 1987.

Shapiro, A.M. Fractured porous media: Equation development and parameter identification, Ph.D. dissertation, Dept. of Civil Engineering, Princeton University, p. 373, 1981.

Shimo, M., and Long, J.C.S. A numerical study of transport parameters in fracture networks, in *Flow and Transport through Unsaturated Fractured Rock*, Geophysical Monograph 42, D.D. Evans et al., eds., 121–131, 1987.

Smith, L., and Schwartz, F.W. An analysis of the influence of fracture geometry on mass transport in fractured media, *Water Resour. Res.*, 20(9) 1241–1252, 1984.

Snow, D.T. A parallel plate model for fractured permeable media, Ph.D thesis, Univ. of Calif., Berkeley, 1965.

Streltsova, T.D. Hydrodynamics of groundwater flow in a fractured formation, *Water Resour. Res.*, 12(3), 405–414, 1976.

Streltsova, T.D. Well hydraulics in heterogeneous porous media, in *Fundamentals of Transport in Porous Media*, J. Bear and M.Y. Corapcioglu, eds. 315–346, 1984.

Streltsova, T.D. Well pressure behavior of a naturally fractured reservoir, *Soc. of Pet. Eng. J.*, 769–780, 1983.

Streltsova-Adams, T.D. Well hydraulics in heterogeneous aquifer formations, *Adv. Hydrosci.*, 11, 357–423, 1978.

Sudicky, E.A., and Frind, E.O. Contaminant transport in fractured porous media: Analytical solutions for a system of parallel fractures, *Water Resour. Res.*, 18(4), 1634–1642, 1982.

Tang, D.H., Frind, E.O., and Sudicky, E.A. Contaminant transport in fractured porous media: Analytical solution for a single fracture, *Water Resour. Res.*, 17(3), 555–564, 1981.

Thomas, L.K., Dixon, T.N., and Pierson, R.G. Fractured reservoir simulation, *Soc. Pet. Eng. J.*, 42–54, 1983.

Torsaeter, O., Kleppe, J., and van Golf-Racht, T.D. Multiphase flow in fractured reservoirs, in *Fundamentals of Transport in Porous Media*, J. Bear and M.Y. Corapcioglu, eds., 551–629, 1987.

Travis, B.J., Hodson, S.W., Nuttall, H.E., Cook, T.L., and Rundberg, R.S. Numerical simulation of flow and transport in fractured tuff, *Material Research Society, Symposium Proceedings*, Vol. 26, 1039–1047, Elsevier, New York, 1984.

van Genuchten, M.Th. A general approach for modeling solute transport in structured soils, in *Proc. 17th Int. Congress, Hydrogeology of Rocks of Low Permeability*, 17, 513–526, 1985.

Wang, J.S.Y., and Narasimhan, T.N. Hydrologic mechanisms governing partially saturated fluid flow in fractured welded units at Yucca Mountain, SAND85-7114, Sandia National Laboratories, Albuquerque, New Mexico, 1986.

Wang, J.S.Y., and Narasimhan, T.N. Hydrologic mechanisms governing fluid flow in a partially saturated fractured porous medium, *Water Resour. Res.*, 21(12), 1861–1874, 1985.

Warren, J.E., and Root, P.J. The behavior of naturally fractured reservoirs, *Soc. Pet. Eng. J.*, 3, 245–255, 1963.

Wilson, C.R., and Witherspoon, P.A. Steady state flow in rigid networks of fractures, *Water Resour. Res.*, 10(2), 328–335, 1974.

Wilson, C.R. An investigation of laminar flow in fractured porous rocks, Ph.D. thesis, Dept. of Civ. Eng., Univ. of Calif., Berkeley, 1970.

Wilson, M.L., and Dudley, A.L. Radionuclide transport in an unsaturated fractured medium, in *Flow and Transport through Unsaturated Fractured Rock*, Geophysical Monograph 42, D.D. Evans et al., eds., 23–29, 1987.

Wu, Y-S., and Pruess, K. A multiple-porosity method for simulation of naturally fractured petroleum reservoirs, SPE Reservoir engineering, 327–336, 1988.

Yamamoto, R.H., Padgett, J.D., Ford, W.T., and Boubeguira, A. Compositional reservoir simulator for fissured systems—the single block model, *Soc. Pet. Eng. J.* 113–128, 1971.

Yeh, G.T., and Luxmoore, R.J. Chemical transport in macropore-mesopore media under partially saturated conditions, in *Symp. on Unsaturated Flow and Transport Modeling*, W.M. Arnold, G.W. Gee, and R.W. Nelson, eds., 267–281, U.S. NRC, Washington, DC., 1982.

9

A SUMMARY OF FIELD TEST METHODS IN FRACTURED ROCKS

P. Jouanna
University Montpellier II, France.

with the collaboration of :
C. Armangau, University Montpellier II, France.
A.S. Batchelor, Geoscience, Falmouth, Cornwall, U.K.
D. Bonazzi, Coyne et Bellier, 5 rue d'Héliopolis, Paris, France.
D. Bruel and E. Ledoux, Ecole des Mines de Paris, Fontainebleau, France.
P. Cheung, and A. Etchecopar, Schlumberger, Etudes et Productions, Clamart, France.
C. Fouillac, Bureau de Recherches Géologiques et Minières, Orléans, France.
P. Louis, Professor, University Montpellier II, France.
P. Mechler, Professor, University Pierre et Marie Curie, Paris VI, France
A. Tabbagh, Professor, University Pierre et Marie Curie, Paris VI, France.
P. Valla, Bureau de Recherches Géologiques et Minières, Orléans, France.

9.1 INTRODUCTION TO THE INVESTIGATION OF FRACTURED ROCKS

In many engineering projects, investigations of natural media are crucial for the design, realization and instrumentation of the prototype. Laboratory tests, which are not examined in detail in this review, can be of some help, but, in the case of large-scale fractured media, they are of limited value, for the following main reasons:

- The scale effect and heterogeneity make it impossible to obtain reasonably sized representative samples, in the sense of a representative elementary volume (REV).

- Sampling fractured rocks is very difficult. If obtaining a good rock sample is not a problem, sampling "undisturbed joints" is undoubtedly difficult, because the smallest disturbance leads to large changes in the rock mass behavior.

- Simulating complex thermo-hydro-mechanical and chemical in-situ conditions is hazardous under laboratory conditions.

Field investigations are consequently a necessary first step in the identification of fractured rock masses. An in-situ investigation program consumes much money and time, and requires a large range of skills. The present review should give general support to engineers in charge of such investigations and should help to answer the following main questions:

- In the case of fractured media, what techniques are available for investigation, and what conceptual approaches are available for rock modeling and test interpretations?

- In each specific case, what are the precise questions to be answered by the investigations, on both qualitative and quantitative aspects, and what are the most efficient tools and techniques to get the answer, at the lowest cost?

The present study approaches these questions by making a distinction between 2D investigations based on direct rock observation, where interpretation is not a problem, and 3D investigations that try to describe the inner structure and properties of the medium by solving the difficult inverse problem.

9.1.1 Investigation Methods (2D)

Preliminary information on a given fractured mass can be obtained directly using 2D investigation methods, which observe the accessible surface of the mass.

As depicted in Fig. 9.1.1, distinctions can be made between :

- Natural surfaces (Σ_1), excavations (Σ_2), pits (Σ_3) or even tunnels or mines (Σ_4). They are observed by eye or remote sensing using reflection and refraction of electromagnetic waves (see Sec. 9.2.1).

- Surfaces observed in wells (Σ_5). Two complementary methods are used: core observation and its negative, well-wall observation. For the latter, techniques are based on close detection using seismic, ultrasonic, electrical or light sensors (see Sec. 9.2.2) with imaging interpretation.

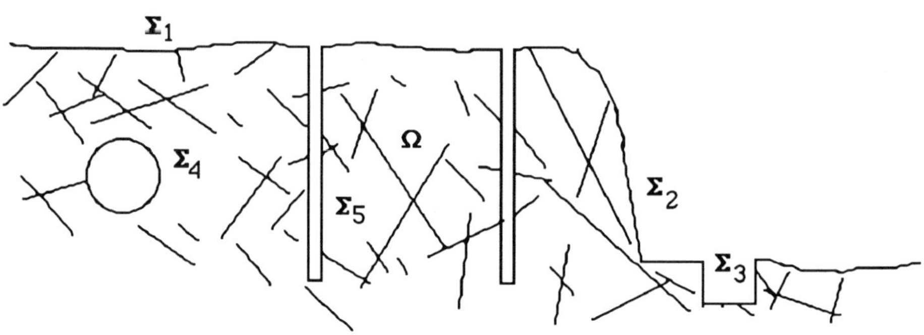

Figure 9.1.1. Different possible investigation domains.

Interpretation of these 2D data is not a problem, because they are easily represented as a mapping of the observed surface. In the case of phenomena invisible to the eye, artificial imaging techniques are now commonly used when scanning of the surface is available.

9.1.2 Investigation Methods (3D)

Although 2D observations may be important, 3D investigation of the inner Ω of the rock domain remains the true question with regard to the difficult inverse problem. The techniques available to investigate the inner characteristics of fractured rock masses can be direct or indirect.

Direct Techniques of Inner Investigation

Techniques are called direct if they refer to fluxes of physical quantities involved in the conservation laws concerning classical thermomechanics. In this chapter,

- Mass transfer of pure water is examined in Sec. 9.3. Mass transfer of water mixed with chemical compounds is examined in Sec. 9.4.
- Heat transfer is treated in Sec. 9.5.
- Momentum transfer, possibly coupled with mass or heat transfers, is treated in Sec. 9.6 essentially for seismic methods. Since it is outside the scope of this book, the static case is treated in overview.

Indirect Techniques of Inner Investigation

In contrast to direct techniques, indirect techniques refer to methods based on physical phenomena. In this chapter,

- Electrical methods are covered in Sec. 9.7.
- Electromagnetic methods are covered in Sec. 9.8.
- Gravimetry and magnetometry methods are covered in Sec. 9.9.

These indirect methods have of course no theoretical link with the required mass or heat transfers. They give essentially qualitative information and require some sort of calibration to give quantitative information on mass or heat fluxes.

The number and sophistication of available tools, especially in indirect techniques, is increasing very rapidly. Most of the techniques were developed first for porous media in general; rare are the methods or tools especially designed for fractured media.

Interpretation of Inner Investigation

Before discussing the identification process itself, it is necessary to place it within a general physical process. Information on the inner Ω of a given system S is obtained by measuring its message or output O, this output being the response to a given excitation or input I. Formally, using the convolution product notation *, one can write $O = S*I$. This formal operation leads to three different types of problems. The convolution problem PB1 consists in finding O, knowing the characteristics of the system S and the input I. This direct problem is the classical forecasting problem in engineering. Its solution is reputed to be unique. Computing the quantity of flow in a given set of fractures, under given boundary conditions, is such a problem. The deconvolution problem PB2 consists in finding the input I, knowing the system S and the output O. Localizing earthquake sources in a fissured media is such a problem. Finally, the reconnaissance problem PB3 is an inverse problem, which involves finding characteristics of a system S, knowing the input I and the output O. Two different approaches to this inverse problem are possible: the so-called black box approach and the identification approach.

Black Box Approach to a Fractured System. The global or black box approach consists of finding relationships between the measured input I and the measured output O, without trying to find any internal information about the system. This pragmatic method assumes that the prototype has been completed. It is thus adapted to the performance tests of an existing prototype. The unknown system S is defined as the fractured medium with its initial conditions and some natural boundary conditions. At the remaining controlled boundaries, instrumentation imposes fluxes in mass, momentum, energy and entropy, or the dual thermodynamic forces. If fluxes are given as an input I, thermodynamic forces are unknown output O, or vice versa. In some circumstances, both I and O are unknown but bound by an imposed relationship. Such a procedure does not depend on the complexity of the system. It is the last chance for characterizing very complex and erratic systems, as in the case of karsts.

Identification Approach to a Fractured System. The identification approach tends to obtain an internal description of the system. The difficulty of this problem increases with the complexity of the medium. When a large number of parameters exist, as in a fissured medium, this inverse problem generally has no unique solution, and many a priori assumptions must usually be made to obtain a solution. The first step consists of determining the distribution of inner boundaries due to fractures. Finding this fracture pattern from outside measurements is by far the most delicate inverse problem in fractured rocks. The second step, assuming the first one to have been completed, consists of measuring the characteristics of discontinuities and matrix.

A number of standard geometrical models have been proposed to emulate the inner structure of a natural medium. The main characteristics of these models are listed as follows:

- **Scale of the Models**. The scale of observation is certainly one of the first parameters to discuss. Different structures can be distinguished at different scales of observation, leading to complex, imbricated structures. Such a description proceeds generally from the larger scale to the microscopic scale. Relationships between the different levels of structure can exist in varying degrees. If the scale does not interfere with the structure, the medium is said to be a fractal; a lot of studies have been carried out recently on such media, but the hopes placed in such descriptions are not yet confirmed.

- **Continuous, Discontinuous and Percolation Models**. At a given scale, even if fractured media are essentially discontinuous, approximation using a so-called equivalent continuous medium is generally attempted, because it allows great simplifications of computation. This equivalent medium is generally considered to be homogeneous and anisotropic in fractured media. The properties of the equivalent medium can be obtained either experimentally or by homogenization techniques. However, this reduction to an equivalent continuous medium is not realistic when large-scale discontinuities exist and no REV can be defined. These discontinuities are assumed to be planar or to be channels. In such cases, discontinuous models have to be adopted. A more critical case appears when discontinuity is unique. The theory of percolation deals with such singular phenomena appearing along a single and critical path.

- **Deterministic or Stochastic Models.** Geometrical description can be performed by a deterministic procedure or a stochastic procedure. Techniques for inverse problems in stochastic descriptions are quite a new field of investigation.

- **Evolutive or Non-Evolutive Models.** A geometrical model should be evolutive when the true fractured medium is modified by the engineering forces. For instance, reversible and irreversible modifications can appear after an injection of a fluid. In some cases, the structure can be entirely modified.

- **Superimposed Model.** Owing to the possibility of superposition of different scale geometries, it is not rare to superimpose different types of models, continuous or discontinuous, either in hydraulic or thermal applications. There is no conceptual limitation to such multiple superposition.

- **Type of Model and Phenomena Involved.** The type of model can depend on the point of view adopted and the phenomena taken as being the most important. For instance, the thermal or mechanical behavior of a fractured mass can be studied with equivalent continuous models, whereas the hydraulic behavior, which is more sensitive to fracture opening, may require discrete models of fractures.

In conclusion, models in complex fractured media present an infinite variety. None of these models can have an intrinsic value; the choice is based on a minimum of information given by the surface investigation as discussed in Sec. 9.2. Part of this choice often remains subjective, as individual engineers will differ on the interpretation of field test results.

There are an infinite number of parameters in natural media. In practice, the number of parameters taken into account is limited by the type of problem and the type of model adopted. Physical parameters related to the phenomena observed can be classified as geometry parameters of the fractured rock structure, coefficients appearing in equilibrium or non-equilibrium relations, and physical constants. Other parameters can appear in indirect testing methods. For example, the electrical conductivity of a fracture is an indirect parameter when heat and mass transfer are the phenomena studied. In such a case, correlating test parameters and the physical parameters needs extra calibration.

9.1.3 Investigation Program

An investigation program requires a set of skills and an excellent leader who can cover the entire background. Strong coordination is necessary to make the decisions and to avoid redundancy, mistakes, and scopeless investigations, which waste time and money. Strong coordination is also necessary to exploit the different results. A diagnosis is not a compilation of the different tests.

A synthesis with correlation of the different investigations should lead to the best possible model of the fractured rock medium. Such multiparameter analysis is more and more commonly used.

Finally, an in-situ investigation program is specific to the field of application considered: geotechnics (dams, tunnels, excavations, mines), energy problems (petroleum engineering, geothermal energy), safety problems (waste disposal, environmental implications) or water resources. The impact of the domain considered is determinant, as can be seen in the two quite different examples given in Sec. 9.10 in the domains of dam engineering and petroleum wire line logging. Other examples can be found in several chapters of this book.

9.2 OBSERVATION OF A FRACTURED SYSTEM BY 2D INVESTIGATION METHODS

9.2.1 Field Observation and Remote Sensing

Direct observation of the surface means that the surface is accessible to the eye or to remote sensing techniques, so-called teledetection. When surface Σ_1 is natural, vegetation and weathering are often a strong barrier to observation of fractures below the surface. Better observation is obtained in excavations Σ_2 or pits Σ_3. Inner excavations Σ_4 allow an excellent visual inspection of fractures but this opportunity generally does not exist for deep works.

Geological surveying by eye is a very old technique. It yields very important information on fractured media when observation of non-weathered rock outcrops is possible or when rock is accessible in tunnels. Field techniques used by the geologist for surveying rock discontinuities at the surface or in tunnels are very simple. They consist of observing directly with the naked eye or a magnifying glass the opening and length of the fractures and their dip and orientation. Although topographic methods are seldom used to describe fracture networks, these methods are becoming much more rapid with the help of computerized tacheometers and localization by satellite, and will be usable in the near future in specific cases.

Today, the eye of the geologist is helped or replaced by teledetection techniques. This section covers 2D observation of the earth's surface. Teledetection in the open air has made tremendous progress in recent years, exploiting an increasing number of electromagnetic techniques that have surprising resolution capabilities. The only electromagnetic phenomena discussed in this chapter are emission and reflection, since they concern 2D observation. The penetration of electromagnetic waves into the rock for the 3D investigation of Ω is discussed in Sec. 9.8.

Principle of Electromagnetic Methods

Emission Phenomena. All bodies radiate energy W depending on temperature T, according to the Stefan-Boltzmann law $W = \varepsilon \sigma T^4$ where emissivity ε is equal to 1 for black bodies and less than 1 for grey bodies, σ being the Stefan constant. The energy density for a given wavelength is given by Planck's law. The principal sources involved in observation methods are the sun, emitting at a temperature T = 5790 K and the earth, emitting at about 300 K.

Absorption and Reflection Phenomena. All bodies receiving energy can absorb it or reflect it. For instance, solar energy can be absorbed or reflected by the earth's surface. Similarly, other types of radiation like radar waves can be absorbed or reflected by the earth's surface.

Observation Principle and Useful Wavelengths. The global energy received by a sensor includes emission from the earth's surface and the reflected energy from outside sources. These sources include the sun and, possibly, artificial sources, such as radar generators, illuminating the earth. The total energy received depends on the influence of the atmosphere between the earth's surface and the receiver sensor. The atmosphere is opaque to all radiation with a wavelength shorter than the visible spectrum; thus only close observation is possible with γ-ray, X-ray and UV radiation. The atmosphere, when clear, is transparent to visible radiation and to some bandwidths, or so-called windows, of IR radiation. For all other wavelengths the atmosphere is always transparent.

Figure 9.2.1 gives a summary of the different domains of electromagnetic phenomena, the observation windows and the corresponding types of sensors related to teledetection.

Remote Sensing Techniques

Observation techniques depend on the wavelength adopted. Two different types of sensors are used:

Passive Sensors. They are used to receive a message given by the observed surface, possibly illuminated by external sources. This is the case for:

- Visible and near-infrared domains, where the ground surface essentially reflects solar energy, the wavelength corresponding to an emission temperature above 6000 K.
- Infrared window from 3 to 5 microns, where the ground surface essentially emits in the wavelength corresponding to an emission temperature around 300 K but also reflects a non-negligible quantity of solar energy.
- Infrared window from 8 to 14 microns, where only emission of the ground surface is observed.

Active Sensors. They are used in radar techniques, where natural reflection and emission can no longer be observed. The method consists in illuminating the target to be observed by an adequate source, the energy being emitted by an antenna. The same antenna receives the reflected energy. It thus plays the role of both transmitter and receiver, the so-called active sensor.

Signal Treatment and Interpretation

The domain of signal treatment and imagery is changing completely and requires special skills. The result is that today the resolution obtained by such techniques can approach a few centimeters or less.

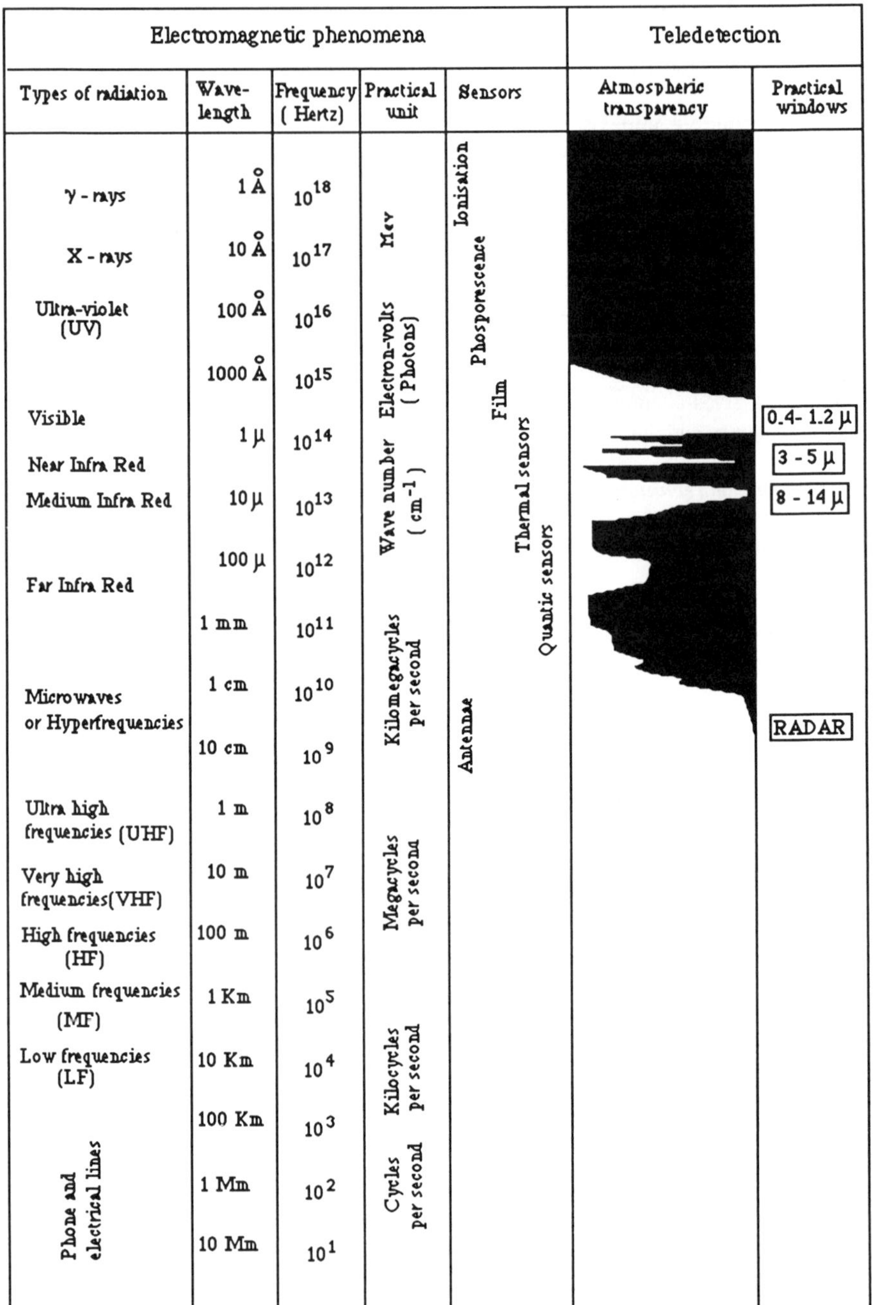

Figure 9.2.1. Teledetection and electromagnetic phenomena.

9.2.2 Well Observation by Logging, Cores and Imaging Methods

Boreholes provide the only practical and direct means of exploring a rock mass at depth, beyond the influence of penetrative weathering and other superficial geological phenomena. Deep wells are costly and, at least in the initial stages of subsurface exploration, typically penetrate far less than one millionth of the volume of the zone under investigation. For these reasons alone the combinations of techniques used for in-situ visualisation of the rock mass should be designed with care. Any one technique can, at best, provide only part of the information that is needed to characterize a rock mass and, in practice, a combination of appropriate, complementary methods must be used and then properly interpreted. A rock mass can be considered as three basic elements: (i) the rock material., (ii) the discontinuities (joints, faults) which intersect the rock material and (iii) the infillings to these discontinuities.

Any attempt at in-situ visualisation can succeed only if the techniques used can in some way measure or otherwise identify distinguishing characteristics of these three basic elements and, therefore, the differences between them.

Geophysical Logging

In-hole logging suites as used by oil service companies can provide much useful information relating to the geophysical characteristics of the rock material, particularly in sedimentary rock formations whose physical characteristics are well known to those responsible for processing the raw data. To detect, measure and visualise discontinuities intersecting the borehole wall, additional logging tools such as the borehole televiewer (BHTV) and formation microscanner (FMS) have been developed. However, infillings to discontinuities, of great significance to fissure flow models, may remain undetected or uncharacterized by geophysical methods, possibly because of the negligible effect that they have on instruments designed expressly to characterize the rock mass rather than the infillings of joints and faults. Detection and identification of infilling materials depend upon the contrast in detectable physical properties over a measurable interval of borehole wall. Soft clays and granular gouge are likely to be eroded during drilling, and typically the aperture will be detected but not the original infill.

Borehole Cores

Geophysical logs alone cannot, therefore, be expected to provide sufficient information to describe the rock in the borehole wall, let alone the mass surrounding it. In the same way that surface geophysics requires calibration, by strategically placed boreholes, to investigate the true nature of the subsurface rock material, so downhole geophysical logging requires calibration by recovery of intact samples of the ground penetrated by the hole, either by coring over its entire length or over carefully chosen intervals. In practice, such intervals cannot be identified without some prior knowledge of the ground penetrated.

Development of the Technique. Fully cored deep boreholes were uncommon until relatively recently, and were reserved almost exclusively for research boreholes used to investigate geological successions in great detail. Even these investigations tend to "open-hole" over long intervals, reserving core drilling for lengths to be investigated in more detail. A need has developed in recent years to examine rock masses in their entirety, with particular emphasis on the nature of joints, faults and induced fractures taking precedence even over the detailed nature of the rock material. Core drilling is the most economical means of recovering complete and intact samples of the rock mass, including the rock material, the discontinuities and their infill materials, an essential preliminary to any attempt at characterizing and modeling fractures.

- **Oil Industry.** The ability to recover high quality core from great depths has developed over recent years from three separate fields of drilling technology, involving the oil, minerals, and civil engineering industries. Deep drilling technology was originally developed for oil exploration and production, as were the geological and geophysical investigation techniques associated with this industry. Sandstone and weak limestone reservoirs predominated, and consequently the properties of the rock material, particularly intergranular porosities and permeabilities, took precedence over the nature of the discontinuities. Rock mass characterization, as we know it, was not an issue.

  Interpretation of subsurface geology from information provided by oil industry boreholes relies, traditionally, on a description of rock cuttings screened from the drilling mud, combined with information provided by in-hole geophysical logging suites. Coring of long intervals of hole was rarely undertaken; instead, short sections of core were, and still are, recovered and tested mainly for porosity, permeability and micropalaeontological purposes rather than for rock mass description.

- **Minerals Industry.** In contrast to techniques traditionally associated with oil exploration and production, continuously cored boreholes have been an accepted method of mineral exploration practice for many years, although the holes are generally much shorter and the core samples of smaller diameter than those used for oil exploration. Wireline core recovery methods, in which the core is recovered by withdrawing the inner tube of the diamond tipped core barrel through the drill rods, or tubulars, were developed to save on time tripping rods and barrel from the hole. Until recently wireline coring was a method more appropriate to holes of intermediate depth. It was unnecessary for shorter, civil engineering holes and not available for the deeper oil industry holes.

  Mineral exploration core is obtained almost exclusively for analytical and lithological purposes. Thin core is weak and is typically highly fractured by the drilling process, making it unsuitable for the evaluation of natural fractures. In contrast, rock core recovered for civil engineering purposes is used specifically for rock mass characterization. Unlike core for oil and mining applications, these samples are preserved for testing physical properties. Descriptions of discontinuities became an important part of rock mass description, particularly where these were related to proposed underground openings, rock slopes and foundations to sensitive structures. Consequently, great care is taken to determine the nature of all natural fractures, recording their orientation, spacing, aperture and also the nature and mechanical properties of the infill material.

- **Civil Engineering Industry.** Techniques of obtaining high quality core for civil engineering purposes have improved dramatically in the last twenty years, with recognizable development milestones that include twin- and triple-tube core barrels held in bearings and designed to prevent rotation of the core, and the development of split barrel liners and removable plastic sleeves to retain and protect poor quality rock or fragile infills to fractures. It has also been recognized that the larger the core the more repeatable the test results, and there is therefore a trend towards core diameters exceeding 100 mm.

Civil engineering investigations seldom require rock boreholes longer than about 200 meters, with the majority between 30 and 60 meters. The demand for high quality cores from greater depth has resulted in a merging of the technologies outlined above, with sophisticated core barrels available as part of wireline systems, routinely coring to 3000 meters and capable of reaching even greater depths. As might be expected, there is a decrease in available core diameters with increasing depth of hole, but typically available diameters range from about 80 mm to 140 mm. The development of sophisticated core recovery techniques for civil engineering purposes was in response to a need to obtain high quality intact core in weak materials and weak zones, such as faulted ground. Similarly, improved oil field techniques that led to downhole tools capable of imaging the borehole wall resulted from the realization that, in some hydrocarbon reservoirs, permeability may be significantly influenced or even dominated by fissure flow.

Aims and Objectives. The aim of core drilling is to produce core of the highest quality, recovered in a condition that is as close as possible to its in-situ state. Incorrect handling, recording and analysis of core will inevitably result, at some stage in a project, in a total loss of critically important information. Specifications for coring programs must be written by those who understand the many difficulties that can arise during and subsequent to coring. Inadequate specifications will lead to loss of data, while overly constrictive specifications, however well intentioned, may prevent those involved from dealing effectively with the many problems that will inevitably arise during the program.

There is an unfortunate but understandable tendency among specialists involved in the planning of exploration boreholes to consider only those aspects of the work of interest to them. This is of little consequence if the mix of specialists is well balanced at the planning stage. However, this is seldom the case, and there is therefore a danger of creating an imbalance in the amount of attention paid to the different aspects of core description, sampling and analysis. Some kinds of information remain available for only a short time after coring; rock cores and fracture fillings can deteriorate rapidly. It is good practice to overestimate the kinds of information required from the core and arrange to obtain as much as possible, although there will undoubtedly be some trade-off. It is also good practice to design a primary logging scheme that will record essential information from the core. Specialized logging and sampling can take place concurrently. An example of generalized core log containing the results from a section of a cored heat flow borehole is shown as Fig. 9.2.2.

In practice, good drilling crews will know how to handle recovered core in its barrel, and experienced geologists will know how to handle, log sample and protect the core. There are a number of general procedural points that should be considered when planning a logging program designed to recover material for fracture studies, for example:

- Core barrels, their associated inner barrels, and sleeves containing core should be handled with extreme care.

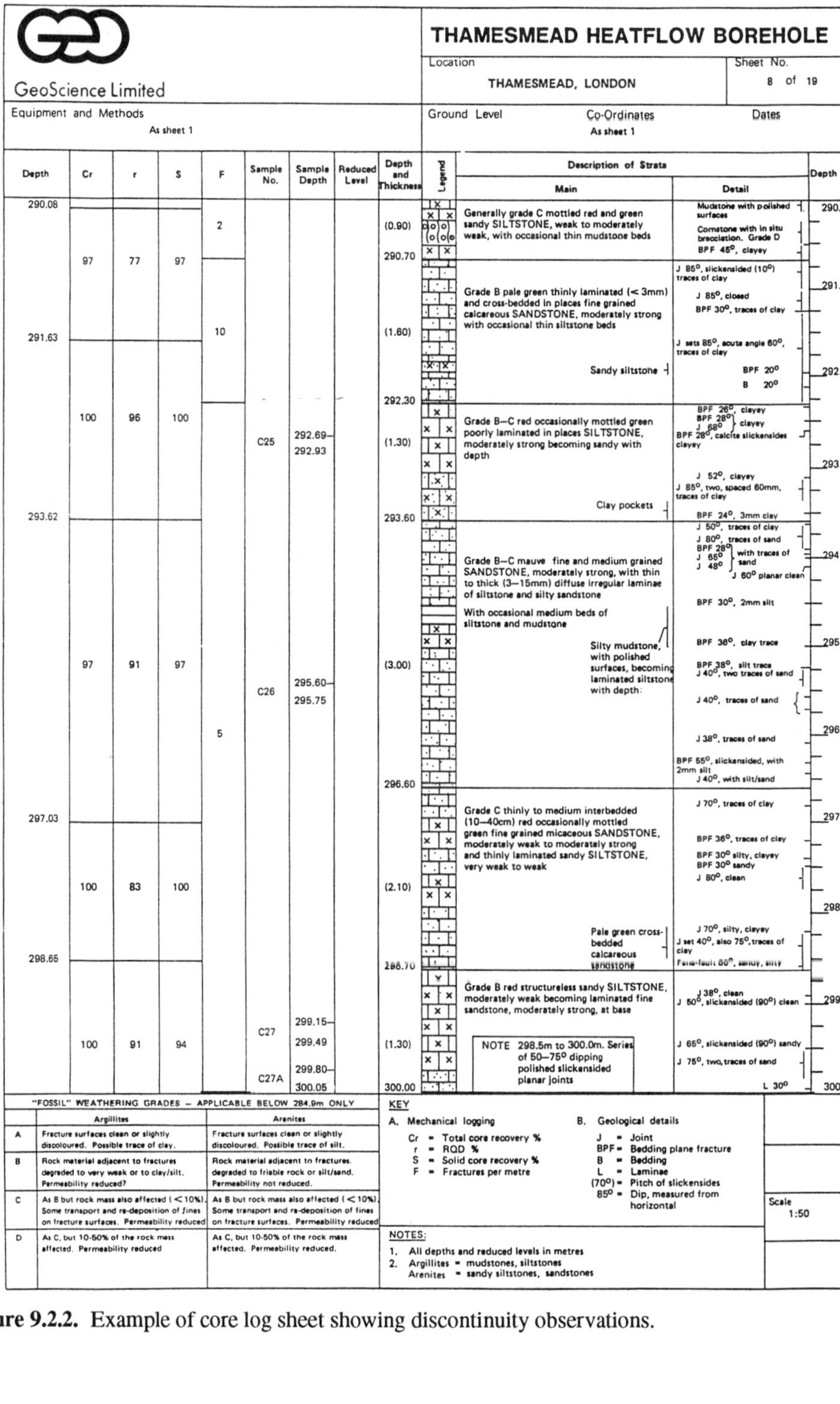

Figure 9.2.2. Example of core log sheet showing discontinuity observations.

- Core should be removed from the barrels under the supervision of structural geologists, with other specialists present if samples are to be taken or specialized core descriptions are to be undertaken. A high-quality video record may provide important information in a later stage in the project when evidence is no longer available.
- Core should be cleaned carefully and gently prior to logging. It should not be hosed and scrubbed.
- Core should be described by the structural geologist and photographed before any samples are taken, before any deterioration can take place, and before the core dries.
- Sampling should take place immediately after description and photography.
- Some forms of analysis and examination, such as palaeontology, can totally destroy core, reducing it to gravel-sized pieces; therefore, sampling programs, especially those that involve core removal, should be programmed with great care.

Logging and Sampling of Core. Core recovered from boreholes for fracture characterization should initially be logged, or at least very closely supervised on site, by a structural geologist who is to be closely involved with the structural interpretation in the rock mass. Initial description should be based on civil engineering practice to ensure that all basic physical characteristics are recorded. Detailed petrographic, lithological and stratigraphic information can be added after more detailed analysis. The following descriptive features should be considered the minimum required, to be used wherever there is a change in rock character: weathering state, structure, color, grain/crystal size, subordinate particle size, texture, alteration state, estimated material strength, mineral type, rock name.

National and International organizations have their own schemes for grading rock strength, particle size, etc.(Anon, 1972, 1978), and these can often be adopted as required. For specialized applications such as fracture logging, it is advisable to adopt such a recognized scheme at the start of the project and to refine it as information accumulates, re-logging earlier core as necessary to conform with the refined scheme.

Discontinuities should be described with great care, recording their dip relative to the axis of the core, their surface properties such as coloration, surface coating, infills, striae and slickensides (and their orientation) and the nature of the surfaces, including roughness and waviness, related to appropriate descriptive scales. Special logging sheets may have to be designed to cope with this level of information. If core orientation has been attempted, then both the dip and dip direction of each discontinuity, relative to the borehole, can be recorded. Calculations to derive absolute orientations should be carried out later, when the borehole has been surveyed accurately.

Several novel mechanical methods exist for orienting the core in boreholes, but even at shallow depths they are notoriously unreliable. The most practical way to orient core from deep holes is to compare fracture descriptions with the results of images of the borehole wall derived from selected downhole geophysical logging data, as will be described later. Whatever method is used to orient core, some means of grading the reliability of the results should be adopted. Recording errors may be suspected but not proven, and it is good practice to document these suspicions and, ideally, to use some form of easily referenced quality coding to pre-empt some of the problems that unreliable orientations can cause at the modeling stage.

Imaging Methods

In recent years a new generation of wireline logging tools have revolutionized visualization of rock mass structure at depth. These tools use different formation responses (sonic, electrical and light) to create images of the borehole wall. These images can then be processed using modern image analysis techniques. Image data from these tools are used for a wide variety of tasks, including fracture studies, hydrofracture stress measurement, assessment of borehole breakout, casing inspection, and stratigraphy. Each group of tools has its advantages and disadvantages and will be described separately before any comparisons are made.[1]

Borehole Televiewer (BHTV). The principle of the BHTV is described by Zemanek et al. (1970). A rotating ultrasonic transducer scans the borehole wall several times a second. As the logging tool moves up the well, the transducer scans a spiral path over the borehole wall. The tool typically produces two sets of measurements, the amplitude of the reflected wave and the time delay between the emission of the ultrasound pulse and its return. The latter measurement can be used to reconstruct the wellbore geometry. When the signal is plotted against depth, the resulting images give a realistic "picture" of the borehole wall, with joints and fractures clearly seen as areas of low amplitude, high transit time data. Digital image processing techniques must be used to enhance the data.

Since its introduction several improvements have been made to the tool and the methods of analyzing the data. Several groups have produced "geothermal" versions of the tool capable of running at temperatures of over 150 °C for a limited time (Dreesen, 1986).

Rambow (1984) describes a BHTV with an automatic gain control facility that allows the tool to see more deeply into fractures. Others have worked on focused transducers to provide better resolution and greater power for viewing through dense muds.

Fracture study work has demanded very high standards of orientation, and BHTV tools can now be run with orientation packages similar to those used with dipmeters. This is especially important for highly deviated wells where use of the original orientation device (a single flux-gate magnetometer) will introduce a systematic error due to the orientation of the earth's magnetic field. Mastin (1985) and Lanyon (1986) have both published formulae for calculating this systematic error.

Since the development of digital image acquisition and storage, another area of progress has been in the development of image analysis methods for BHTV data (Lloyd et al., 1986; Paillet et al., 1985). An illustration is given in Fig. 9.2.3 (Barton et al., 1989). The BHTV IMAGE package, developed at Stanford University, runs on an Apple Mac II workstation and provides facilities for generating false color images, 3D borehole geometry reconstructions, fracture orientation, and aperture measurements. The major disadvantages of the BHTV are its relatively slow logging speed and limited temperature range. Image quality will be degraded when running in heavy muds, but tools with better transducers are being developed. The quality of data suffers in highly elliptic wellbores, or where the tool is not well centralized. Georgi (1985) describes these geometrical problems, which lead to dark stripes of poor image quality on the log.

---

[1]Such comparisons, with complementary information and detailed examples, are available in Sec. 9.10.3, which covers wireline logging in the oil industry.

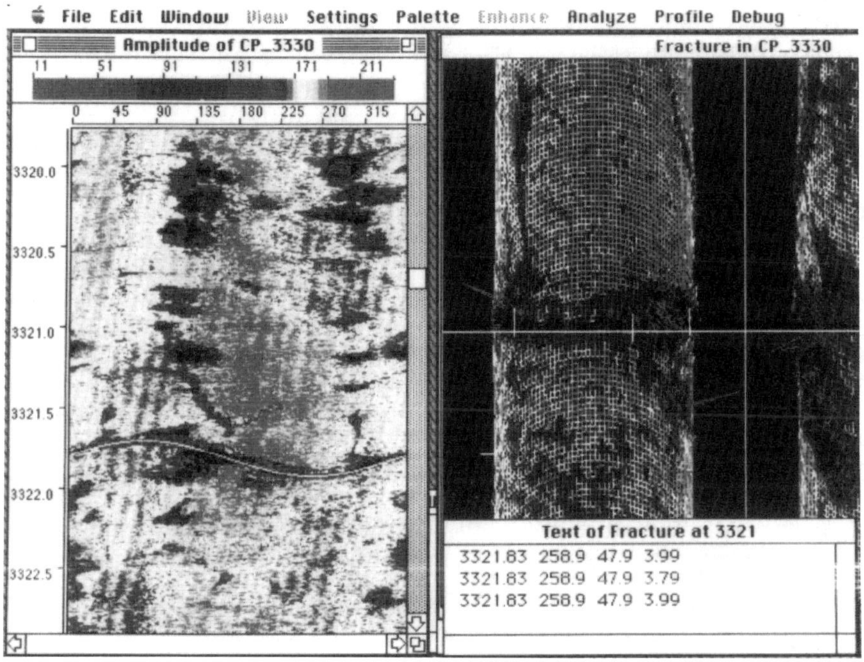

Figure 9.2.3. Borehole Televiewer Image Program (after Barton, C.A., 1989).

Formation Microscanner (FMS). The principle of the FMS tool is described in Lloyd et al. (1986) and further discussed by Cull (1988) and Burke (1989). The tool is a development of the dipmeter, where one or more of the dipmeter pads carries an array of button electrodes. These electrodes measure the conductance of the formation. Current versions of the tool have either two or four imaging pads. The four-pad tool has an array of 16 image buttons on each arm pad. The two pad tool is a hybrid dipmeter, where two of the four dipmeter pads carry an array of 27 small image buttons. The four-pad tool provides greater coverage of the wellbore at the cost of slightly lower resolution. In an 8.5-inch hole the four-pad tool provides approximately 40% coverage of the well, compared with 20% for the two-pad version. Typically, measurements from each image electrode are taken every .1 of an inch as the tool moves up the well.

The resolution of the tool is currently limited by telemetry and data processing considerations. The four-pad version makes over 7500 conductivity measurements over each foot of formation and, for this reason alone, detailed FMS logging is usually carried out on limited depth zones of specific interest. The tool has been used in fracture studies to estimate fracture orientation density and to make qualitative aperture estimates, and work has also been carried out to distinguish between natural and induced fracturing. Other uses include estimation of bedding thickness, and evaluation of sedimentary structure and rock texture. The major disadvantages of the tool are the limited lengths of borehole that can be covered in practice and the high cost of data processing.

Borehole Cameras. Camera systems may be used for providing detail about single fractures and for observing fracture dilation during and after hydraulic testing. Additional uses include inspection of casing and viewing well blockages. In general, camera tools can be used only where the wellbore contains clean water or gas. They are also limited to low temperature applications, although work has been done on both active and passive cooling systems.

Several different forms of borehole cameras have been developed, using systems based on film or television image transmission. Typically they have provided either continuous coverage of an axial view down the well or snapshot radial views of part of the borehole wall. In general, they have not produced a log output, although researchers have used records (usually videotape) of camera runs to reconstruct log-type images, similar to those of the BHTV. These images are normally produced from continuous axial recording and have to be corrected for the viewing geometry.

Very high resolution pictures can be obtained of small segments of the borehole wall. Batchelor (1984) reports images of 75 × 100 mm showing natural fractures. These provide excellent data on both rock and fracture texture, but the small field of view makes it impossible to measure fracture orientation. Similarly, while single fractures can be examined in great detail, comprehensive fracture studies are rarely possible. The major disadvantage of the camera tools has been that they do not produce digital output, which could be processed and integrated with other in-hole measurements.

9.3 MASS TRANSFER: HYDRAULIC METHODS

9.3.1 Introduction

Methods involving fluid transfer are direct methods for recognizing heat and mass transfer properties of fractured rocks. This fact explains their importance among all other recognition methods. Fluid existing inside the fractures and/or pores may be a gas, a liquid or a mixture of gas and liquids, possibly with solid particles. Table 9.3.1 presents the different elementary cases. Tests presented in this chapter refer essentially to situations where fractures and pores are saturated with one liquid. Tests involving one pure liquid are developed in the present section. These tests sometimes include coupling with heat or momentum transfers. Tests involving a fluid mixed with tracers are developed in Sec. 9.4. Thermal tests are mainly described in Sec. 9.5.

Types of Hydraulic Tests

Most hydraulic in-situ tests are well tests. A very large number of different tests have been devised. Tests can be realized in one single hole or between two or more holes, with different boundary conditions. Input signals are quantity of flow, pressure (or head), or a combination of both. Experiments are generally carried out in time space, but recent tests have been developed in frequency space. Even if techniques are apparently different as determined by the domain of application, the large variety of tests can be classified in all circumstances as given in Table 9.3.2.

Table 9.3.1. Different situations in fluid testing

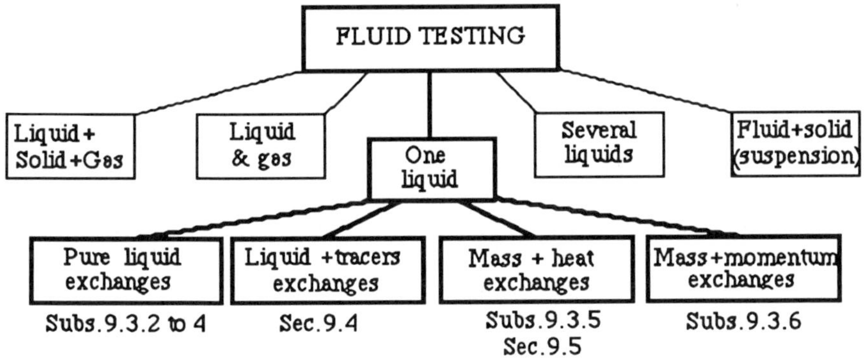

Table 9.3.2. Different types of hydraulic well tests.

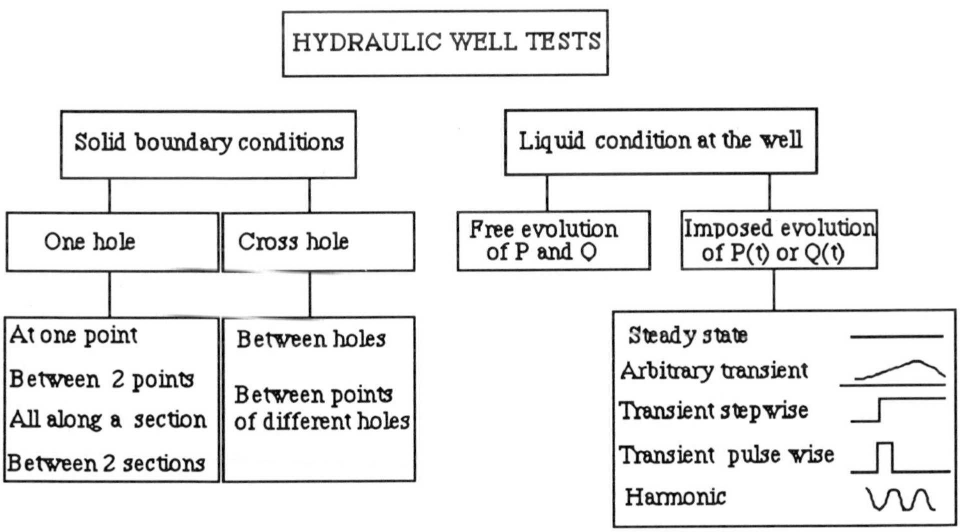

Steady-state or quasi-steady-state hydraulic tests were derived in the early days for water resource research in soils. These methods were extended in the 1970s to fractured rocks, with different improvements according to the execution and interpretation of the tests (Louis, C., 1977).

Transient tests were derived during the same period in petroleum engineering (Earlougher, R.C., 1977). In classical oil field tests, the basic test is a transient stepwise flow test in one well; a well section of length h is isolated between two packers, and a constant oil flow Q is imposed at the head of the well. The evolution, due to pumping or injection, of the pressure $P(t)$ versus time t is measured at the formation level. When these tests are interrupted, the variation in $P(t)$ can give extra information. If the time of aperture of the well is very short, it is a so-called pulse-test. Several successive pulse-tests can be considered. In more complex tests or in historical cases, the imposed quantity of flow can vary according to an arbitrary law. Interference tests between two wells can be performed under similar conditions; the production well, initially closed, is open under a constant quantity of flow, and the pressure variations are measured in observation wells, closed and initially stable.

All the above transient tests are limited to an order of magnitude of one second because the step between shut-in and producing is not perfect due to inertia, viscosity and/or compressibility of the fluid between the bottom and the head of the well. The principle of harmonic tests was proposed for identification purposes using highly transient effects (Jouanna et al., 1979). It consists of creating, in an open hole or between two packers, a sinusoidal quantity of flow, alternatively injecting a liquid into the formation and pumping it out of the formation. The resulting pressure in the well is also alternating and depends on the parameters of the formation and, accessorily, of the injection device. The test is first carried out at a given frequency and then repeated at different frequencies. The frequency can vary from 0.1 Hz up to 100 Hz or even more. In case of linear phenomena, results are expressed directly in the frequency space by a complex relationship $P^*(\omega) / Q^*(\omega)$ versus $\omega = 2\pi f$, $P^*(\omega)$ and $Q^*(\omega)$ being the Fourier transforms of P and Q.

Types of Interpretation of Fluid Testing

For direct prediction problems, different approaches are listed in Table 9.3.3.

The black box approach (1) consists of finding the relationship between pressure and quantity of flow, leading to a gross estimation of the mass transfer properties. In some cases, when the fracture medium is erratic, in karsts for instance, only this approach is possible.

A more refined phenomenological approach (2) can be based on equivalent continuous models, if fractures are small enough to be assimilated to pores. Filtration laws in such a continuous medium can be found experimentally or be assessed by homogenization techniques.

A more precise approach (3) takes explicitly into account discontinuities of the rock by some description of the fractures. This approach can be either deterministic or, as in more recent work, stochastic. The new filtration theory, based on fractal representation, is another promising tool for singular percolation situations.

These different approaches can be combined in a superposition approach (4), to interpret complex situations.

Theoretically, all these models could be used for inverse identification problems. However, models with too many parameters become impractical. Simplified models, where an analytical solution exists, are generally preferred as a first step in studying the inverse problem.

Table 9.3.3. Different interpretation approaches.

INTERPRETATION OF HYDRAULIC TESTS					
TYPE OF APPROACH	FRACTURED ROCK IDEALIZATION				
(1) Black box approach	Natural medium + its boundary conditions				
(2) Identification by continuous approach	One fracture	One set of fractures	One system of fractures	Several sets of fractures	One path
Experimental laws	(shaded)				(shaded)
Periodic Homogenization	(shaded)				(shaded)
Stochastic Homogenization	(shaded)				(shaded)
(3) Identification by discontinuous approach	One fracture	One set of fractures	One system of fractures	Several sets of fractures	One path
Deterministic					(shaded)
Stochastic	(shaded)				(shaded)
Percolation theory	(shaded)	(shaded)	(shaded)	(shaded)	
(4) Superposition approach					

Case Studies

Even in the case of a pure liquid (water or oil), Tables 9.3.2 and 9.3.3 show the infinite possibilities of combining different test situations with different interpretation approaches. Three interpretation points of view are presented in Secs. 9.3.2, 9.3.3, and 9.3.4: black box, continuous equivalent and discontinous approaches. Two examples of coupling are then presented: coupling between mass and heat and coupling between mass and momentum.

9.3.2 Black Box Approach

Conceptualization

Whatever the hydraulic test may be, in one-hole or cross-hole situations crude experimental results can always be presented as relationships between input and output. These inputs and outputs, generally functions of time t, are here the quantity of flow $Qi(t)$ in the vicinity of points i and pressure or head $Pj(t)$ at points j.

Simulation in Time Space

To represent possible fracture paths between two boreholes or different points of one borehole, black box modeling is the first degree of interpretation. It consists in finding mathematical expressions able to represent the experimental relationships between vector *Qi* and vector *Pj*. The proposed mathematical model has generally no physical bases and is said to be a simulation model. Linear relationships can be used in a first attempt. Such linear relationships assume that flow is laminar and that the fractured system is invariant. Engineers often refer to impedance of a fractured system, impedance coefficients *Cij* being defined by the matrix coefficient coupling vectors *Qi* and *Pj* by the relationship *Qi* = [*Cij*] *Pj*.

However, linear relationships between *Qi* and *Pj* are rare in fractured media, owing to turbulent flow phenomena and coupled effects modifying the fractured rock structure. If turbulent flow becomes predominant, a relationship such as $(\boldsymbol{Qi})^2 = [Cij]\,\boldsymbol{Pj}$ should be more pertinent. In fact, in practice the flow is turbulent around the well but laminar far from it. A global quadratic expression could be the signature of channeling because in such channels the flow is generally turbulent. For quasi-steady-state tests these relationships still apply to functions *Qi*(*t*) and *Pj*(*t*), coefficients *Cij* remaining unchanged.

Simulation in Frequency Space

For transient tests, this expression becomes time dependent; *Qi*(*t*) and *Pj*(*t*) are related by a convolution product. If phenomena can be linearized, performing a Laplace or Fourier transform enables the writing of a relation in the form *Qi** = [*Cij**] *Pj**; the transformed quantities are frequency-dependent and complex, as a phase angle now exists between flow and pressure.

Validation of Simulation

The domain of validity of simulation models is restricted to systems similar to the prototype for which measurements have been done. Moreover, a simulation model is valid within given experimental limits. It is important to test this domain of validity by back analysis of different history cases.

9.3.3 Continuous Approach

Conceptualization

In the most simple conceptualization, as used classically in petroleum engineering, fissured rock formations are considered as isotropic porous layers of porosity ϕ_1, permeability k_1, and thickness h, and as perpendicular to the well reduced to a line (line source model). If the well is considered with its true radius r_w, the effect of the liquid column is introduced by a coefficient C (wellbore storage). Around the well, a zone disturbed by perforation or stimulation may be taken into account either by a coefficient S (skin) or by a zone around the well with modified porosity and permeability (composite radial model). Physical parameters are the total compressibility c_t of the formation, including fluid and bulk compressibilities, and the dynamic viscosity μ of the liquid.

When porosity is expected to be created by fractures, these fractures may be assumed to be stochastically distributed in space, and their presence may be represented by an equivalent porous medium. Generally this fictitious porous medium is superimposed on the matrix porous medium, resulting in a "double porosity medium." In this superposition process, the flow of two fictitious fluids is studied, one being assigned to the pores and the other to the fractures. Parameters are allocated to each fictitious liquid together with cross parameters governing exchanges of mass, momentum, energy and/or entropy between these two liquids. Because the process is assumed to be quasi-static and isothermal, only the coefficient governing the mass exchange versus the difference of head between matrix and fractures is considered. When a block size in the true fractured medium is assessed, this parameter can be considered as a phenomenological coefficient or as resulting from a homogenization procedure. In a more sophisticated conceptualization, two horizontal layers (double permeability models) or more complicated boundary limits can be introduced.

Interpretation in Time Space

For each kind of test and for each kind of conceptualization, modeling leads to reference curves, the so-called "type curves." According to the matching procedure, comparison is made between type curves and experimental curves. Identification of unknown parameters or groups of parameters requires special training in each case. In the most sophisticated models, the number of parameters increases and leads to difficult interpretations, especially concerning unicity of the possible parameters. This identification process becomes very intricate when analytical solutions of the models are no longer available. A very recent account of the state of the art is given on double porosity and double permeability systems by Chen (1989). A complete description of double porosity type curves will be found in Daviau (1986, Ch. 3, Annex 5, charts 5 to 8). For heterogeneous formations, a review is given in Streltsova (1988).

The more general analytical solutions lead to an expression of the dimensionless pressure P_D depending on four dimensionless parameters. As it is impossible to draw such type curves with four parameters, approximations have been proposed. For instance, Bourdet and Gringarten (1980) proposed an approximation for intermediate times, the type curves for short or long times being similar to the type curves obtained for a homogeneous medium. Thus only two dimensionless parameters remain, making it possible to draw sets of curves. For the purpose of matching, dimensional results plotted on log-log axes are superimposed on the type curves, as shown in Fig. 9.3.1. When coincidence is obtained, the coordinates of the match point allow correlation between dimensional and dimensionless variables, leading finally to the engineering parameters, here $k_f h$ (product of the fracture permeability by the height of the formation) and the skin coefficient S.

Interpretation in Frequency Space

For harmonic tests, the same double porosity case has been developed in frequency space by Alami (1983) and Bonnet (1985). In these new developments, the type curves are given in modulus and phase; they depend on new dynamic terms introducing complex coefficients such as complex Darcy's coefficient or complex skin coefficient S. Such analysis yields new groups of parameters resulting from transient terms and inertia that give complementary information. This helps to solve theoretical inverse problems, leading, for instance, to unicity where this was not

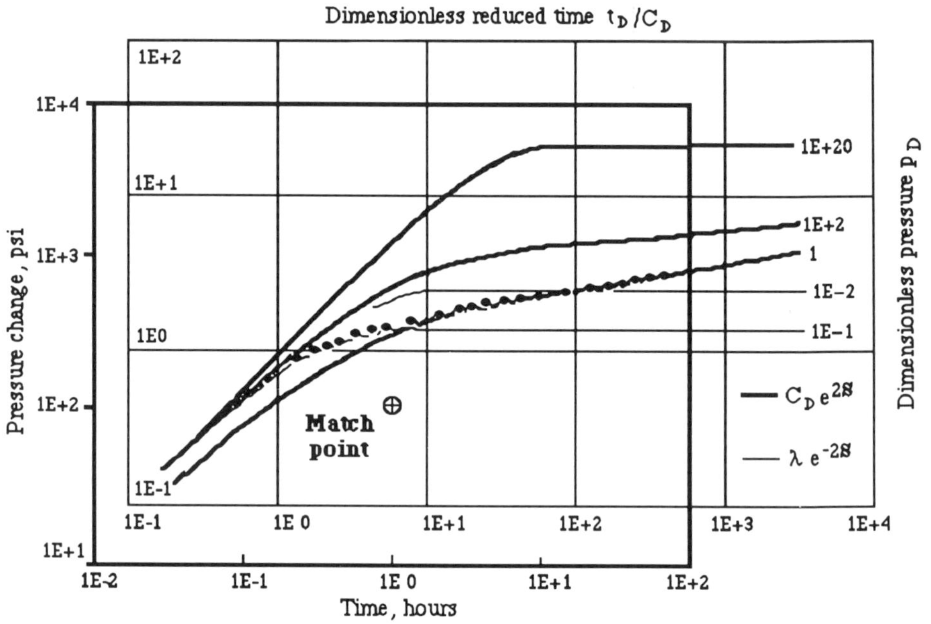

Figure 9.3.1. Drawdown type curves and matching in the double porosity case (after Bourdet and Gringarten, 1980).

previously possible. In addition, it is a self-checking method, because the fit has to be achieved on the curves for both modulus and phase. Moreover, sensitivity may be better for phase when it is poor for modulus, and vice versa.

If these transient tests are executed at high frequencies, the classical dynamic equation used in oil interpretation (Earlougher, 1977) has to be modified. Using homogenization techniques or Biot's theory, it appears that two equations instead of one are necessary to describe the movement of the liquid and the movement of the bulk (Bonnet, 1985). To achieve this, all the work done in time space, for the double porosity case, can be developed in frequency space. Alami (1983, p. 64) gives, for a double porosity medium, the dimensionless expression in Fourier frequency space of the pressure P_w at the well, as follows:

$$P^*_{wD}(\omega_D) = \frac{(1+iIL\omega_D)[K_o(F(\omega_D)) + S'F(\omega_D)K_1(F(\omega_D))]}{C_D(iL\omega_D - IL^2\omega_D{}^2)K_o(F(\omega_D)) + [1 + C_D S'(iL\omega_D - IL^2\omega_D{}^2)]F(\omega_D)K_1(F(\omega_D))}$$

with the following expressions, (Q_w^* being the harmonic quantity of flow imposed at the well):

$$\omega_D = \frac{\mu \Phi_f c_{tf} r_w^2}{L\, k_f}\omega \qquad L = \frac{\Phi_f c_{tf}}{\Phi_m c_{tm} + \Phi_f c_{tf}} \qquad P_{wD}^* = \frac{2\pi\, k_f\, h}{\mu\, Q_w^*} P_w^*$$

$$I = \frac{\rho_{22f} k_f^2}{\mu^2 \Phi_f^3 c_{tf} r_w^2} \qquad F(\omega_D) = \left[(i\omega_D - IL\omega_D^2) \frac{L(1-L)i\omega_D + \lambda}{(1-L)i\omega_D + \lambda} \right]^{1/2}$$

$$C_D = \frac{C}{2\pi \Phi_f\, c_{tf}\, h\, r_w^2} \qquad S' = \frac{S}{1 + iIL\omega_D} \qquad \lambda = \alpha \frac{k_m}{k_f} r_w^2 \qquad i^2 = -1$$

where f, m, D, w refer to "fissure", "matrix", "dimensionless", "well." ω is the pulsation, μ the dynamic viscosity, c_t the total compressibility, k the permeability, h the well section length, Φ the porosity, P^* the Fourier transform of the pressure P, Q^* the Fourier transform of the quantity of flow Q, α a shape coefficient, r_w the well radius, C the wellbore storage, S the standard skin coefficient, ρ_{22} a Biot's dynamical coefficient, K_0 and K_1 modified Bessel functions.

This expression presents evident similarities with Mavor and Cinco's solution (1979) given in Laplace space, but exhibits complementary terms due to the dimensionless inertia parameter I. This parameter is typical of transient tests such as harmonic pumping tests. The inverse transform into time space is no longer necessary because the experimental curves are obtained directly in frequency space and the matching is done in frequency space. Moreover, plotting the frequency type curves is much easier, because no approximation is necessary to obtain an analytical expression of the frequency type curves. The following example (see Alani, 1985, Fig. 23-3, p. 169 and Fig. 23-4, p. 170) is given just to illustrate the matching procedure in frequency space. The tested material is a standard porous medium with a limited extension. Matching can be carried out simultaneously in phase and modulus, as shown respectively in Figs. 9.3.2(a) and 9.3.2(b). The coefficients C_D and D refer to the adimensional wellbore storage and to a caisson coefficient.

Validation

To verify the experimental information given by a test and/or to check the values of the parameters obtained by this inverse process, it is very important to validate these results on history cases. Such back analysis is a necessary final step in any parameter determination by matching.

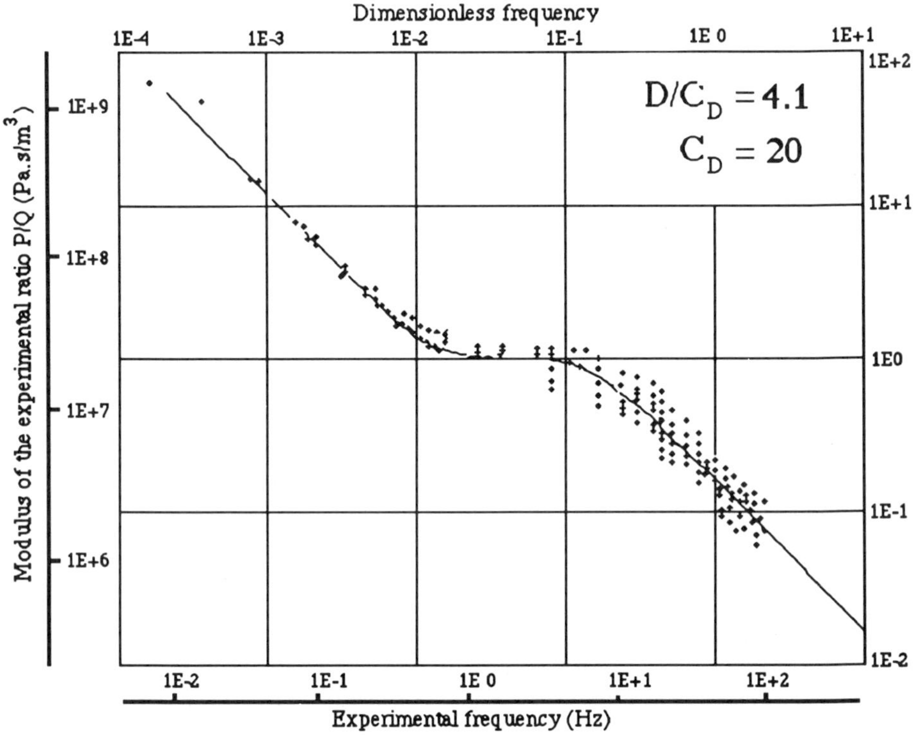

Figure 9.3.2 (a). Harmonic pumping test. Modulus matching (after Alani, 1985, p. 169).

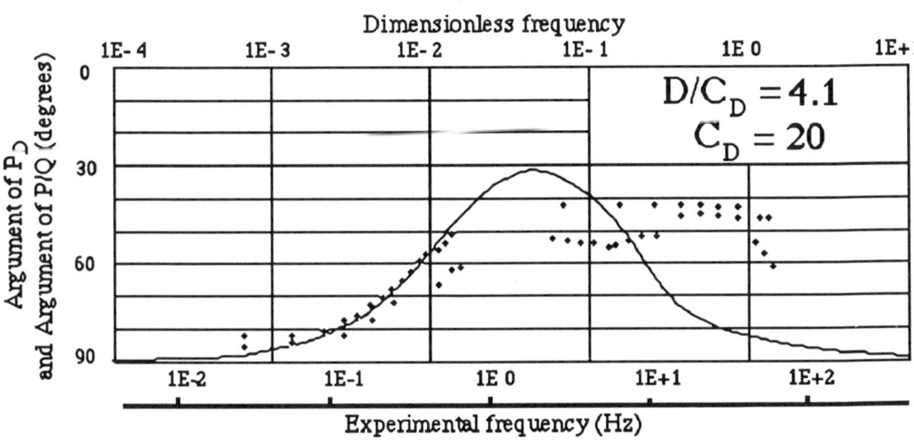

Figure 9.3.2 (b). Harmonic pumping test. Phase matching (after Alani, 1985, p. 170).

9.3.4 Discontinuous Approach

Conceptualization

Fractured rocks can no longer be considered as a continuum when the fractures have relatively large dimensions compared to the size of the rock mass considered. Such circumstances are common in surface rock formations, in mining and in artificial fractures created for stimulation in deep formations (oil or geothermal engineering). Predicting the hydraulic behavior of a given fracture or set of fractures embedded in a pervious or impervious matrix is a direct problem, extensively studied in the 1970s. By contrast, progress on the inverse problem was very slow; cases as simple as finding the characteristics of a single limited fracture intersected by one well remained unanswered because appropriate tests were lacking.

Such questions were the source of development of the harmonic technique for individual fractures (Jouanna et al., 1979), which has been reviewed recently in Crosnier (1987). In porous or equivalent porous media, this method (described in Sec. 9.3.3) gives more information than other existing methods. In the case of isolated and closed fractures, it is often the only procedure available for understanding the hydraulic behavior of small fractures, especially in a single-well investigation. Harmonic tests are also used in cross-hole tests.

Interpretation of harmonic tests in isolated fractures depends, of course, on the type of fracturation model. Choosing among all possible models can be helped by visual well reconnaissance or some sort of logging.

To guide interpretation, the two following elementary models are fundamental:

- One closed fracture in an impervious and rigid matrix, where the liquid movement is essentially governed by inertia and compressibility forces.

- One open fracture in an impervious and rigid matrix, where the liquid movement is essentially governed by inertia and viscous forces.

More sophisticated models can be exploited for refining interpretation, taking into account inertia, compressibility and viscosity forces, permeability and/or deformation of the matrix, etc.

For complex situations, it is also possible to combine the preceding elementary emulation models by using the theory of hydraulic quadrupoles under linear assumptions. Combining these quadrupoles as electrical quadrupoles leads to a great number of models (Crosnier, 1987, pp. 122-134).

Interpretation in Time Space: Example of One Closed Fracture

A great deal of work has been done in petroleum engineering for the case of one well with a vertical fracture embedded in a porous formation. A review of the different available cases is given in Daviau (1986, Ch. 4, pp. 61-72). Analytical solutions assume that the well is reduced to the fracture (no wellbore storage) and the skin coefficient is zero. The fracture can have a uniform flow per surface, have an infinite hydraulic conductivity, or be filled with a porous medium.

Interpretation in Frequency Space: Example of One Closed Fracture

The case of a harmonic pumping test in one vertical well with a vertical fracture could also be studied using the previous analytical models developed for common petroleum engineering. However, other cases may be encountered in deviated wells, in geothermal engineering, and in geotechnics applications, where fractures can have any inclination on the well axis. The simpler case is a limited circular fracture of extension R_f, which can have a constant thickness, different inclinations θ or different shapes around the well of radius R_p as shown in Fig. 9.3.3 (a) to (c). The harmonic quantity of flow Q is imposed between two packers placed at the fracture entrance; the resulting harmonic variation of pressure P between these two packers is measured. The average static pressure P_o does not interfere when the matrix has no deformation; in other cases, the fracture opening depends on the static pressure P_o but does not depend on the harmonic variation P, which is in any case much lower than P_o.

In case (a), with a constant thickness e of the fracture, $Q(t)$ and $P(t)$ are related in frequency space by the transfer function $Z(R_p, R_f, \omega, c)$ as given below, with V_o maximum radial velocity, $\omega = 2\pi f$ the pulsation, f the frequency, ρ_o the liquid mass density, c the wave velocity in the fluid and J_0, J_1, Y_0, Y_1 the Bessel functions of the first and second kind, of order 0 and 1:

$$Q(t) = 2\pi\ R_p e\ V_o \cos\left(\omega t - \tfrac{\pi}{2}\right)$$

$$P(t) = \rho_o\ V_o\ c\ Z(R_p, R_f, \omega, c)\ \cos(\omega t)$$

$$Z(R_p, R_f, \omega, c) = \frac{J_1\left(\frac{\omega}{c} R_f\right) Y_0\left(\frac{\omega}{c} R_p\right) - Y_1\left(\frac{\omega}{c} R_f\right) J_0\left(\frac{\omega}{c} R_p\right)}{J_1\left(\frac{\omega}{c} R_f\right) Y_1\left(\frac{\omega}{c} R_p\right) - Y_1\left(\frac{\omega}{c} R_f\right) J_1\left(\frac{\omega}{c} R_p\right)}$$

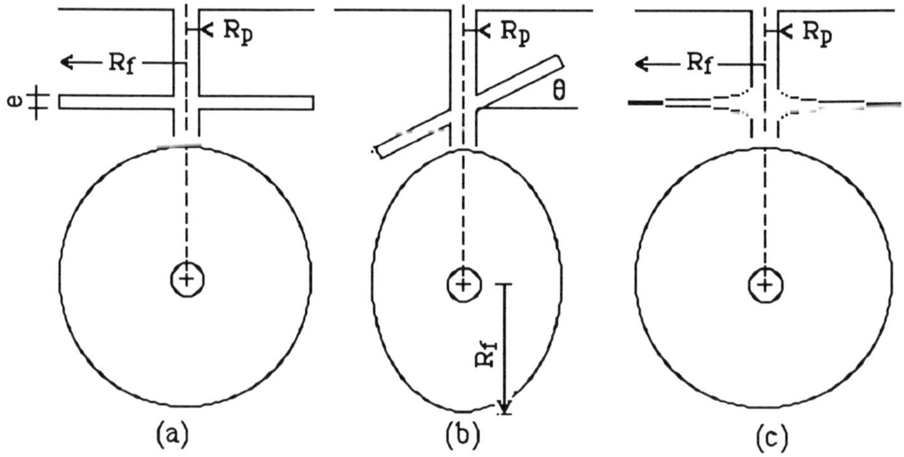

Figure 9.3.3. Possible fracture situations in a harmonic pumping test (after Crosnier, 1987, p. 46).

The graph of the transfer function Z versus f is given in modulus in Fig. 9.3.4 and in phase in Fig. 9.3.5. The forces of inertia and compressibility create in the fluid a resonance phenomenon, which explains the very simple shape of the phase graph.

- The resonance frequency f_o gives the extension R_f of the fracture (Fig. 9.3.6).
- The shape of the modulus curve gives the thickness e of the fracture (Fig. 9.3.4). The sensitivity of the method is evident, the two extreme curves corresponding to thicknesses of 0.2 mm and 2 mm.
- Other studies show that the transfer function does not depend in practice on the inclination θ of the fracture up to angles of 80° (Fig. 9.3.7).
- The shape of the fracture does not interfere with the determination of its extension, since it can be derived from Fig. 9.3.8.

Interpretation in Frequency Space: Example of Open Fractures

Let us consider now several sets of parallel fractures such that the hydraulic head H_o can be supposed constant at a distance R_f from the well of radius R_p, as shown in Fig. 9.3.9. The set (i) is made up of n_i fractures of the same thickness e_i. The fluid is assumed to be Newtonian, incompressible, and have a kinematic viscosity ν and laminar flow. The input signal is the flow $Q(t)$, and the output is the difference between the instantaneous head $H(t)$ and the constant head H_o, their difference being referred to as $\Delta H(t)$.

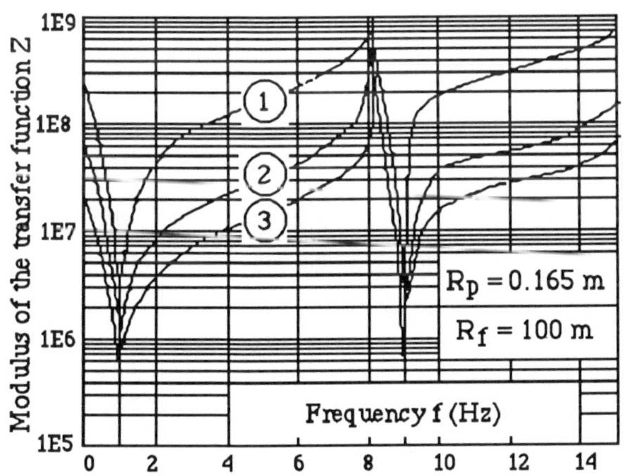

Figure 9.3.4. Theoretical modulus of the transfer function Z (after Crosnier, 1987, p. 49).

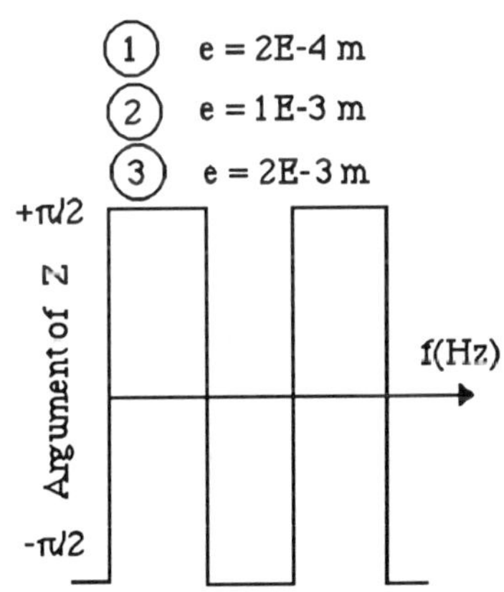

Figure 9.3.5. Theoretical phase of the transfer function Z (after Crosnier, 1987, p. 49).

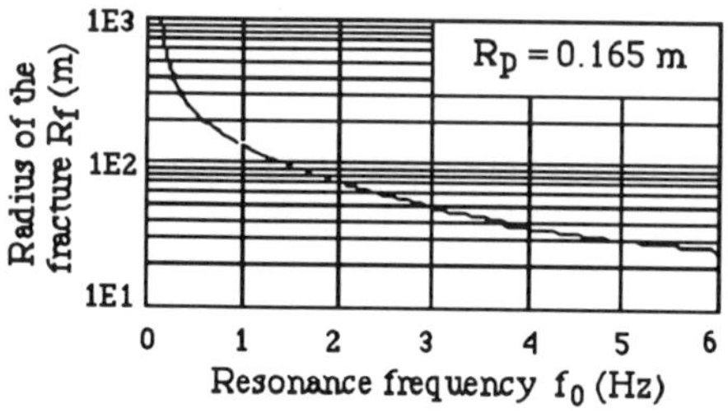

Figure 9.3.6. Fracture extent vs. resonance frequency (after Crosnier, 1987, p. 51).

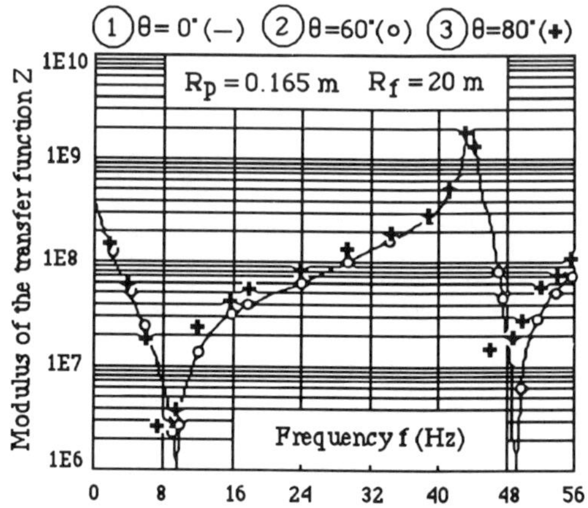

Figure 9.3.7. Influence of the inclination of the fracture on the modulus of the transfer function Z (after Crosnier, 1987, p. 54).

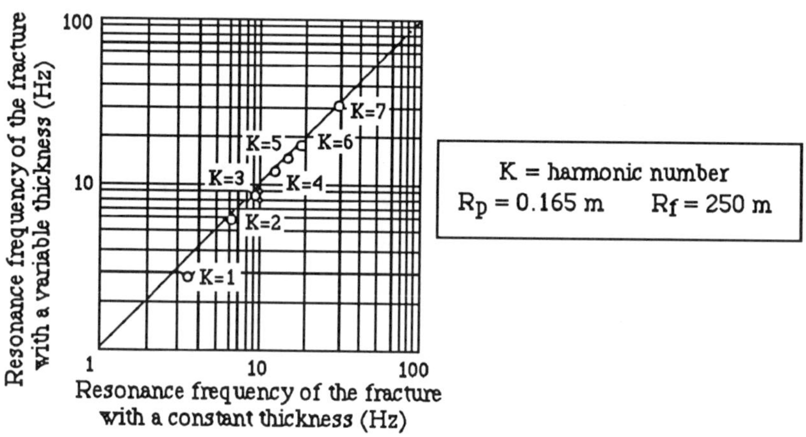

Figure 9.3.8. Influence of the shape of the fracture on resonance frequencies, from order 1 up to order 7 (after Crosnier, 1987, p. 56).

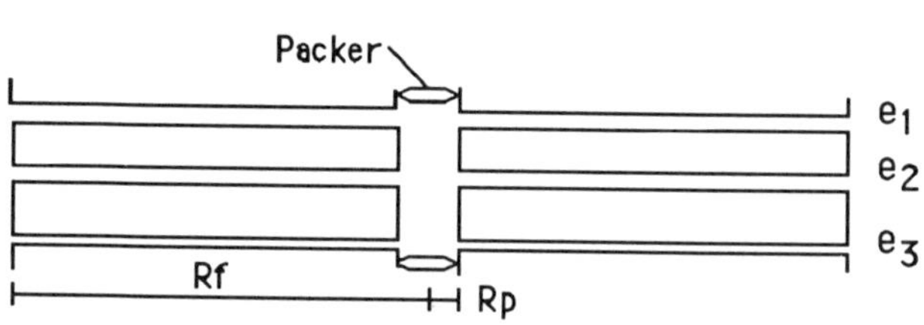

Figure 9.3.9. Representation of a horizontal open fracture family.

- Modulus of the transfer function F:

$$|F| = \frac{|\Delta H^*|}{|Q^*|} = \frac{K\,\omega}{\left[\left(\sum_i n_i e_i \frac{A_i C_i + B_i}{2\beta_i(1+C_i^2)}\right)^2 + \left(\sum_i n_i e_i \left(-1 + \frac{A_i - B_i C_i}{2\beta_i(1+C_i^2)}\right)\right)^2\right]^{1/2}}$$

with: $K = \text{Ln}(R_f / R_p)/(2\pi g)$

$$\beta_i = \frac{e_i}{2}(\omega / 2\nu)^{1/2}$$

$$A_i = \tanh\beta_i + \tan\beta_i$$

$$B_i = \tanh\beta_i - \tan\beta_i$$

$$C_i = \tanh\beta_i \cdot \tan\beta_i$$

- Argument of the transfer function F:

$$\text{Arg } F = \text{Arctan}\frac{\sum_i n_i e_i \left(1 - \frac{A_i - B_i C_i}{2\beta_i(1+C_i^2)}\right)}{\sum_i n_i e_i \frac{A_i B_i + C_i}{2\beta_i(1+C_i^2)}}$$

Analysis of the transfer function F shows groups of parameters that are identifiable by adjustment of the real spectral signature to the theoretical type curves. The first step consists of obtaining as much information as possible from the behavior of the transfer function, first at low frequencies and then at high frequencies, using the following limited and asymptotic expansions:

- Low frequencies:

$$|F| = \frac{6\nu\, Ln(R_f / R_p)}{\pi g} \frac{1}{\sum_i n_i e_i^3} \qquad \text{Arg } F = \text{Arctan}\left[\frac{1}{10\nu} \frac{\sum_i n_i e_i^5}{\sum_i n_i e_i^3}\,\omega\right]$$

- High frequencies:

$$|F| = \frac{Ln(R_f / R_p)}{2\pi g} \frac{\omega}{\sum_i n_i e_i} \qquad \text{Arg } F = \text{Arctan}\left[(\omega/2\pi)\frac{\sum_i n_i e_i}{\sum_i n_i}\right]$$

Knowing the values of these expressions, comparison with experimental curves makes it possible to determine the four parameters p_1, p_2, p_3, p_4 linked to the unknown physical parameters by the following relations:

$$p_1 = \sum n_i \qquad p_2 = \sum n_i\, e_i \qquad p_3 = \sum n_i\, e_i^3 \qquad p_4 = \sum n_i\, e_i^5$$

Here the determination of p_1, p_2, p_3, p_4 is possible and unique, the inverse problem leading to uniqueness for the mathematical parameters. Finding the physical parameters requires an inversion of the above formulas, which can lead to one or several solutions. If several solutions are possible, more information to achieve uniqueness can be obtained by considering the behavior of the transfer function for intermediate frequencies between these two extreme low and high frequency behaviors. In the case of two sets of fractures, the characteristics (n_1, e_1, n_2, e_2) of these two sets can be obtained without referring to the intermediate behavior of the transfer function F. In the case where e_1 is much larger than e_2, an analytical approximation is given by

$$n_1 = \frac{(p_3)^{5/2}}{(p_4)^{3/2}} \qquad e_1 = \frac{(p_4)^{1/2}}{(p_3)^{1/2}}$$

$$n_2 = p_1 - \frac{(p_3)^{5/2}}{(p_4)^{3/2}} \qquad e_2 = \frac{p_2 - (p_3)^2 / p_4}{p_1 - (p_3)^{5/2} / (p_4)^{3/2}}$$

Figure 9.3.10 (a) provides an example of comparison between experimental points, giving the modulus of the transfer function as recorded on a laboratory model, and corresponding type curves for 1 fissure of 4 mm + 2 fissures of 2 mm. Figure 9.3.10 (b) gives the corresponding argument type curves. A detailed analysis is given in Crosnier et al. (1985).

In Fig. 9.3.10 (a) it can be observed that the fit becomes poor above 5 Hz. This fact is due to turbulence in the flow regime. It is possible to avoid such a discrepancy by reducing the piston displacement, the quantity of flow being proportional to the frequency and to the piston displacement. In practice, harmonic tools possess an adjustable piston displacement, large displacements being used at low frequencies and low displacements at high frequencies.

9.3.5 Coupling of Mass and Heat Transfers

Conceptualization

A large number of injection tests, coupling heat and mass transfers, are carried out at medium or high depths for geothermal purposes or other applications such as environmental investigation. These tests can be performed using any of the preceding pure hydraulic tests by adding temperature and/or heat flux measurements. Injection or pumping can be carried out in one well or between two producing wells, monitoring being performed in these wells or using adjacent piezometers.

For pure fluids the black box, continuous, or discontinuous approach may be used in modeling mass and heat exchanges. The black box approach, not specific to the fractured medium, is not presented here. The discontinuous approach is analyzed in detail in Sec. 9.5, which covers thermal methods. An example of a hydraulic and thermal continuum approach is given in the present section.

Heat and mass transfers in simultaneously porous and fractured formations are treated here using the double porosity concept. The flow of two fictitious fluids is studied, one in the pores and the other in the fractures, each fluid having its own temperature. Each fictitious liquid possesses its own phenomenological parameters; moreover, cross parameters govern exchanges of mass and energy between these two fictitious media.

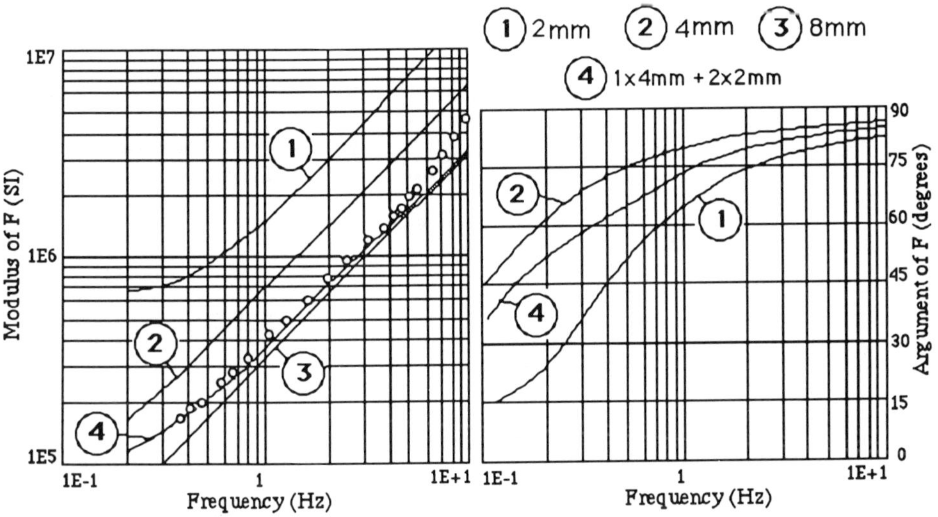

(a) Modulus of F. (b) Argument of F.

Figure 9.3.10. Transfer function F for a set of two families of fractures (after Crosnier et al., 1985).

Under given mass and heat boundary conditions, integration of the coupled heat and mass differential equations leads to a set of type curves, which give, for instance, temperature variations versus time. Comparison with experimental results by means of a matching procedure leads to parameters coupling heat and mass transfers. Validation of modeling through a case history is a necessity.

Example of a Coupled Hydraulic and Thermal Continuum Approach

The proposed example is a pilot scheme for an industrial project situated near Dieppe, in northern France (Mary, 1987). In-situ tests consisted of injecting and then extracting, in a single well, hot water in a porous and fissured pervious limestone zone situated between a compact clay and a deeper impervious limestone. Temperatures and quantities of flow of injected or extracted water were recorded in the well; formation water temperatures were observed by means of three piezometers, as shown in Fig. 9.3.11 (Mhamdi Alaoui, 1988, Vol. 2, p. 18).

A first test was conducted to analyze hydraulic and thermal properties of the double porosity limestone formation. Temperatures and quantities of flow were imposed by industrial requirements during a first injection period (Jan. 14–Feb. 10). Then after a storage period (Feb. 10–March 9), hot water was pumped out (March 9–April 11). A second test, simulating industrial exploitation, consisted of injecting hot water for a week and pumping it out during the weekend.

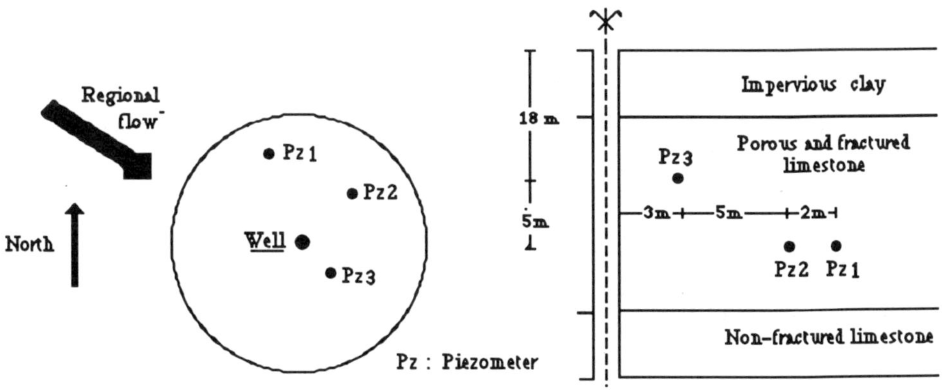

Figure 9.3.11. Heat injection well and temperature observation piezometers (after Mhamdi Alaoui, 1988).

Creating the Model. Based on geological observations, the storage formation was considered a double porosity medium, the two other zones being regarded as pervious for heat transfers but impervious for mass transfers. The set of differential equations for the double porosity model is given below (Mhamdi Alaoui, 1988, p. 84), the permeability of the matrix being assumed to be much lower than the permeability of the fractures.

The state variables are as follows: H_2, T_1, T_2

The mass equations are as follows:

- Medium 1, equivalent to the saturated porous matrix (equation not necessary here).
- Medium 2, equivalent to the fluid in fractures.

$$\operatorname{div}\left(\frac{k_2}{\mu_2}\operatorname{grad} H_2\right) = 0$$

The thermal equations are as follows:

- Medium 1, equivalent to the saturated porous matrix:

$$\rho_t\, c_t \frac{\partial T_1}{\partial t} = \operatorname{div}(\lambda_t \operatorname{grad} T_1) + H'(T_2 - T_1)$$

- Medium 2, equivalent to the fluid in fractures:

$$\rho_2 c_2 \frac{\partial T_2}{\partial t} = \mathrm{div}(\lambda_2\, \mathrm{grad} T_2) - \rho_2 c_2 V_2\, \mathrm{grad} T_2 + H'(T_1 - T_2)$$

with:

$$V_2 = -\frac{k_2 \rho_f g}{\mu_2 \phi_2}\, \mathrm{grad} H_2$$

H_2 = water head of the liquid in fractures

T_1 , T_2 = temperatures of medium 1 and medium 2, respectively

k_2 = permeability of the fractured equivalent medium 2

μ_2 = dynamic viscosity of the liquid in fractures

ρ_t , ρ_2 = apparent mass density of medium 1 and medium 2, respectively

c_t, c_2 = specific heat of the saturated matrix and of the liquid in fractures, respectively

t = time

λ_t , λ_2 = apparent thermal conductivity in medium 1 and medium 2, respectively

H' = heat exchange coefficient between medium 1 and medium 2

V_2 = average water velocity in the fractures, medium 2

ρ_f = mass density of the water

g = gravity acceleration

ϕ_2 = porosity of the fractures

Identification of Unknown Parameters. A first approach entails carrying out a sensitivity study to isolate the most important parameters. Such a study shows that heat exchanges due to conductivity λ_2 in medium 2 are probably not important and that fracture porosity ϕ_2 influences heat transfers in the early stages only. On the other hand, influence of the heat exchange coefficient H' is essential.

To determine H', a homogenization approach can be attempted, as given for instance in Mary (1987, pp. 113-138), but this approach is possible only under very simple assumptions. Moreover, it requires knowledge of the geometry of fractures, this geometry being precisely the main unknown of the problem.

Given these conditions, identification of a mean value of H' only was sought in this test. Assuming other parameters to be estimated, the inverse process consists of comparing the experimental evolutions of temperature with a set of type curves plotted for different values of H'.

This comparison is made on graphs of fluid temperatures T_2 in the fractures, direct measurement of temperatures T_1 being difficult in the saturated matrix.

The series of results at the producing hole, given by the first test mentioned above, was compared to type curves during both storage and pumping periods. Computation assumptions are given in Fig. 9.3.12. The results are given in dimensionless temperature $T+$ defined as $T+ = (T - T_o) / (T_{in} - T_o)$ with T_o = initial temperature of the medium and T_{in} = maximum temperature of the medium. Taking as reference the maximum injection flow, the dimensionless quantity of flow (injection or exhaust) is also plotted, the reference being the maximum injection flow; this flow imposed by industrial requirements is essentially variable, except during the storage period.

The matching procedure is shown in Figs. 9.3.13 to 9.3.15. For a very low value of $H' = 1$ $W/(m^3.K)$, in Fig. 9.3.13, the temperature in the fractures is too low in comparison with the recorded temperature. The fit seems to be realized for values of H' between 10 $W/(m^3.K)$ in Fig. 9.3.14 and 100 $W/(m^3.K)$ in Fig. 9.3.15. Nevertheless, the shapes of the curves are not entirely satisfactory, and complementary investigations should be made on different values of λ_2 and ϕ_2. Moreover, a regional flow and a general anisotropy created by large fractures were not considered in this first study.

Checking the Validity of the Model. The third step of the procedure consists of verifying how the identified model is able to emulate the hydraulic and thermal behavior of the porous and fractured formation when subjected to other boundary conditions. Such a validation process has been carried out on a weekly injection and pumping cycle, simulating industrial exploitation. The predicted temperature curves are compared to the recorded industrial measurements in Figs. 9.3.16 for $H' = 10$ $W/(m^3.K)$ and in Fig. 9.3.17 for $H' = 100$ $W/(m^3.K)$. Other comparisons have been made at observation points having piezometers. Moreover, influence of the regional flow was tested, showing that it has no significant effect on the weekly storage cycle (Mhamdi Alaoui, 1988, p. 152).

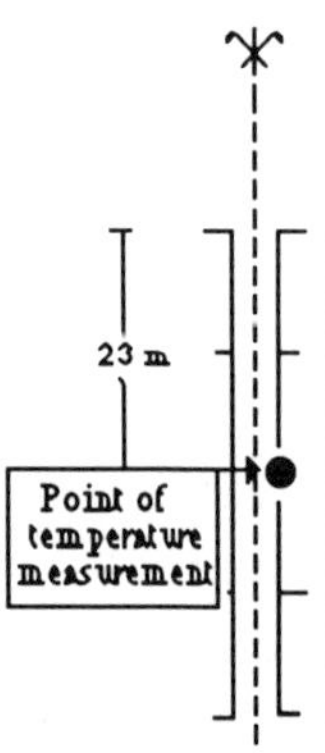

		Thermal conductivity W / (m.K)	Specific heat J / (kg.K)	Porosity (%)	Permeability (m/s)	Volumic mass kg / m^3
Impervious clay		1.5	750	45	10^{-9}	2800
Pervious & fractured limestone	Medium 1 (Matrix)	1.0	1200	35	10^{-7}	2650
	Medium 2 (Fractures)	0.6	4200	5	$2 . 10^{-3}$	1000
Non fractured limestone		1.25	900	35	10^{-7}	2650

Figure 9.3.12. Parameters of the heat and mass modeling (after Mhamdi Alaoui, 1988, Tome 2, p. 71).

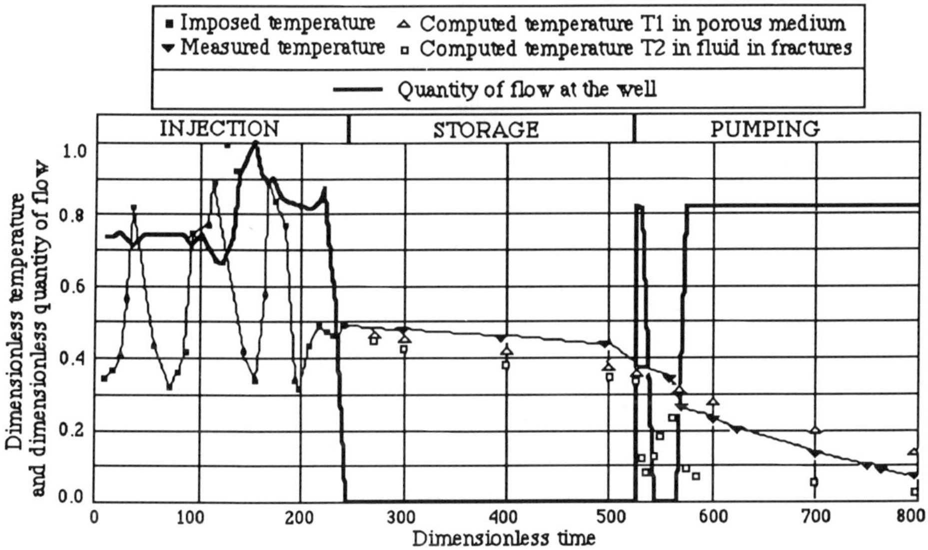

Figure 9.3.13. Matching on type curves for $H' = 1\ W/(m^3.K)$ (after Mhamdi Alaoui, 1988, Tome 2, p. 70).

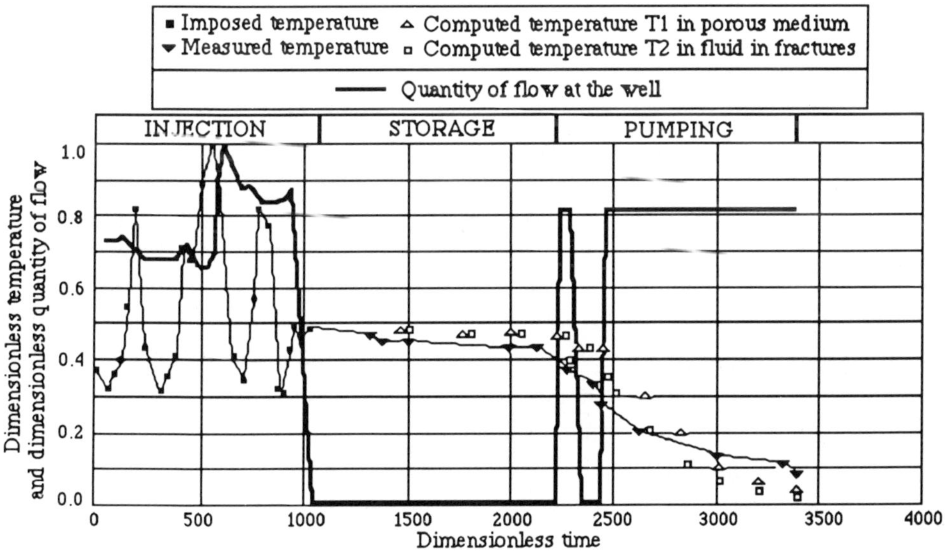

Figure 9.3.14. Matching on type curves for $H' = 10\ W/(m^3.K)$ (after Mhamdi Alaoui, 1988, Tome 2, p. 80).

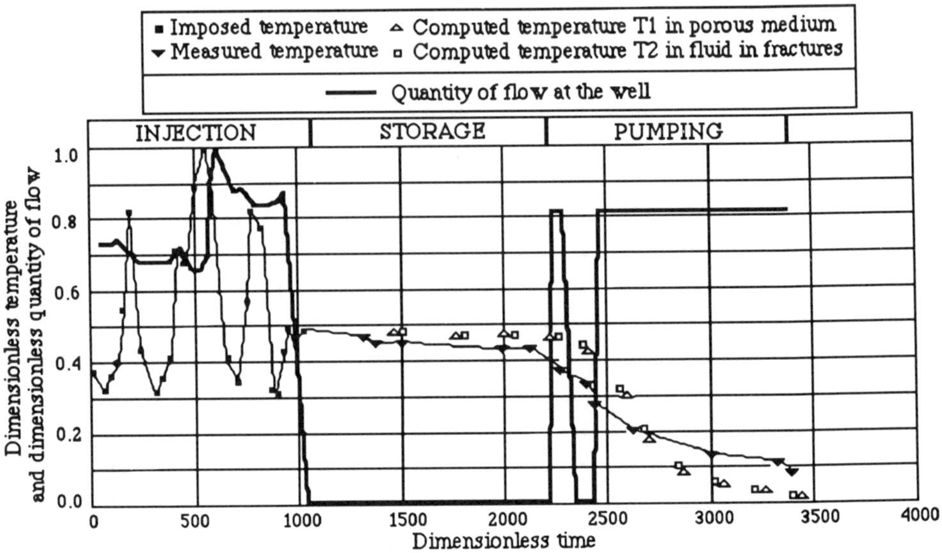

Figure 9.3.15. Matching on type curves for $H' = 100\ W/(m^3.K)$ (after Mhamdi Alaoui, 1988, Tome 2, p. 90).

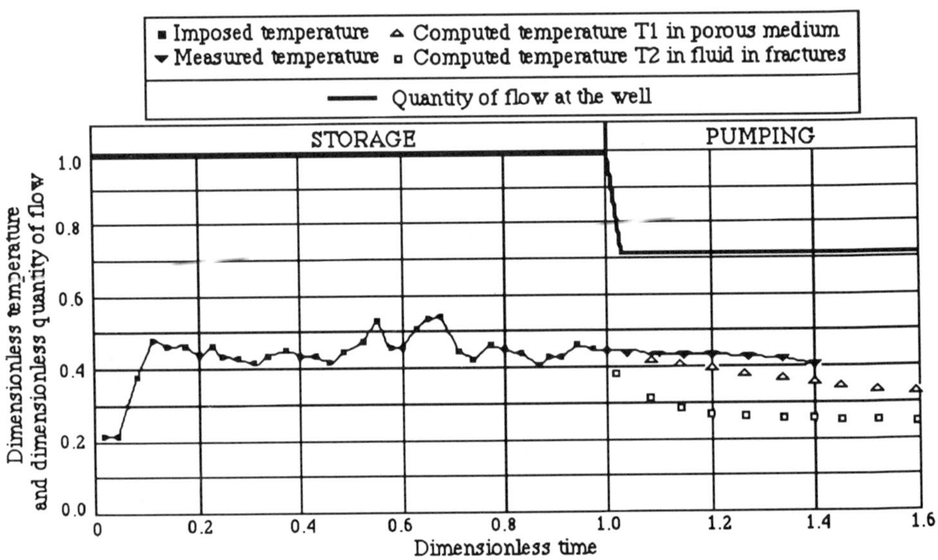

Figure 9.3.16. Validation process for $H' = 10\ W/(m^3.K)$ (after Mhamdi Alaoui, 1988, Tome 2, p. 106).

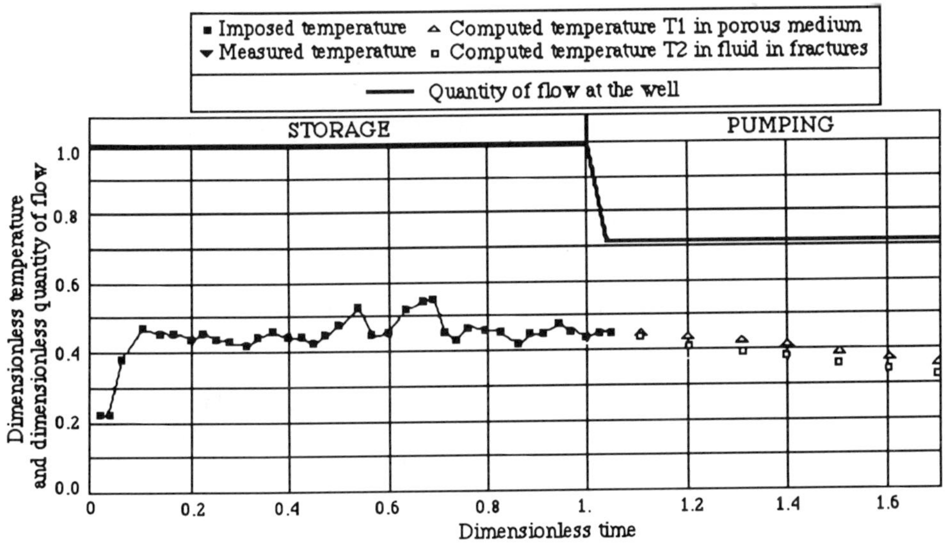

Figure 9.3.17. Validation process for $H' = 100$ $W/(m^3.K)$ (after Mhamdi Alaoui, 1988, Tome 2, p. 114).

9.3.6 Coupling of Mass and Momentum Transfers

It was recognized in the 1970s that permeability of rocks could depend on stresses, but that the rock was assimilated to a porous medium (Bernaix, 1967). These studies were carried out to explain the failure of the Malpasset dam. The first studies coupling mass transfers and stresses on fractured rocks appear to be those of Jouanna (1972). In this study, two points of view were considered:

- The continuum equivalent point of view was adopted for studying mass and stress interactions in finely fissured rocks, for instance a mica schist, experiments being conducted on large samples in the laboratory and also in-situ.
- The discontinuous approach was used for studying mass and stress interactions in rock presenting individual fractures. Experimental studies were carried out in the laboratory on large samples of fractured limestone.

Since that time, although a few results have been obtained in continuous terms, many studies have adopted the second point of view, especially in geotechnical and geothermal in-situ investigations. An illustration of these two points of view follows.

Continuum Approach

The following experiments (Jouanna, 1972) were performed on a mica schist, as shown in Fig. 9.3.18. This picture explains why the rock has to be considered as an equivalent anisotropic

Continuum Approach

The following experiments (Jouanna, 1972) were performed on a mica schist, as shown in Fig. 9.3.18. This picture explains why the rock has to be considered as an equivalent anisotropic continuum. The impact on the flow of a stress normal to the average direction of the foliation is reported in the paragraphs that follow for an in-situ experiment. The influence on the flow of the shear stress along the foliation also is reported from a laboratory experiment, no in-situ experiment being available under such circumstances. Such an influence explains the possibility of modifying percolation of water by hydrofracturing in the presence of shear stresses.

Normal Stress Influence. In this in-situ experiment, a piece of rock one meter in diameter was perforated by a series of adjacent holes as shown in Fig. 9.3.19 (a). A vertical stress was applied perpendicularly to the foliation by a plate and three pistons, the reaction being created by a central bar anchored in the deep rock, as shown in Fig. 9.3.19 (b). An example of the variations of the quantity of flow versus the stress normal to the schistosity is given in Fig. 9.3.20.

These variations show a hysteretic phenomenon; the decrease in permeability when normal stress n increases is greater than the increase in permeability when n decreases. The relation of the quantity of flow versus hydraulic head depends on the stress normal to the foliation, as shown in Fig. 9.3.21. It should be mentioned that in this test this law is global, because the hydraulic gradient varies with the distance to the well. In similar tests carried out in the laboratory, but under constant gradients, such relations between hydraulic laws and normal stresses have been very clearly demonstrated. Another noteworthy result obtained in the laboratory showed that a reduction of flow can occur under creep of the rock. Such an influence is rarely reported in the literature.

Shear Stress Influence. Precise demonstration of shear stress influence on rock permeability is rare in the field or even in the laboratory. In the laboratory experiment reported, applying a shear stress on the schistosity plane was produced by means of two variable stresses n_1 and n_2 acting on both faces of a mica schist sample (Fig. 9.3.22). The quantity of flow measured along the third principal direction, in the schistosity plane, shows a drastic diminution of flow when the shear stress is increased. This fact corresponds to a closure of the water paths, understandable here for a soft mica schist. For harder rocks, dilatancy leads to an inverse effect, the quantity of flow increasing with shear displacements. It is interesting to note in the same graph a hysteretic effect on the flow variations resulting from the irreversibility of shear strains.

Discontinuous Approach

In individual fractures, mass exchanges are very often strongly coupled with mechanical effects. This fact is so commonly encountered that hydraulic tests are used to measure in-situ stresses. These mechanical effects are also commonly found in stimulation, where natural fracture opening varies with pressure. Fractures, of course, can be induced by flow, but that effect is beyond the scope of this chapter.

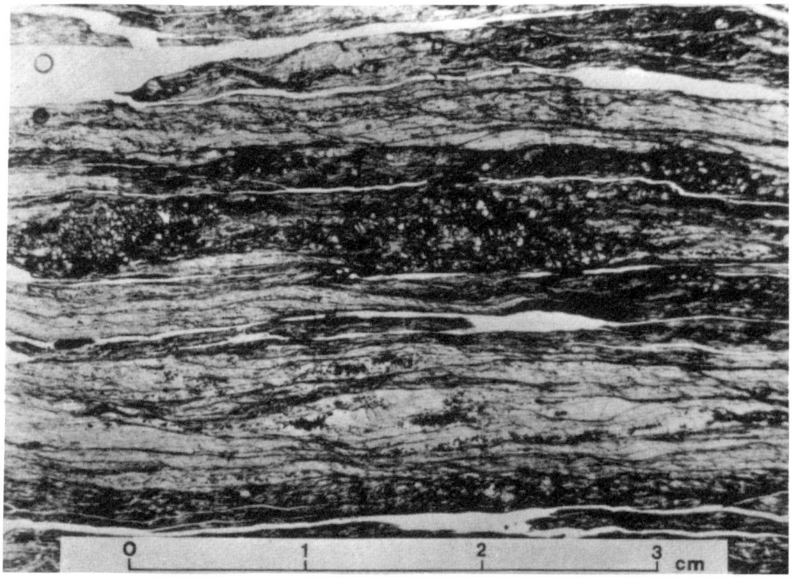

Figure 9.3.18. Thin cut of a typical mica shist sample (after Jouanna, 1972, p. 83).

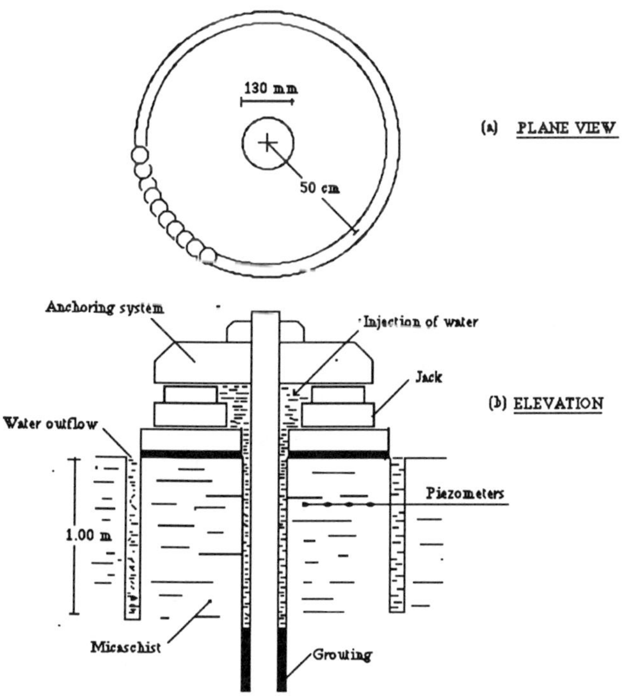

Figure 9.3.19. In-situ experiment on interaction between flow and normal stress (after Jouanna, 1972, pp. 148-149).

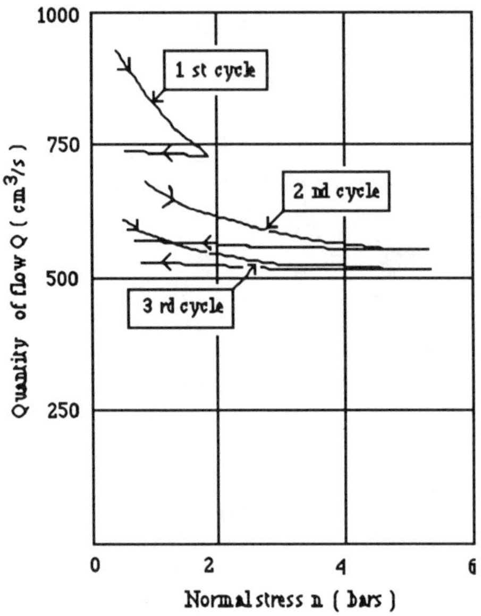

Figure 9.3.20. Influence of a normal stress on quantity of flow (after Jouanna, 1972, p. 173).

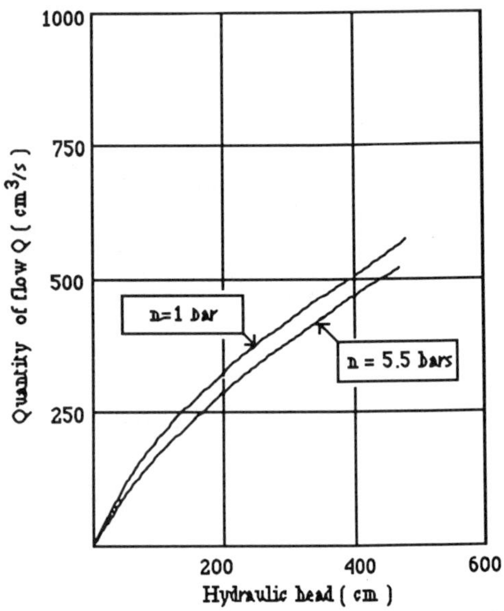

Figure 9.3.21. Influence of hydraulic head on quantity of flow (after Jouanna, 1972, p. 161),

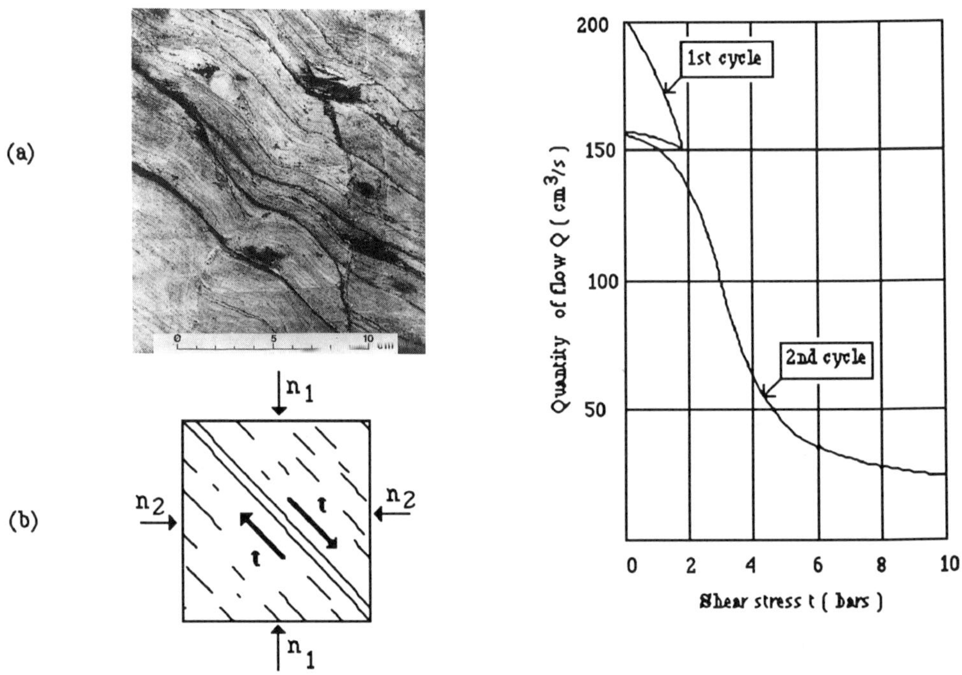

Figure 9.3.22. Influence of shear on quantity of flow: (a) View of rock before applying shear, (b) normal and shear stress device, (c) quantity of flow vs shear stress (after Jouanna, 1972, (a) p. 123, (b) and (c) p. 133).

The two main mechanisms are:

- Opening the fracture by playing on the deformability of the surroundings; in principle, a highly reversible effect.

- A discontinuous shear displacement of the faces of a fracture, induced by natural shear stresses, that creates an irreversible variation of thickness of the fracture, generally an expansion due to a positive dilatancy effect. This mechanism is fundamental in hydraulic stimulation.

Numerous examples are now available in the literature to illustrate these effects of interaction between mass transfer and stresses. The most pertinent are geothermal experimental studies in Los Alamos, Camborne, Urach, Falkenberg, Maillet de Montagne, Soultz, etc. Two up-to-date publications are: Parker, R.H., editor (1989), and Cornet F.H., coordinator (1989).

For instance, Jung (1987) gives a typical example, referring to an injection and shut-in test with several boreholes intersecting a fracture created in a granite by hydraulic fracturing. Opening and closing of the fracture during the test was investigated by three independent methods: mechanical caliper, effective fracture width deduced from fracture transmissivity, and volume stored in the fracture. The relation found in this case study between effective fracture width versus average fluid pressure is shown in Fig. 9.3.23. The evolution with time of the correlated fracture width, well pressure and injected or recovered quantity of flow at the injection borehole is shown in Fig. 9.3.24. The average pressure in the fracture is measured in a closed observation borehole.

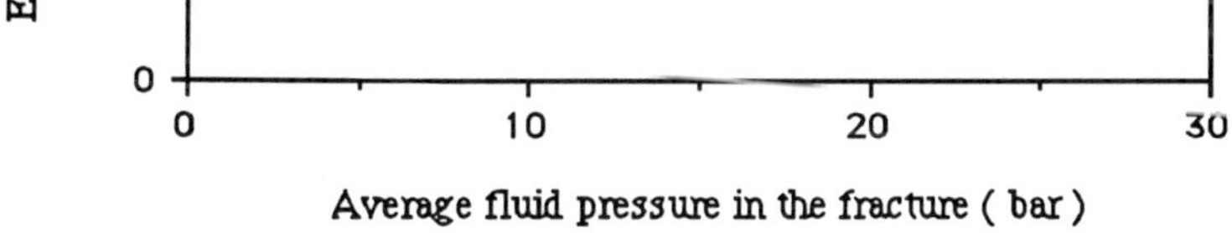

Figure 9.3.23. Correlation between average fluid pressure and opening of a hydraulic fracture (after Jung, Fig. 10, 1987).

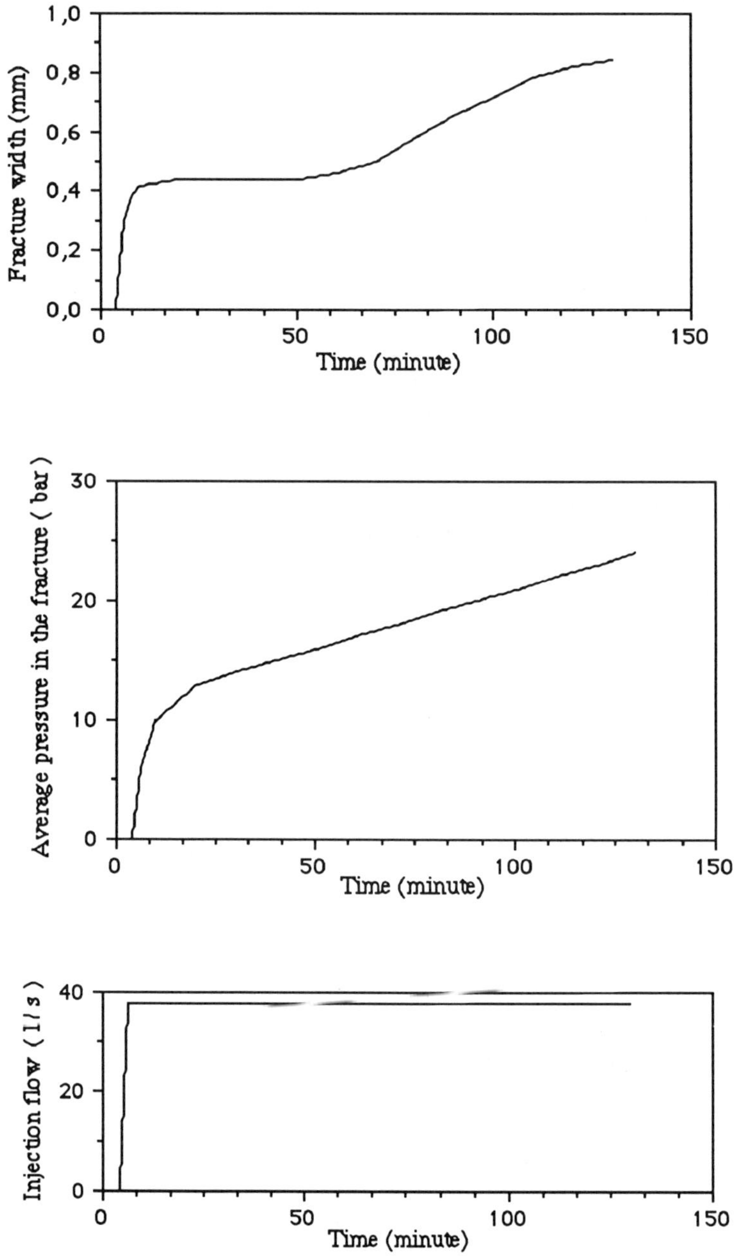

Figure 9.3.24. Correlated variations of width of a fracture with pressure and quantity of flow during an injection test (after Jung, Fig. 8, 1987).

9.4 MASS TRANSFER: CHEMICAL METHODS

9.4.1 Introduction

In mass transfer phenomena, physical chemical phenomena are to be added to the preceding considerations on pure water flow. These considerations can come from two directions:

- Chemistry of natural water.
- Water tagged by chemical components.

Considerations on natural water are not really field tests in the sense of this chapter. The chemical field tests considered here consist of monitoring the chemical composition of injection and/or returning fluid in one or several wells. Chemical substances may either be added before injection or produced by water rock interaction during the injection-production sequence. Such chemical components involved in these two situations are called tracers. A tracer is a substance that will follow the injected water and will experience the same physical phenomena. When added to water, the tracer molecule should be similar to the molecules included in the original fluid; however, the tracer must be distinguished from the original fluid components. The choice of such a tracer requires the best balance between these two contradictory requirements. The resolution of this apparent conflict depends on the hydraulic and thermal environment. Moreover, the easier the analysis of the tracer, the smaller the quantity of tracer.

Theory

The behavior of solutes in water in contact with rock involves a variety of time and temperature dependent phenomena, such as:

- mechanical phenomena (transport, mixing);
- physico-chemical phenomena (phase changes, diffusion, adsorption);
- chemical phenomena (reactions).

If there is no theoretical limit for studying such complex and interacting phenomena by thermodynamics of irreversible processes, in practice field tests are constructed in such a way that coupling phenomena can be neglected. Thus the measure consists in analyzing and interpreting the return of a given chemical species, versus time or recovered volume.

Different Kinds of Tests and Classification

The parameters to be considered in a classification of the chemical tests, as presented in Secs. 9.4.2 to 9.4.5, can be as follows:

- Choice of the tracers, which can be artificial or natural, inert or reactive
- Choice of the injection procedure, in one well or multiple wells, steady-state or transient.

Interpretation of Chemical Tests in Fractured Rocks

To obtain information on the rock structure by such chemical procedures, the general identification theory of systems applies. The identification procedure consists of:

- Obtaining experimental tracer response curves
- Building a set of theoretical tracer response curves, assuming some hypotheses on the structure of the fractured system
- Comparing both in order to identify the most critical parameters of the fractured medium.

9.4.2 Natural Inert Tracers

One Hole

In some situations, one can take advantage of certain chemical particularities of the injected water. For example, in some areas the available surface or ground water has a rather high nitrate (NO_3^-) content, compared with the interstitial formation fluid. Injecting this water corresponds in fact to a continuous injection of a nitrate tracer. By contrast, one can also use the chemical properties of the fluid filling the formation, which, in the case of deep holes, is frequently very different from the injected water. Such favorable but rather rare situations can reduce the cost of tracer tests. It is only necessary to monitor the chemical concentration at the outlet; no particular device for injection is required. This situation was encountered during injection backflow testing on a research well in a European hot dry rock project in Soultz-sous-Forêts in the Rhine Graben (Pauwels et al., 1989). There, the chloride content of the formation fluid was very high (around 60 g/l) whereas the chloride content was practically zero in the injected water.

The return profile curve of chloride concentration during backflow is shown on Fig. 9.4.1. Such experiments can only be used to investigate a limited rock volume in the immediate vicinity of the well. It is easy to calculate the mixing volume between the injected water and the fluid formation.

Multiple Hole

The situation is practically the same as in a single hole, except that the chemical monitoring is carried out at the wellhead of the producing well. The volume swept by the tracer can be much larger than in a single-well test. The duration of the test is also longer and more costly. The practical implementation of such tests is summarized in Fig. 9.4.2.

Theoretically, the results are interpreted by comparing the return profile curve to analytical response curves corresponding to an assumed fractured reservoir geometry. In practice, the use of natural inert tracers can only give information on mixing phenomena and on intensity and direction of the fluid movement through the total quantity of tracer returned.

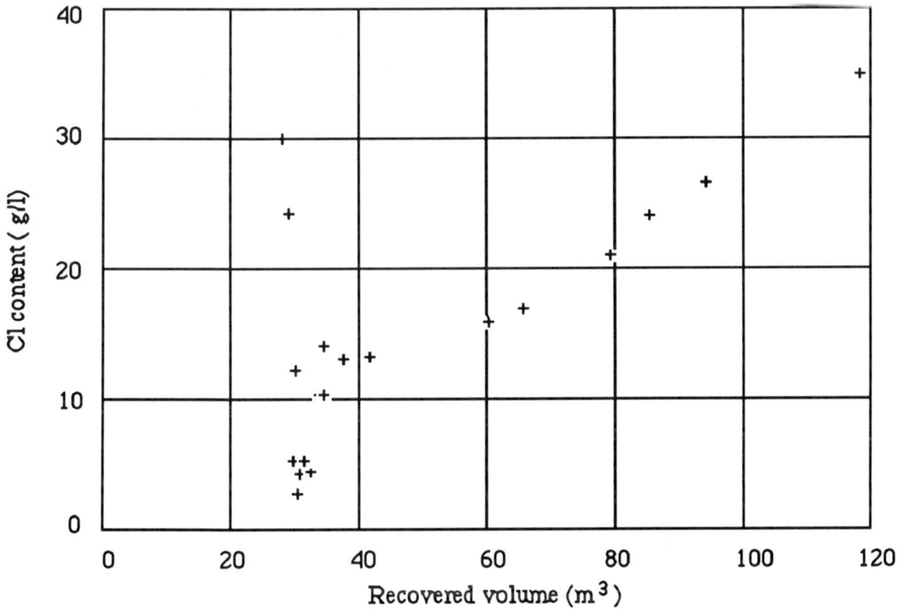

Figure 9.4.1. Return profile curve of chloride concentration during backflow.

9.4.3 Natural Reactive Tracers

One Hole

Unlike natural inert tracers, a natural reactive tracer is a species that is picked up through water-rock interaction during the transit in the formation of the injected water. Silica is a very good example of such a class of tracers. Silica can be used to trace hot, fractured formations. Its concentration in solution is strictly related to the temperature. Normally it is at a very low concentration in surface water (< 30 mg/l), but it can reach a high concentration at higher temperatures in short periods of time (only a few hours are needed to raise the silica content from

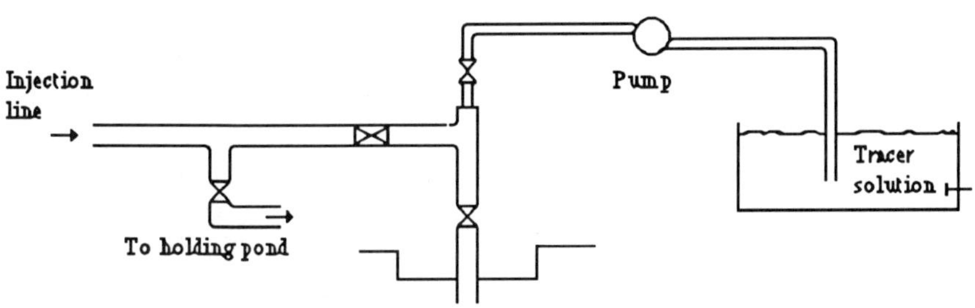

Figure 9.4.2. Tracer injection method using pump and injection line bypass to holding pond.

a few mg/liter to hundreds of mg/liter). Figure 9.4.3 gives an example of the increase of silica caused by rock dissolution during an injection backflow experiment in Soultz-sous-Forêts, in the Rhine Graben (Pauwels et al., 1989). Such experiments can give a rough idea of the quantity of rock dissolved during the interaction. The injection procedure is the same as described in Fig. 9.4.2.

Multiple Hole

For multiple well testing, the situation is exactly the same, except that the chemical changes are monitored at the producing well. Because the volume swept by tracers is larger in multiple hole experiments, it has to be noted that mixing phenomena can proceed to a greater extent and consequently shadow the behavior of natural reactive tracers, whose use could in fact be restricted to injection backflow testing.

9.4.4 Artificial Inert Tracers

One Hole

This test consists of injecting water labeled with tracers into a well for a predetermined period of time and then withdrawing this water from the same well, either immediately or after a short period of time. The tracer concentration must be monitored at the wellhead to obtain information

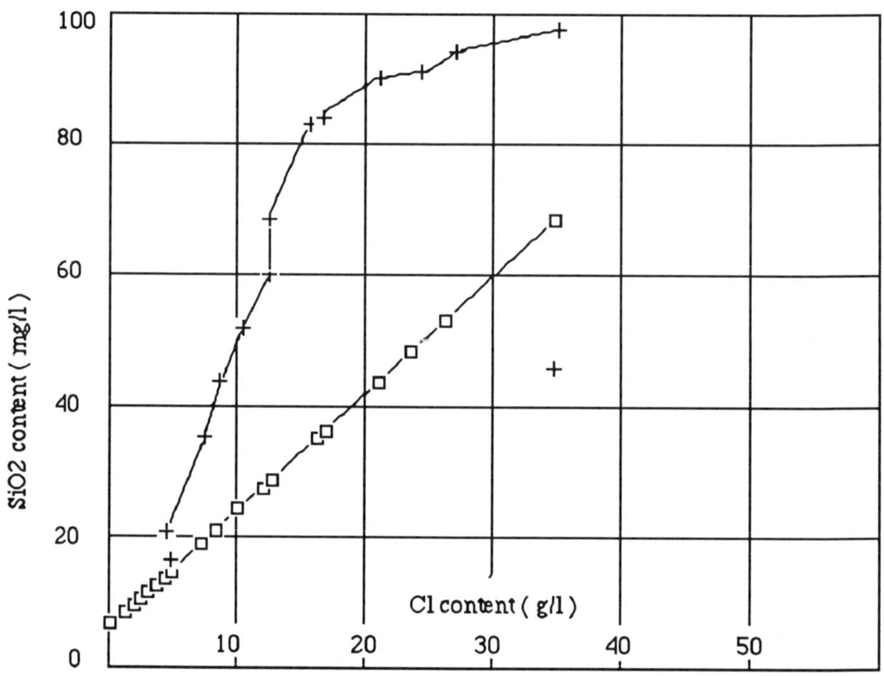

Figure 9.4.3. Increase of silica caused by rock dissolution (after Pauwels et al., 1989).

about the fractured medium. The quantity of artificial tracer required must be sufficient to be recognizable above any background concentration in the reservoir. Moreover, its concentration at the output must be sufficient for accurate measurements. Conversely, since the cost of the tracer is usually a significant expense in a tracer test, the quantity needed has to be kept as small as possible. Hence, it is common to use "slug" or "spike" injection, since it results in a greater increase in the concentration, simplifying the analytical problem and lowering the cost for materials. A schematic diagram for a "spike" tracer injection is shown in Fig. 9.4.4.

Good examples of injection backflow testing of a fractured reservoir can be found in Capuano et al. (1983). In the fractured reservoir at Raft River (Idaho), artificial tracers were used, including halogenide (Cl^-, Br^-, I^-), thyocianate (SCN^-) and some cations. The tests produced useful information on the mixing mechanism and on the permeability of the formation.

Attempts to analyze these data were made by Russell et al. (1983). They compare the actual response of the tracer test to the analytical solution of the advection-dispersion equation of Ogata and Banks (1961) for continuous injection and one-dimensional flow in a single fracture. They demonstrate that the field data collected at Raft River were not consistent with a single-fracture model. They show that flow is dominated by a major fracture with some form of dual porosity, probably resulting from secondary fractures. The injection backflow test has helped to refine the conceptual reservoir model. A complete theoretical study of the return profiles from injection backflow tracer tests was published by Kocabas and Horne (1987).

Multiple Hole

This test consists of injecting water labeled with tracers into a well and then withdrawing it from another well. Such tests are particularly well-suited for looking at the presence of fractures intersecting both wells.

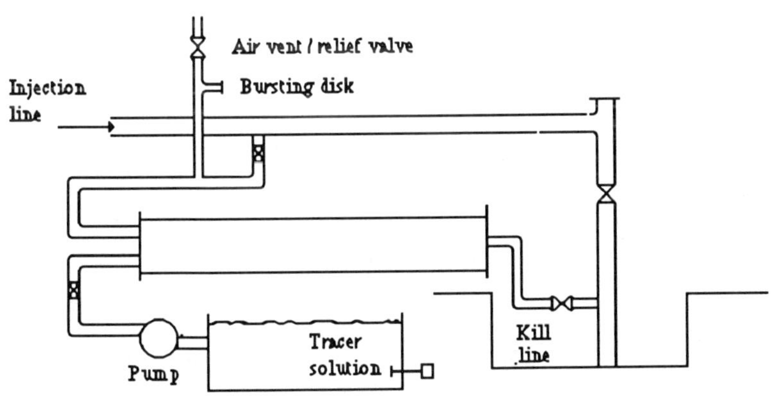

Figure 9.4.4. Tracer injection method using flow bypass through a high pressure tracer vessel.

9.4.5 Artificial Temperature-Dependent Tracers

There are soluble molecules whose concentration in solution is not only a function of the dispersion nature of the fluid flow but also of the temperature field within the reservoir. This class of tracers was proposed for the study of fractured geothermal reservoirs by Robinson et al. (1984).

A complete study of several types of molecules showing the required thermomechanical properties was studied in detail by Robinson, and their application to the granite, fractured, hot, dry, rock reservoir of Fanton Hill was published by Tester et al. (1987).

This type of tracer is at an early stage of development, and the theoretical treatment of the tracer response is currently modeled simply by using the one-dimensional form of the convective dispersion equation:

$$D\frac{\partial^2 c}{\partial x^2} - V\frac{\partial c}{\partial x} - kc = \frac{\partial c}{\partial t}$$

where x is the flow direction, V the mean velocity in the x direction, c the tracer concentration, D the effective dispersivity and k the first order rate constant of the thermal decomposition of the chemical species.

In theory, the analysis of the return profile curve of this type of tracer can provide information on both the temperature and the fluid velocity distributions. In practice, as the kinetics of thermal decomposition of such molecules is also very frequently pH dependent, additional efforts are needed before practical use of these tracers can become effective.

9.4.6 Conclusion and Trends

The full range of chemical tracers represents a powerful tool box for investigating fractured media, but different parameters have to be taken into account in the choice of tracer. Practical implementation has to be carried out under a controlled procedure. Finally, to be complete, a tracer test requires correct interpretation.

Tracer Choice

The selection of the tracer is site-specific. Successful implementation of a tracer test requires a good knowledge of the natural environment for both phases present, fluid and rock, and of the temperature of both.

Sufficiently good knowledge of the chemistry of the fluid filling the rock mass is necessary in order to investigate the mixing phenomena with the tracer itself. This should be done by sampling and analyzing the in-situ fluid.

Good knowledge of the mineralogy of the rock reservoir is necessary to obtain information on adsorption tracer behavior. Knowledge of properties governing adsorption of tracers by the rock material is currently the most critical point in selecting the tracer. Such information can be obtained from core or cuttings under downhole temperature conditions.

Finally, a tracer test depends on temperatures of both fluid and rock. With respect to thermally reactive tracers, basic research efforts are still needed to know whether decomposition of molecules is purely temperature dependent or also pressure dependent.

Compared to such in-situ difficulties resulting from the uncertainty of natural conditions, problems of pure analytical chemistry or those related to the frequency of sampling are much less troublesome.

Test Procedure

In addition to these chemical aspects, the tracer injection procedure is also of great importance. Injection at the wellhead has been widely used in geothermal industry. In this way, the total thickness of the formation (open hole testing) is investigated. If packers are used, selected portions of the fractured zones can be investigated separately. However, it can be of great benefit to use downhole injectors to limit the quantity of tracer and the mixing phenomena that occurs before the labeled water enters the formation. Such tools have been used to good effect in hot dry rocks for investigating fractured reservoirs (Tester et al., 1987).

It also seems desirable for the future to be able to simultaneously monitor the chemical changes at different depths of the production well. Research and development on downhole chemical probes are still needed for such tests to become routine.

Test Interpretation and Consequences on Rock Modeling

Compared to the engineering aspects, the mathematical tools used for interpretation seem well developed. If, at a first step of investigation, the exact geometrical description of the fractured medium is not precise, simplified models make it possible to derive from tracer tests very useful information on the following parameters:

- Average fluid velocity in the formation
- Quantity returned
- Mixing volumes.

This information can be used by the reservoir engineer to apply constraints to reservoir modeling. However, an effort has to be made in inverse problems to obtain more specific information in the case of fractured media.

9.5 HEAT TRANSFER: THERMAL METHODS

9.5.1 Heat Rock Parameter Identification

Three main methods seem to be used to approach thermal characteristics of fractured media, the choice of methods depending on ease of application. Parameters are the thermal conductivity λ_f ($Wm^{-1}K^{-1}$) and the diffusivity α_f (m^2s^{-1}). The first method is derived from the "line source" method used in the laboratory, the measure being performed in an open hole. The second method is the so-called "single well" method, requiring a double packer at the fracture level. The third is a cross-hole method.

Line Source Method: Open-Hole Test

Principle. This method is derived from the classical measure of conductivity used in the laboratory (Van der Held and van Drunen, 1949). For a steady flow Q of water in a vertical hole, a uniform geothermal gradient $g = \partial T_i / \partial z$ is equivalent to a heat source, existing at a large distance from the well; this heat source provides, to the flow Q of water, a constant radial heat flux $q_0 = \rho_w c_w Q g$ per depth meter of the rock formation, with:

$T_i(z)$ = temperature in the formation at depth z, far away from the well.

ρ_w = mass density of water.

c_w = specific heat of water.

At depth z in the well, water temperature varies versus time t. Starting from an initial thermal equilibrium established at temperature T_i without any flow in the well, the water temperature variations, after establishing a constant quantity of flow Q at $t = 0$, lead to an estimation of the rock thermal conductivity λ_f and the rock diffusivity $\alpha_f = \lambda_f / (\rho_f c_f)$, ρ_f being the mass density of the rock formation.

Test Realization. A steady state flow is created at the top of an open hole and measurements of temperature are made using sensors placed at different depths. Thermal equilibrium has to be reached before the test. Because it is simple, this method is particularly adapted to deep wells.

Interpretation. Experimental water temperature results are interpreted by comparing them with type curves giving the temperature $T(z,t)$ of the well water by an analytical approach. The analytical solution $T(z,t)$ is derived from the coupled problem of heat exchange between the fluid circulating in the well and the rock formation, expressed as:

$$\rho_w c_w Q \frac{\partial T}{\partial z} + \pi r_w^2 \rho_w c_w \frac{\partial T}{\partial t} = q \tag{9.5.1}$$

with:

$$q = \frac{2\pi\lambda_f (T_i - T)}{f(t_D)} \qquad t_D = \frac{\alpha_f t}{r_w^2} \tag{9.5.2}$$

where:

r_w = radius of the well.

q = heat flux exchange at the wall of the well.

t_D = dimensionless time.

$f(t_D)$ is a continuous function, its asymptotic behaviors being:

$$t_D << 1, f(t_D) = \sqrt{\pi t_D}$$

$$t_D >> 1, f(t_D) = \ln\left(\sqrt{t_D}\right) + 0.405$$

The key point is the estimation of the heat flux exchange q at the wall of the well versus the dimensionless time parameter t_D. An approximation of the solution is given by Ramay (1962). Detailed developments are given by Murphy and Lawton (1977).

In the case $t_D >> 1$, the second term of Eq. 9.5.1 becomes negligible. In that case, substitution of Eq. 9.5.2 in simplified Eq. 9.5.1, assuming a constant geothermal gradient g, leads by integration to:

$$T_i(z) - T(z,t) = \frac{\rho_w c_w Q g f(t_D)}{2\pi\lambda_f} + \left[T_i(0) - T_0 - \frac{\rho_w c_w Q g f(t_D)}{2\pi\lambda_f}\right] \exp\left[\frac{-2\pi\lambda_f z}{\rho_w c_w Q g f(t_D)}\right] \qquad (9.5.3)$$

with: T_0 = temperature of the entering fluid at depth $z = 0$.

The last term of Eq. 9.5.3 becomes negligible when the following "short term criterion" is satisfied:

$$\frac{-2\pi\lambda_f z}{\rho_w c_w Q g f(t_D)} > 3$$

Expression of the water temperature, taking the transient storage effect into account, is derived from a dimensionless analysis, the dimensionless fluid temperature T_D being defined as:

$$T_D = \frac{2\pi\lambda_f (T_i - T)}{\rho_w c_w Q g} \qquad (9.5.4)$$

Figure 9.5.1 gives, in log-log coordinates, T_D versus t_D varying with the dimensionless parameter $\beta = \rho_w c_w / (2\rho_f c_f)$. The use of such theoretical log-log type curves is classical, and translation establishes correspondence between measured points (ln T, ln t) and computed points (ln T_D, ln t_D). Knowing quantities Q, g, ρ_w, c_w and r_w and the expressions bounding T_D, t_D to T, t, it is easy to obtain values of λ_f and α_f from matching.

Single Well Method: Double Packer Test

Principle and Measurements. First, a fracture situated between two packers is injected with a constant quantity of flow. Then pressure at the top of the well is released and part of the water is allowed to flow out of the fracture.

Interpretation of the test is related essentially to the temperature measurement during the shut-in phase, which characterizes heat exchanges between the fluid and the rock. It also requires knowledge of the initial temperature map in the rock, which can be obtained by an initial thermal log.

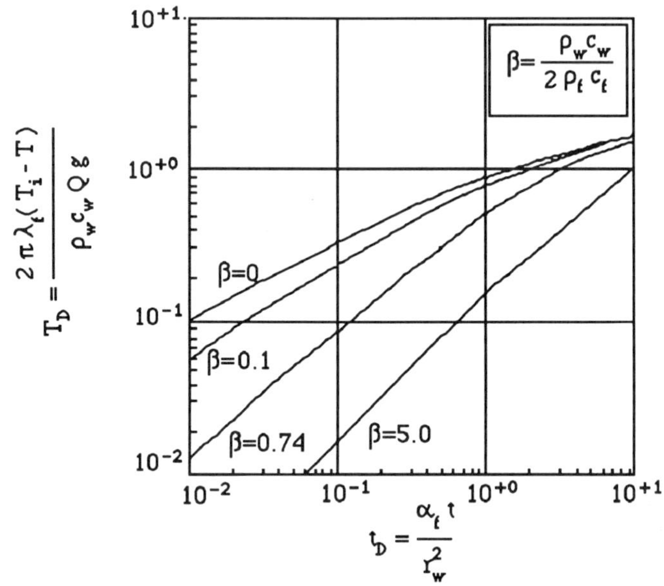

Figure 9.5.1. Theoretical type curves (after Murphy, 1977).

Temperature evolution depends on discharge, which is related to mechanical properties of the rock and also to possible interconnection with other fractures. This discharge generally decreases rapidly; when it is low, heat exchanges with the well by conduction are not negligible when compared to the heat flux coming from the fracture wall. Working with high discharges thus simplifies interpretation.

Theoretical Analysis and Treatment of the Data. In this analysis, the fracture is assumed to be flat, rigid and infinite. The rock is assumed to be impervious. Flow is assumed to be radial and in a steady state. In current cases, heat capacity and conduction of the fluid in the fracture can be neglected (Hosanski and Ledoux, 1982). If no temperature difference is assumed between the rock and the fluid in the fracture and if heat exchanges are assumed to be essentially perpendicular to the fracture, then the problem is governed by heat conduction in the rock.

The heat equation for conduction in the rock becomes:

$$\frac{\partial \theta_f(r,z,t)}{\partial t} = \alpha_f \frac{\partial^2 \theta_f(r,z,t)}{\partial z^2} , \qquad (9.5.5)$$

$\theta_f(r,z,t)$ being the rock temperature at distance r of the well axis, at depth z and time t.

The boundary condition at the fracture wall is obtained by equalizing conduction heat flux in the rock to heat transported by convection in the fracture fluid:

$$c_w \left(\frac{Q(t)}{2\pi r} \right) \left(\frac{\partial T_w(r,t)}{\partial r} \right) - 2\lambda_f \left(\frac{\partial \theta_f}{\partial z} \right)_{z=0} = 0 \qquad (9.5.6)$$

with:

T_w = fluid temperature in the fracture.

t = time elapsed from the beginning of the injection phase.

$Q(t)$ = inlet or outlet quantity of flow depending on time.

A numerical procedure has to be used to optimize the determination of parameters λ_f and α_f, because of the time variations of the quantity of flow $Q(t)$. This procedure leads to the determination of the temperature field versus time, in the rock and in the fluid. At the entrance of the fracture in particular, type curves of the fluid temperature versus time can be obtained that depend on λ_f and α_f. Comparison with experimental curves leads to an estimation of λ_f and α_f. An illustration is given in Fig. 9.5.2.

Thermal Cross-Hole Test

In a thermal cross-hole test, injection over a long period is performed in one well while observations of flow and temperature are made in the other one. Temperature evolution depends on the fracture network, and, through observation, enables one to estimate a so-called thermal exchange surface.

Interpretation methods are in general based on analytical solutions that couple a bidimensional flow in a plane fracture, where advection is preponderant, with heat conduction in the rock formation. More sophisticated configurations can be analyzed by placing fractures in parallel or in series. Type curves that depend on a shape coefficient are available for estimating the exchange surface of a doublet after some months of heat transfer, as shown in Fig. 9.5.3.

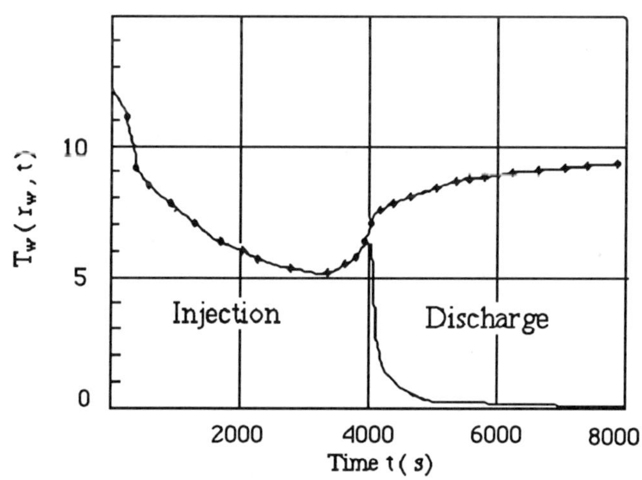

Figure 9.5.2. Example of thermal "single well" test interpretation (after Hosanski and Ledoux, 1982).

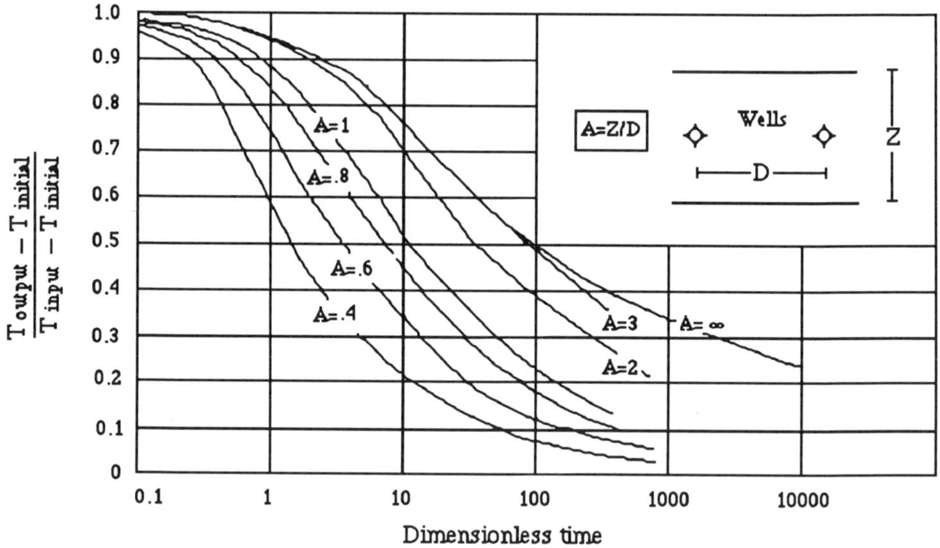

Figure 9.5.3. Thermal cross-hole type curves (after Hosanski et al., 1984).

Such tests require a long period of time and large fluid volumes that may appear prohibitive, often with results to which chemical tracer testing bears favorable comparison. However, exchange surfaces estimated by chemical means do not necessarily relate directly with economic heat exchange estimations.

9.5.2 Application of Thermal Logs to Outflow Estimation of Fractures

Assuming λ_f and α_f are known, it is now possible to estimate quantities of flow coming out of a fractured formation or individual fractures. An illustration is given by an adaptation of the line source method in an open hole flow test.

This method has been proposed by Michel and Haenel (1984), who assume the flow to be constant at the top of the well. If quantity of flow, conductivity and diffusivity remain constant in a given well length, Eq. 9.5.3 shows that the temperature gradient $g(z,t)$ remains independent of time as soon as the "short-term criterion" is fulfilled. Thus a simple relation can be derived between flow and thermal gradient variations with depth.

At the level of some fluid-conducting fracture, the quantity of flow changes, inducing a change in the slope of the temperature log. The fracture level is thus indicated by the knee of the temperature log, as illustrated in Fig. 9.5.4.

In practice, the invariance with depth of the ratio $\Delta T / Q\, g_{eff}$ is used, g_{eff} being defined by:

$$g_{eff} = \sqrt{g(t)\frac{1}{t}\int_0^t g(\tau)d\tau} \tag{9.5.7}$$

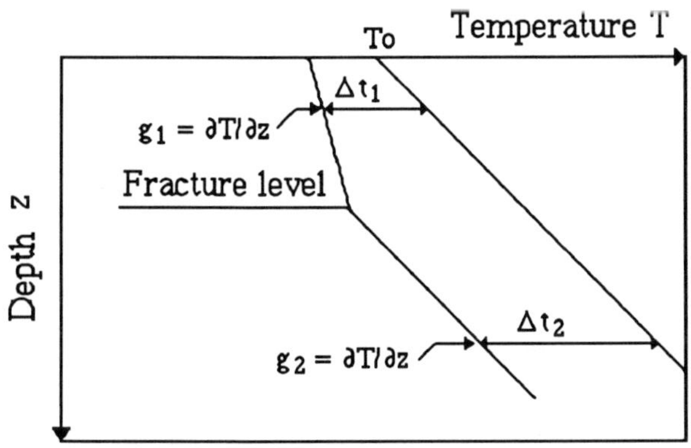

Figure 9.5.4. Temperature gradient changes induced by a conducting fracture in an open-hole flow test (after Schellschmidt and Haenel, 1987).

Temperature profiles at different times are necessary for estimation of g_{eff}. The diagram giving the different step values $\Delta T / Q\ g_{eff}$ versus depth z shows the position of the fractures inducing a flow (Fig. 9.5.5).

Knowing the flow at the top of the well, it is possible to estimate the different quantities of flow at all depths by using the following recurrent relation:

$$\frac{Q_i}{Q_j} = \frac{\Delta T_i / g_{eff\ i}}{\Delta T_j / g_{eff\ j}} \tag{9.5.8}$$

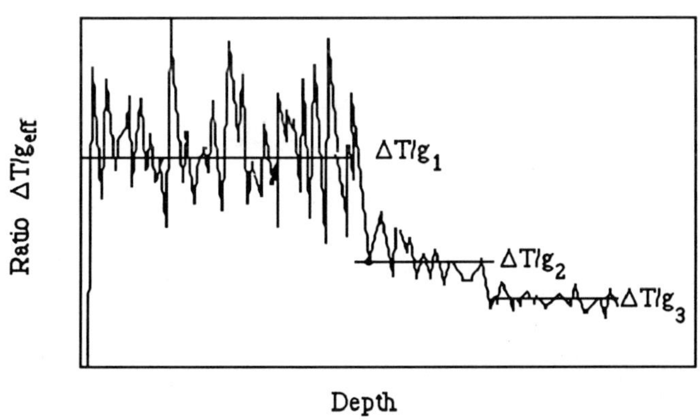

Figure 9.5.5. Ratio $\Delta T\ /\ g_{\text{eff}}$ and conducting fracture depth (after Schellschmidt and Haenel, 1987).

9.6 MOMENTUM TRANSFER METHODS

Momentum transfer methods of studying fractured rocks consist of dynamic and static methods. Dynamic methods are represented essentially by active and passive seismic, as described in Secs. 9.6.1 and 9.6.2. Static methods essentially concern the mechanical behavior; their link with heat and mass transfer in fractured rock is indirect, as shown in Sec. 9.3, and they are not further discussed here.

9.6.1 Active Seismic Methods

Active seismic methods use artificial sources to create waves. These sources can be placed at the soil surface or in wells. Similarly, reception of waves is measured either at the surface or in wells. Transfer functions between sources and receivers are used to obtain information on fractured rocks.

Surface Seismic

Theory and Principle of Measurement. Seismic waves, produced by a shot set off at a surface source S, travel at velocity v. When they encounter a discontinuity at depth z, assumed here to be subhorizontal, these waves are reflected and can be recorded at the surface by a receiver R at distance x from S (see Fig. 9.6.1). The travel time is given by:

$$t = \frac{\sqrt{x^2 + 4z^2}}{v} \tag{9.6.1}$$

Knowing x, and assuming a value of v, the principle of measurement consists in finding z by measuring t. Theoretically it would be best to measure time t_0 necessary for the wave to travel vertically from ground back to ground. But it is impossible to set off a shot in S and place a receiver R at the same surface point O.

To avoid this difficulty, the following procedure is used around the surface point O under consideration, the "midpoint." For a shot S at distance $x/2$ from the midpoint, the travel time t is recorded at a receiver R placed at $x/2$ on the other side of the midpoint (Fig. 9.6.1). A series of travel times t corresponding to different values of x can be considered and plotted on the "stack" diagram (Fig. 9.6.2). On this diagram, extrapolating the curve $t(x)$, when x tends to zero gives the

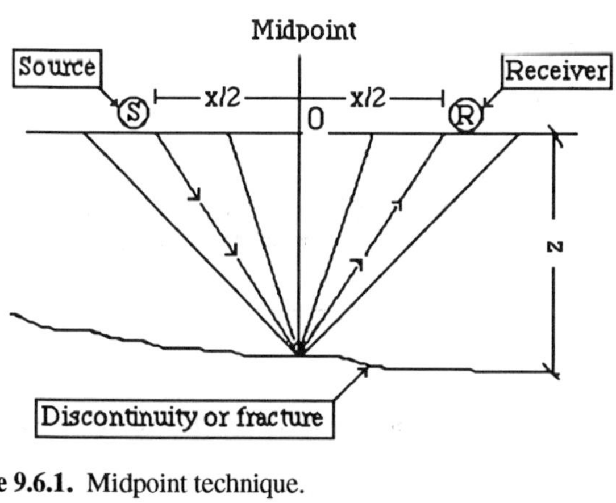

Figure 9.6.1. Midpoint technique.

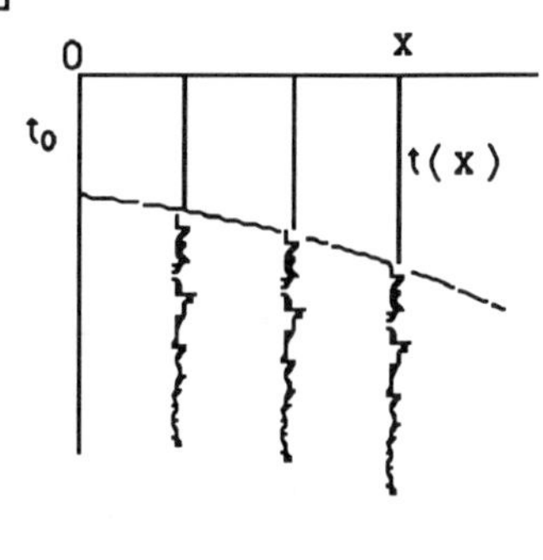

Figure 9.6.2. Stack diagram.

theoretical vertical transfer time t_0 for $x = 0$. Assuming v, obtained eventually by a "velocity analysis", gives $z = v.t_0/2$. If the discontinuity surface is subhorizontal, this procedure can be applied at different midpoints. The diagram, or "film cut," giving theoretical arrival times t_0 at different midpoints, represents an image of the geological cut (see the left hand side of Fig. 9.6.3).

Test Realization and Treatment of the Data. In the past, shots were created by about one kilogram of explosive. Today in crowded areas, either a mass percussion is applied on the ground or a frequency modulated wave is generated for about 15 seconds (Vibroseis^R), this emission being equivalent after correlation to an explosive shot. In any case, recording at the surface is by vertical geophones with a resonance frequency of about 10 Hz.

Comprehensive treatment of the data can be found in specialized textbooks (e.g., Sheriff and Geldart, 1982). Classical seismic treatments use 4 ms time space recording, more recent treatments going below 1 ms for a higher resolution. For a velocity of about 3000 m/s, the corresponding bandwidth of 20-90 Hz gives a limitation on precision of about 20 m. To overcome the difficulty created by the weathered zone (WZ), several geophones are used in parallel to obtain a more precise arrival time. Nevertheless, a distortion is present at high frequencies.

"Multiple cover" is used with the midpoint technique. It consists of a series of shots, with arrival times recorded conventionally at 48 points on each side of a shot, the distance between geophones being of the order of 100 m.

Interpretation and Detection of Faults. In practice, *surface seismic* studies use only the wave arrival time, the shape of waves being reserved for improving the time arrival estimate by deconvolution. The surface seismic method is used to obtain a preliminary general idea of the underground. It gives an excellent idea of boundaries between layers in the case of subhorizontal sedimentary layers or large faults, but subvertical layers or faults, frequent in the basement formation, cannot be detected. A more detailed observation of the fracture pattern is not possible with this technique.

Vertical Seismic Profiling (VSP)[1]

Theory and Principle of Measurement (Hardage and Toksoz, 1983). This method consists of generating waves at the top of a well (*VSP* method) or at a distance from it (offset *VSP* method) and measuring arrival times by geophones placed along the hole. Moreover, recent results of Mari (1989) show the existence of tube waves, which travel at low velocity along the well. Tube waves result from a partial conversion of P waves when a fractured zone is encountered.

Test Realization and Treatment of the Data. The seismic waves are generated by a classical Vibroseis^R. Geophones are similar to those used in surface seismic surveying but have 3D components. Spacing of geophones is generally about 15 m but can be reduced to 7 m for precise recording.

The first treatment consists in separating downgoing waves and upgoing waves. Assuming sufficient continuity of geological properties, the first waves or direct waves are very important and are easily distinguished from the upgoing waves by using about ten reception levels. The

---

[1] The principle of *sonic* logging is identical to that of *VSP* in the conventional tool layout (see Secs. 9.10.3, b3). For this method, emitter and receiver are both placed in the well, at a distance of around 1 meter, giving more detail. However, *sonic* logging can cause systematic errors in integrating data along the well.

upgoing wave is similar to the wave computed at the end of the multiple cover treatment in surface seismic surveying but is not affected by the weathered zone (WZ), because it is recorded in the better conditions found close to reflection planes. However, this wave cannot give lateral information. Three-dimensional geophones enable measurement of the angle of incidence of the propagation direction in case of offset *VSP* but essentially allow the separation of longitudinal P waves and transverse S waves (Naville, 1986), as discussed below for *tomography.*

Interpretation and Detection of Fractures. VSP can be used to interpret, around the hole, surface seismic data given by the multiple cover. Figure 9.6.3 gives an example for detecting fractured zones in a basement formation, showing the correlation between surface seismic measurements on the left hand side of the figure and offset VSP on the right hand side of the figure. Distances are shown horizontally, and vertical double time t is shown vertically. Depth z is deduced assuming a wave velocity $v = 3000$ m/s.

Interpretation of tube waves also seems promising for the study of fractured media.

Tomography and Cross-Hole Methods

Theory and Principle of Measurement (Nolet, 1987). In tomography techniques, shots and sensors are both down the hole, with spacing between wells of 50 to 500 m. The term cross-hole is sometimes used when the shot and the receiver are at the same depth. For a wave amplitude A_o at the source, the wave amplitude A at a distance x is given by :

$$A(x) = A_o \exp\left[-\frac{\pi f x}{Q v}\right] \tag{9.6.2}$$

with :

Q = dimensionless damping factor or so-called "quality factor" Q.

v = velocity of the waves.

f = wave frequency.

The "quality factor" Q, during one inelastic cycle, is related to the ratio between the irreversible loss of energy ΔE and the total energy E involved in one cycle by the expression:

$$\frac{1}{Q(\omega)} = \frac{\Delta E}{2\pi E} \tag{9.6.3}$$

The quality factor should depend on the pulsation ω, but in practice Q remains constant throughout the bandwidth used.

The decay of amplitude $A(x)$ is also affected by a geometrical spreading factor. However, when comparing pre- and post-stimulation data on the same geometry, this factor is canceled out. Hence :

$$\ln\frac{A_{pre}}{A_{post}} = \pi f \frac{x}{v}\left[\frac{1}{Q_{post}} - \frac{1}{Q_{pre}}\right] \tag{9.6.4}$$

Knowing f, x and v, the measurement of A_{pre} / A_{post} leads to Q_{post}, assuming Q_{pre} to be infinite.

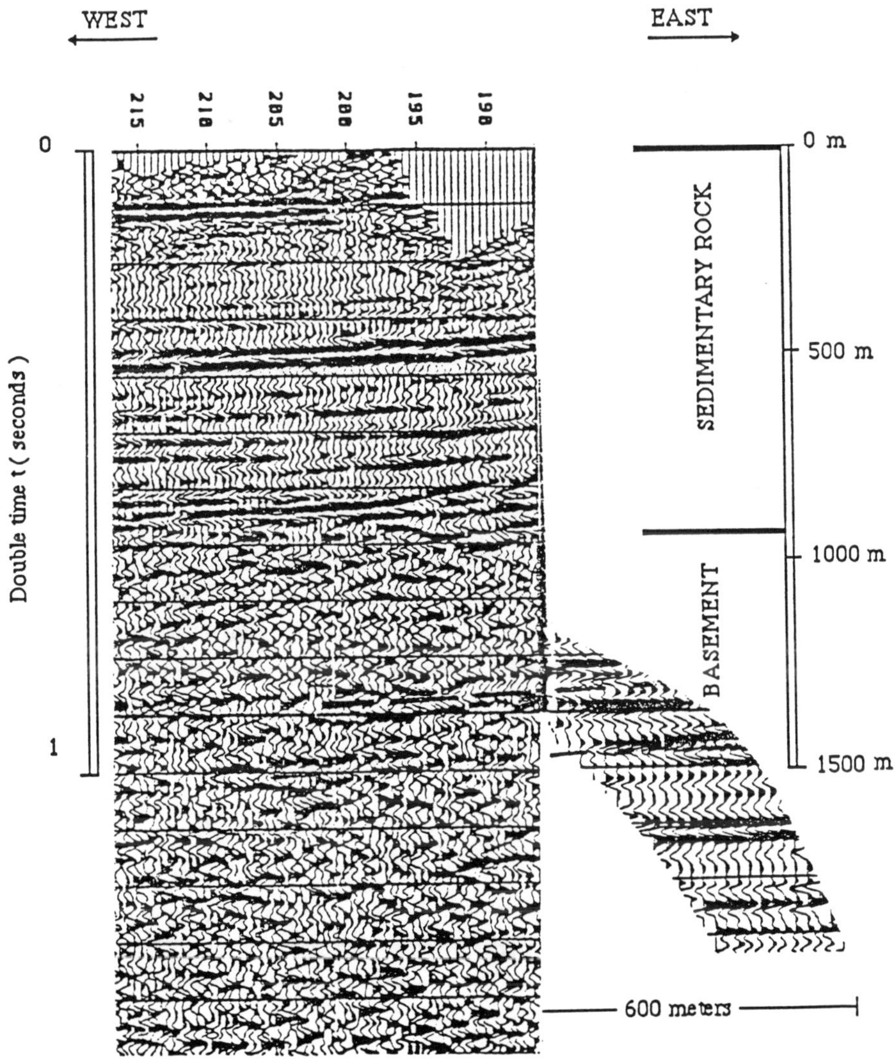

Figure 9.6.3. Correlation between surface seismic measurements and offset VSP (after Paul et al., 1988).

Test Realization and Processing of the Data. The main difficulty consists, especially in sedimentary rocks, in setting off shots without destroying the well. This difficulty is overcome using small shots (blasting caps), gun shooting (using a lateral cutter) or a sparker adapted to the hole. Energy emission is low.

Sensors are 3D geophones similar to those used in VSP, but with a much wider bandwidth of several kHz. In the present state of the technique, tomography is time consuming due to use of individual geophones, as streamers are not yet operational, and also because the gun must be recharged after each shot as multishot devices are not yet used.

Interpretation of the Data and Detection of Fractures. For P waves, Fig. 9.6.4 (a) and (b) give an example of *cross-hole* signals obtained at the Camborne site before and after hydraulic stimulation (Camborne School of Mines, 1988, ETSU G 137-P6). Comparison between both figures shows that:

- Arrival times *B* and *C* are larger after stimulation due to decrease in the velocity *v*.

- Frequencies are lower after stimulation as coefficient *Q* is lower.

As it compares data before and after injection of a formation, tomography can map the Q_{post} values (Lagabrielle, 1986, Ch. 4, pp. 135-219).

Secondary arrivals can be interpreted to determine the difference in velocities of SH and SV waves, leading to an estimate of the anisotropy of the medium. As the response of geophones depends on coupling with the soil, it is important to use only 3D geophones placed in parallel positions in the well; the direction of the geophones is deduced by interpretation of the measures. This is also possible in VSP tests.

9.6.2 Passive Seismic Method

The passive seismic method can localize seismic events produced during artificial fracturation and follow their evolution.

Theory and Principle of Measurement (Aki and Richards, 1980)

The passive seismic method is a special case of classical seismology, applied on a reduced scale with small earthquakes. Starting from the arrival times of waves at different stations, the origin time of the seismic event and its location are calculated, assuming that the velocity distribution in the rock is known.

Test Realization and Processing of the Data

The extension of the seismograph array at the surface will be of the order of magnitude of the depth of seismic events, i.e. a few kilometers. Theoretically 4 stations are sufficient to estimate the origin time *t* and the 3 location parameters *x, y, z* of one seismic event. However, redundancy is necessary in practice.

In homogeneous geological structures, about 15 seismographs placed in small wells to avoid the weathered zone are sufficient. If the sedimentary cover is important, the seismographs are

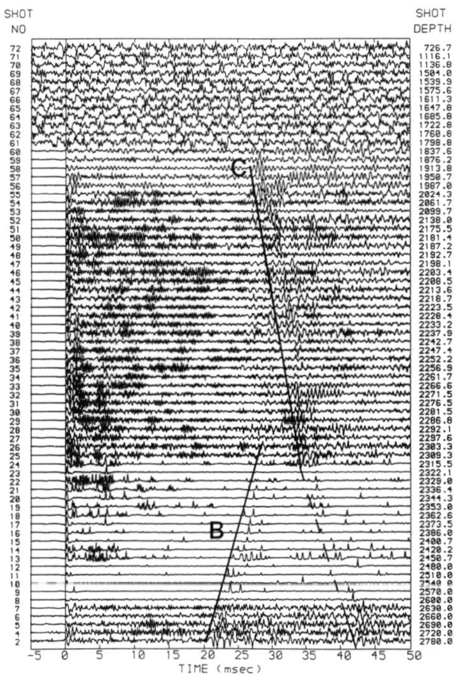

(a) Cross-hole before stimulation.

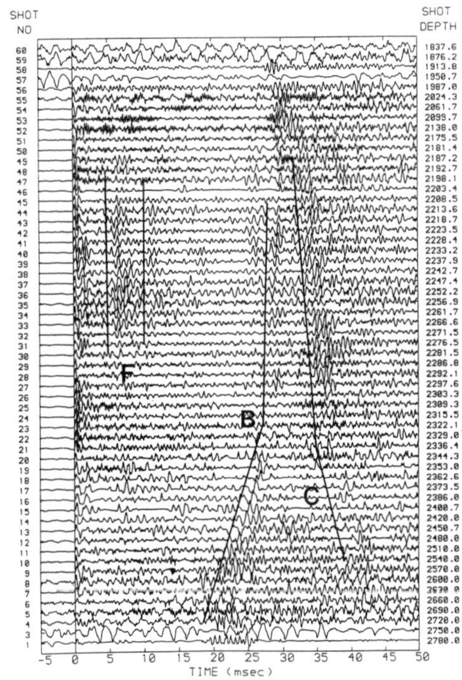

(b) Cross-hole after stimulation.

Figure 9.6.4. Cross-hole interpretation before and after hydraulic stimulation (after Camborne School of Mines, ETSU G 137-P6, 1988).

placed in deep wells to avoid damping of small-energy events. The number may be reduced for cost reasons.

Interpretation of the Data and Detection of Fractures

The passive seismic method is able to give the source mechanism; i.e., the location, direction of failure plane, and direction of sliding along this plane. An example of results has been taken from a report of the Camborne School of Mines (ETSU G 137-P7, 1988). An example of mapping the seismic events during viscous stimulation is given in Fig. 9.6.5 (a) in vertical view and (b) in horizontal view. Interpretation giving azimuth and dip of two fault planes involved in the shear mechanism is given in Fig. 9.6.6.

If this failure plane is assumed to be related to the fracture pattern, passive seismicity leads to cartography of active fractures. Similar results obtained in Camborne with the passive seismic method and tomography favor such a relation, at least in hard rocks.

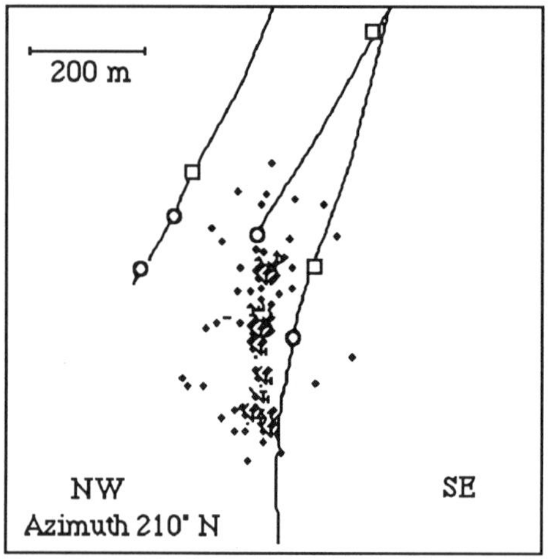

Figure 9.6.5 (a). Mapping seismic events by passive seismic method. Vertical view (after CSM, ETSU G 137-P7, 1988).

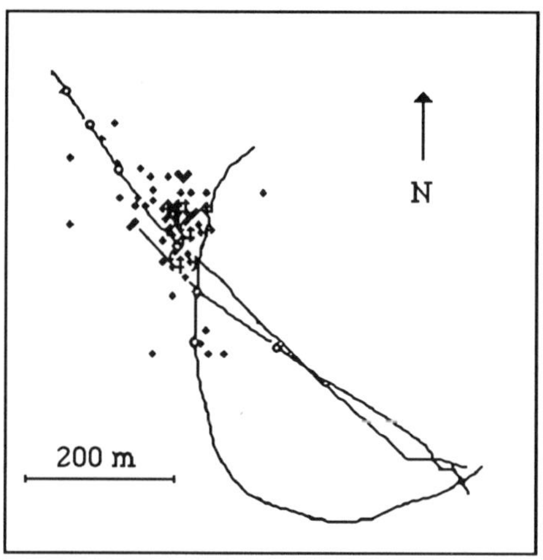

Figure 9.6.5 (b). Mapping seismic events by passive seismic method. Plan view (after CSM, ETSU G 137-P7, 1988).

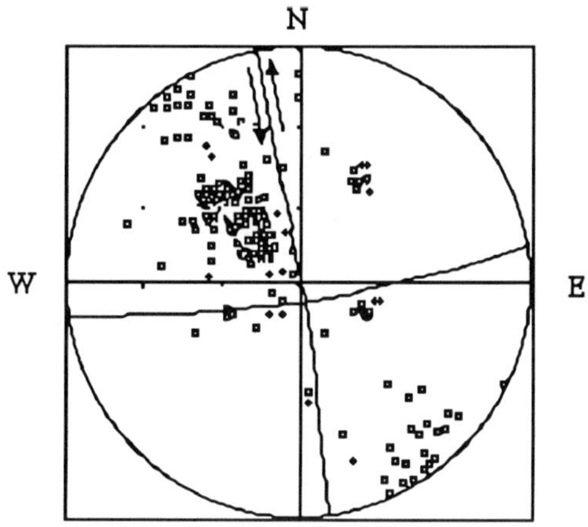

Figure 9.6.6. Interpretation of the passive seismic method. Fault planes and shear mechanism (after CSM, ETSU G 137-P7, 1988).

9.7 ELECTRICAL METHODS

The aim of electrical prospecting is to determine either the distribution of electrical conductivity (active procedure) or the electrical current flow distribution in the ground (passive procedure). In both cases, anomalies may be related to the fracture pattern. When penetration is very low, such methods can identify the boundary Σ of the rock; this is essentially the case of electrical imaging in a well, as described in Sec. 9.7.2. When penetration is greater, the reconnaissance of the inner rock domain Ω is possible from the surface and/or from one or more wells. Geoelectric methods are based upon the analysis of steady electrical currents near the surface of the rock or in the ground. In active methods, these currents are operator-controlled (direct current or DC methods). In passive methods, naturally occurring currents are exploited (spontaneous potential or SP methods).

9.7.1 Electrical Properties of Rocks and Relation with Fracture Pattern

Theory

The equations describing the flow of electrical currents within a continuous medium are:

- Ohm's law, $J = \sigma E$, where J is the current density (A/m^2), E is the electrical field (V/m) and σ is the electrical conductivity (S/m), the reciprocal of the electrical resistivity ρ ($\Omega.m$).
- The law of current conservation, div $J = \delta$, where δ is the source distribution.
- Rot $E = 0$, from which it follows that the electric field E is the gradient of a scalar potential U, such that E = – grad U.

Principle of Measurement

The electrical conductivity of most rocks, as given in Table 9.7.1, is linked to the mineralogical composition, porosity, water content and water salinity. A fractured zone, having a higher porosity than the massive rock, will show a higher conductivity. For a water-saturated non-argillaceous rock, Archie's empirical law $\sigma = w \ \Phi^n$ (Archie, 1942) relates the bulk conductivity of the rock σ to the resistivity of the water w, the porosity of the rock Φ, n being a formation factor generally lying between 1.3 and 2.3.

The aim of electrical prospecting is to determine the distribution of electrical conductivity under the ground, or at least to detect conductivity anomalies linked to the studied feature of the rock, here the fracture pattern.

Table 9.7.1. Resistivity of common rocks.

Waters and rocks	Resistivity ($\Omega.m$)
Sea water	0.2
Ground water	10 to 100
Spring water	10 to 100
Dry sand and gravel	100 to 10,000
Sand and gravel with fresh water	50 to 500
Sand and gravel with salted water	0.5 to 5
Claystone	2 to 20
Marlstone	20 to 100
Limestone	150 to 10,000
Argillaceous sandstone	50 to 150
Massive sandstone or quartzite	300 to 10,000
Volcanic tuffs	5 to 100
Lavas	100 to 10,000
Graphitic shales	0.5 to 5
Argillaceous or altered slates	100 to 300
Massive slates	100 to 1,000
Altered gneiss or granite	50 to 300
Massive gneiss or granite	1,000 to 10,000

9.7.2 Boundary Electrical Investigation in Wells

Boundary investigation using electrical methods has been used mainly in wells where possible contact with rock is excellent.

Active Procedure: Electrical Imaging in Wells

Electrical imaging is a direct current (DC) method similar to resistivity profiling at the surface or in wells, as described in Sec. 9.7.3, but the depth of investigation of the resistivity properties is limited to the near-surface of the well wall.

The borehole wall electrical imaging technique has progressed in the past few years. The works to be mentioned are those of the Schlumberger company (Ekstrom et al., 1986) on the formation microscanner (FMS) and those of other teams (Mosnier, 1981; Straub, 1988). The principle is to perform conductance measurements through a set of microelectrodes placed on the hole circumference. After various corrections, results are displayed as images of the borehole wall; these images may be processed by various means to optimize the observation scale of the fractures. A fracture is seen as a sinusoidal shape, as shown in Fig. 9.7.1. As the vertical spreading of the sinusoid is related to the dip and as the FMS is orientated, the dip and orientation of the fracture can be determined.

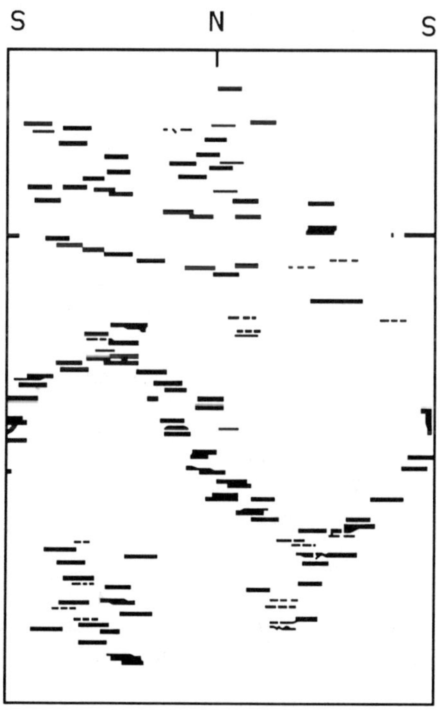

Figure 9.7.1. Microelectrode ELIAS imaging of a fracture in granite (after Straub, 1988).

Passive Procedure: Spontaneous Potential in Wells

Spontaneous potential possibilities have been exploited essentially at the surface, as developed in the sections that follow. However, recent developments have been achieved in SP logging, as reported in Sec. 9.10.

9.7.3 Inner Electrical Investigation

Active Procedure: The Direct Current (DC) Methods

Surface Electrical Prospecting: Test Realization and Treatment of the Data. The various electrical arrays used at the ground surface are all of the four-pole type, with two current injection electrodes A and B and two potential electrodes M and N. The injected current is on the order of a few milliamperes to a few amperes; the measured voltage difference is on the order of a few microvolts to a few volts. For each position of the array, an apparent resistivity is computed by multiplying the ratio V/I, voltage difference over injected current, by a geometric factor. The apparent resistivity ρ_a is the resistivity that a homogeneous half-space would have in order to give such results. A given array may be used in two modes: the profiling mode, which consists of moving the entire array along a profile, keeping the array geometrically unchanged, and the sounding mode, which consists of increasing the size of a fixed center array.

The most common setup is the Schlumberger array, as shown in Fig. 9.7.2, for which measurements are made over an MN dipole at the center of the AB injection line. Data plotted bilogarithmically show the apparent resistivity ρ_a as a function of a reference length $AB/2$.

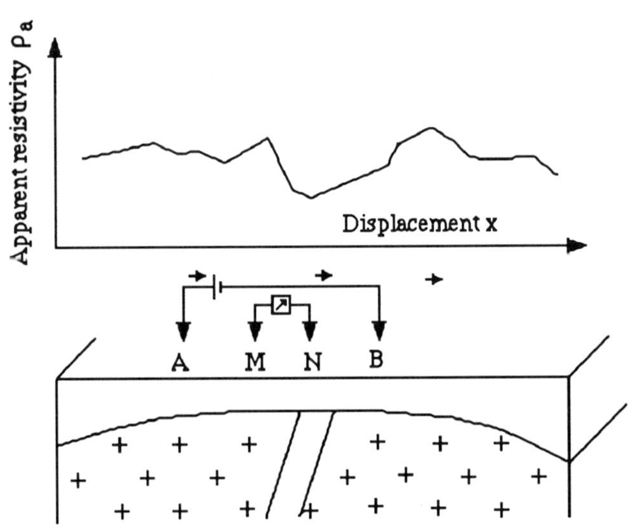

Figure 9.7.2. Resistivity profiling by a Schlumberger array.

A related method, called the gradient array, consists of mapping the central zone of the AB line. Another type of array, the square array, is made of a current bipole and a potential bipole located on opposite sides of a square, measurements being made in two perpendicular directions. This array can be used to study anisotropy and gives promising results in locating anomalies when used in the profiling mode (Darboux-Afouda and Louis, 1989).

Use of Surface Electrical Prospecting in the Case of Fractured Rocks. Electrical profiling with a Schlumberger array is a very commonly used method to localize boreholes for water supply in basement areas. The problem usually consists of specifying the localization of a fractured zone suggested by aerial photography. This procedure is commonly used for optimizing borehole locations in the most fractured zones of the basement for water resources purposes.

Ideally, electrical profiling should not be used alone as is too often the case. Comprehensive electrical prospecting should include the following steps :

i. Preliminary electrical sounding to define the A_oB_o optimum length for basement investigation under the weathered overburden. As shown in Fig. 9.7.3 (a), this length is about one decade greater than that corresponding to the start of the ascending branch of the curve; the dashed curve corresponds in this figure to a 500 $\Omega.m$ basement.

ii. Electrical profiling, with the optimum length A_oB_o, in a direction roughly perpendicular to the expected fracture zone. Fig. 9.7.3 (b) shows an electrical profile in the basement with indication of a conductive axis, perpendicular to the profile. This axis indicates a possible highly fractured zone.

iii. Electrical gradient mapping located on the conductive axis to search laterally for a possible higher fractured zone, as shown on the resistivity map in Fig. 9.7.3 (c). The gradient array AB in this figure is of 300 m. This array optimized the location of the well H with a water yield of 2.7 m^3/hour in the basement.

iv. Final electrical sounding on the minimum resistivity zone to detect a possible superficial cause of the anomaly. This is illustrated in Fig. 9.7.3 (d), where the dashed line indicates a shallow conductor of apparent conductivity 50 $\Omega.m$, situated over a 5000 $\Omega.m$ basement.

Electrical Prospecting in One Well. Resistivity prospecting in one well is the borehole counterpart of resistivity profiling on the surface, but with a much smaller measurement step. It enables identifying conductive zones at a certain distance from the well and quantifying fracture-linked porosity (see Pezard et al., 1988) if the assumption of a constant facies is possible. However, because resistivity logs are nondirectional, they give no information about fracture *orientation.*

Cross-Hole Electrical Prospecting: Test Realization and Processing of the Data. In the cross-hole setup, a current injection electrode is moved along one borehole, and a potential measurement electrode is moved independently along another borehole.

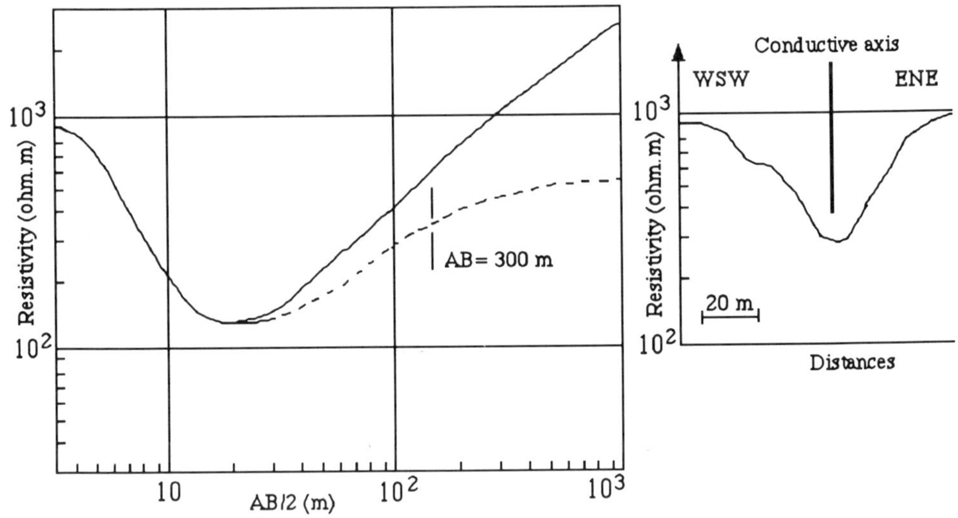

(a) Preliminary electrical sounding.

(b) Electrical profiling.

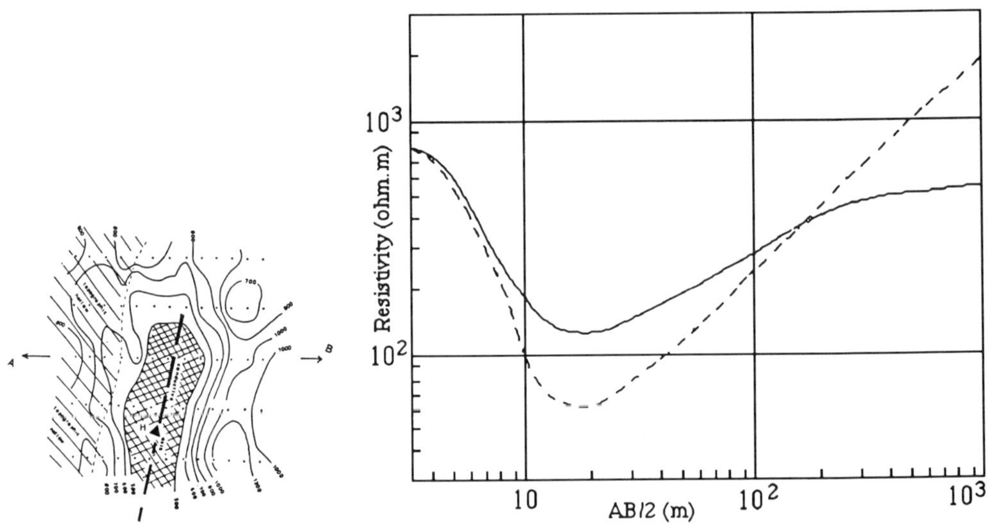

(c) Mapping of the anomaly.

(d) Final electrical sounding.

Figure 9.7.3. Different steps of comprehensive electrical prospecting.

Cross-Hole Electrical Prospecting: Interpretation and Usefulness in Fractured Rocks. Improvements in numerical modeling methods which have occurred in the last decade have made possible a renewal of studies on cross-hole electrical methods. Most of these studies are made in view of fracture pattern studies over potential waste disposal sites. From the theoretical point of view, the inverse problem is approached either by 2D modeling or 3D modeling (Poirmeur and Vasseur, 1988).

Such measurements are used on the one hand to study electrical links between fractures intersected by the holes (Pottecher and Poirmeur, 1989) and on the other hand to detect possible fractured zones in the vicinity of the holes; this is called resistivity tomography (Shima and Sakayama, 1987)

Passive Procedure: The Spontaneous Potential (SP) Method

Principle. The spontaneous potential (SP) method for reconnaissance of the inner rock domain Ω consists of measuring the electric potential at the earth's surface resulting from natural currents underground. These natural currents are created by different phenomena :

i. The most widely known is the electrochemical oxidation-reduction effect in sulfide or graphitic bodies intersected by the water table.

ii. The electrokinetic effect or electrofiltration creates a negative potential anomaly at the source of a flow along an interface between two media presenting a contrast in streaming potential coefficients (Fitterman, 1978).

iii. The thermoelectric effect is caused by the coupling of the thermal gradient and the electrical field at the interface between two media having different thermoelectrical coupling coefficients.

Interpretation in Fractured Rocks. The interpretation of spontaneous potential (SP) measurements is directly linked to the different coupled phenomena creating underground electrical currents.

The first effect (i) has no link with fracture pattern. The two other effects (ii) and (iii) are linked to fractures when, for different reasons, an electrical contrast exists between the rock on both sides of a fracture. Thus some sets of fractures can be detected, others remaining undetected. These two effects explain the *SP* anomalies observed in geothermal fields, as shown in Fig. 9.7.4 (Corwin and Hoover, 1979). Electrofiltration may also be used for detecting possible leakages within or under dams.

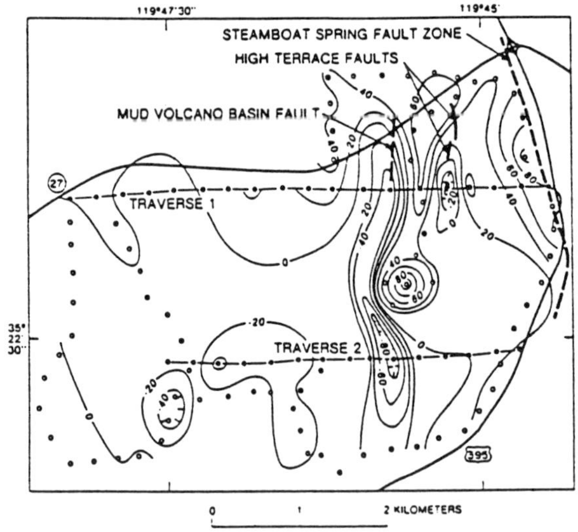

Figure 9.7.4. Self potential map in the Steamboat Hills area, Nevada (after Corwin and Hoover, 1979).

9.8 ELECTROMAGNETIC METHODS FOR INNER INVESTIGATION

Electromagnetic methods have been described in Sec. 9.2.1 for investigation of the rock surface Σ by teledetection, using the reflection and diffraction properties of electromagnetic waves. The present section, devoted to the inner inspection of the rock domain Ω, is based on the penetration properties of the same waves, the tools being placed essentially in wells.

9.8.1 Theoretical Considerations

The term "electromagnetic" applies in the presence of coupled time-varying magnetic and electric fields. Electromagnetic phenomena are governed by Maxwell equations. The electric field E and the magnetic field H are related by:

$$\nabla \cdot (\mu H) = 0$$

$$\nabla \cdot (\varepsilon E) = q$$

$$\nabla \times E = -\mu \frac{dH}{dt}$$

$$\nabla \times H = \sigma E + \varepsilon \frac{dE}{dt}$$

q being the volumetric density of electrostatic charges, generally equal to zero at a macroscopic scale. Three physical properties of the medium appear in these equations:

i. The electrical conductivity σ of rock and filling of fractures, as given in Table 9.7.1.

ii. The modulus of the relative dielectric permittivity ε_r, which has a very high value 80 for water and varies between 3 to 5 for a rock matrix. This property explains why dielectric permittivity can be an excellent indicator of the presence of water in the case of a saturated fracture.

iii. The magnetic permeability μ of a rock, related to the vacuum permeability μ_0 by :

$$\mu = \mu_0(1+\kappa)$$

where κ is the magnetic susceptibility, which varies for rock between 10^{-2} (basalt rocks) to 10^{-5} (general value for rock particles). In most cases, $\mu = \mu_0$.

To a first approximation, magnetic permeability does not interfere with the fracture pattern. The two other parameters appear in the same equation, but their relative influence depends on time variations. For a sinusoidal variation, assuming $\sigma = 2.10^{-3}\ S/m$ and $\varepsilon_r = 10$, both terms have the same effect for a frequency of 3.6 MHz. In the low frequency domain, below 36 kHz for instance, the terms including ε can be ignored, and only the rock conductivity will have an influence; this is the induction domain. In the high frequency domain, well above 3.6 MHz, permittivity ε governs the velocity of electromagnetic signal propagation, conductivity creating the attenuation of the signal. Both low and high frequency domains can be exploited in the study of fractured rocks.

9.8.2 Low Frequency Domain

Principle of Measurement

Skin Effect. In the low frequency domain, and for harmonic variations, the electromagnetic field obeys the following differential equation:

$$\Delta E - i\,\sigma\mu\omega\,E = 0$$

In the most simple case which may be considered, with a uniform periodic electrical field E at the surface of a homogeneous soil, the field decreases with depth according to an exponential law with a phase variation as given by:

$$E = E_0\, e^{-\sqrt{\frac{\sigma\mu\omega}{2}}z}\; e^{-i\sqrt{\frac{\sigma\mu\omega}{2}}z}$$

The skin depth p is defined as $p = \sqrt{\frac{2}{\sigma\mu\omega}}$

This depth p is inversely proportional to the square root of the conductivity and frequency, the induction currents being concentrated at the surface. Therefore, the investigation zone is situated at the vicinity of the surface; nevertheless, by changing the frequency, it is possible to explore different depths.

Source Types and Signals. The primary field, which would exist in only one medium, the vacuum, can have different distributions depending on the type of source. In practice, the following types of sources are used:

- Sources at a large distance from the zone to be studied, with a uniform primary field.
- Long cables, creating a cylindrical primary field varying as $1/r$, r being the distance to the cable.
- Dipoles, in which the primary field depends on $1/r^3$. They are created either by two electrodes close to each other or by an easily movable coil.

All these sources can deliver either a single or a superposition of sinusoidal signals or pulses. In the frequency domain, things are simple; in the time domain, a larger bandwidth and a higher power may be used.

Whatever the source, the response of the medium with its particularities is directly proportional to the conductivity σ. A fracture of conductivity σ_2, located in a rock of conductivity σ_1, exhibits a response proportional to $(\sigma_2 - \sigma_1)$. This difference, written $\sigma_1(\sigma_2/\sigma_1 - 1)$, depends not only on the ratio but also on the absolute values of the conductivities. This situation is different from that encountered in electrical prospecting, where the response of a structure is related only to the ratio of the conductivities, by the expression $(\sigma_1 - \sigma_2)/(\sigma_1 + \sigma_2)$. For a fracture in a rock of low conductivity, the low frequency method would be less favorable because of proportionality between signal and conductivity, but would be more favorable in terms of signal penetration.

Interpretation and Detection of Fractures

In the low frequency domain, measurements of the three components give the orientation of the electrical or magnetic field. Thus, orientation of structures like fractures, which produce the orientations of these fields, can be determinated.

In a one-well configuration, Fig. 9.8.1(a) shows the example of a transmitter consisting of one vertical coil and a receiver made of two coils, transmitter and receiver being placed on both sides of a fracture of dip θ. Theoretical vertical and horizontal components H_x and H_z of the magnetic field at the receiver are given in Fig. 9.8.1(b) versus inclination θ of the fracture. The dimensionless ratio $H_x(\theta)/H_z(\theta = 0)$ is very sensitive to the dip θ, when dimensionless ratio $H_z(\theta)/H_z(\theta = 0) = 0)$ does not vary significantly with the dip θ.

In a cross-hole configuration, the transmitter is placed in one well and the receiver in another. Fig. 9.8.2 (a) shows the simple case of a fracture, with a dip θ, situated between the two wells and perpendicular to the plane of the two well axes. Fig. 9.8.2 (b), in dimensionless variables, shows a great sensitivity to the dip θ for the component H_y perpendicular to the axes plane, its variation being symmetrical around $\theta = 45°$. The vertical component E_z of the electrical field is also very sensitive to the dip θ. Finding the dip θ of the fracture would require two parameters, but in practice the four parameters H_x, H_y, H_z and E_z are available.

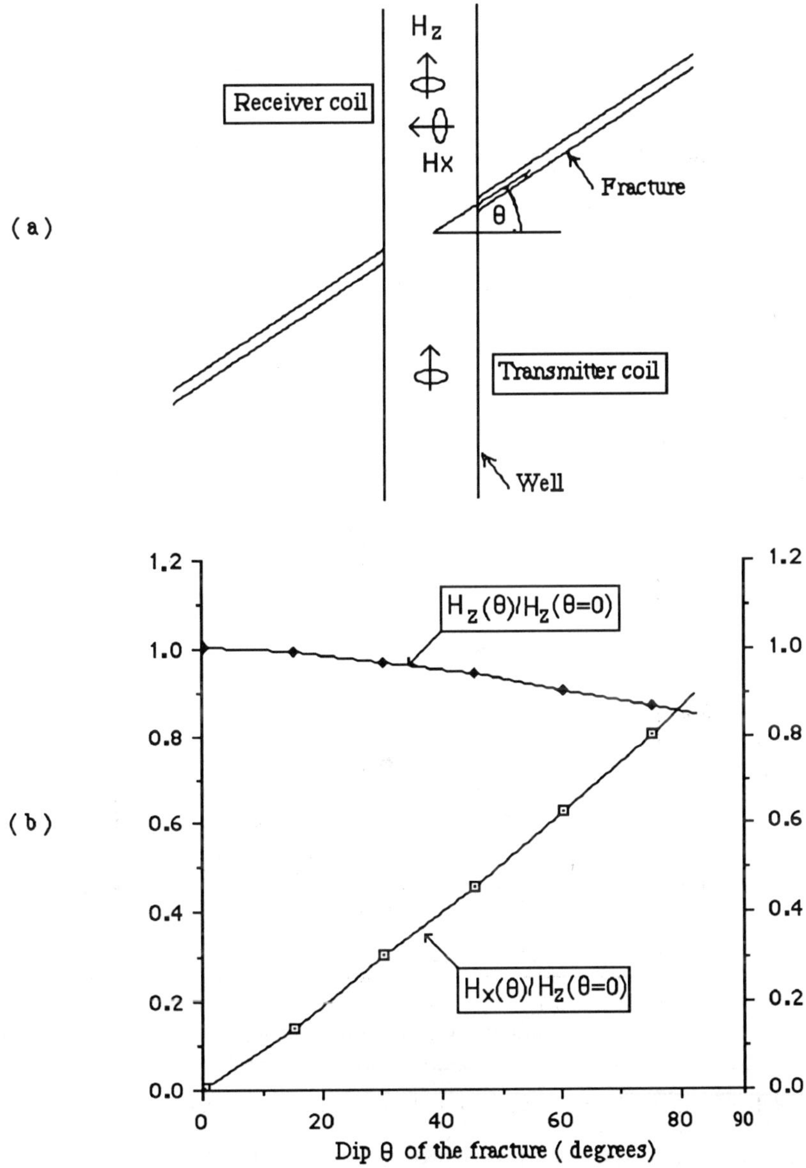

Figure 9.8.1. Detection of a fracture in a one-hole electromagnetic test.

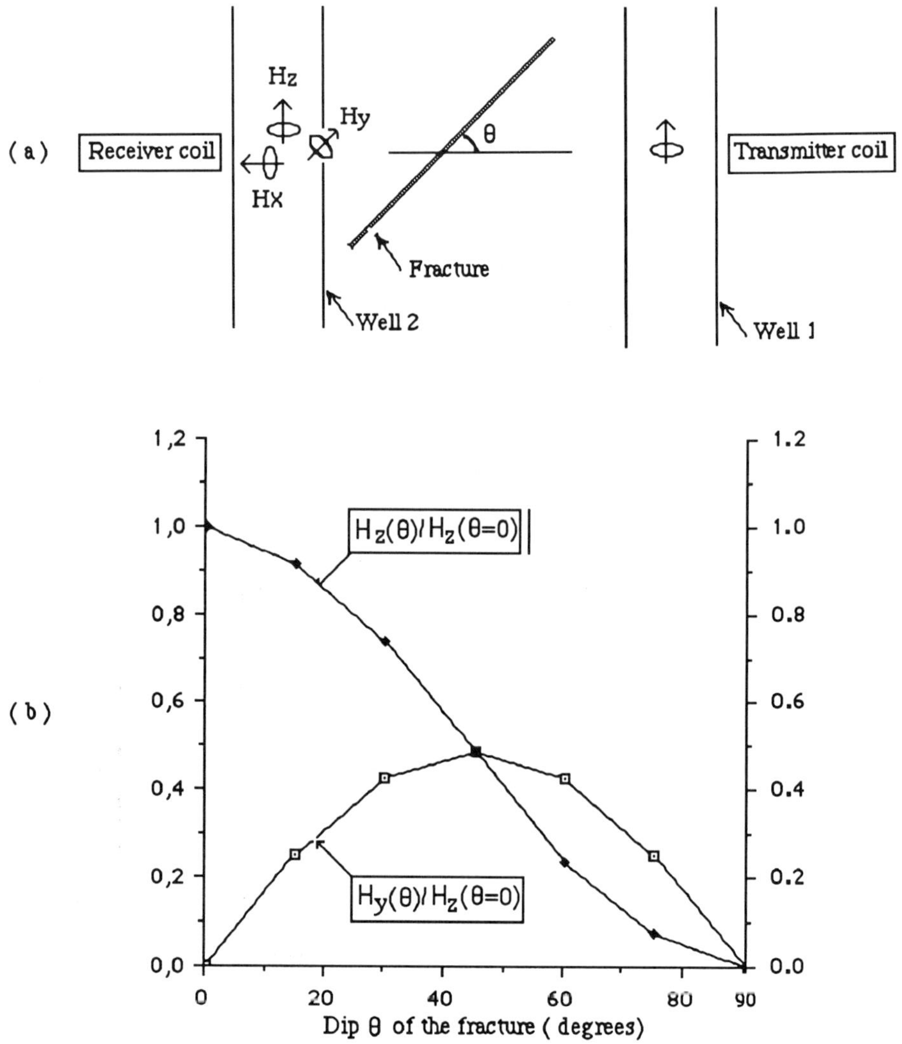

Figure 9.8.2. Detection of a fracture in a cross-hole electromagnetic test.

9.8.3 High Frequency Domain

Theory

At high frequencies, the electromagnetic field obeys the following differential equation:

$$\Delta E - \sigma\mu\frac{\partial E}{\partial t} - \varepsilon\mu\frac{\partial^2 E}{\partial t^2} = 0$$

Thus the field is propagated with a velocity that depends on the dielectric permittivity of the medium in question. This method has a strong analogy with seismic reflection (see Sec. 9.6), but four important differences have to be underlined:

i. In electromagnetism, all phenomena are much more rapid; for example, the travel time for 1 meter is about 10 nanoseconds, while it is about 300 microseconds in the seismic domain.

ii. Water has a strong influence on the electromagnetic propagation velocity and is the first parameter when studying velocity contrasts.

iii. Coupling the transmitter with the ground is not difficult, and the source is mobile and can be placed in water.

iv. Damping in soils is important in electromagnetism, and attenuation is the limiting factor for ground investigation. Attenuation and resolution both increase with frequency, sometimes leading to an impossible compromise.

For harmonic variations the governing equation becomes:

$$\Delta E - i\,\sigma\mu\omega\,E + \varepsilon\mu\omega^2\,E = 0$$

The propagation factor γ is defined by:

$$\gamma^2 = i\,\sigma\mu\omega - \varepsilon\mu\omega^2 \qquad \text{or} \qquad \gamma^2 = \varepsilon_o\mu_o\,\omega^2\left(\varepsilon_r' - i\left(\varepsilon_r'' + \frac{\sigma}{\varepsilon_o\omega}\right)\right)$$

where ε_r' and ε_r'' are the real and imaginary parts of the relative permittivity. A loss angle δ and a skin depth p are defined as follows, λ_o being the wavelength in free space:

$$tg\,\delta = \frac{\varepsilon_r'' + \frac{\sigma}{\varepsilon_o\omega}}{\varepsilon_r'} \qquad p = \frac{\lambda_o}{2\pi}\cdot\frac{1}{\sin\frac{\delta}{2}\sqrt[4]{\varepsilon_r'^2 + \left(\varepsilon_r'' + \frac{\sigma}{\varepsilon_o\omega}\right)^2}}$$

From the value of p, it is possible to define an amplitude of the attenuation in dB/m, $\alpha = 8.686/p$. Thus the importance of the attenuation can be expressed by $tg\delta$, p or α. In practice, above 1 GHz, attenuation is very high. Under 50 MHz, resolution is poor; the transmitter and receiver antennae become too large. Measurements for rocks and soils are made between these two frequencies. Attenuation is greatly dependent on the type of rock; thus, at 100 MHz the coefficient α is equal to 100 dB/m for a clay, 0.5 dB/m for a dry, sound granite, and even less for a salt.

Principle of Measurement and Signal Treatment

The principle of measurement for detecting a fracture considered as a reflector consists of measuring the time necessary for a signal emitted by a source to come back to the source after reflection on the fracture (double time). Due to attenuation, the distance between the source and the reflector is rarely larger than the wavelength. Travel times can be approximated by the very simple ray theory; for large reflectors, such an approximation is excellent as compared to travel times obtained by a complete computation. However, an indetermination remains, as in seismic, between velocity and distance, and the same methods are used to try to solve this problem. The polarization of the signal can be used to try to separate the reflection due to a precise reflector from backscattering due to inhomogeneities of the medium; however, this method is not yet commonly used in rock studies.

The signals used are generally a simple pulse, but it is possible to use a continuous wave transmission with a frequency modulation in sawtooth (FMCW) or any kind of signal adapted to more sophisticated treatments.

Test Realization. Radar for Rock Studies

From the ground surface, radar has been used for various investigations in salt domes, coal mines, etc. Ground penetrating radars were developed in the seventies, essentially for permafrost investigation.

At the moment, two commercial apparatus are commonly used for low depth geophysical problems and tunnel wall inspection. The following example in Fig. 9.8.3 gives an illustration of such a fracture detection, along the floor of a tunnel. The frequency of the central antenna was 120 MHz. The depth scaling is deduced from the double time scale using a value of $\varepsilon'_r = 7$.

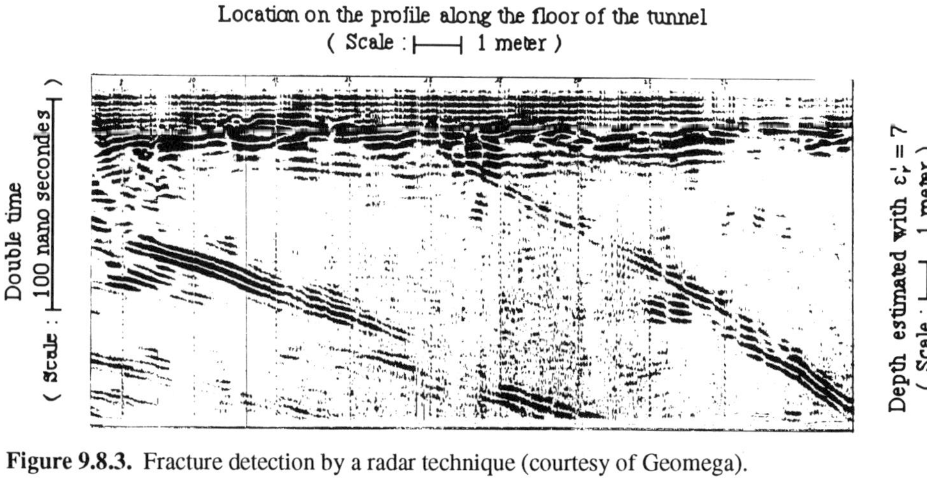

Figure 9.8.3. Fracture detection by a radar technique (courtesy of Geomega).

In wells, two experimental radar apparatus are being tested with a large distance—about 10 meters—between emission and reception antennae, with a relatively low frequency bandwidth, between 20 and 60 Hz. These devices should allow great penetration and should be adapted to the detection of important fractured zones.

In conclusion, detection of fractures in rock by electromagnetic methods, at low or high frequencies, is fairly recent, the consequence being the lack of tested methods or examples. However, current theoretical and experimental studies for investigating fractures by electromagnetic methods are promising for fractures filled with water (high frequency domain) or with mineralized water (low frequency domain).

9.9 POTENTIAL METHODS

Potential methods, including gravimetry and magnetometry, are not generally considered to be methods of choice for detection of fractured rocks, and no exhaustive study has been done in this area. Nevertheless, they can give information on fractures in special cases.

9.9.1 Gravimetry

Principle

Gravimetry is sensitive to mass anomalies: high density minerals, low density minerals (salt, gypsum) or complete lack of mass due to artificial (quarries) or natural (karst) cavities. Thus, such a method should be able to detect fractured zones that have a lower density than the surrounding rock.

Test Realization and Treatment of the Data

The value of the gravity acceleration g is measured at the surface, using a gravimeter. After rather complicated processing, profiles or maps of Bouguer's anomaly are drawn in milligals (1 milligal is equal to 10^{-5} $m.s^{-2}$). Measurement density is quite different according to the fractured zone to be investigated; the distance between stations can vary from some meters to some hundreds of meters. Very high precision is required due to the weakness of the anomalies; the station elevation has to be determined very precisely by topography, increasing the cost of measurements considerably.

Interpretation and Detection of Fractures

Numerous models give a quantitative interpretation, and they all necessitate complementary information given by geophysics or wells to limit the number of possible solutions. An example of reconnaissance by gravimetry of a fractured zone, at a kilometric scale, is given in the region of Couy, in the Paris basin (Lorenz and Millon, 1990). The purpose was to determine the location of a 3500 m well, to study a well-known large magnetic anomaly. This survey, presented in Fig. 9.9.1, shows a positive gravimetric anomaly, interpreted as a horst, which is cut by a fractured zone N-90° to N-110°.

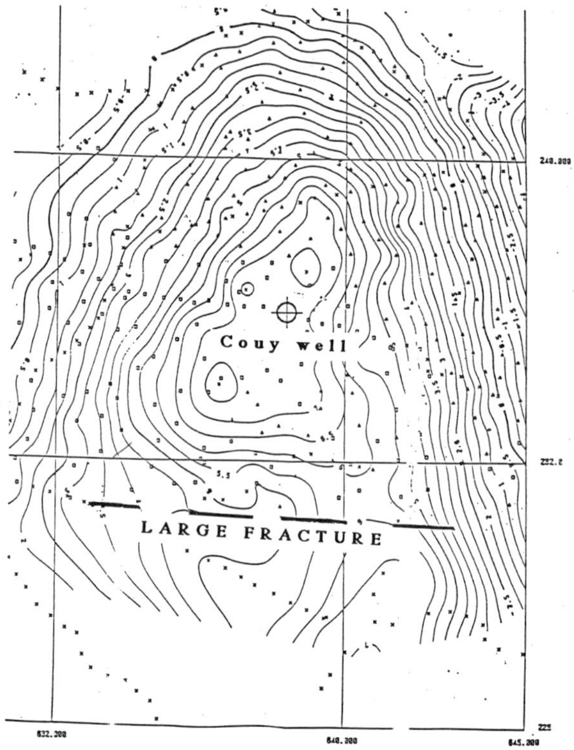

Figure 9.9.1. Large accident determination by gravimetry in the Paris Basin (courtesy of BRGM, Carte gravimétrique de la zone de Couy, 1988).

9.9.2 Magnetometry

Principle

Reconnaissance of fractured zones by magnetic prospecting is possible when they are injected by high-magnetic-susceptibility rocks, such as dolerite and, in general, basic rocks.

Test Realization and Treatment of the Data

The magnetic field is measured using a proton magnetometer, the unit being the gamma equal to a nanostela. Measures on the ground with only one operator are easy and rapid. On a larger scale, kilometric or decakilometric, aeromagnetic maps can be realized.

Interpretation and Detection of Fractures

Qualitative interpretation of magnetic measurements is relatively simple and is very efficient in locating basic rocks and, subsequently, fractured zones injected by such rocks. In such cases, magnetometry is of great help, especially when coupled with electrical methods. However, other types of fractures are not detected by magnetometry. Two types of applications of this method can be cited in West Africa. The first application concerns the search for water in fractured zones along massive dolerite intrusions. By contrast, the second one concerns the detection of fractured dykes. A magnetic profile of such a fractured dyke, which is an excellent water drain, is presented in Fig. 9.9.2.

9.10 CASE STUDIES

9.10.1 Fractured Rock Investigation and Engineering Fields

The first engineering domain where the importance of fracturation in rocks has been realized is certainly geotechnics (dams, tunnels, excavations, mines). In this field, a lot of fundamental studies were carried out in the sixties and early seventies, especially under the impact of some famous collapses. The number of engineering fields interested by the fractured aspect of rock suddenly increased in the early eighties, motivated either by energy problems (petroleum engineering, geothermy), safety problems (waste disposals, environment implications) or resources problems (water needs). Among all these domains, we have chosen two case studies, opposite in many respects:

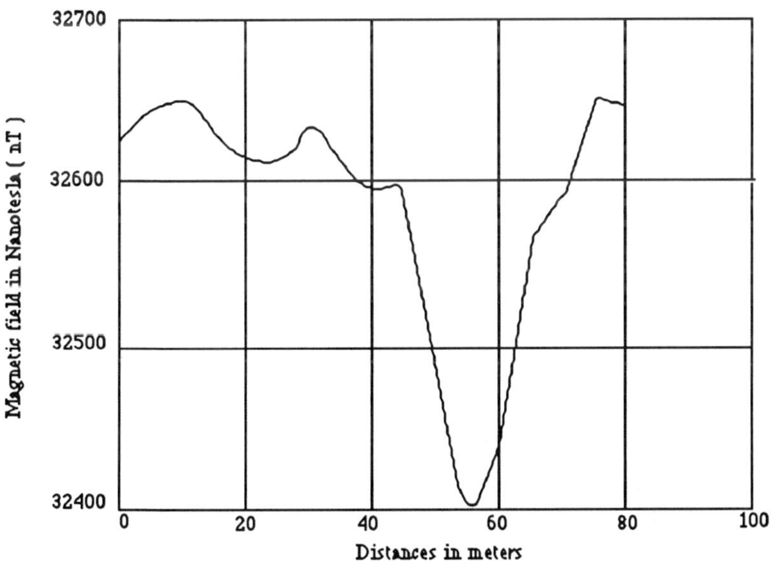

Figure 9.9.2. Magnetic profile on a fractured doleritic dyke in Mali (courtesy of Dubreuil, G., BRGM).

- The first domain is dam engineering, which is concerned with surface rock, and for which new techniques have a relatively small impact as compared to more traditional considerations. This fact probably results from the feasibility of qualitatively visualising the rock at a shallow depth. However, it is difficult to convert a mass of sophisticated quantitative data into synthetic information useful in the project computations. In this field, the art of the engineer remains very important for data interpretation.

- The second domain, logging in petroleum engineering, has produced the majority of the most recent high tech methods and tools for studying rock fracturing at great depth. Petroleum engineering is a relatively old discipline, in which the rock has been considered for a long time as a porous medium. Taking fractures into account is quite recent, and this is partly the consequence of the poor resolution of traditional logging tools. The qualitative detection of fractures has made decisive progress recently, and the problem is now one of how to obtain quantitative data.

9.10.2 Fractured Rock Investigation in Dam Engineering

The depth of rock that is of concern for purpose of dam construction is small in comparison to mines or oil engineering. This depth is about the height of the dam, the first meters of the rock being of special interest for determining foundation excavation. This rock will have been generally submitted to natural low stresses and weathering, exacerbating the role of potential fractures existing in confined or sound deep rock. From the mechanical and hydraulic points of view, the importance of these fractures is vital for dams. Since temperature plays a secondary role, investigation in dam engineering is thus devoted essentially to isothermal fractured rock.

Geologists, geotechnicians and engineers involved in a dam project should be able to describe rocks, their mechanical and hydraulic properties, and their discontinuities and filling material, which determine the mechanical properties of the joints.

Of course, the case of an arch dam is the most critical, because the dam abutments are subjected to a concentration of stresses and high hydraulic gradients. The case of earth and rock fill dams is less critical from this point of view; for these dams, investigations concern the dam abutments, and borrow pits and quarries.

Observation Methods

Surface Observations. Dam investigations begin with careful regional and local geological investigation. With the help of geological maps and outcrops, the geologist should investigate the site formation, nature of rock and major fractures existing on the site. Such surface investigation leads to a first structural sketch, indicating the main different rock zones and discontinuities with their directions and dip. For high dams and seismic regions, specialists in seismo-tectonics must evaluate the potential for earthquakes. This first geological examination will result in a program of more precise investigations, in deeper rock, using different techniques as described below.

Observation in Galleries, Adits and Pits. Observation in galleries or adits is typical of investigations for concrete dams; it is also common for high earth and rock fill dams. Observation in pits is particularly interesting for horizontal stratified formations. The importance of such observation lies in the possibility of a direct rock characterization. In particular, fracture directions and thicknesses can be analyzed and filling material can be sampled for laboratory identification.

Core Observation. Boreholes are an old investigation procedure, still currently used for dam site study, giving large amounts of information. Cores are generally taken throughout drilling. Diameters vary between 25 and 120 mm but are generally between 50 and 90 mm. Cores give information on the different geological units and allow petrographic observation on thin rock cuts and representative tests for quarry products. Identification of soft materials in joints and fractures depends on their strength, the diameter of the hole and the skill of the operator. The quality of the fractured rock can be assessed as a whole using the RQD index, which indicate, for each meter of borehole, the cumulative length of elements larger than 10 cm.

However, this technique is inadequate for detecting large fractures or faults. In particular, because the orientation of the cores is undefined around the axis of the borehole, which can be vertical or inclined, the cores generally do not reveal the directions of joints. This fact demonstrates the importance of galleries for dam foundation investigation, especially for arch dams over 50 m high.

Well Observation. Borehole logging tools (see Sec. 9.10.3) are rarely used for dam purposes, as pits or adits are required for direct rock observation.

Hydraulic Tests

One-Well Hydraulic Tests. The Lugeon test is still commonly used, which involves increasing (in principle) the pressure by steps up to 1 MPa. The Lugeon unit, i.e., the absorption of 1 liter per meter of hole per minute, is assumed to correspond to a global permeability of the fractured rock of 10^{-7} m/s. Directional coefficients of permeability are not commonly determined because of the numerous joint and fracture systems. Studying the influence of stresses on permeability is not common but seems to be a tendency for the future (see Sec. 9.3.6).

Cross-Hole Tests. They are not common, although piezometers are generally installed in boreholes.

Cores. On cores, permeability tests are a common practice.

Mechanical Tests: Dynamic Testing

Surface Seismic. Rapid and cheap, this procedure is used almost systematically at the beginning of investigations. It gives a first site classification, especially when dealing with concrete dams, and leads to a first approximation of the excavation depth. If P-wave velocities are between 2500 and 4000 m/s, a concrete dam is possible. Above these values, an arch dam can be considered when other site characteristics, especially topography, are suitable for this type of structure.

Seismic Method in Galleries. Small seismic tests can be performed also in galleries, as proposed by Schneider in the 1960s (Bienawski, 1978; A.I.G.I., 1979). This test records the total signal between two points some 10 m apart and correlates this measure empirically with the static deformation modulus.

Seismic Tests in One Borehole or in Cross-Holes (Tomography). They are not commonly used in dam engineering.

Seismic Tests on Cores. The dynamic Young's modulus of the rock matrix is determined by measuring the velocity of sound.

Mechanical Tests: Static Testing

In Galleries. Mechanical tests, all of which are time-consuming and costly, can be performed in galleries:

- Pressure tests can be realized in pressurized rooms.
- Plate tests consist in applying jacks on the opposite wall of a gallery, in different directions according to the fracture directions indicated on the Schmidt diagram.
- It is current practice to try to determine in-situ stresses in the rock foundation for very high dams (above 100 m). The method used is the relaxation method in a saw cut on the adit walls.
- Large scale in-situ tests on joints can be performed in special rooms in the rock, as in Vouglans, France (C.I.G.B., 1967) or in Longtan or Lijiaxia, China (Carrère et al., 1987).

In Boreholes. In boreholes, the overcoring technique can be used at the bottom of boreholes, and interpretation done as previously by the relaxation technique.

On Cores. On cores, mechanical tests are currently done by direct compression and traction on both wet and dry samples.

Generally speaking, all these measures are of interest for the geologist studying the site history. However, they generally are of little import for the dam engineer, except for projects where the in-situ state of stress is required, for instance, for the lining of pressurized conduits.

Mechanical Tests: Measurement while Drilling (MWD)

The use of drilling parameters, even if not a common practice in dam engineering, should become a help, either for dam foundation or rock quarry investigation (Lutz, 1981).

9.10.3 Fracture Evaluation from Wireline Logs in Petroleum Engineering

In this section, the use of different wireline logs for fracture evaluation is reviewed. Logging techniques for fracture analysis may be divided into two types:

- Indirect methods, which are based on the observation of phenomena induced or modified by the presence of fractures.
- Imaging methods, in which fractures are observable in an image of the borehole wall.

Most conventional logging techniques are of the indirect type; imaging methods rely on ultrasonic or electrical sensors. We shall describe both indirect and imaging techniques in turn. We will then discuss the possibility of obtaining quantitative estimates of fracture density, orientation and aperture. The principles of the logging tools will be described only in passing, and standard texts (Serra, 1984; Labo, 1987; Ellis, 1987) should be consulted for details. Fracture detection techniques are also described in these same references, and in some detail in Kent (1987).

In general, indirect methods alone can only indicate the probable occurrence of fractured zones. Imaging techniques, on the other hand, greatly enhance the probability of detection, since they allow individual fractures to be observed. However, since imaging methods have a very shallow depth of investigation, they require some of the 'deeper' conventional tools to identify the fractures which extend deep into the formation. As all methods are limited to some degree, either in the type of information they yield or by environmental factors, this review will be concluded by a table summarizing the advantages and limitations of each technique.

Indirect Methods

Many indirect methods have been developed over the years for fracture detection. They all suffer from at least one of the following drawbacks:

i. the technique is non-unique in that phenomena other than fractures will produce a 'positive' response.

ii. the technique has 'blindspots' so that it fails to detect some fractures because of their unfavorable orientation or for some other reason.

iii. the number of closely spaced fractures is not resolved.

iv. little or no information is obtained on fracture orientation.

v. vertical resolution, of the fracture location and extent, is poor.

Nevertheless, these methods are often the only ones available, and they can prove to be extremely effective when used in the appropriate context and in combination. Among the indirect techniques, the sonic and resistivity methods deserve special mention because they can provide quantitative information on fracture aperture.

Resistivity Measurements (Figure 9.10.1). When fractures intersecting the borehole are invaded by borehole fluid (mud), which is much more conductive than the rock matrix, they are detected as zones of abnormally low resistivity by resistivity (laterolog) tools. In the conventional laterolog tool configuration, several measurements can be made simultaneously, with depths of investigation varying from a few millimeters to about three meters and more.

The response of the microlaterolog, with a depth of investigation of a few millimeters, can fluctuate dramatically in front of fractures, but its value as a fracture indicator is limited by similar behavior in the presence of bad borehole conditions and heterogeneous formations.
The shallow (LLS) and deep (LLD) laterologs, with depths of investigation of about one and three meters, respectively, respond to fractures differently according to their dip. With fractures dipping 50 degrees or less, both deep and shallow laterologs see fractures and formation

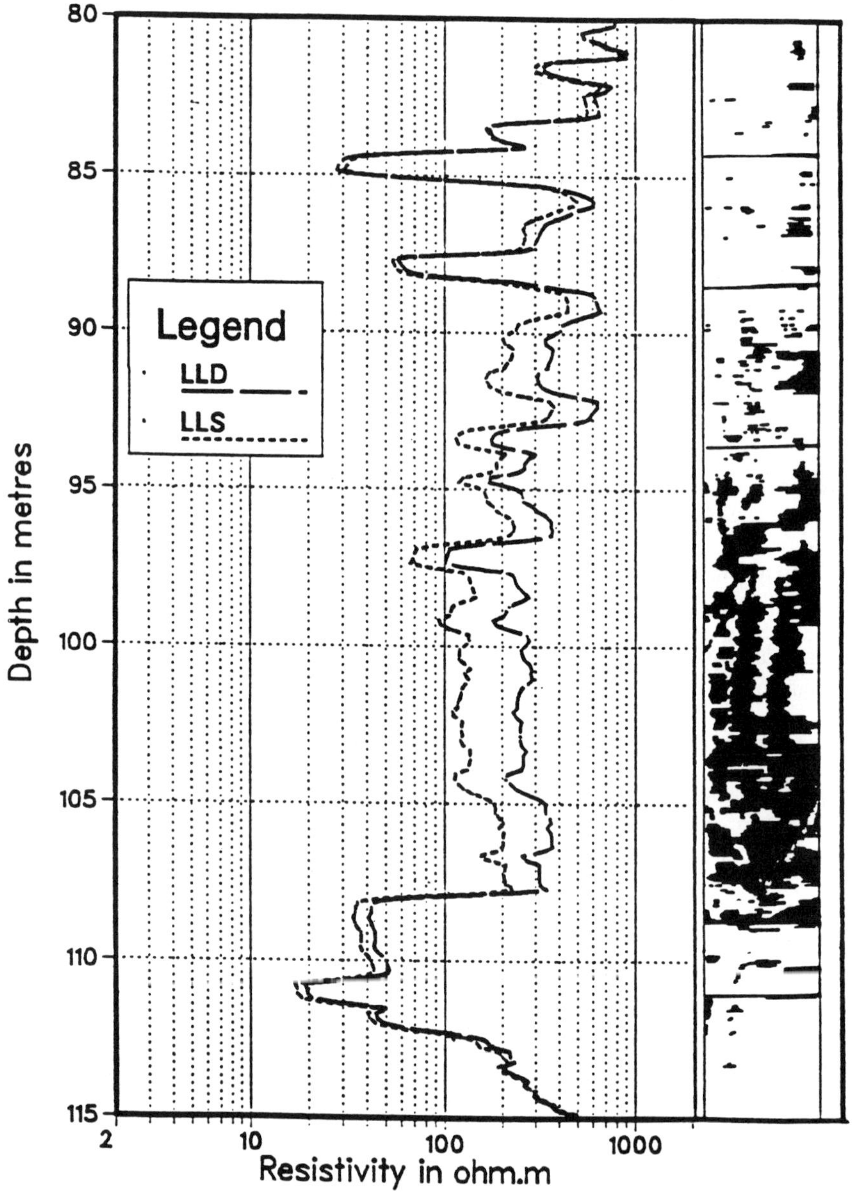

Figure 9.10.1. Dual laterolog response (LLD = deep, LLS = shallow) showing low dip fractures at 85, 88 and 111 ft and high dip fractures between 90 and 107 ft.

resistivities in series. Consequently, both measurements respond with a low resistivity anomaly (see Fig. 9.10.1 at 85, 88 and 111 ft). High dip fractures short-circuit laterolog currents sent into the formation and distort the normally symmetrical field surrounding the tool. The shallow laterolog, due to the proximity of the current return electrode, is more affected than the deep one, and records lower resistivity (see Fig. 9.10.1, 90-107 ft).

These patterns of behavior of the deep and shallow laterologs can be exploited to detect low and high dip fractures. They are most effective in low-porosity, high resistivity formations (100 ohm-meters or more), and perform the worst at detecting horizontal fractures in high-porosity, low-resistivity formations.

Recently, more and more wells have been drilled with oil-based muds that preclude use of the laterologs. They are replaced by *induction tools* with vertical coils, which respond to currents induced in horizontal loops surrounding the borehole. When a vertical fracture filled with non-conductive fluid, such as hydrocarbon, cuts one of these loops, the tool records a decrease in conductivity. However, the tool is insensitive to both horizontal and conductive fractures.

Dipmeters. Dipmeters are tools with pads that come into contact with the borehole wall. Each of these pads carries buttons that measure on a few square millimeters the microresistivity of the formation, which is recorded as a curve. The dip in the formation is determined by correlating the different curves. For perfectly horizontal beds, the curves would be identical. For dipping beds, the correlated points on the curves are shifted in depth by a value depending on the dip (see Fig. 9.10.2). In the presence of fractures, the curves will exhibit local differences that do not follow the bedding correlations.

Unfortunately, not all anomalies can be attributed to fractures. Sources of confusion include shale breaks, vugs, pebbles, borehole irregularities, etc. However, when the anomalies on the different pads correlate with a plane whose dip and strike are different from those of the bedding, the confidence in identifying the anomalies as a fracture is greatly enhanced (see Fig. 9.10.2). Use of dipmeters is generally limited to fracture detection in relatively thick, massive formations where the borehole is in good condition.

Sonic Methods. The waveforms recorded by sonic tools consist of the superposition of compressional and shear headwaves and borehole modes, including the Stoneley wave (White, 1983). The principal effects of fractures on the different acoustic waves are reflection, mode conversion and attenuation. All these effects have been exploited for fracture detection.

For the early generation of sonic tools, cycle skips, i.e., the failure to detect an arrival in the expected time window, are taken to be indicative of fractures on the assumption that the expected arrival has attenuated below the detection threshold. When the full waveform is digitally recorded for processing, many other measurements with superior reliability become available: the attenuation of compressional and shear waves, the mode conversion between compressional and shear waves, slowness anomalies, and the attenuation and reflection of Stoneley waves (Kent, 1987 and references therein).

It is known that for headwaves traveling vertically in the direction of the borehole axis, as in the conventional tool layout, the compressional and shear waves are relatively unattenuated by low and high dip fractures, respectively. In the circumferential tools where acoustic waves are propagated in a tangential direction around the borehole, vertical fractures are more easily detected. With transmitter-receiver pairs covering the four quadrants of the borehole circumference, orientation information is also obtained.

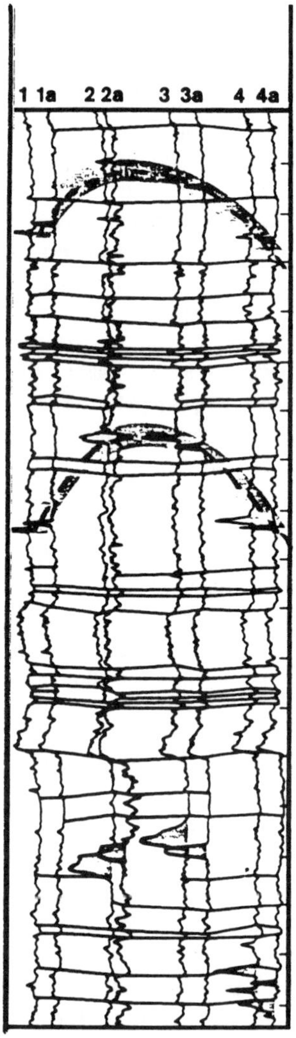

Figure 9.10.2. Dipmeter curves showing events corresponding to bed boundaries (connected by solid lines) and high dip fractures (connected by dashed lines).

With modern tools incorporating transmitters with low frequency output (typically extending below 1 kHz), Stoneley wave methods come to the fore. Stoneley wave measurements, as compared to compressional and shear, have the advantage of being relatively unaffected by changes in the acoustic impedance of the formation, and respond mostly to open fractures. They also offer the possibility of estimating fracture aperture. Of the attenuation and reflection measurements, the latter has the superior signal-to-noise performance. Fig. 9.10.3 from Hornby et al. (1989) illustrates the method. In the recorded waveform (shown for a receiver - transmitter spacing of 11 feet = 3.36 m), the reflected Stoneley waves are observed as slanted arrivals against a background of nearly vertical direct arrivals. The reflection coefficient (RC) can be estimated from the stacked amplitudes of the direct and reflected arrivals at each depth. An effective fracture width is then estimated either by assuming a horizontal fracture (open bars) or using the fracture dip as determined from a BHTV image (solid bars). Note that in this example, the 'fractures' with width exceeding a millimeter usually correspond to zones which comprise several fractures or in which the reflection coefficient is affected by borehole enlargement. In general, all sonic techniques are adversely affected by bad boreholes. Vugs, laminations and thin beds can also lead to false indications.

Natural Gamma Ray Spectrometry. Rocks contain naturally radioactive elements, of which uranium, thorium and potassium are detected by gamma ray tools. Uranium salts are well known for their solubility in water and they are frequently deposited in the fracture network. When the concentration of uranium is high relative to thorium and potassium, it is suggestive of a fracture zone, probably cemented. This technique helps to build evidence for fractures, but is far from confirmatory. It is most commonly used in carbonates, where uranium tends to be incorporated into calcite infilling the fractures. A variation of the technique is to log the gamma ray tool twice using a radioactive mud in the second pass. Invasion of fractures by mud will be indicated by additional high count rates in the second pass. This technique has also been used to investigate the hydraulic connectivity between wells by injecting radioactive markers into one well and measuring the gamma ray response in another.

Photoelectric Effect. In photo-electric logging, low energy gamma rays emitted from a Co 60 or Cs 137 source are absorbed by atoms in the formation and electrons re-emitted as a result are detected. The photo-electric effect, being the strongest for high atomic number elements, allows the detection of such elements. When barite mud is used in a well, spikes will appear on the photo-electric log at places where the barium-rich mud has invaded the fractures. Note, however, that the detectors are small, covering approximately 10% of the borehole circumference.

Spontaneous Potential. The value of the electrical potential between an electrode in the well and a reference at the surface is commonly recorded during logging. It is generally admitted that this spontaneous potential is due to electrode-filtration of fluids into the porous rocks. So this measure can indicate fracturing when the invasion of the formation by the flow of mud filtrate ions causes a streaming potential. However, streaming potentials can also occur from permeable beds. So this technique is again an inconclusive but useful diagnosis for fracture detection.

Caliper. The diameter of the borehole is usually monitored by two or more pairs of mechanical calipers (arms) or by the transit time of a radial ultrasonic beam, as used in the borehole televiewer (BHTV). Caliper logs have also been used for fracture detection, since chipping at the

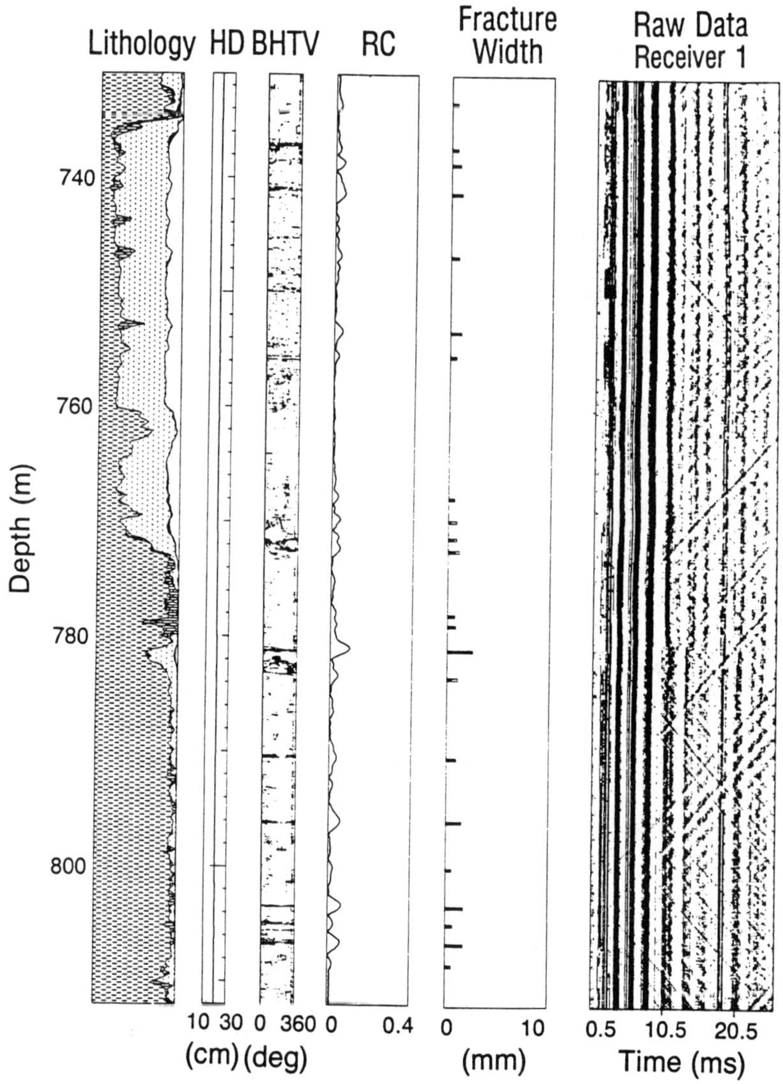

Figure 9.10.3. Fracture evaluation using Stoneley wave reflections (from Hornby et al., 1989). From left to right, the tracks show:

1) the results of a volumetric analysis.
2) a borehole televiewer image.
3) the reflection coefficient (RC) of Stoneley waves.
4) the effective fracture aperture computed, assuming zero dip (open bars) and dip as determined from the borehole televiewer image (solid bars).
5) the recorded waveforms for an 11-ft transmitter-receiver spacing.

intersection between fractures and the borehole wall will produce a sharp localized increase in the hole diameter. However, this is an unreliable indicator because borehole breakouts are more usually associated with the failure of the borehole, either as a result of the drilling process (e.g., the "key-hole" effect arising from motion of the drill stem, found especially in deviated wells) or of the imbalance of stresses around the borehole. Indeed, breakouts have been analyzed extensively for the determination of the direction of the stress anisotropy.

Nevertheless, caliper logs are indispensable for fracture analysis for two reasons. First, a knowledge of the borehole geometry is essential to correctly interpret nearly all log responses. Second, identification of the direction of stress anisotropy, and zones of borehole elongation, are necessary to interpret drilling-induced fractures.

Imaging Techniques

Because all the indirect methods can lead to confusion between fractures and other phenomena, different techniques were tried to provide images of the borehole wall in which fractures can be 'directly' observed. Images allow fractures to be counted and their orientation determined with precision. They also allow an experienced interpreter to decide if fractures are natural or borehole induced and, in general, place the fractures in their geological context.

Borehole Televiewer. Zemanek et al. (1969) were the first to propose a tool providing an image of the well borehole on which fractures can be observed. This tool, the borehole televiewer (BHTV), is an acoustic scanner. As it is pulled up the hole, it scans the borehole wall with a rotating transducer that emits a pulsed ultrasonic beam. Two measurements are recorded: the reflectivity, which is the ratio of the amplitudes of the first peak of the signals transmitted and received, and the transit time between the tool and the borehole wall. When these measurements are displayed as if the borehole wall were split vertically and laid flat, then a fracture intersecting the borehole appears on the images as a sinusoid from which the azimuth and dip can be computed (see Fig. 9.10.4).

Usually the acoustic reflectivity provides a more accurate image of the fractures than the transit time. Because the amplitude of the signal is drastically affected by any damage to the wellbore, the fractures are highlighted on reflectivity images, provided that the mudcake is not too thick. The transit time measurements, on the other hand, can be used to detect post-drilling movements of fractures, as well as deformation of the borehole, as seen in Fig. 9.10.5. The BHTV gives coverage of the borehole wall and can be used with oil- or water-based muds, provided the mud is not too heavy.

Formation Microscanner. More recently, an electrical imaging tool, called the formation microscanner (FMS), has been developed (Serra, 1989). It is a tool with 2 or 4 pads pressed against the borehole wall. In each of the pads, an array of micro-electrodes is inserted, covering a width of about 7 cm per pad. The buttons have a depth of investigation of less than or equal to one centimeter. The current intensity is recorded for each button at a very high sampling rate (0.1 inch = 2.5 mm). Images are obtained by displaying the resistivity using grey levels. Low resistivity zones appear dark and high resistivity features appear white. Open fractures invaded by conductive mud will appear as dark lines on the images (see Fig. 9.10.6). On the other hand, cemented fractures infilled by resistive minerals are white.

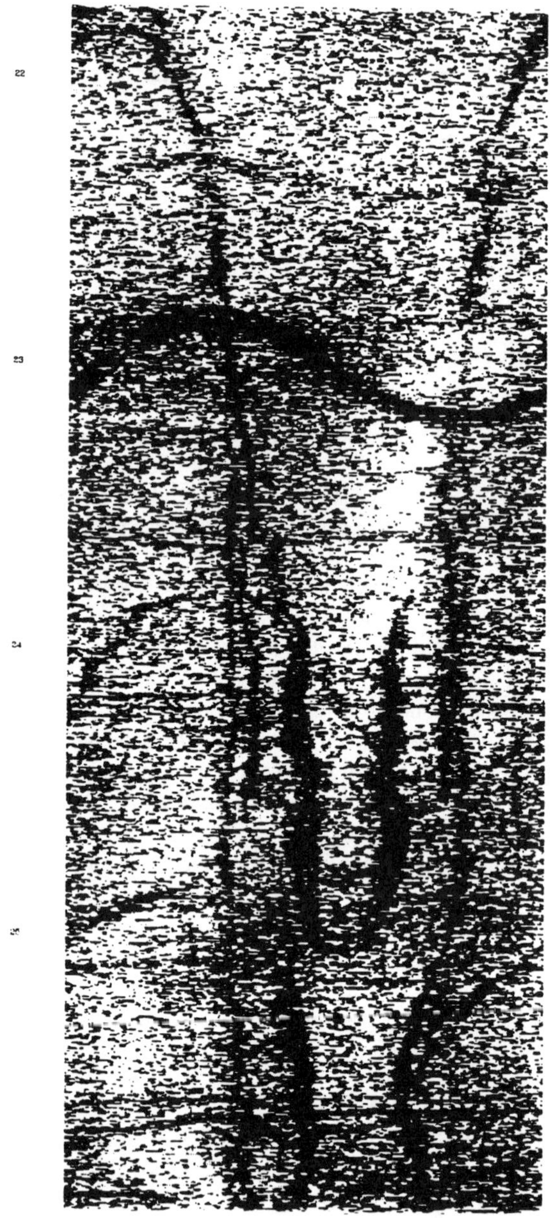

Figure 9.10.4. Borehole televiewer image (amplitude) of a granite section showing planar fractures as sin
and non-planar, near-vertical fractures as U-shaped traces.

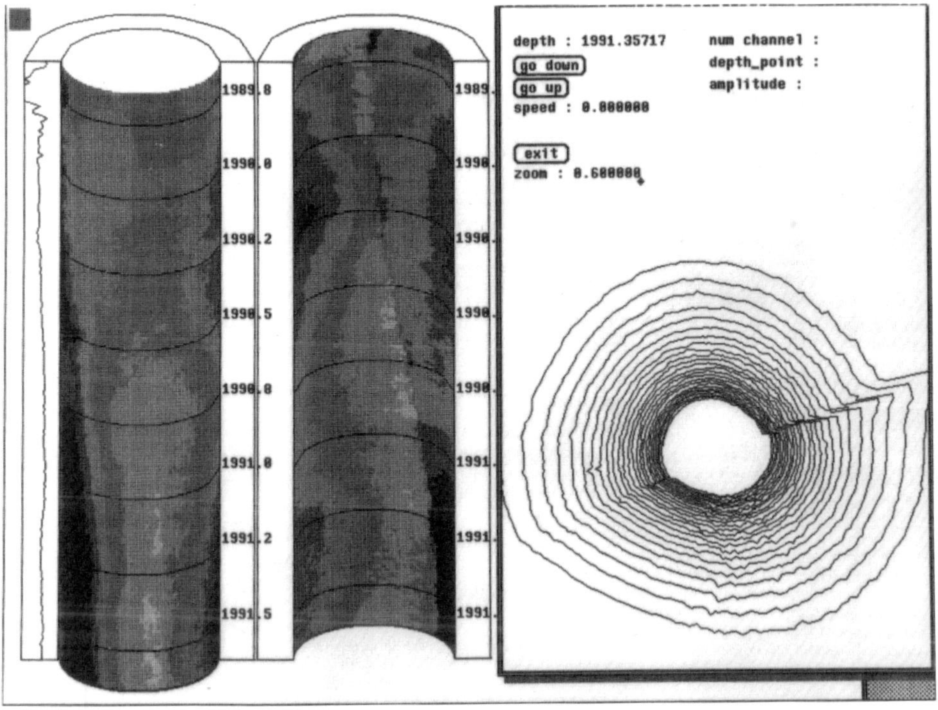

Figure 9.10.5. Transit time measurements from a borehole televiewer showing shear movement on a fracture after drilling. The drawing on the right is a perspective view of the borehole wall looking downwards from a point inside the borehole.

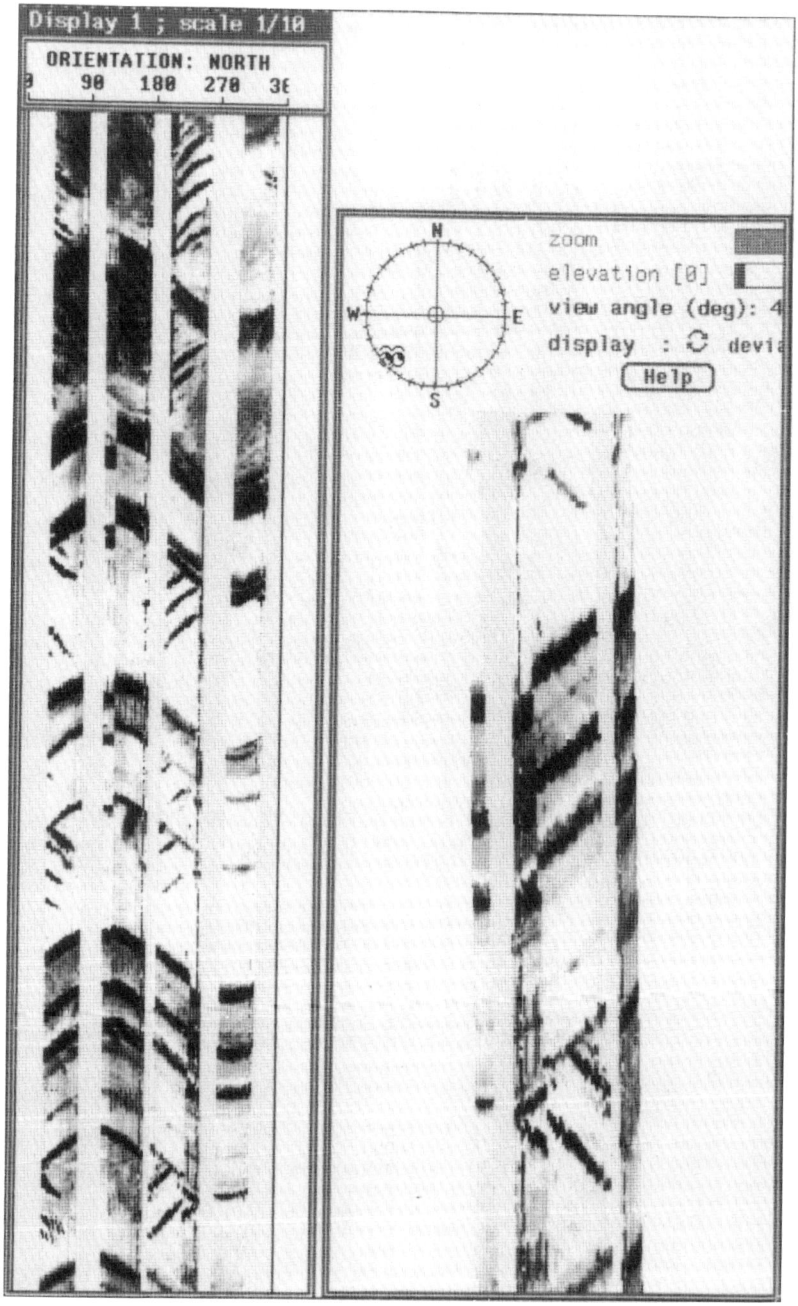

Figure 9.10.6. Formation microscanner images made from three passes of a two-pad tool. These show thin conductive beds dipping 45 degrees to WNW and strata-bounded microfractures normal to the beds. The microfractures have been reopened in the NE-SW direction as a result of drilling. The left image shows the borehole wall unfolded onto a plane and the right image shows the same data in 3D (core presentation).

The resolution of the tool is between one-half and one centimeter, as shown by comparison with core, but, of course, it can detect much thinner fractures, depending on the conductivity contrast between the fracture fluid and the matrix. For example, if the contrast is 1000, a fracture 5 microns wide crossing the well can be detected. Obviously, the FMS must be used with conductive mud.

Azimuthal Laterolog. The azimuthal laterolog (Mosnier, 1987) is a centered tool with a girdle of 24 or more focused electrodes. These electrodes measure, directly ahead, an "azimuthal" resistivity with a depth of investigation on the order of a few centimeters. Images can be obtained by displaying these resistivity curves as wiggle lines or using color levels (see Fig. 9.10.7). The tool has been introduced between packers to observe the appearance of fractures on the borehole wall during hydrofracturing operations.

Quantification

Going beyond the mere detection of fractures, well logs can provide quantitative measures of the fracture system, such as the density, orientation and aperture of the fractures.

Fracture Density. Imaging techniques provide the obvious means to obtain the number of fractures intersecting the borehole and their orientations. Interactive computer software has rendered this a routine task. However, the fracture population thus sampled is heavily biased against high dip fractures. To correct for this bias is a well-known problem which affects core measurements as much as logs (Narr and Lerche, 1984).

Orientation. Most fractures are more or less planar. Their planarity can be quantified by fitting a plane to the fracture trace as seen on borehole images. The dip and azimuth can then be computed, using the directional information provided by inclinometry devices fitted to logging tools to effect the transformation from borehole (tool) to earth coordinates.

Aperture and Extent. From some resistivity and sonic logs, a fracture aperture can be estimated by comparing the measurements against the results of mathematical modeling. Usually, the mathematical model is a planar, perfectly smooth fracture of infinite extent. Thus the aperture obtained is always understood to be an effective value, equal to that of an equivalent model fracture.

Stoneley Wave Reflectivity. The details of this technique are described by Hornby et al. (1989). Briefly, an aperture can be obtained from the measured reflection coefficient if the dip of the fracture is known, and provided the extent of the intersection of the fracture with the borehole wall is small compared to the wavelength of the Stoneley wave (roughly two meters at 600 Hz). The depth of investigation is approximately half a wavelength.

Deep/Shallow Laterologs. Finite-element modeling by Sibbit and Faivre (1985) shows that the deep and shallow laterolog measurements can be used to infer fracture aperture, provided the conductivity of the fracture fluid is known and sufficiently large compared to the matrix conductivity.

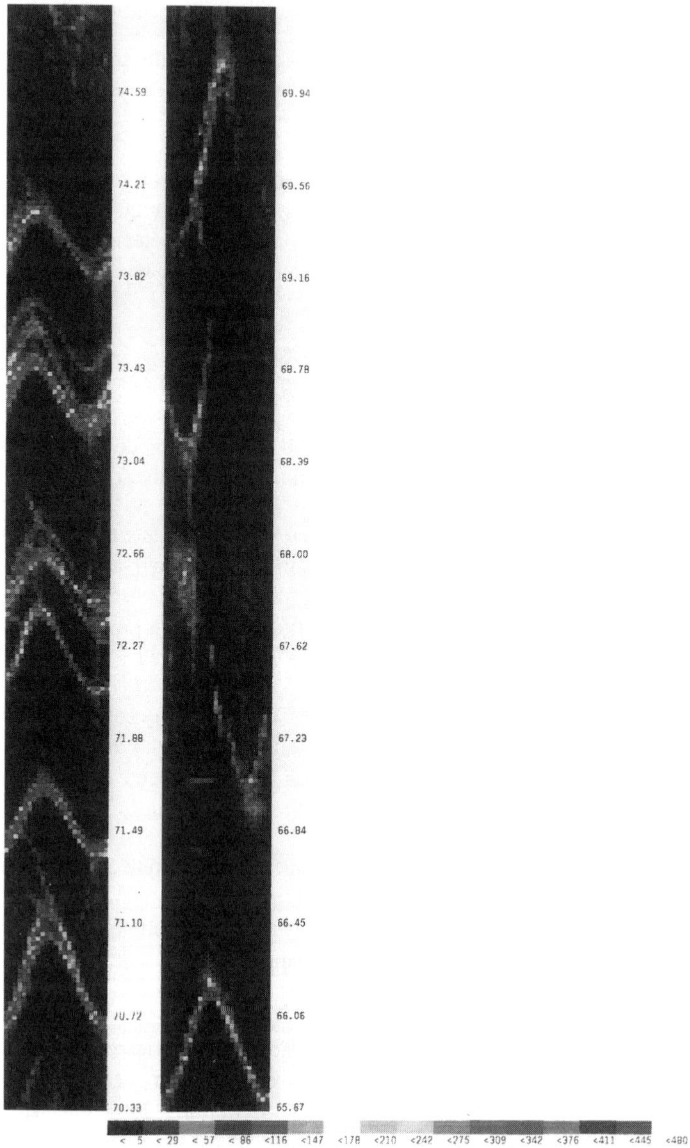

Figure 9.10.7. Azimuthal laterolog obtained by a 24-electrode tool, showing fractures on the wall of a borehole in granite (Mayet de Montagne experimental station). Depths are in meters. Resolution is 4 cm in the vertical dimension. The image was obtained in real time without any processing of the data.

Formation Microscanner (FMS). Similarly, Luthi and Souhalte (1990) havc shown that it is possible to deduce the aperture of a fracture from the electrical current response of the FMS tool. Briefly, the excess current measured in the vicinity of a fracture over the unfractured background is found to be a function of the fracture aperture and mud resistivity. Since the mud resistivity is routinely logged, even though only in the borehole, the aperture can be estimated.

Because the FMS has a very shallow depth of investigation, the aperture thus determined cannot be assumed to be representative at distances exceeding a few millimeters from the borehole wall. In general, the aperture obtained from the FMS must be confirmed by other measurements.

Synthesis

Table 9.10.8 gives an up-to-date comparison between the different possibilities of logging tools, in view of fracture detection in oil open boreholes.

Table 9.10.8 Comparison of the different possibilities of logging tools, for fracture detection in open boreholes.

	Imaging Techniques			**Indirect Measurements**
	Borehole Televiewer	**Azimuthal Laterolog**	**Formation MicroScanner**	**Compressival and Shear Waves**
Can it detect • Open fractures? • Healed fractures?	Open fractures most of the time, healed fractures sometimes. Calcite-filled fractures have a high reflectance and may appear as open fractures if the background has a low impedance.	Both usually, if resistivity contrast is sufficient. If the mudcake is thick, current is defocused and the resolution drops drastically.	Both usually. Short fractures parallel to the borehole may be missed. Healed fractures will have a lower con-ductivity contrast than open fractures. Clay filled fractures may appear as open ones.	Both
How thin a fracture can it detect ? (sensitivity)	0.5 mm	0.1 mm with a good resistivity contrast	On the order of microns, given a sufficient mud / formation resistivity contrast.	Undetermined
Aperture estimation	No	Yes	Yes	No
Distinguish natural / induced	Possibly	Possibly	Possibly	No
Vertical resolution	• 1 cm at 100 m/h logging speed. • 3.5 cm at 300 m/h	4 cm	1 cm	20 cm
Features mistakeable for fractures	Drilling damage, high-impedance bedding	Drilling damage	Drilling damage	Washouts, thin beds, vugs, breccia, boundary of high acoustic contrast
Can it provide strike and dip ?	Yes	Yes	Yes	No
Depth of investigation	Surface to a few mm depending on signal techniques	1-2 cm	1 cm	One wave length (30 cm)
Borehole coverage	100 %	100 %	7 cm / pad (2 to 8 pads)	100 %
Mud limitations	May be used in muds lighter than about 14 lb/gal	Very high contrast in resistivity is required between mud and formation	Usable only in conductive muds; will work in mud with up to 20 % oil	None

Indirect Measurements				
Stoneley waves Attenuation / Reflection	**Dipmeter**	**Laterologs Deep / Shallow**	**Photoelectric Effect**	**Natural Gamma Ray**
Mainly open	Both usually, but small coverage of the borehole limits the picture	Open only	Open only	Detects emplacement of radioisotopes in either open or healed fractures
0.5 mm	10 microns, assuming a large resistivity contrast between formation and drilling fluids	10 microns, given a 1,000 ohm-m resistivity contrast between formation and drilling fluids	0.5 to 1 cm given a fracture of sufficient depth	Not applicable; detection threshold 2 to 5 ppm uranium
Yes	Yes	No	No	No
No	No	No	No	No
20 cm	2 cm if the conductivity of each fracture is about equal	1 m	15 cm	60 cm
Washouts, as p- and s- waves, but to a lesser degree	Breakout zones can appear as fractured zones	Deep invasion in permeable intervals; conductive thin beds	Washout, drilling damage, vugs	Carbonate cementation in formerly permeable zones; kerogen-rich shales; phosporites
No	Yes	No	No	No
One wavelength (>1 mm at < 1 KHz)	1 cm	For a 500 micron fracture containing a 0.1 ohm-m mud in a 10,000 ohm-m formation • 70 cm (shallow laterolog) • >500 cm (deep laterolog)	90 % of measurement within 10 cm of borehole	20 to 30 cm
100 %	Point coverage from button on four pads	100 %	10 % in 8.5 in borehole	100 %
None	Usable only in conductive muds; salty muds favor fracture detection	Usable only in conductive muds	Effective only in baryte mud	Correction required for heavy or potassium rich muds

NOTATION

A	Wave amplitude.
c	Wave velocity. Specific heat. Tracer concentration.
c_t	Total compressibility (liquid+ bulk) of a saturated rock.
C	Wellbore storage coefficient.
Cij	Hydraulic impedance coefficient.
Cij*	Fourier transform of Cij.
D	As subscript, symbol denoting "dimensionless."
D	Caisson coefficient. Effective dispersivity.
e	Thickness of a fracture.
e_i	Thickness of a fracture belonging to set (i).
E	Electrical field.
f	As subscript, symbol denoting "fissure."
f	Frequency.
f_o	Resonance frequency.
g	Gravity acceleration. Thermal gradient.
g_o	Geothermal gradient.
h	Well section length.
H	Hydraulic head. Magnetic field.
H'	Heat exchange coefficient between fluid in fracture and matrix.
i	Complex symbol; $i^2 = -1$.
I	Input. Current intensity. Dimensionless inertia parameter.
J	Current density.
J_0, J_1	Bessel functions, of the first kind, of order 0 and 1.
k	Permeability. Thermal decomposition constant of a chemical specie.
K_0, K_1	Modified Bessel functions, of order 0 and 1.
m	As subscript, symbol denoting "matrix."
n_i	Number of fractures of thickness e_i belonging to set (i).
O	Output.
P	Pressure.
P_o	Static pressure.
P_w	Pressure at the well.
$P^*(\omega)$	Fourier transform of P(t).
q	Heat flux. Volumetric density of electrostatic charges.
Q	Quantity of flow. Damping factor or "quality factor."
$Q^*(\omega)$	Fourier transform of Q(t).
r	Radial coordinate.
r_w or R_p	Well radius.
R_f	Circular fracture radius.
S	System. Skin coefficient.
t	Time.
T	Temperature.
U	Electrical potential.

v	Seismic wave velocity.
V	Velocity of a fluid. Voltage difference.
w	As subscript, symbol denoting "well," "water" or "fluid."
w	Electrical resistivity of water.
W	Energy radiated by bodies.
x	Horizontal coordinate.
y	Horizontal coordinate.
Y_0, Y_1	Bessel functions of the second kind, of order 0 and 1.
z	Depth.
Z	Transfer function.

Greek Letters

α	Shape coefficient. Diffusivity = $\lambda/\rho c$.
β	Dimensionless parameter = $\rho_w c_w / (2\rho_f c_f)$.
γ	Electromagnetic propagation factor.
ΔH	Hydraulic head difference.
δ	Electrical source term.
ε	Emissivity. Dielectric permittivity.
ε_r	Modulus of the relative dielectric permittivity.
θ	Inclination of a fracture or dip.
$\theta_f(r,z,t)$	Rock temperature at a distance r of the well axis, at depth z and time t.
λ	Thermal conductivity.
μ	Dynamic viscosity of a fluid. Magnetic permeability.
μ_o	Vacuum magnetic permeability.
ν	Kinematic viscosity of a fluid.
ρ	Liquid mass density. Electrical resistivity.
ρ_a	Apparent electrical resistivity.
ρ_{22}	Biot's dynamical coefficient.
σ	Stefan constant. Electrical conductivity of the rock.
Σ	Natural rock surface.
ω	Pulsation = $2\pi f$.
Φ	Porosity.
κ	Magnetic susceptibility.
Ω	Inner rock domain.

Special Symbols

*	Convolution product. Fourier transform.
Arg F	Argument of a complex quantity F.
$\lvert F \rvert$	Modulus of a complex quantity F.
K	Kelvin temperature unit.
ln	Decimal logarithm.
Ln	Neperian logarithm.
$\nabla .$	Divergence operator.

$\nabla \times$	Curl operator.
Δ	Laplace operator.

ABBREVIATIONS

2D ; 3D	Two-dimensional. Three-dimensional.
BHTV	Borehole televiewer.
DC	Direct current.
FMCW	Frequency modulation in saw tooth.
FMS	Formation microscanner.
IR	Infrared.
LLD	Deep laterolog.
LLS	Shallow laterolog.
MWD	Measurement while drilling.
REV	Representative elementary volume.
RQD	Cumulative length of elements larger than 10 cm (on a 1 meter core).
SH wave	Horizontal shear wave.
SP	Spontaneous potential.
SV wave	Vertical shear or transverse wave.
P wave	Longitudinal wave.
RC	Reflection coefficient.
UV	Ultraviolet.
VSP	Vertical seismic profiling.
WZ	Weathered zone.

REFERENCES

A.I.G.I. (Association Internationale de Géologie de l'Ingénieur). Reconnaissance des massifs rocheux par la 'Petite Sismique', *Association Internationale de Géologie de l'Ingénieur,* Bulletin No. 20, p. 121–123, Krefeld, 1979.

Aki, K., and Richards, P.G. *Quantitative Seismology, Theory and Methods,* Vol. I and II, W.H. Freeman and Company, San Francisco, 1980.

Alami, M. Reconnaissance de formations poreuses saturées par analyse harmonique, Thèse de 3ème cycle, Université Paul Sabatier, Toulouse, 1983.

Alani, M. Phénomènes dynamiques en milieux poreux saturés. Incidence du couplage solide-liquide en écoulement harmonique, Thèse de 3ème cycle, Université des Sciences et Techniques du Languedoc, Montpellier, 1985.

Anon. Suggested methods for the quantitative description of discontinuities in rock masses, *Int. J. Rock. Mech. Min. Sci.*, Vol. 15, 1978.

Anon. The logging of rock cores for engineering purposes, *Quarterly Journal of Engineering Geology*, V3, p. 1–94, 1972.

Archie, G.E. The electrical resistivity log as an aid in determining some reservoir characteristics, *AIME Technical paper 1422, Petroleum Technology*, Jan. 1942, p. 54–62, 1942.

Barton, C.A., Tesler, L.G., and Zoback, M.D. Interactive Image Analysis of Borehole Televiewer data, *Automated Pattern Recognition in Exploration Geophysics Palaz I*, Sengupta SK (eds), Geophysical Press, 1989, (in press).

Batchelor, A.S. Hot Dry Rock reservoir stimulation in the UK: an extended summary, *IEE 4th International Conference on Energy Options*, April 1984.

Bernaix, J. *Etude géotechnique de la roche de Malpasset*, Dunod, Paris, 1967.

Bienawski, Z.T. The 'petite sismique' technique, A review of current developments, Second Conference on acoustic emission, Microseismic activities in geological structures and materials, *The Pennsylvania State University*, Nov. 1978.

Bonnet, G. Contribution à l'étude de milieux poreux saturés en régime dynamique. Application à la reconnaissance par pompage harmonique et à la reconnaissance sismique, Thèse de Doctorat d'Etat, Université des Sciences et Techniques du Languedoc, Montpellier, 1985.

Bourdet, D., and Gringarten, A.C. Determination of Fissure Volume and Block Size in Fractured Reservoirs by Type Curve Analysis, *Paper SPE 9293 presented at the 55th Annual Fall Technical Conference and Exhibition of the Society of Petroleum Engineers of AIME, held in Dallas, Texas*, September 21–24, 1980.

Burke, L. et al. Using formation microscanner images, *The Technical Review*, p. 16–40, January 1989.

C.I.G.B. (Commission Internationale des Grands Barrages). Essais et calculs de mécanique des roches appliqués à l'étude de la sécurité des appuis d'un barrage-voûte, Exemple de Vouglans, Groupe de travail du Comité National Français, *9ème Congrès des Grands Barrages*, Vol. I, p. 793 et al., Istambul, 1967.

Camborne School of Mines. Camborne Geothermal Energy Project, Microseismic Results (1984–1986), ETSU G 137-P7, 1988.

Camborne School of Mines. Camborne Geothermal Energy Project, Microseismic Results (1984–1986), ETSU G 137-P6, 1988.

Capuano, R.H.C., Adams, M.C., and Wright, P.M. Tracer recovery and mixing for two geothermal injection backflow studies, *Proceedings 9th Workshop on Geothermal Reservoir Engineering,* Stanford University, Stanford, p. 299–304, 1983.

Carrère, A., Nury, C., and Pouyet, P. Un modèle tridimensionnel non linéaire aux éléments finis contribue à évaluer la convenance du site géologiquement difficile de Longtan en Chine au support d'une voûte de 220 m de hauteur, *6ème Congrès International de Mécanique des Roches, Montréal,* p. 305 et al., 1987.

Chen, Z.X. Transient Flow of Sligthly Compressible Fluids Through Double-Porosity, Double-Permeability Systems. A State-of-Art Review, *Transport in Porous Media* 4, Editor J. Bear, Haïfa, Israël, p. 147–184, 1989.

Cornet, F.H. Issue Coodinator, Forced Fluid Flow Through Fractured Rock Masses, *Int. J. Rock Mech. Min. Sci. & Geomech. Abstr.*, Vol. 26, No. 3/4, p. 169–349, July 1989.

Corwin, R.F., and Hoover, D.B. The self-potential method in geothermal exploration, *Geophysics*, Vol. 44, No. 2, p. 226–245, 1979.

Crosnier, B., Fras, G., and Jouanna, P. Reconnaissance of fractured media with several systems of fractures by means of harmonic techniques, *Rock Mechanics and Rock Engineering*, Vol. 18, p. 77–105, 1985.

Crosnier, B. Reconnaissance harmonique en milieux rocheux fissurés et/ou poreux, Thèse de Doctorat d'Etat, Université des Sciences et Techniques du Languedoc, Montpellier, 1987.

Cull, R. Formation Microscanner tool imaging and laterolog resistivity measurements in interpretation of crystalline terrains, *Deep Drilling in Crystalline Bedrock,* V.2, Springer-Verlag, 1988.

Darboux-Afouda, R., and Louis, P. Contribution des mesures de l'anisotropie électrique à la recherche des aquifères de fracture en milieu cristallin au Bénin, *Geophysical Prospecting*, Vol. 37, No. 1, p. 91–106, 1989.

Daviau, F. Interprétation des essais de puits. Les méthodes nouvelles, Ecole Nationale Supérieure du Pétrole et des Moteurs, Centre d'études supérieures de développement et d'exploitation des gisements, Edition Technip, Paris, 1986.

Dreesen, D. Fracture Detection: Interpretation of well logs to select packer seats and locate injection intervals, *11th Workshop on Geothermal Reservoir Engineering, University of Stanford, Stanford,* 1986.

Earlougher, R.C. Advances in Well Test Analysis, Soc. Pet. Eng. J. AIME, New York, Dallas, 1977.

Ekstrom, M.P., Dahan, C.A., Chen, M.Y., Lloyd, P.M., and Rossi, D.J. Formation Imaging with Microelectrical Scanning Arrays, *Transactions of the 27th Annual Logging Symposium, Houston,* June 9–13, paper BB, 1986.

Ellis, D.V. *Well Logging for Earth Scientists,* Elsevier, Amsterdam, 1987.

Fitterman, D.V. Electrokinetic and Magnetic Anomalies Associated with Dilatant Regions in a Layered Earth, *Journal of Geophysical Research*, Vol. 83, No. B12, p. 5923–5928, 1978.

Georgi, D.T. Geometrical aspects of borehole televiewer images, *SPWLA 26th Annual Logging Symposium,* June 10–13, 1985.

Hardage, B.A., and Toksoz, M.N. *Vertical Seismic Profiling*, Parts A and B, Geophysical Press, The Netherlands, 1983.

Held(van der), E.F.M., and van Drunen, F.G. A method of measuring conductivity of liquids, *Physica* 15, No. 10, p. 865–881, 1949.

Hornby, B.E., Johnson, D.L., Winkler, K.W., and Plumb, R.A. Fracture evaluation using reflected Stoneley-wave arrivals, *Geophysics*, Vol. 54, p. 1274–1288, 1989.

Hosanski, J.M., Bernaudat, F., Ledoux, E., and Ribstein, A. Transferts thermiques en milieu fissuré, *Annales des Mines*, p. 93–102, mai-juin, 1984.

Hosanski, J.M., and Ledoux, E. In situ determination of the hydrothermal properties of a deep fractured medium by a single-well technique, *J. of Hydrology* 56, p. 39–47, 1982.

Jouanna, P. Effet des sollicitations mécaniques sur les écoulements dans certains milieux fissurés, Thèse de Doctorat d'Etat, Université Paul Sabatier, Toulouse, 1972.

Jouanna, P., and Fras, G. Introduction à la reconnaissance harmonique, dans l'espace des fréquences, de milieux fissurés par essais d'eau transitoires. Notion de signatures spectrales et application au cas d'horizons fissurés reconnus par pompage harmonique. Unicité des solutions, *C. R. Acad. Sc. Paris*, t. 288, série B, p. 33–36, 15 janvier 1979.

Jung, R. Propagation and Hydraulic Testing of a Large Unpropped Hydraulic Fracture in Granite, Geol. Jb., E39, p. 37–65, Hannover, 1987.

Kent, J.M., Editor. Fracture detection with logs, *Technical Review*, Vol. 35, No. 1, Schlumberger Educational Services, Houston, Texas, 1987.

Kocabas, I., and Horne, R.N. Analysis of Injection-Backflow Tracer Tests in Fractured Geothermal Reservoirs, *Proceedings 12th Workshop Geothermal Reservoir Engineering*, Stanford University, Stanford, California, 1987.

Labo, J. A Practical Introduction to Borehole Geophysics, Geophysical References, Vol. 2, ed. T.N. Crook, Society of Exploration Geophysicists, 1987.

Lagabrielle, R. Contributions à la mise au point de quelques méthodes géophysiques appliquées à la reconnaissance en Génie Civil, Thèse de Doctorat d'Etat, Université des Sciences et Techniques du Languedoc, Montpellier, 1986.

Lanyon, G.W. Interactive computer analysis of borehole televiewer data, In: Society of Professional Well Log Analysts, *10th European Formation Evaluation Symposium, Aberdeen*, April, 1986.

Lloyd, P.M., Dahan, C., and Hutin, R. Formation imaging with micro-electrical scanning arrays. A new generation of stratigraphic high resolution dipmeter tool, *SPWLA 10th European Formation Evaluation Symposium*, April 22–25, 1986.

Lorenz, C., and Millon, R. A major structural feature deduced from the geological and the geophysical surveys related to the hole drilled on the magnetic anomaly of Paris basin, *Bull. Soc. Geol. France*, (8), VI, n° 5, p. 715–721, 1990.

Louis, C. Suggested methods for determining hydraulic parameters and characteristics of rock masses, Report, Int. Soc. Rock. Mech., 1977.

Luthi, S.M., and Souhaite, P. Fracture apertures from electrical borehole scans, *Geophysics*, Vol. 55, No.7, p. 821–833, 1990.

Lutz, J. Enregistrement des paramètres de forage, *Travaux*, No. 552, février 1981.

Mari, J.L. Unpublished personal communication, *Institut Français du Pétrole*, Rueil-Malmaison, France, 1989.

Mary, J.P. Thermodynamique d'aquifères à double porosité. Application au stockage de chaleur, Thèse de Doctorat, Institut National des Sciences Appliquées, Toulouse, 1987.

Mastin, L. Effect of borehole deviation on breakout orientation, *Journal of Geophysical Research*, Vol. 93, No. 88, p. 9187–9195, 1985.

Mavor, M.J., and Cinco, H. Transient Pressure Behavior of Naturally Fractured Reservoirs, *Paper SPE 7977, presented at the 1979 California Regional Meeting of the Society of Petroleum Engineers of AIME, Ventura, California,* April 18–20, 1979.

Mhamdi Alaoui, M. Contribution à la modélisation thermodynamique de milieux poreux et fissurés, Thèse de Doctorat d'Etat, Tomes 1 et 2, Université des Sciences et Techniques du Languedoc, Montpellier, 1988.

Michel, W., and Haenel, R. Quantitative Bestimmung von Wasserinjektionen und Extraktionen in Bohrungen mit Hilfe von Temperaturmessungen im Hot Dry Rock Projekt, Urach, *Ber. Arch. NLFB*, 95824 Hannover, 1984.

Mosnier, J. Localisation en profondeur et en azimuth de fractures conductrices dans un encaissant résistant, *Actes du quatrième colloque annuel de diagraphies, communication 20, S.A.I.D., Paris*, 21-23 octobre 1981.

Mosnier, J. Méthode électrique de détection des fractures sur la paroi d'un forage, *Bull. Soc. Geol. France*, III-6, p. 1049–1054, 1987.

Murphy, H.D., and Lawton, R.G. Downhole measurements of thermal conductivity in geothermal reservoirs, *J. Pressure Vessel Tech.*, Vol. 99, 1977.

Narr, W., and Lerche, I. A method for estimating subsurface fracture density in core, *American Association of Petroleum Geologists Bulletin*, 68, p. 637–648, 1984.

Naville, C. Detection of anisotropy using shear waves splitting in VSP surveys requirements and applications, *Expanded abstracts SEG, Houston, Texas,* p. 391 et al., 1986.

Nolet, G., Editor. *Seismic tomography, with applications in global seismology and in exploration geophysics*, D. Reidel Publishing Company, Dordrecht, Holland, 1987.

Ogata, A., and Banks, R.B. A solution of the Differential Equation of Longitudinal Dispersion in Porous Media, *U.S. Geological Survey Professional Paper* 411-A, p. A1–A7, 1961.

Paillet, F.L., Keys, W.S., and Hess, A.E. Effects of lithology on televiewer-log quality and fracture interpretation, *SPWLA 26th Annual Logging Symposium*, June 10–13, 1985.

Parker, R.H., Editor. *Hot Dry Rock Geothermal Energy, Phase 2B Final Report of the Camborne School of Mines Project,* Vol. 1 and 2, Pergamon Press, Oxford, New York, Beijing, Frankfurt, Sao Paulo, Sydney, Tokyo, Toronto, 1989.

Paul, A., Jegouzo, M., Millon, R., Track, A., Vezat, R., *Forage de Sancerre-Couy, Signification des réflecteurs sismiques dans le socle, in: G.P.F., Forage scientifique de Sancerre-Couy. Données géophysiques préliminaires,* Document B.R.G.M. No. 138, Editions BRGM-BP 6009–45060 Orléans Cedex 2, France, 1988.

Pauwels, H., Criaud, A., Brach, M., Vuataz, F-D., and Fouillac, C. Geochemical study at Soultz-sous-Forêts, *Proceedings of Hot Dry Rock geothermal energy, International Conference 27–30 juin 1989, Camborne School of Mines*, Roy Baria editor, p. 136–145, 1989.

Pezard, P.A., Anderson, R.N., Howard, J.J., and Luthi, S.M. Fracture Distribution and Basement Structure from Measurements of Electrical Resistivity in the Basement of the Cajon Pass Scientific Drillhole, California, *Geophysical Research Letters* 15, August 1988 supplement, p. 1021–1024, 1988.

Poirmeur, C., and Vasseur, G. Three-dimensional modelling of a hole-to-hole electrical method: Application to the interpretation of a field survey, *Geophysics*, Vol. 53, No. 3, p. 402–414, 1988.

Pottecher, G., and Poirmeur, C. The MIMAFO direct current cross-hole method: a support to hydraulic investigations, in Detection of Subsurface Flow Phenomena, *FRG, March 16–18, 1988, Karlsruhe, Lecture Notes in Earth Sciences*, Vol. 27, Springer-Verlag, 1989.

Ramay, H.J. Wellbore Heat Transmission, *J.P.T.*, Vol. 14, p. 427–435, 1962.

Rambow, F.H.K. The borehole televiewer: some field examples, *SPWLA 25th Annual Logging Symposium*, June 10–13, 1984.

Robinson, B.A., Tester, J.W., and Brown, L.F. Using chemically reactive tracers to determine temperature characteristics of geothermal reservoirs, *Geothermal Resource Council Transactions* 8, 26–29 August, p. 337–341, Reno, Nevada, 1984.

Russell, B.F., Hull, L.C., and Downs, W.F. Response of East Mesa and Raft River Reservoirs to Injection-Backflow Testing, *Proceedings 9th Workshop Geothermal Reservoir Engineering, SCP-TR-74, Stanford University, Stanford, California*, p. 311–315, 1983.

Schellschmidt, R., and Haenel, R. Influx and water loss in open hole sections of boreholes determined by temperature measurements, *Geol. Jb.*, E-39, p. 101–108, Hannover, 1987.

Serra, O. *Formation MicroScanner image interpretation,* Schlumberger Educational Services, Houston, Texas, 1989.

Serra, O. Fundamentals of well log interpretation. Volume 1: The acquisition of logging data, *Developments in petroleum science, 15A*, 440 p., Elsevier, Amsterdam, 1984.

Sheriff, R.E., and Geldart, L.P. *Exploration seismology*, Vol. 1 and Vol. 2, Cambridge University Press, Cambridge, U.K., 1982.

Shima, H., and Sakayama, T. Resistivity tomography: an approach to 2-D resistivity inverse problem, *Expanded abstracts of the 57th annual international SEG meeting*, New Orleans, p. 59–61, 1987.

Sibbit , A.M., and Faivre, O. The dual laterolog response in fractured rocks, *Transactions of SPWLA 26th Annual Symposium,* Dallas, June 17–20, paper T, 1985.

Straub, A. Panorama et exemples de méthodes géophysiques appliquées à la fracturation, *Hydrogéologie*, No.4, p. 235–244, Editions BRGM-BP 6009-45060 Orléans Cedex 2, France 1988.

Streltsova, T.D. *Well Testing in Heterogeneous Formations*, John Wiley, 1988.

Tester, J.W., Robinson, B.A., and Ferguson, J. The theory and selection of chemically reactive tracers for reservoir thermal capacity production, *Proceedings 12th Workshop Geothermal Reservoir Engineering,* Stanford University, Stanford, California, 1987.

White, J.E. *Underground sound*, Elsevier, Amsterdam, 1983.

Zemanek, J., Caldwell, R.L., Glenn, E.E., Jr., Holcomb, S.V., Norton, L.J., and Straus, A.J.D. The borehole Televiewer - A new logging concept for fracture location and other types of borehole inspection, *Journal of Petroleum Technology* 21, p. 762–774, 1969.

Zemanek, J., Glenn, E.E., Norton, L.J., and Caldwell, R.L. Formation evaluation by inspection with the borehole televiewer, *Geophysics*, Vol. 35, No. 2, p. 254–269, 1970.

BIBLIOGRAPHY

Aki, K., Fehler, M., Aamodt, R.L., Albright, J.N., Potter, R.M., Pearson, C.M., and Tester, J.W. Interpretation of seismic data from hydraulic fracturing experiments at Fenton Hill, New Mexico, hot dry rock geothermal site, *J. Geophys. Res.*, Vol. 87, N B2, p. 936–944, 1982.

Baria, R., Hearn, K., and Batchelor, A.S. The monitoring of induced seismicity during stimulation of the potential hot dry rock geothermal reservoir, *4th Conf. on acoustic emission microseismic activity in geological structures and material,* Pennsylvania State University, 1985, ed. Hardy, R.H., Trans Tech Publications, Clausthal, Germany, 1986.

Bayar, M., and Okandan, E. Tracer flow in a fractured geothermal reservoir model, *Proceedings 12th Ann. Work. Geother. Res. Eng.,* Stanford University, Stanford, California, 1987.

Berest, P. *La thermomécanique des roches*, Editions du Bureau de Recherches Géologiques et Minières (B.R.G.M.), Orléans la Source, 1988.

Bernaudat, F. Contribution à l'étude de l'exploitation de la chaleur des roches chaudes et sèches, Thèse de Docteur-Ingénieur, Ecole des Mines de Paris, 1983.

Bruel, D. Exploitation de la chaleur des roches chaudes et sèches. Etude des phénomènes hydrauliques, mécaniques et thermiques au moyen d'un modèle à fracture discrète, Thèse, Ecole des Mines de Paris, déc. 1990.

Cash, D., Homuth, E.F., Keppler, H., Pearson, C., and Sasaki, S. Fault plane solutions for microearthquakes induced at the Fenton Hill Hot Dry Rock Geothermal Site: implications for the state of stress near a quaternary volcanic center, *Geophys. Res. Let.*, Vol. 10, No. 12, p. 1141–1144, 1983.

Constant , O., and Bouchon, M. Diffraction of seismic waves by fluid-filled fractures and cracks, *European Association of Exploration Geophysicists, Den Haag*, 6–10 June, 1988.

Crampin, S. Evaluation of anisotropy by shear-wave splitting, *Geophysics*, Vol. 50, p. 142–152, 1985.

Elsworth, D. Theory of Thermal Recovery From a Spherically Stimulated Hot Dry Rock Reservoir, *J. Geophys. Res.*, Vol. 94 (B2), p. 1927–1934, 1989.

Fehler, M., and Pearson, C. Cross-hole seismic surveys: applications for studying subsurface fracture systems at a hot dry rock geothermal site, *Geophysics*, Vol. 49, p. 37–45, 1984.

Fehler, M. Using dual-well seismic measurements to infer the mechanical properties of a hot dry rock geothermal system, *J. Geophys. Res.*, Vol. 87, p. 5485–5494, 1982.

Héliot, D., Etchecopar, A., and Cheung, Ph. New developments in fracture characterization from logs, *Rock at Great Depth,* Maury & Fourmaintraux (eds), Balkema, Rotterdam, 1990.

Horne, R.N., and Rodriguez, F. Dispersion in tracer flow in fractured geothermal systems, *Geophysical Research Letters*, Vol. 10, No. 4, p. 289–292, April 1983.

Hudson, J.A. Wave speeds and attenuation of elastic waves in material containing cracks, *Geophys. J. R. Astr. Soc.*, Vol. 64, p. 133–150, 1981.

Luthi, S.M. Sedimentary structures of clastic rocks identified from electrical borehole images, *Geological Applications of Wireline Log,* Hurst, A., Lovell, M.A. & Morton, A.C. (eds), *Geological Society Special Publication*, No. 48, p. 3–10, 1990.

McCann, D.M., Baria, R., Jackson, P., and Green, A.S.P. Application of cross-hole seismic methods in site investigation surveys, *Geophysics*, Vol. 51, No. 4, p. 914–929, 1986.

McKenzie, D.P. The relation between fault plane solutions for earthquakes and the directions of principal stresses, *B.S.S.A.*, Vol. 56, No. 2, p. 591–601, 1969.

Molnar, P., Tucker, B.E., and Brune, J.N. Corner frequencies of P and S waves and models of earthquake sources, *B.S.S.A.*, Vol. 63, No. 6, p. 2091–2104, 1973.

Talebi, S., and Cornet, F.H. Analysis of the seismicity induced by fluid injection in a granitic rock mass, *Geophysic. Res. Let.*, Vol. 14, No. 3, p. 227–230, 1987.

INDEX

N

V

W

ISBN 0-12-083980-6

9 780120 839803 90090